ADVANCED MICROPROCESSOR AND MICROCONTROLLERS

ADVANCED MICROPROCESSOR AND MICROCONTROLLERS

By

Prof. S.K. VENKATA RAM

ME (Eln) MIF, MISTE, Fellow IETE
Principal
Karnatak Law Society
Sri Vasantrao Potdar Polytechnic,
Tilkwadi, Belgaon
Karnataka

UNIVERSITY SCIENCE PRESS

(An Imprint of Laxmi Publications Pvt. Ltd.)

An ISO 9001:2015 Company

BENGALURU • CHENNAI • GUWAHATI • HYDERABAD
KOCHI • KOLKATA • LUCKNOW • RANCHI
NEW DELHI

Printed and bound in India

First Edition : 2002; *Reprint* : 2003, 2004, 2005, 2006, 2008, 2013 ; *Edition* : 2025

ISBN **:** 978-81-31806-50-0

Branches

✆	Bengaluru	080-26 75 69 30
✆	Chennai	044-24 34 47 26
✆	Guwahati	0361-254 36 69
✆	Hyderabad	040-27 55 53 83
✆	Kochi	0484-405 13 03
✆	Kolkata	033-40 04 77 79
✆	Lucknow	0522-430 36 13
✆	Ranchi	0651-224 24 64

Published in India by

UNIVERSITY SCIENCE PRESS

(An Imprint of Laxmi Publications Pvt. Ltd.)

An ISO 9001:2015 Company
113, GOLDEN HOUSE, GURUDWARA ROAD, DARYAGANJ,
NEW DELHI - 110002, INDIA
Corporate Office : B-33, Sector-5, Noida,
Gautambuddha Nagar (U.P.) - 201 301, INDIA
Telephone : +91-120-44 46 900
www.laxmipublications.com
info@laxmipublications.com

PREFACE

Microprocesor-8086/8088 is a 16 bit Microprocessor Architecture and Semiconductor technology and is an extremely powerful microprocessor. The 8-bit microcontroller series 8048, 8051 etc. are more powerful microcontrollers in the form of a single chip microcomputer.

This book covers the details of 16 bit microprocessor 8086 and 8 bit microcontroller for the architecture, interface, instruction set etc. 8087 co-processor is also discussed for its architecture and features, and the book is titled "**Advanced Microprocessor and Micro-controller**."

The book is divided into 15 chapters to fulfill the exhaustive information on 16 bit microprocessor and MCS-51, and 8-bit micro controller for developing number of applcations. After 15 chapters, in the Appendix instruction set summary of 8086 microprocessor and MCS - 51 microcontrollers are added. Care is also taken to fulfill the syllabi requirement for E & C branch students of VTU, Karnatka.

I am sure that the readers will be benefited with this book and feedback from them are welcome. After publishing few text books as per VTU syllabus good feedback and complements received inspired me to bring out this book i.e. "**Advanced Microprocessor & Micro-controller**."

I am highly indebted to the Management of Karnatak Law Society, Tilakwadi, Belgaum for their inspiration and motivation for writing this book. The compliments and motivation received from the Chairman/Secretaries/Members, Karnatak Law Society, Belgaum for my earlier publications has inspired me to continue and contribute such publication work to help the student community in the field of Technical Education.

I thank my wife Smt. S.V. Aruna Ram for her co-operation while writing and editing this book. Finally I thank M/S Laxmi Publications, New Delhi for undertaking the printing and publishing work of this book.

S. K. VENKATARAM

ABOUT THE AUTHOR

Prof. S.K.Venkata Ram is native of Chitradurga in Karnataka. He has done his B.E.(E & C) from Karnataka Regional Engg College, Suratkal and M.E.(Electonics) from University VIsvesvaraya College of Engineering, Bangalore. Prof.S.K.Venkata Ram has already published few Text Books and study materials to cater the needs of Engineering and Diploma students as per the revised syllabus in Visweswaraiah Technological University and Director of Technical Education Bangalore, Karnataka. From his publications one of the book entitled "Bio Medical Electronics & Instrumentations" is recommended by the Board of Technical Education, Banagalore to the final year diploma in E & C branch students, in Karnataka. The author has also published few technical papers in International and national level technical journals, among which few papers are using Micro controllers. He actively participates in Indian Society for Technical Education activities and he is invited as resource person to deliver lectures on technical and non technical topics under staff development and student development programme at different institutions. The author is having about 23 years of teaching experience in different levels in Polytechnics and engineering colleges. At present he is principal in Karnatak Law Society's Sri Vasantrao Potdar Polytechnic, Tilakwadi, Belgaum, in Karnataka.

Dedicated to my Beloved Parents

SMT. S.K. SUSHEELAMMA
SRI S.R. KADANDA RAMA SETTY
Chitrdurga-577501
&
Parents-in-law
SMT. M.K. SUBBARATNAMMA
SRI M.C. KRISHNIAH SETTY
Tumkur

CONTENTS

1 INTRODUCTION TO ADVANCED MICROPROCESSOR

1.0 OBJECTIVE

At the end of this chapter, you can, know the history of computer, draw block diagram/ Architecture of 8086, draw functional find diagram of 8086, know different flag, segmentation and list difference between 8086 and 8088 microprocessor.

1.1 INTRODUCTION

In this section we briefly discuss the history of events which lead to the Microprocessor development. It is the way which started from mechanical age to microprocessor age, and we discuss following topics :

1. Mechanical age
2. Electrical age
3. Advancement in programming
4. Microprocessor age

1.1.1 Mechanical Age

The first practical geared, mechanical machines used for computing information automatically was in early 1800 *(before humans invented light bulb)*. An early pioneer of mechanical computing machinery was Charles Babbage, who was aided by Augusta Ada Byron, the Countess of Lovelace. Babbage was commissioned in 1823 by the Royal Astronomical Society of Great Britain to produce a programmable calculating machine. The machine use to generate navigational tables for the Royal Navy. He accepted the challenge and began to create what he called his **Analytical Engine**. That engine was a mechanical computer which can store 1,000 20-digit decimal numbers and a variable program that could modify the function of the machine so it could perform various calculating tasks. Input to that engine was through punched cards, much as computers in the 1950s and 1960s used punched cards. The idea of using punched cards were obtained from Joseph Jacquard, a Frenchman who used punched cards as input to a weaving machine he invented in 1801, which is today called Jacquard's loom. Jacquard's loom used punched cards to select intricate weaving patterns in the cloth that it produced. *(The punched cards programmed the loom)*.

1.1.2 Electrical Age

In 1800s advent of electric motor with multitude of motor-driven added the machine. They all based on mechanical calculator developed by Blaise Pascal. These electrically driven mechanical calculators were common pieces of office equipment until well into the early 1970s, when the small hand-held electronic calculator, first introduced by Bomar, appeared. Monroe was also a leading pioneer of electronic calculators, but their machines were desktop, four-function models the size of cash registers. Next in 1889, Herman Hollerith developed the punched card for storing data. Like Babbage, he too apparently borrowed the idea of a punch card from Jacquard. He also developed a mechanical machine–driven by one of the new electric motors–that counted, sorted and collated information stored on punched cards. *(The idea of calculating by machinery intrigued the United States government so much that Hollerith was commissioned to use his punched-card system to store and tabulate information for the 1890 census).*

The Mechanical machines driven by electric motors continued to dominate the information processing world until the advent of the first electronic calculating machine in 1941 by a German inventor named Konrad Zuse. His calculating computer, the Z3, was used in aircraft and missile design during World War II for the German was effort. Had Zuse been given adequate funding by the German government, he most likely would have developed a much more powerful computer system. Zuse is today finally receiving some belated honor for his pioneering work in the area of digital electronics and for his Z3 computer system.

First general-purpose programable electronic computer system was developed in 1946 at the University of Pennsylvania. This first modern computer was called the ENIAC ***(Electronics Numerical Integrator and Calculator)***. The ENIAC was a huge machine containing over 17,000 vacuum tubes and over 500 miles of wires. This massive machine weighed over 30 tons, yet performed only about 100,000 operations per second. ENIAC thrust the world into the age of electronic computers. The ENIAC was programmed by rewiring its circuits—a process that took many workers several days to accomplish. The workers changed the electrical connections on plug-boards that looked much like early telephone switchboards. One problem with the ENIAC was the life of the vacuum tube components, which required frequent maintenance.

1.1.3 Advancement in Programming

As programmable machines developed programs and programming languages started to appear. As mentioned, first truly programmable electronic computer system was programmed by rewiring its circuits, which proved too cumbersome for practical application.

Therefore early in the evolution of computer systems, computer languages began to appear to control the computer. The first such language was machine language, which was constructed of ones and zeros using binary codes that were stored in the computer's memory system in groups called programs. This was more efficient than rewiring a machine to program it, but it was still extremely time consuming to develop a program because of the sheer number of codes required. John von Neumann, the mathematician, was the first to develop a system that accepted instructions and stored them in a memory. Computers are often called von **Neumann machines** *(in honor of John von Neuman)*.

First truly successful and widespread programming language for business applications was ***COBAL (Computer Business Oriented Algorithmic Language)***. Although COBAL usage has diminished somewhat in recent years, it is still a major player in many large business systems. Another fairly popular business language is RPG ***(Report Program Generator)***, which allows programming by specifying the form of the input, output and calculations.

In early days of programming, additional languages have appeared, the more common are BASIC, C/C++ (OOP), PASCAL and ADA. The BASIC and PASCAL, languages were both designed as teaching languages, but have escaped the classroom and are used in many computer systems. The BASIC language is probably the easiest of all to learn. Some estimates indicate that the BASIC language is used in the personal computer for 80 per cent of programs written by users. Recently, a new version of BASIC called VISUAL BASIC has appeared, so programming in the Windows environment has become easier, *(VISUAL BASIC language may eventually supplant C/C++ and PASCAL).*

ADA language is used heavily by the Department of Defense. The ADA language was named in honor of Augusta Ada Byron, Countess of Lovelace. The Countess worked with Charles Babbage in the early 1800s in the development of this Analytical Engine.

1.1.4 Microprocessor Age

Intel 4004, a 4-bit microprocessor was the world first microprocessor. It was a programmable controller on a chip, which merges by today's standards.

The intel 4004 addressed a mere 4,096 4-bit wide memory locations. *(A bit is binary digit with a value of one or zero. A 4-bit wide memory location is often called a nibble).* The 4004 instruction set contained only 45 instructions. It was fabricated with the then-current state-of-the-art P-channel MOSFET technology that only allowed it to execute instruction at the slow rate of 50 KIPs ***(kilo-instructions per second)***, which was slow when compared to the 100,000 instruction executed per second by the 30-ton ENIAC computer in 1946. The main difference was that the 4004 weighed much less than on ounce.

The 4-bit microprocessor debuted in early video game systems and small microprocessor-based control systems. One such early video game, a shuffleboard game, was produced by Balley. The main problems with this early microprocessor were its speed, word width, and memory size. Evolution of the 4-bit microprocessor ended when Intel released the 4040, an updated version of the earlier 4004. The 4040 operated at a higher speed, although it lacked improvements in word width and memory size. Other companies, particularly Texas Instruments *(TMS - 1000)*, also produced 4-bit microprocessors. The 4-bit microprocessor still survives in low-end applications such as microwave ovens and small control systems and is still available from some microprocessor manufactures. Most calculators are also still based on 4-bit microprocessor that process 4-bit BCD *(binary-coded decimal)* codes.

In 1971, realizing that the microprocessor was a commercially viable product, Intel Corporation released the 8008, an extended 8-bit version of the 4004 microprocessor. The 8008 addressed an expanded memory size *(16K bytes)* and contained additional instructions *(a total of 48)* that provided an opportunity for its application in more advanced systems. *(A byte is generally an 8-bit wide binary number and a K is 1,024 bytes).*

As engineers developed more demanding uses for 8008 microprocessor, they discovered that its somewhat small memory size, slow speed and instruction set limited its usefulness. Intel recognized these limitations and, in 1973, introduced the 8080 microprocessor - the first of the modern 8-bit microprocessors. About six months after intel released the 8080 microprocessor, Motorola Corporation introduced its MC6800 microprocessor. The floodgates opened the 8080 and, to a lesser degree, the MC6800 ushered in the age of the microprocessor. Immediately other companies began to introduce their own versions of the 8-bit microprocessor. Table 1. lists these early microprocessors and their manufacturers.

Table 1. 8-Bit Microprocessors

Manufacturer	Part Number
Fairchild	F-8
Intel	8080
MOS Technology	6502
Motorola	MC6800
National Semiconductor	IMP-8
Rockwell International	PPS-8
Zilog	Z-8

. Amongst Intel and Motorola continue successfully to introduce newer and improved versions of the microprocessor. Zilog still manufactures microprocessors, but has remained in the background, deciding to concentrate, fairly successfully, on microcontrollers and embedded controllers instead of general-purpose microprocessors. *(Rockwell has all but abandoned microprocessor development in favour of modern circuitry).*

The 8080 address more memory and execute additional instructions, but it executed them 10 times faster than the 8008. An addition that took 20 μs *(50,000 instructions per second)* on an 8008-based system required only 2.0 μs *(500,000 instructions per second)* on an 8008-based system. Also, the 8080 was compatible with TTL (transistor-transistor *logic*). Whereas the 8008 was not directly compatible. This made interfacing much easier and less expensive.

In 1977, Intel Corporation introduced an updated version of the 8080 *i.e.,* the 8085. This was to be the last 8-bit general-purpose microprocessor developed by Intel. Although only slightly more advanced than an 8080, the 8085 executed software at an even higher speed. An addition that took 2.0 μs *(500,000 instructions per second)* on the 8080 required only 1.3 μs *(769,230 instructions per second)* on the 8085. Main advantages of the 8085 were its internal clock generator, internal system controller, and higher clock frequency. This higher level of component integration reduced the 8085's cost and increased its usefulness. Intel has managed to sell well over 100 million copies of the 8085, including the 8085s manufactured by other licensed second sources, such as Advandel Micro Devices *(AMD);* there are over 200 million of these microprocessors in existence *(Applications that contain the 8085 are still being used and designed and will likely continue to be popular well into the future).*

1.2 ADVANCED MICROPROCESSOR

Intel corporation further developed 16 bit and 32 bit microprocessors after 8-bit processors. The main processors so developed are :

1. 8086/8088 microprocessors
2. 80286 microprocessor
3. 32-bit microprocessor

4. 80486 microprocessor
5. Pentium microprocessor

1.2.1 8086/8088 Microprocessor

Intel corporation released 8086 microprocessor in 1978, later after a year it released 8088 microprocessor. The two devices were 16-bit microprocessors, which executed instructions in as little as 400ns *(2.5 MIPs or 2.5 million instructions per second).* This represented a major improvement over the execution speed of the 8085. In addition, the 8086 and 8088 addressed 1M bytes of memory, 16 times more memory than the 8085. ***(A 1M byte memory*** *contains 1,024K byte-sized memory locations, or 1,048,576 bytes).* Such a higher execution speed and larger memory size allowed the 8086 and 8088 to replace smaller minicomputers in many applications. One other feature found in the 8086/8088 was a small 4- or 6-byte instruction cache or queue that prefetches a few instructions before they are executed. The queue sped the operation of many sequences of instructions and proved to be basis for the much larger instruction caches found in modern microprocessors.

Increase in memory size and additional instructions of the 8086 and 8088 have led to many sophisticated applications for microprocessors. Improvements to the instruction set included a multiply-and-divide instruction, which were missing on earlier microprocessors. Also, the number of instructions increased from 45 on the 4004, to 246 on the 8085, to well over 20,000 variations on the 8086 and 8088 microprocessors. Note that these microprocessors were called CISC *(complex instruction set computers)* because of the number and complexity and instructions. The additional instructions eased the task of developing efficient and sophisticated applications even though their number was at first overwhelming and time consuming to learn. Such 16-bit microprocessor also provided more internal register storage space than the 8-bit microprocessor, *(additional registers allowed software to be written more efficiently).*

1.2.2 808286 Microprocessor

Intel introduced 80286 microprocessor, which is updated 8086 in the year 1983. The 80286 microprocessor *(which is also a 16-bit architecture microprocessor)* was almost identical to the 8086 and 8088 except it addressed a 16M byte memory system instead of a 1M byte system. The instruction set of the 80286 was also almost identical to the 8086 and 8088 except for a few additional instruction that managed the extra 15M bytes of memory. The clock speed of the 80286 was increased, so it executed some instructions in as little as 250ns *(4.0 MIPs)* with the original release 8.0 MHz version. Few changes also occurred in the internal execution of the instruction that led to an eight-fold increase in speed for many instructions when compared to 8086/8088 instructions.

1.2.3 32-bit Microprocessor

Application began to demand faster microprocessor speeds, more memory, and wider data paths, led to the arrival of the 80386, in 1986, by Intel Corporation. The 80386 represented a major overhaul of the 16-bit 8086-80286 microprocessor's architecture. The 80386 was Intel's first practical 32-bit microprocessor that contained a 32-bit data bus and a 32-bit memory address. *(Intel produced an earlier, although unsuccessful, 32-bit microprocessor called the iapx-432).* Through these 32-bit buses, the 80386 addressed up to 4G bytes of memory. *(1G **of memory** contains 1,024M or 1,073,741,824 locations)* A 4G byte memory can store an astounding 1,000,000 typewritten, double-spaced pages of data. The 80386 was also available in a few modified versions such as the 80386SX, which addressed 16M bytes of memory through a 16-bit data and 24-bit address bus, and the 80386SL/80386SLC, which addressed 32M bytes of memory through a 16-bit data and 25-bit address bus. An 80386SLC version contained an internal cache memory that allowed it to process data at even higher rate. In 1995, Intel released the 80386EX microprocessor. 80386EX is called an **embedded PC,** containing all the components of the AT class personal computer on a single integrated circuit, *(80386EX also contains 24 lines for input/output data, a 26-bit address bus, a 16-bit data bus a DRAM refresh controller, and programmable chip selection logic).*

The 32-bit microprocessor is needed because of the size of its data bus, which transfers real *(single-precision floating-point)* numbers that require 32-bit wide memory. In order to efficiently process 32-bit real numbers, the microprocessor must efficiently pass them between itself and memory. If they pass through an 8-bit bus, it takes our read or write cycle; when passed through a 32-bit data bus, however, only one read or write cycle is required. It significantly increases the speed of any program that manipulates real numbers. Most high-level languages, spreadsheets, and database management systems use real number for data storage. Real numbers are also used in graphical design package that use vectors to plot images on the video screen. These include such CAD *(computer aided drafting/design)* systems as AUTOCAD, ORCAD etc.

Note. The instruction set of the 80386 microprocessor was upward compatible with the earlier 8086, 8088 and 80286 microprocessors. Additional instructions referenced the 32-bit registers and managed the memory system and memory management instructions and techniques used by the 80286 were also compatible with the 80386 microprocessor. These features allowed older, 16-bit software to operate on the 80386 microprocessor.

1.2.4 80486 Microprocessor

Intel released the 80486 microprocessor, in 1989 which incorporated an 80386-like microprocessor, an 80387-like numeric coprocessor and an 8K byte cache memory system into one integrated package. Even though the 80486 was not radically different from the 80386, it did include one substantial change. The internal structure of the 80486 was modified from the 80386 so about half of its instructions executed in one clock instead of two clocks. Because the 80486 was available in a 50 MHz version, about half of its instructions executed in 25ns *(50 MIPs)*. Average speed improvement for a typical mix of instruction was about 50 percent over the 80386 operated at the same clock speed. Later versions of the 80486 executed instructions at even higher speeds with a 66 MHz double-clocked version *(80486DX2)*. The double-clocked 66 MHz version executed instruction at the rate of 66 MHz, with memory transfers executed at the rate of 33 MHz. A triple-clocked version from Intel, the 80486DX4, improved the internal execution speed to 100 MHz with memory transfers at 33 MHz. *(Note that the 80486DX4 executed instructions at about the same speed as the 60 MHz Pentium).* It also contained an expanded 16K byte cache in place of the standard 8K byte cache found on earlier 80486 microprocessors. Recently, Advanced Micro Devices *(AMD)* has produced a triple-clocked version that runs with a bus of 40 MHz and a clock speed of 120 MHz. The future promises to bring microprocessors that internally execute instructions at rate of up to 250 MHz or higher.

Now Table 2 lists many microprocessors produced by Intel and Motorola with information about their word and memory sizes as below :

TABLE 2. Intel and Motorola Microprocessors

Manufacturer	Processor	Data Bus Width	Memory Size
Intel	8048	8	2K internal
	8051	8	8K internal
	8085A	8	64K
	8086	16	1M
	8088	8	1M
	8096	16	8K internal
	80186	16	1M
	80188	8	1M
	80251	8	16K internal
	80286	16	16M
	80386EX	16	64M
	80386DX	32	4G
	80386SL	16	32M
	80386SLC	16	32M + 1K cache
	80386SX	16	16M
	80486DX/DX2	32	4G + 8K cache
	80486SX	32	4G + 8K cache
	80486DX4	32	4G + 16K cache
	Pentium	64	4G + 16K cache
	Pentium Overdrive		4G + 16K cache
	(P24T)	32	
	Pentium Pro processor	64	64G + 16K LI cache + 256K L2 cache
Motorola	6800	8	64K
	6805	8	2K
	6809	8	64K
	68000	16	16M
	68008Q	8	1M
	68008D	8	4M
	68010	16	16M
	68020	32	4G
	68030	32	4G + 256 cahce
	68040	32	4G + 8K cahce
	68050	32	4G + 8K cahce Proposed, but never released
	68060	64	4G + 16K cache
	PowerPC	64	4G + 32K cahce

1.2.5 Pentium Microprocessor

Pentium was introduced in 1993, which was similar to the 80386 and 80486 microprocessors. This microprocessor was originally labelled the P5 *(80586)* but Intel decided not to use a number because it appeared to be impossible to copyright a number. The two introductory versions of the Pentium operated with a clocking frequency of 60 MHz and 66 MHz and a speed of 110 MIPs, with a higher frequency 100 MHz one and one-half clocked version operating at 160 MIPs. The double-clocked Pentium, operating at 120 MHz and 133 MHz, was also available, as were higher speed versions *(200 MHz).* One more difference was that the cache size was increased to 16K byte from the 8K cache found in the basic version of the 80486. Pentium contains an 8K bytes instruction cache and an 8K byte data cache. This allowed a program contained up to 4G bytes, with the data bus width increased from the 32-bits found in the 80386 and 80486 to a full 64-bits. The data bus transfer speed was either 60 MHz, or 66 MHz depending on the version of the Pentium. This wider data bus width accommodated double precision floating-point numbers used for high-speed vector-generated graphical displays. It also transferred data between the memory system and microprocessor at a higher rate. It should allow virtual reality and software to operate at more realistic rates on current and future Pentium based platforms. The widened data bus and higher execution speed of the Pentium should also allow full-frame video displays that operate at scan rates of 30 Hz or higher-comparable to commercial television.

Note. Intel has released recently the Pentium Overdrive *(P24T)* for older 80486 systems that operate at either 63 MHz or 83 MHz clock. The 63 MHz version upgrades older 80486DX2 50 MHz systems, and the 83 MHz version upgrades the 80486DX2 66 MHz systems. The upgraded 83 MHz system performs at a rate somewhere between a 66 MHz Pentium and a 75 MHz Pentium. *(If older VESA local bus video and disk caching controllers seem too expensive to toss out, the Pentium OverDrive represents an ideal upgrade path from the 80486 to the Pentium).*

1.3 THE COMPUTER

A computer is defined as a system comprising of hardware and software designed to provide an effective tool for good, computation, communication and imaging characteristics of such computers are high speed of operation, accuracy, flexibility, reliability, and economy with efficient processing data. The computer can be classified into these types on the basis of the principle of operation viz. (*a*) analog, (*b*) digital, and (*c*) hybrid computers.

A computer may be a single user or multiuser type. In single user computer it has less processing power and only one user can operate at a time with single task execution, such as DOS *(Disk operating system)* ex : microcomputer. In multiuser computer it has large processing power and storage capacity, such as UNIX, LINUX, VAX, ex : minicomputer main frame computer and supercomputer. Next computers can be classified based on the technology evolution as, zero, first, second, third, fourth and fifth generation as shown in Table 3 below :

Table 3. Computer Generation

Generation	Year	Example Computers	Computer Hardware	Computer Software	Computer Performance
Zero	Pre 1946	Abacus Difference Engine Analytical Engine	Counter Wheels wit Mechanical Components		Numerical Operations at slow Speed
First	1946 - 1956	Enivace Udvac Univac IBM 650	Vacuum Tubes Magnetic Drums CRT Screen	Stored Programs Machine Code	2KB Memory 10 Kilo Instructions Per Second (KIPS)
Second	1957 - 1963	IBM 7094 CDC 6600 Honeywell 400 Burroughs 500	Transistor, Magnetic Code Memory	HLL, FORTRAN, COBOL, ALGOL	32 KB Memory 200 Kilo Instructions Per Second (KIPS)
Third	1964 - 1981	IBM 360/370 Honey well 2000 CRAY - I, CYBER 205	Integrated Circuits (ICs), Semit-Conductor Memory, Magnetic Disk Micro and Mini Computers	Pascal, C, Computer Graphics, IS, Time Sharing Systems	2 MB Memory 5 Million Instructions Per Second (MIPS)
Fourth	1982 - 1989	CRAY-XMP IBM 305	Distributed Computing System, VLSI, Bubble Memory, Optical Disk	Widespread of packaged software OOP Languages C++, Small Talk	8MB Memory 30 Million Instru-ctions Per Second (MIPS)
Fifth	1990	Extensive development in distributed computing PARAM 9000, C-DAC, India CRAY-II, Cray Corp. NAl Flosolver, India	Advance packaging and Integration Technology, parallel Architecture 3D-IC Design, Gallium Arsenide Technology Optical and Neutral Computing	Concurrent languages, Object Oriented OS, Micro-Kenrel based OS (C-DAC PARAS) Multithreaded parallel Programming Model	1 Giga to 1 Tera Instructions Per Second

1.3.1 A Digital Computer

Fig. 1.1 shows the block diagram of a digital computer. A computer can perform a task by processing a sequential list of instructions.

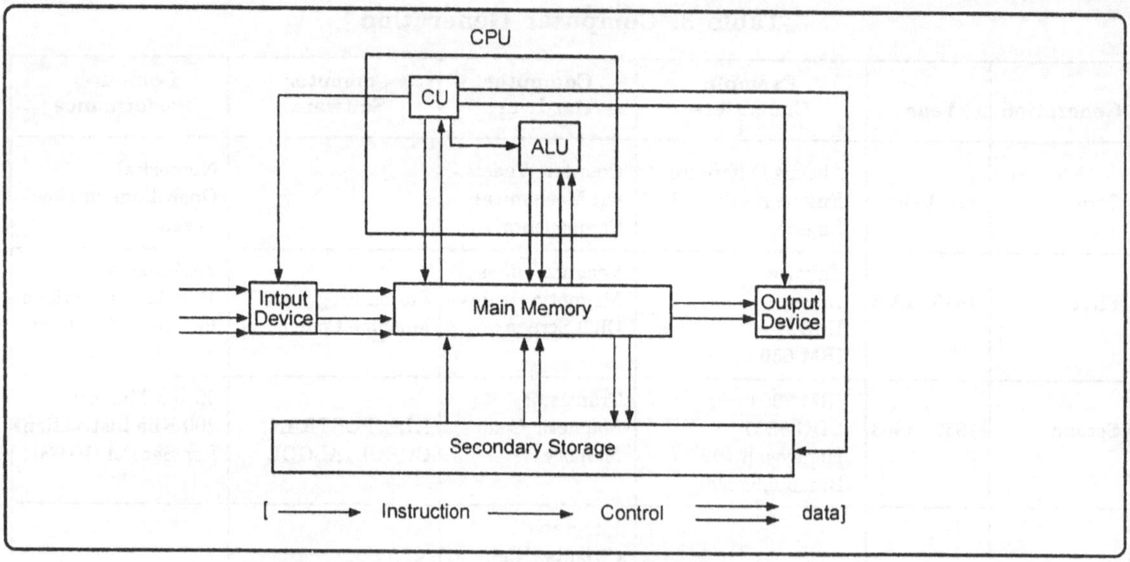

Fig. 1.1 Block Diagram of a Digital Computer

A list of instruction constitute a program. A set of programs to achieve a desired objective is known as software. The physical components of a computer constitute the hardware. A computer consists of five basic hardware blocks, as under ;

(*a*) Central Processing Unit *(CPU)*
(*b*) Main Memory Unit *(MMU)*
(*c*) Secondary Storage Unit *(SSU)*
(*d*) Input Unit *(IU)*
(*e*) Output Unit *(OU)*

(*a*) Central Processing Unit

In Fig. 1.1 the Control Unit *(CU)* and Arithmetic and Logic Unit *(ALU)* constitute the Central Processing Unit *(CPU)*, the brain of the computer, *i.e.*, CPU = CU + ALU. The CU coordinates and directs the operations of a computer. The ALU performs arithmetic and logical operations. The microprocessor functions as the CPU of the computer system. The unit of the processing capacity of CPU is defined in terms of Million Instructions Per Second *(MIPS)* of Floating Point Operations Per Second *(FLOPS)*.

Few microprocessors used in digital computers are Intel's x86 family processors *i.e.*, 8086, 80186, 80286, 80386, 80486, Pentium *(80586)*, P6 *(8-686)*, P7 *(80786)*, 1860, and Motorola's 68000 series processors *(68000, 68020, 68030, 68040)*. Some computers have coprocessors *(Intel 8087 series)* which help the main processor *(Intel 8086 series)* in speeding up the floating point operations.

(*b*) Main Memory Unit

It consists of contiguous memory cells used to store data and instructions during the execution of programs. The main memory can be random-access memory or read only memory. Random Access Memory *(RAM)* is an ultra-fast memory that provides a fast and uniform access to any storage location in the memory. It is a volatile memory that gets erased when powered off. The Read Only Memory *(ROM)* is a nonvolatile memory that will not be erased even when powered off. RAMs are of two types, static RAM and dynamic RAM. The static RAM stores the data in Flip-Flops and need not be refreshed, whereas dynamic RAM stores the data in the form of the capacitor charge and therefore must be refreshed periodically.

(*c*) Secondary Storage Unit

The secondary storage unit is used to store programs and data. Programs and data stored in the secondary memory are loaded into the main memory for execution by the operating system and removed from the main memory on execution of the program. Information stored in memory remains until it is removed from the secondary storage. Secondary storage devices are characterised by their large memory storage capacity, show speed of access and low cost. The time required to read or write data from the secondary storage is more than the time that is required for primary memory, examples of secondary storage devices are like floppy diskettes and hard disk.

(*d*) Input Unit

A computers serve purpose without the capability to communicate with the external world. Input unit is used to enter program and data into the computer. A read command in a program directs the computer to read the data stored or entered from the input computer. A read command in a program directs the computer to read the data stored or entered from the input device and transfers data to CPU or memory unit. *Eg.* Keyboard, Mouse, Scanner and so on.

(*e*) Output Unit

Output unit will cause computer to transfer data from the CPU or the main memory unit to an output device that records or displays the data. An output devices can be used to display character or graphics image. The secondary storage device can also be used as input and output unit under the control of a program, but there will be no direct interaction of the user in the input and output processes. *Eg.* CRT, Printer, Plotter and so.

1.4 THE MICROPROCESSORS

The microprocessor is the controlling element in a computer system and is sometimes referred to as the CPU *(central processing unit).* A microprocessor can control memory and I/O through a series of connections called buses. Buses select an I/O or memory device, transfer data between an I/O device or memory and the microprocessor, and control the I/O and memory system. Memory and I/O are controlled through instructions that are stored in the memory and executed by the microprocessor.

Microprocessor performs three main tasks for the computer system :

(1) data transfer between itself and the memory or I/O systems

(2) simple arithmetic and logic operations, and

(3) program flow via simple decisions. Albeit these are simple tasks, but through them, the microprocessor performs virtually any series of operations or tasks.

Power of the microprocessor is in its ability to execute millions of instructions per second from a program or software ***(group of instructions)*** stored in the memory system.

This stored program concept has made the microprocessor and computer system a very powerful device. *(Recall that Babbage also wanted to use the stored program concept in his Analytical Engine.)*

The Table 4 shows arithmetic and logic operations executed by the Intel family of microprocessors. These operations are very basic, but through them, very complex problem are solved. Data are operated upon from the memory system or internal registers.

Table 4. Arithmetic and Logic Operations

Operation	Comment
Addition	
Subtraction	
Multiplication	
Division	
AND	Logical multiplication
OR	Logical addition
NOT	Logical inversion
NEG	Arithmetic inversion
Shift	
Rotate	

The Data widths are variable and include a **byte** *(8-bits)*, **word** *(16 bits)*, and **doubleword** *(32-bits)*. Note that only the 80386 through the Pentium directly manipulate 8-, 16- and 32-bit numbers. The earlier 8086-80286 directly manipulate 8- and 16-bit numbers, but not 32-bit numbers The 80486DX and Pentium also contain a numeric coprocessor that allows them to perform complex calculations using floating-point arithmetic, *(numeric coprocessor was an additional component in the 8086-through the 80386-based personal computer)*.

One more feature that makes the microprocessor powerful is its ability to make simple decisions. Decisions are based upon numerical facts. For instant, a microprocessor can decide if a number is zero, if it is positive, and so forth. Such simple decisions allow the microprocessor to modify the program flow so programs appear to think through these simple decisions. Table 5 lists the decision-making abilities of the Intel family of microprocessors, as under.

Table 5. Decisions Observed in 8086 - 80486 Microprocessors

Decision	Comment
Zero	Test a number for zero or not-zero
Sign	Test a number for positive or negative
Carry	Test for a carry after addition or a borrow after subtraction
Parity	Test a number for an even or an odd number of ones
Overflow	Test for an overflow that indicates an invalid signed result after addition or subtraction

1.4.1 Buses

A **bus** is a common group of wires that interconnect components in a computer system. The buses that interconnect the sections of a computer system transfer address, data and control information between the microprocessor and its memory and I/O systems. In the microprocessor based computer system, three buses exist for this transfer of information : address, data and control. Fig. 1.2 depicts how these buses interconnect various system components such as the microprocessor, read/write memory *(RAM)*, read-only memory *(ROM)*, and a few I/O devices.

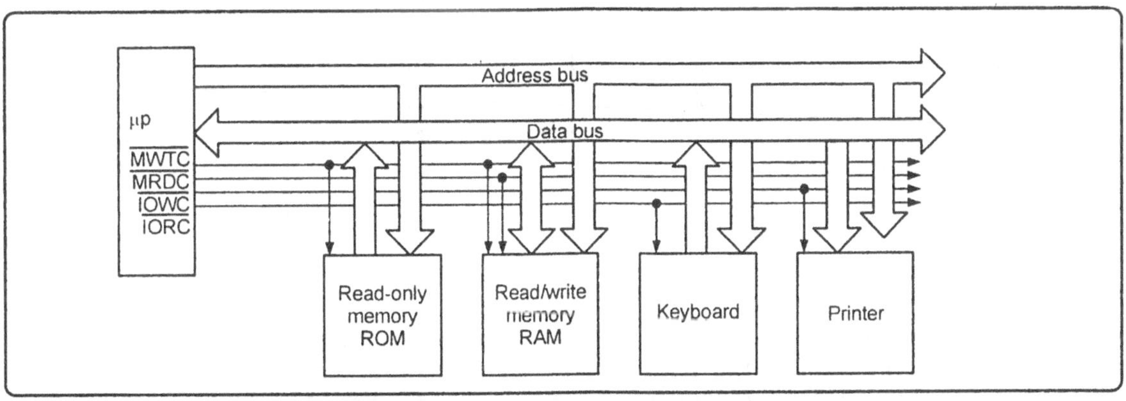

Fig. 1.2. Block diagram of a computer system showing the address, data and control bus structure

As observed in Fig. 1.2 the address bus requests a memory location from the memory or an I/O location from the I/O devices. If I/O is addressed, the address bus contains a 16-bit I/O address from 0000H-FFFFH. The 16-bit I/O address or port number selects one of 64 K different I/O devices If memory is addressed, the address bus contains a memory address. The memory address varies in width with the different versions of the microprocessor. The 8086 and 8088 address 1M byte of memory using a 20-bit address that selects locations 00000H-FFFFFH. The 80286 and 80386SX address 16M bytes of memory using a 24-bit address that selects locations 000000H-FFFFFFH. The 80386SL, 80386SLC, and 80386EX address 32M bytes of memory using a 25-bit address that selects locations 0000000H-1 FFFFFFH. The 80386DX, 80486SX, and 80486DX address 4G bytes of memory using a 32-bit address that selects locations 00000000H-FFFFFFFFH. The Pentium also addresses 4G bytes of memory, but it uses a 64-bit data bus to access up to 8 bytes of memory at a time. *(The Pentium Pro processor has a 64-bit data bus and a 32-bit address bus that addresses 4G bytes of memory from location 00000000H-FFFFFFFFH or a 36-bit address but address bus that addresses 64G bytes of memory at locations 000000000H-FFFFFFFFFH depending on its configuration)* Table 6 gives complete listing of bus and memory sizes on the Intel family of microprocessors, as below :

Table 6. Intel Family of Microprocessor Bus and Memory Sizes

Microprocessor	Data Bus Width	Address Bus Width	Memory Size
8086	16	20	1M
8088	8	20	1M
80186	16	20	1M
80188	8	20	1M
80286	16	24	16M
80386SX	16	24	16M
80386DX	32	32	4G
80386EX	16	26	64M
80486	32	32	4G
Pentium	64	32	4G
Pentium Over Drive	32	32	4G
Pentium Pro	64	32	4G
Pentium Pro	64	36	64G

Data bus transfers information between the microprocessor and its memory and I/O address space. Data transfers vary in size from 8-bits wide to 64-bits wide in various members of the Intel microprocessor family. The 8088 contains an 8-bit data bus that transfers 8-bits of data at a time. The 8086, 80286, 80386SL, 80386SX, and 80386EX transfer 16-bits of data through their data buses; the 80386DX, 80486SX, and 80486DX transfer 32-bits of data; and finally, the Pentium and Pentium Pro transfer 64-bits of data.

Note. Advantage of a wider data bus is speed in applications that use wide data. For instant, if a 32-bit number is stored in memory, it takes the 8088 multiprocessor four transfer operations to complete because its data bus is only 8-bits wide. The 80486 accomplishes the same task with one transfer because its data bus in 32-bits wide.

1.5 ARCHITECTURE OF 8086 MICROPROCESSOR

Fig. 1.3 depicts the architecture *(internal block diagram)* of 8086 central processing unit *(microprocessor is called as CPU)*. It is observed in Fig. 1.3 that the CPU is divided into two independent functional units as :

1. Bus Interface unit *(BIU)*
2. Execution Unit *(EU)*.

1.5.1 Bus Interface Unit

Bus Interface Unit is responsible for transfer of instructions, address and data on the system bus to the execution unit. In handles the transfer of data between the processor. memory, and I/O devices. It includes instruction fetch, data fetch, address transfer, and computation of effective address of the memory. In Bus Interface Unit, there are three functional parts, they are :

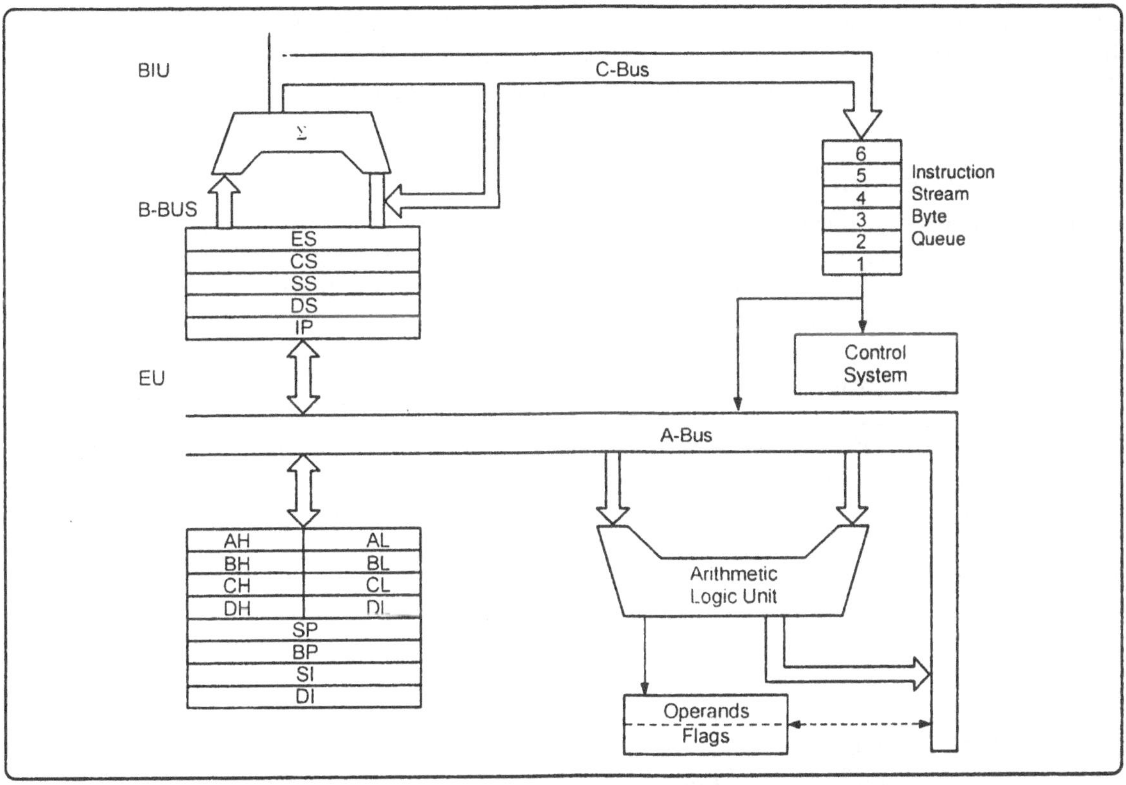

Fig. 1.3 8086 Microprocessor internal block diagram (Architecture)

a. Instruction Queue *(IQ)*

b. Segment Registers

c. Instruction Pointer *(IP)*

(*a*) Instruction Queue *(IQ)*

Instruction queue is of six bytes length which is used to speed up the execution of programs, by prefetching six instruction bytes in advance from the memory. The prefetched instruction are stored in a group of high speed registers known as the instruction queue. The BIU works in parallel with the EU, and BIU fetches the instruction bytes while the EU is executing an instruction. The simultaneous operations of BIU and EU are possible only when the EU does not require the system bus. The process of fetching the next instruction in advance while the EU is executing the current instruction, is known as *pipelining.*

(*b*) Segment Registers

Bus Interface Unit contains four 16-bit segment registers. They are,

1. Code Segment *(CS)* register
2. Data Segment *(DS)* register
3. Stack Segment *(SS)* register
4. Extra Segment *(ES)* register

These registers are used to store the 16-bit starting address of the four memory segments. The BIU generates a 20-bit address using the segment and the offset components of an address. BIU can address the memory locations starting from 00000H to FFFFFH *(0 to 1 Mb)*.

(c) Instruction Pointer (IP)

Instruction pointer is designated as register IP. It always holds the address of memory location *(offset)* of the next instruction to be executed. As the instruction is executed, the IP is advanced to point to the next instruction in the memory. Instruction pointer is also called as the program counter in other microprocessors.

1.5.2 Execution Unit

Execution Unit *(EU)* works in parallel with the BIU which informs the BIU about the location at which the next instruction or data is to be fetched. The phases of execution of the instruction are Fetch, Decode, Execute, and Write. The fetch phase performs fetching of the instruction from the instruction queue. The decode phase performs the decoding of the instruction. The execute phase performs real *(actual)* operations on the data. The write phase performs the operation of storing the computer result at the destination. Functional parts of the Execution Unit are,

(*a*) Control Circuitary and Instruction Decode,
(*b*) Arithmetic and Logic Unit *(ALU)*,
(*c*) Flag Register,
(*d*) General Purpose Registers,
(*e*) Stack Pointer Register and
(*f*) Pointer and Index Registers

(a) Control Circuitary and Instruction Decode

Control circuit of the EU directs all the internal operations of the processor. The instruction in the EU translates the instruction fetched from the memory into a series of actions carried out by the execution unit.

(b) Arithmetic and Logic Unit (ALU)

Arithmetic and logic unit is one of the component of the execution unit. It performs 8-bit or 16-bit mathematical operations such as addition, subtraction, multiplication, division, data conversion and logical operations like logical *NOT, OR* or *AND*. It also performs register increment, decrement, and shift operations.

(c) Flag Register

8086 processor has nine, 1-bit flags to reflect the status of the CPU. The 16-bit flag register of 8086 stores the information about the status of the processor and the status of the instruction executed most recently.

(d) General Purpose Registers

The 8086 processor has four 16-bit general purpose registers, which are AX, BX, CX and DX. Each of such 16-bit registers can be considered as two 8-bit registers distinguished as high and low order bytes of the respective 6-bit registers and referenced as AH, AL, BH, CH, CL and DH, DL.

(e) Stack Pointer Register

Register SP points to the current top of the stack, while the register BP is used as the stack pointer. The SI and DI registers are the source and destination index registers respectively used primarily for manipulating strings.

(f) Pointer and Index Registers

The 8086 has one more pointer register known as base pointer *(BP)* register in addition to the stack pointer *(SP)*. The register BP can be used as a pointer to a memory location, which is similar to the usage of registers BX, SI and DI. This register is generally used for accessing the parameters from the stack in the procedures.

8086 processors has two index registers known as the source index *(SI)* register and the destination index *(DI)* register. The SI register is especially used as a pointer to a memory location, which is similar to the usage of the BX register. It is suitable for accessing the elements of an array from the memory. The use of DI register as a pointer to a memory location, is similar to the usage of BX and SI registers. It is suitable for accessing the elements of an array from the memory.

1.6 PIN DESCRIPTION OF 8086 MICROPROCESSOR

Fig. 1.4 depicts functional pin diagram of 8086 microprocessor, where only functions of various pins are shown, without actual pin numbers.

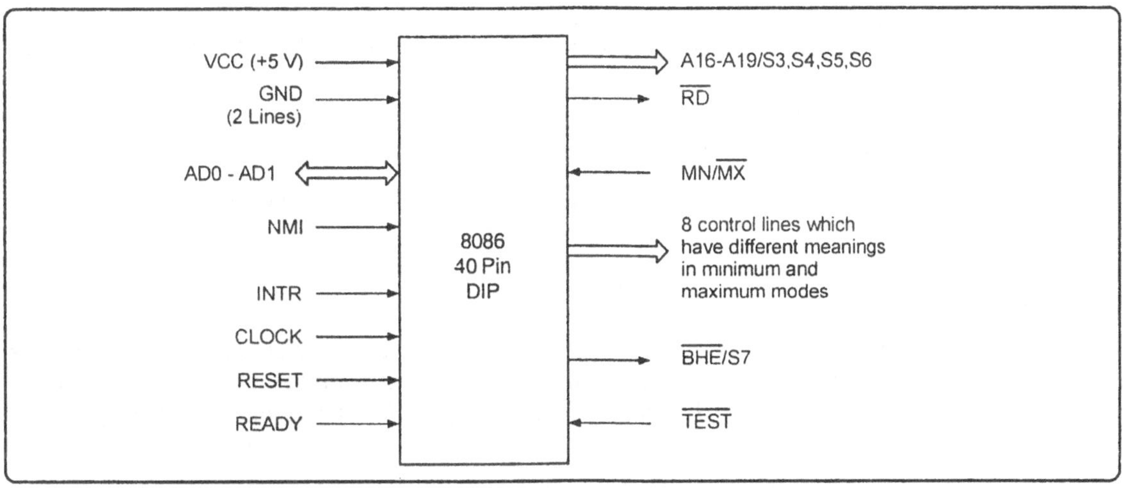

Fig. 1.4 Functional pin diagram of 8086 Multiprocessor

The 8086 processor is packaged in a 40 pin, Dual In-line package *(DIP)* which requires +5 volts power supply. There are 2 pins which are required to be connected to ground. Now let us brief about important functional pins such as :

(*a*) Clock
(*b*) Ready
(*c*) AD 15-0
(*d*) A19-16/S6-3
(*e*) $\overline{RD}$
(*f*) $\overline{BHE}$ / *S*7
(*g*) NMI and INTR
(*h*) Reset
(*i*) *MN* / $\overline{MX}$

(*a*) Clock

All microprocessors require a clock for timing purpose. For 8086, it requires to generate clock externally with the help of an external clock generator/driver chip like 8284. However, 8086 is available is different clock speeds, without change in its internal architecture. Now for instant, 8086-1 works at 10 MHz, 8086-2 at 8 MHz and 8086 works at 5 MHz and so on. The Intel 8254 is divided by 3 circuits internally to provide clock input for 8086. A 30 MHz crystal is to be connected to 8284 for 8086-1 to work at 10 MHz.

(*b*) Ready

The 8086 generally needs 4 clock cycles for writing/reading information to/from memory I/O ports. These clock cycles are termed T1, T2, T3 and T4. These 4 clock cycles in which the μP performs a read or write operations is called a machine cycle. The Ready pin is checked in the middle of T2 by 8086. The Ready pin is an active high input pin to the 8086. If it is a 0, the μP inserts a Wait state between T3 and T4.

When Ready signal is still 0, at the end of the Wait state, then one more Wait state is introduced. The μP enters T4 state only if Ready pin is active at the end of T3, or a Wait state.

Wait state lasts for a clock cycle, and during this time, the various signals sent out by the μP remain the same as they were at the start of the Wait state.

Note : Ready pin receives the input from 8284, which synchronizes the ready signal with the clock.

(*c*) AD 15-0

Since 8086 is a 16-bit μP, with 16 data pins, D15-0, which are bidirectional. 8086 receives 16-bits of information on these pins from memory or I/O ports, manipulates the information internally, and then sends out 16 bit result to memory or I/O ports on the same pins. *(If it is required 8086 can also work on 8 bit information).*

Note. 8086 is provided with twenty address pins. With 20 address pins it can address 2^{20} = 1 million memory locations. In order to save pins, the data and the least significant (LS) 16 bits of address are multiplexed. So, we have in the diagram the pins AD 15-0, which are used for data as well as the LS 16 bits of address. During T1 state of a machine cycle the μP sends out address on AD 15-0. In the later part of a machine cycle the μP sends out data or receives data on AD 15-0.

(*d*) A19-16/S6-3

Pins A19-16 provide the most significant *(MS)* 4 bits of memory address during the first clock cycle, T1, of a machine cycle. In case, the 8086 is sending out address for accessing I/O ports, the information on these pins will be zeros during T1. This is because, I/O port addresses in 8086 are only 16 bits. In subsequent clock cycles these pins send out S6-3 status information. The memory in a 8086 based microcomputer can be visualized as shown below, in Fig. 1.5.

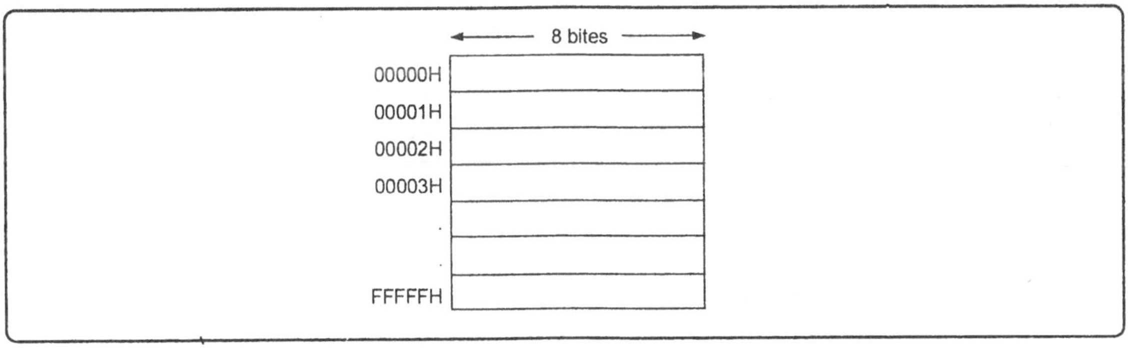

Fig. 1.5. Memory in a 8086 based microcomputer

From Fig. 1.5 it is observed that, the number of bits stored in every memory location is not 16 bits, but only 8 bits. In other words, every byte has a separate address. This is because, sometimes it is desired to access only a byte of information from memory. To facilitate this, the memory is having the ability to address an individual byte.

Thus two locations are needed to store a word of information. Therefore the 1 million locations, 0.5 million words can be stored. Word 0 information is contained in memory locations 0 and 1. Next, contents of locations 2 and 3 constitute word 2, and not word 1. Next, contents of locations 2 and 3 constitute word 2, and not word 1. That is, Word 1 is the contents of locations 1 and 2.

In few computers, like PDP-11, words with odd addresses, like word 1 or word 3, are illegal. However in 8086, we can access words with odd addresses also. But accessing of odd addressed words takes twice the time needed to access an even addressed word.

For word 4, the LS byte of the word is at byte location 4, and the MS byte of the word is at byte location 5. Similarly, for word, 7, the LS byte of the word is at byte location 7, and the MS byte of the word is at byte location 8. Thus, for any word in memory, the 8086 stores the LS byte at a lower byte address, and the MS byte at the next higher byte address. *(Which is similar to that with 8085 μP).*

(e) $\overline{\text{RD}}$

$\overline{RD}$ pin is an active low output pin. Whenever the μP needs to get information from a memory location or an I/O port, it activates the $\overline{RD}$ pin.

(*f*) $\overline{\text{BHE}}$ / S7

During the first clock cycle, T1, of a machine cycle, the μP uses this pin to send out $\overline{\text{BHE}}$ *(Bus High Enable).* In subsequent clock cycles the μP sends out a 0 on S7 as status information. It is a spare status line. It is used by 8087 numeric coprocessor to determine whether the CPU is a 8086 or a 8088.

Physically in 8086 based microcomputer, the main memory can be viewed as two banks, each of 512 K bytes maximum. The banks are called the *'Lower bank'* and the *'Upper bank'*. To select a location in these banks, a 19 bit address (*as* 2^{19} = 512*K*) is sent by the 8086 on A19-1.

Lower bank contains bytes with only even addresses, like 0, 2, 4 etc. In other words, location 0 in this bank will have the address 0, location 1 will have the address 2, location 2 will have the address 4 etc. The data lines of the lower bank is connected to D7-0 of 8086. The lower bank is selected by A0. The upper bank contains bytes with only odd addresses, like 1, 3, 5 etc. In other words, location 0 in this band will have the address 1 location 1 will have the address 3, location 2 will have the address 5 etc. Data lines of the upper bank is connected to D15-8 of 8086 and the upper bank is selected by $\overline{BHE}$.

(*g*) NMI and INTR

All microprocessors (μP), will have some interrupt pins. In a microcomputer system whenever an I/O port wants to communicate with the μP urgently, it interrupts the μP. Then, the μP completes the instruction it is presently executing. Then, it saves the address of the next instruction on the stack top, and branches to an Interrupt Service Subroutine (ISS), to service the interrupting I/O port. After completing the ISS, the 8086 returns to the original program, by making use of the address that was saved on the stack top. The Stack and the stack operations are described in the later chapter.

In 8086 there are 2 interrupt pins, INTR and NMI.

Where, INTR stands for Interrupt. It is an input pin to the 8086, and is an active high signal. Whenever an external device activates this pin, the μP will be interrupted only if interrupts are enabled using a STI *(set interrupt flag)* instruction.

If interrupts are disabled using CLI *(clear interrupt flag)* instruction, the μP will not get interrupted even if INTR line gets activated by an external device. In other words, INTR can be masked.

INTR is a non vectored interrupt. This means, the 8086 does not know where to branch, to service the interrupt. The 8086 has to be told by an external device like a Programmable Interrupt Controller regarding the branch to be made.

Next, NMI stands for Non Maskable Interrupt. It is an input pin to the 8086, and is an active high signal. Whenever an external device activates this pin, the μP will be interrupted. CLI and STI instructions have no effect on NMI. *(In other words, this signal cannot be masked).*

Note. NMI is a vectored interrupt. This means, the 8086 knows where to branch, to service the NMI interrupt. When both NMI and INTR are activated at the same time, NMI will be serviced first. Thus NMI has a higher priority than INTR.

(*h*) Reset

When a 1 is sent into the 8086 on Reset pin, for at least 4 clock cycles, the 8086 μP gets reset. That is, it starts from a clearly defined initial state.

The CS register contents will become FFFFH and IP register contents changes to 0000H. After reset, the first instruction to be executed will be fetched from FFFF0H. The other segment registers viz. DS, ES, SS will be initialized to 0000H, and the Flags register contents become 0000H.

Next, the processor gets into disable interrupt state. Only NMI can interrupt the 8086 after a reset. If INTR is required to interrupt the 8086, 'enable interrupt' instruction has to be executed.

Reset pin of 8086 receives the input from the Reset out of 8284, which synchronizes the reset signal with the clock. Also, the Reset out signal of the 8284, resets the rest of the circuitry in the microcomputer.

There are 2 ways of resetting the 8086, microprocessor

(*i*) When the Reset on the microcomputer is momentarily pressed and then released, the Reset pin is activated for a few clock cycles.

(*ii*) Whenever the microcomputer is switched on, the Power on reset circuit will activate the Reset pin for at least 50 μ seconds.

(*i*) MN / $\overline{\text{MX}}$

In simple systems with a single μp, all the control signals needed in the microcomputer system can be directly generated by the 8086.

But, in more complex systems, especially those containing arithmetic coprocessor like Intel 8087, I/O processors like Intel 8089, and multiple processors, more number of control signals are required to be generated for proper interaction among them *(However, 8086 has only 40 pins).*

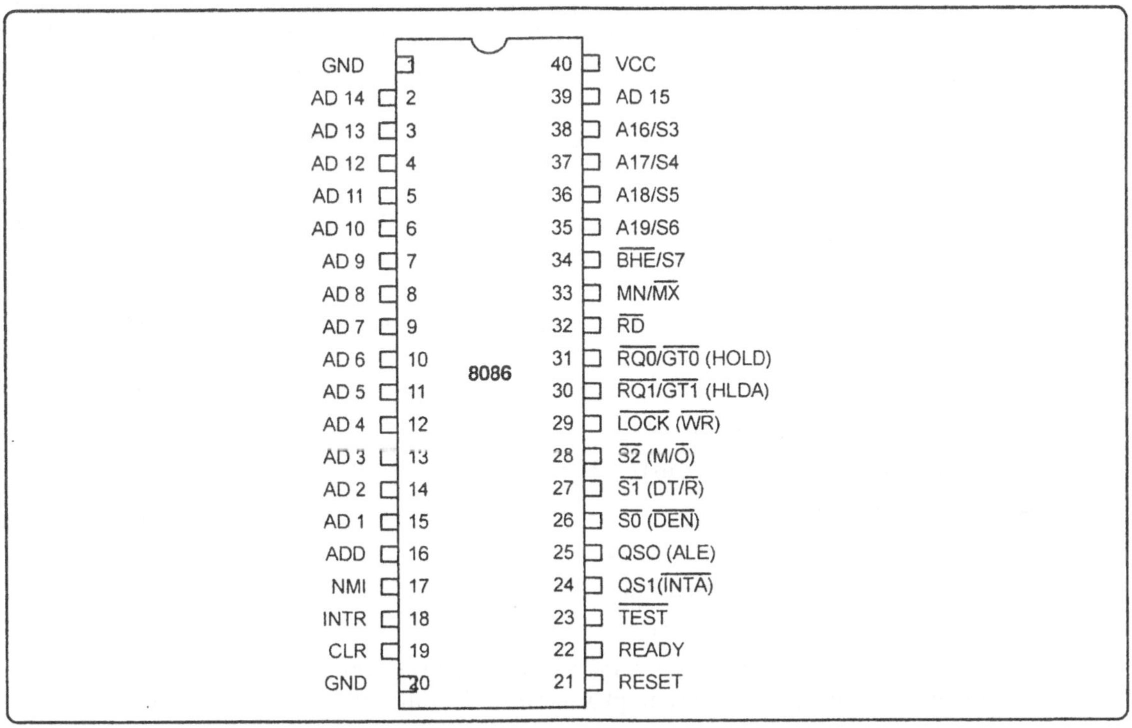

Fig. 1.6 Pin Diagram of 8086 Microprocessor

This problem is solved by allowing 8086 to work in 2 different modes, called the **'Minimum'** and **'Maximum'** modes. This is indicated by the input pin $MN/\overline{MX}$. In simple systems, where 8086 is the only processor in the system, we connect $MN/\overline{MX}$ pin to +5 volts. In case the 8086 directly generates all the control signals needed in the system. In complex systems, which for instant uses a coprocessor, we connect $MN/\overline{MX}$ pin to Ground. In this case, pins 24 to 31, a total of 8 pins will have different functions in minimum and maximum modes.

Now Fig. 1.6 depicts the pin diagram of 8086 microprocessor, where all pins of the processor are shown. However, detail of all other pins are not discussed here.

1.7 WORKING PRINCIPLE OF 8086 MICROPROCESSOR

8086 CPU consists of two separate processing units, *i.e.,* Execution Unit *(EU)* and the Bus Interface Unit *BIU* connected by a 16-bit ALU data bus and an 8-bit queue bus. The EU obtains instructions from the instruction prefetch queue *(IQ)* maintained by the BIU and executes the instruction using a 16-bit ALU. Execution of the instruction involves maintenance of the CPU status, the control logic, and the manipulations of segment and offset address within the 16-bit limits. The EU accesses the memory and peripheral devices through requests to the BIU, which is the second processing unit, performing the bus operations for the EU on a demand basis. It involves generating physical addresses from the segment registers, the offset values and writing out the results. The BIU is also responsible for prefetching instructions from the IQ whenever possible, to keep the EU busy with prefetched instruction under normal condition and for resetting the IQ when EU transfers control to another location *(say JUMP). (The BIU and EU operate independently enabling the 8086 to overlap the instruction fetch phase and the execution phase so as to speed up the execution of instructions).*

1.7.1 Instruction set of 8086

8086, supports a wide array of instruction to perform the movement of data *(between memory, CPU, I/O devices),* arithmetic and logical operations *(addition, subtraction, multiplication, division, data conversion, logical instructions),* branch *(jump instructions),* and control of processor operations *(enable interrupt, set direction flag),* string operations, and protection control.

In 8086 microprocessor instructions are classified into the following categories :

- Data Transfer Instructions
- Arithmetic and Logical Instructions
- Branch Instructions
- Processor Control Instructions
- String Operation Instructions
- Protection Control Instructions

Instruction set of 8086 is upward compatible, *i.e.,* programs written for lower version processor can be run on higher version processors without further modification.

1.7.2 Instruction Types

8086 processor instructions are classified into four groups depending on the number of operands explicitly specified in the instructions. They are zero, single, double and three operand instructions.

■ Zero Operand Instructions

The instructions in which the operands are not explicitly specified are known as zero operand or address instructions. The implicit operands are assumed to be in the registers.

Examples :

Instruction		Explanation
CMC	;	Complement carry
STC	;	Set carry
CLD	;	Clear direction flag

The above instructions operate on an implicitly assumed 8086 flag register.

■ One Operand Instructions

The instruction in which a single operand is explicitly specified are known as one operand or address instructions.

Examples :

Instruction		Explanation
INC AX	;	AX ← AX + 1
DEC CX	;	CX ← CX –1, decrement CX content by 1
POP BX	;	pop BX from the stack

The above instructions operate on the single operand specified in the instruction.

■ Two Operand Instructions

The instructions in which two operands are explicitly specified are known as two operands or address instructions.

Examples :

Instruction		Explanation
MOV AX, 100	;	AX ← 100
ADD AX, BX	;	AX ← AX + BX
SUB CX, 1	;	CX ← CX –1

The above instructions operate on the two operands specified in the instruction.

■ Three Operands Instructions

The instructions in which three operands are explicitly specified are known as three operands or address instructions. The three operands instruction IMUL is supported by the 80186 processor only.

Examples :

Instruction	Explanation
IMUL BX, CX, 10	; BX ← CX * 10
IMUL AX, BX, 1024	; AX ← BX * 1024

The above instructions operate on the three operands specified in the instruction.

1.7.3 Segment Registers

8086 uses memory segmentation. It treats the 1 Mega byte of memory as divided into segments, with the maximum size of a segment as 64 K bytes. Thus a location within a segment can be addressed using 16 bits. *(This is convenient in view of the 16 bit word size of 8086).*

A segment always starts at a memory location with 0H as the LS hex digit, *(for a example : 23450H)* A memory address with 0H as LS hex digit is called a paragraph boundary. This constraint was put on the starting address of a segment so that it is enough to specify only the MS 16 bits, with the LS 4 bits always taken as zeros. And working with 16 bits comes naturally to 8086.

There are 4 different types of segments in memory. They are the Code Segment, Data Segment, Data Segment, Stack Segment, and the Extra Segment.

1.7.4 CS (Code Segment) Register

Code Segment holds the program code. Where exactly the Code Segment starts in memory is indicated by the contents of a register called CS *(Code Segment)* register. Once again, we may think that the CS register should be a 20 bit register, to indicate the starting address of Code Segment in memory. But CS register is only 16 bits wide. But, we can imagine it to be a 20 bit register where the LS hex digit is 0H *i.e.,* the LS 4 bits are always zeros. This 20 bit value provides the starting address of the Code-Segment in memory.

For instant, if the contents of CS is 2345H, it means that the Code Segment starts from memory location 23450H, which is a 20 bit address. Thus we find that the Code Segment always starts at a paragraph boundary in memory. For example, the Code Segment can not start at 23456H. *(The 16 bit contents of a segment register, like CS, is called the Segment Offset.)* Let us see the generation of 20 bit address.

■ Generation of 20 bit Address

Let us observe how 20 bit address is generated by the BIU using the 16-bit IP contents. Let us say, IP content is 123CH. This is called the Offset value, or Effective Address *(EA)*. Let the contents of CS be 2345H. This is called the Segment Base value. This means that the Code segment starts at 23450H, and we want to fetch an instruction from a memory location which is 123CH bytes away from the beginning of the Code Segment. Thus the BIU adds up 2345H and 123CH using the adder in the BIU, and sends out the sum 2468CH as the Physical Address *(PA)* on the 20 address pins.

Segment Base value in CS = 2345H

20 bit starting address of Code Segment = 23450H

16 bit EA or offset in IP = 123CH

20 bit PA = 2468CH

A 20 bit physical address of 2468CH is quite often represented in Segment Base : Offset form as 2345 : 123C, as this form clearly indicates the Segment Base value and the EA.

Note. The segment and offset addressing scheme appears quite complex. But it provides a very important advantage. It allows programs to be relocated in memory. A relocatable program can be placed in any area of memory and executed without any changes. Similarly, relocatable data can be placed in any area of memory and used without any change to the program. To enable proper relocation the only thing that must be changed are the contents of the segment registers

1.7.5 Instruction Queue

If CS contains 2345H and IP contains 123CH, the BIU sends out 2345H : 123CH = 2468CH as the address. Hex digit 2 is sent out on A19-16, and 468CH is sent out on AD 15 - 0. To indicate that address is sent out on the multiplexed pins, the 8086 activates the ALE signal during T1. Using the ALE signal, an external latch will latch on to the address information sent out by the 8086. That means, even though the 8086 stops sending out the address on its pins after T1 in a machine cycle, the memory or I/O ports continue to receiver the address from the latch for whole machine cycle.

The 8086 activates $\overline{RD}$, and outputs a 1 on $M / \overline{IO}$, indicating that it wants to read from the addressed memory location. In the subsequent clock cycles of the machine cycle, the 8086 receives from memory 16 bit instruction code on the multiplexed address data pins. Instruction length depends on the instruction. It is a minimum of 1 byte and a maximum of 6 bytes long. With segment override prefix, to be discussed later, it uses another extra byte to make the maximum size of an instruction as 7 bytes long. An instruction code enters the 6 byte instruction queue in the BIU. It works on first in first out basis.

For instant assume, we have received from memory into the instruction queue, the 2 byte code for the instruction MUL BX. This instruction multiplies the contents of AX and BX registers *(to know about these registers, let us have a slight digression).*

1.7.6 General Purpose Registers

In all μP there will be a number of registers for temporary storage of frequently required data. Accessing data in a register is always much faster than accessing a memory location. In 8086, processor there are 4 registers, which are AX, BX, CX and DX. Each of them can hold 16 bit information. They are 16 bits wide because 8086 is a 16 bit μP. Which normally manipulates 16 bit information.

Normally, in many instructions we require 8086 to work on 8 bit data. To facilitate this, these registers can be thought of as divided into two 8 bit portions. AX can be thought of as divided into AH and AL, which respectively hold the High and Low bytes of AX. Similarly, BX is divided into BH and BL, CX is divided into CH and CL, and DX is divided into DH and DL.

In addition to serving as general purpose registers, they play the following special roles.

- In few instructions, like the 16 bit Multiply, one of the operands must be in AX. The other operand can be a word in memory location or in another register. The resulting 32 bit product is stored in DX and AX, with DX storing the MS word.

■ In few instructions, like the 8 bit Multiply, one of the operands must be in AL. The other operand can be a byte in memory location or in another register, like BH. The resulting 16 bit product is stored in AX, with AH storing the MS byte.

■ In instructions where we need to specify in a general purpose register the 16 bit EA of a memory location, the register has to be BX.

■ In LOOP instructions to be discussed later, CX register will be always used as the implied counter.

■ In I/O instructions, the 8086 receives into or sends out data from AX or AL, depending on a word or byte operation. In these instructions, the port address, if greater than FFH, has to be given as the contents of DX register.

Note. *(On Instruction Queus)* Earlier we digressed to discuss about general purpose registers. Let us say, we have in the instruction queue the 2 byte opcode for MUL BX. This instruction is now decoded by the instruction decoder in the EU. Thus the EU understands that it has to multiply the contents of AX and BX, and store the product in DX and AX.

Now, Control unit in the EU, which is responsible for controlling all the operations in the microcomputer, directs the ALU to perform the required multiplication. In this operation, all the operands needed for the instruction execution are within the 8086 registers. Still, the instruction needs more than 100 clock cycles for the execution. This is because, multiplication has to be internally performed using a number of shift and add operations. *(There are a number of instructions in 8086 which need a quite a large number of clock cycles).*

During the time the EU is busy executing the instruction, the BIU instead of simply idling, fetches the next instruction or instructions from memory into the instruction queue, which is done as long as the queue is not full. That means the time an instruction like MUL BX is executed, the next instruction would have arrived in the queue. All that the EU has to do is, to decode the instruction in the queue and start executing it. This way, the instruction fetch time is eliminated. Except in the case of branch group of instructions, where the queue should be newly formed, this scheme greatly speeds up the processing. *(This feature of fetching the next instruction while the current instruction is being executed is called Pipelining).*

1.7.7 DS (Data Segment) Register

During program execution, the µP accesses a data memory location for a read or a write operation. In 8086 based microcomputer, unlike a 8085 based microcomputer, there will be a separate Data Segment, where the data will reside.

Data Segment holds the data, constants, and work areas needed by the program. Where exactly the Data Segment starts in memory is indicated by the contents of a register called DS *(Data Segment)* register. As before, we may think that the DS register should be a 20 bit register, to indicate the starting address of Data Segment in memory. But DS register is only 16 bits wide. But, we can imagine it to be a 20 bit register where the LS hex digit is 0H which provides the starting address of the Data Segment in memory.

1.7.8 SS (Stack Segment) Register

A Stack is a portion of memory set aside to store addresses and data, when the µP branches to a subroutine. Stack is a Last in First out *(LIFO)* data structure implemented in the RAM. During program execution, the μP accesses the stack for a pop or a push operation. These operations are discussed later. In the case of 8086 based microcomputer, unlike a 8085 based microcomputer, there will be a separate, Stack Segment, where the stack will reside.

Where exactly the Stack Segment starts in memory is indicated by the contents of a register called SS *(Stack Segment)* register. But SS register is only 16 bits wide. But, we can imagine it to be a 20 bit register where the LS hex digit is 0H. This 20 bit value provides the starting address of the Stack Segment in memory.

For instant, if the contents of SS is A200H, it means that the Stack Segment starts from memory location A2000H, which is a 20 bit address. The Stack segment ends at A2000H + FFFFH = B1FFFH.

The top most filled location in the stack is indicated by the SP *(Stack Pointer)* register, SP contains a 16 bit EA, and is in the EU of 8086. Suppose SP content is 3456H, that means, A2000H + 3456H = A5456H is the PA of the top most filled location in the stack. SP contents makes use of SS contents to generate 20 bit PA. In 8086, the stack operations are on words only.

BP, which stands for Base Pointer, is a 16 bit register in the EU. It is used for storing a 16 bit offset value or EA. To generate the 20 bit PA it uses the SS register. That means both SP and BP, are used to access data in the Stack Segment. BP can also be used for holding a 16 bit data temporarily. *(But, as the main function of BP is to hold 16 bit offset, this register cannot be thought of as divided into two 8 bit portions. BX and BP are called the pointer registers).*

1.7.9 ASCII Codes

ASCII stands for American Standard Code for Information Interchange. In computer programs which are in Assembly language, or a High level language, there is a need to represent characters like A, B,.. Z, and a, b,.. z, and numerals like 0, 1,.. 9, and punctuation marks like ? . , ‘, ”, etc. as well as symbols like +, -, *, / etc. But everything inside a computer has to be in the form of 0s and 1s only. Such problem is solved by assigning a code or a numerical value to each of the above using only 0s and 1s. A very popular code is the 7 bit ASCII code. With 7 bits, it is possible to have 128 distinct numbers *(00H to 7FH),* each of which corresponds to a character. For example, the ASCII code for ‘A’ is 1000001 = 41H. ASCII code for ‘0’ is 011 0000 = 30H.

IBM PC uses extended ASCII characters. This means it uses 8 bit codes. It appends a 0 at the MS bit position for the standard ASCII code. Thus, extended ASCII code for ‘A’ is 0100 0000 = 41H. Hence, standard ASCII characters will have codes from 00H to 7FH.

With 8 bit code we can have 256 distinct numbers *(00H to FFH).* Codes having values from 80H to FFH represent the extended ASCII character set which is used to represent graphic characters like ‘ - ’, Greek letter like *‘m’* etc.

Ex : The ASCII code for ‘MUL BX’ will be 4D 55 4C 20 42 58. This 6 type ASCII code is translated by the assembler to a 2 byte operation code.

Note. One more popular character code, the Extended Binary Decimal Interchange code *(EBCDIC, pronounced as ebb’see dick)* represents each character by an eight bit code.

1.7.10 ES (Extra Segment) Register

A string is a series of bytes or words in sequential memory locations. A string generally consists of ASCII character codes.

The 8086 μP has a number of instructions for manipulation of strings. If we want to move a string from one memory area to another, the string must be in the Data segment, with the offset value or EA specified in SI register. SI stands for Source Index and is a 16 bit register in the EU. SI can also be used for holding a 16 bit data temporarily. But, as its main function is to hold 16 bit offset, this register cannot be thought of as divided into two 8 bit portions.

The destination where the string will be moved must be in another segment called the Extra Segment. Where exactly the Extra Segment starts in memory is indicated by the contents of a register called ES *(Extra Segment)* register. ES register is only 16 bits wide. But, we can imagine it to be a 20 bit register where the LS hex digit is 0H, which provides the starting address of the Extra Segment in memory.

Address for the destination of the string must be in the Extra segment, with the offset value or EA specified in DI register. DI stands for Destination Index, and is a 16 bit register in the EU. DI can also be used for holding a 16 bit data temporarily. But, as its main function is to hold 16 bit offset, this register cannot be though of as divided into two 8 bit portions. SI and DI registers are called as the Index Registers.

For accessing a data memory location, using a register to provide the EA, in operations other than string manipulations, the DS register provides the Segment base value and the offset is provided in one of BX, SI or DI.

Note. With only 4 segments, we may feel that the total amount of memory that can be used by 8086 is only 4 × 64 K = 256 K Bytes. However there may be any number of segments of each type. For example, to address another Data Segment, it is only required to change the contents of DS register.

1.7.11 Flag Register

A Flag is a flip-flop which will be in the 1 state or 0 state. In other words, a flag is only one bit of information. Flags are of two types, which are the *'Status flags and the Control flags'*.

Flag is set to 1 or reset to 0 by the 8086 to indicate to the programmer some situation like whether the execution of an instruction produced a negative, positive or zero result etc. Such flags are called the Status flags, and few flags are deliberately set to 1 or reset to 0 by the programmer to control the working of 8086. Such flags are called the Control flags.

Flags register is a 16 bit register in the EU. But only 9 bits are meaningful. The other 7 bits do not contain any useful information. As working with 16 bits is convenient for 8086, this register is also 16 bits wide.

There are six Status flags : which are Z(zero), S(sign), P(parity), C(carry) A(auxiliary carry)_, and O(overflow) flags. The remaining 3 are the Control flags : which are D (direction), I(interrupt enable), and T(trap) flags. The contents of the Flags register is a shown below in Fig. 1.7 (x stands for don't care). There is no need to memorize the position of the various flag bits. There are instructions in the branch group, which will perform a branch depending on the values of the various flags.

Bit no.	15	14	13	12	11	10	9	8	7	6	5	4	3	2	1	0
Flags	x	x	x	x	O	D	I	T	S	Z	x	A	x	P	x	C

Fig. 1.7 Flag format in 8086

Now after looking into the flag format in Fig. 1.7, let us discuss about

(*a*) Status flags

and (*b*) Control flags

(*a*) Status Flags : The flags can be will understood with examples as under :

Example 1. Let us consider the execution of the instruction ADD BX, CX. It adds the contents of BX and CX, and stores the result in BX. Depending on the result in BX, the status flags as mentioned below :

	0 1 0 1
(BX)	1234H
(CX)	7F2EH
(BX)	9162H = 1001 0001 0110 0010

■ **C Flag** C (carry) flag is set to 1 if a carry resulted in the overall addition of the two numbers. In this example, the result is 9162H with a carry of 0. As such, the C flag is reset to 0. In the case of a subtract operation the C flag is set to 1, if we had to borrow in the overall subtract operation.

■ **A Flag** A (auxiliary carry) flag is set if a carry resulted during the addition of the LS hex digits. In this example, the addition of 4 and E in LS hex position resulted in a carry. As such, A flag is now set to 1. In the case of a subtract operation the A flag is set to 1, if we had to borrow in the subtraction of LS hex digits.

The A flag is not of any use to programmer. The programmer is not interested in knowing about the generation or otherwise of this auxiliary carry. But the 8086 makes use of this flag in Decimal arithmetic instructions like DAA.

■ **Z Flag** Z (zero) flag is set to 1 whenever the result stored in a register or a memory location is zero. Otherwise it is reset to 0. In this example, the result stored in the register BX is 9162H, which is non zero. As such, the Z flag is reset to 0.

■ **S Flag** Positive numbers are represented in a 8086 based computer with a 0 in the MS bit position, and the remaining bits provide its magnitude. For instant, +3 is represented using 16 bits as

0000 0000 0000 0011 = 0003H

Then negative numbers are represented in 2's complement notation. Thus, -3 is represented as 2's complement of the above number, as shown below.

1111 1111 1111 1101 = FFFDH

Observe that negative numbers always start with a 1 in the MS bit position. However, if we are working with unsigned numbers, FFFDH corresponds to the number 65533, and not - 3.

For deciding the value of S flag the 8086 treats the result as a signed number. S flag is set to 1 if the MS bit of the result stored in a register or a memory location is a 1. In other words, S flag is to 1 if the result is negative. It is reset to 0 if the result is positive. Of course, this flag is meaningful only if we are working with signed numbers. In case we are working with unsigned numbers, we just ignore this flag.

In the above example, for addition the result stored in BX is 9162H, which starts with a 1 in the MS bit position. As such, the S flag is set to 1.

■ **O Flag.** If we are working with unsigned numbers. Then, in the addition example above, we have got the correct answer in BX later. If we happen to add 2345H and F256H, we get 159BH with C flag = 1. The result can then be interpreted as 1159BH, which is also correct. With unsigned number arithmetic we always get the correct answer.

Overflow indicates a wrong answer. The processor gives a wrong answer sometimes when it is working with signed numbers. In the example of ADD BX CX is we treat the numbers as signed numbers, the numbers 1234H and 7F2Eh are positive numbers *(they start with a 0 in the MS bit position)*. But the result 9162H is negative, as it starts with a 1 in the MS bit position. This is obviously a wrong answer. Thus in the above example, the O *(overflow)* flag will be set to 1.

Note. O flag value is decided by the 8086 assuming that it is working with signed numbers. The O flag is meaningful only if we are working with signed numbers. If we are working with unsigned numbers, we just ignore this flag.

(Why did overflow occur in the above example ? The largest positive number using 16 bits is 7FFFH, and the sum is required to be larger than this value, which is not possible. So, we get a wrong answer. It may be noted that overflow is a possibility, only when we are adding numbers of the same sign, or subtracting numbers of the opposite sign).

■ **P Flag.** After an arithmetic operation, the 8086 counts the number of 1s in the result for a byte operation, or in the LS byte of result for a word operation.

Now, if there are even number of 1's, the *(parity)* flag is set to 1. If there are odd number of 1s, the P flag is reset to 0. This flag is not of much use, except to check for errors in data received from a communication link.

In the above example the result 9162H has 3 number of 1's in the LS byte. As the result has an odd number of 1's, the PO flag is reset to 0.

***Example 2** : Now let us consider the execution of the instruction ADD BL, CH. It adds the contents of BL and CH, and stores the result in BL. Depending on the result in BL, the status flags are affected as below :*

```
        1 1
(BL)    04H
(CH)    FCH
        ----
(BL)    00H  =  0000 0000
```

As a carry resulted in the overall addition, C flag is set to 1. Since there was a carry generated in the addition of LS hex digits, the A flag is also set to 1. As the 8 bit result stored in BL is 00H, the Z flag is set to 1. The S flag is reset to 0, as 00H starts with a 0 in the MS bit position. The P flag is set to 1 as 00H has 0 number *(an even number)* of 1s.

The O flag is also reset to 0, as the answer is correct. The number in CH register is -4, if FCh is treated as a signed number. Thus the sum of +4 and –4 is 00H, which is correct.

(*b*) Control flags. In control flags the 3 flags can be analysed as under :

- **I Flag** INTR is a maskable interrupt in 8086. If the 1 (interrupt enable) flag is reset to 0, the 8086 does not respond to the request on INTR. Only if I is set to 1, the 8086 responds to the request on INTR pin. The I flag bit has no effect on NMI, which is a non maskable interrupt. The I flag is set to 1 by executing the STI instruction, and reset to 0 by executing the CLI instruction.

- **T Flag** Generally, any program will have errors or *'bugs'* to start with. In such a case, it is necessary to run the program one instruction at a time, and see the contents of various registers and memory variables at the end of every instruction. This process is called *'single steeping'* through a program.

An 8086 μP provides a facility for single stepping through a program. All we have to do is to set the T (trap) flag to 1. But, there is no instruction is 8086 to directly set the T flag to 1, or reset the T flag to 0. So, to set the T flag to 1, the following instructions are executed, *i.e.*,

```
PUSHF
MOV BP, SP
OR WORD PTR [BP + 0], 0100h
POPF
```

That means the above instructions save the Flags register information on the Stack top, change the value at T bit position (bit 8) to 1 in the word at the stack top, and then transfer this word to the Flags register.

If the T flag is a 1, the 8086 gets automatically interrupted at the end of every instruction. The interrupt request does not come to 8086 onm NMI or INTR. IT is internally generated by the 8086. The 8086 then branches to an Interrupt Service Subroutine (ISS), which should be written to display the contents of the various registers and memory variables on the CRT. At the end of the ISS the 8086 returns to the program to be checked, and executes the next instruction. Once again, at the end of this instruction, the 8086 branches to the single step ISS and displays the register contents. This way, the entire program can be run, an instruction at a time, and the errors can be easily traced and corrected.

However, the instructions in the ISS should not be executed in single step. To facilitate this, the T flag should be made 0 in the beginning of this ISS, using the instructions mentioned are :

```
PUSHF
MOV BP, SP
AND WORD PTR [BP + O], OFEFFH
POPF
```

Note. At the end of the ISS, once again T flag is set to 1, so that our program is interrupted at the end of every instruction.

- **D Flag.** Assume that, we have the string 'Move this' starting at an offset of 1231H in the Data segment. We want to move this string, of size 9 characters *(including the blank character)* to another memory area, starting at an offset off 3451 in the Extra segment.

The 8086 provides MOVSB *(stands for Move String Byte)* instruction to perform this movement a byte at a time. This instruction should be executed 9 times in a loop to perform the desired string movement. MOVSB instruction expects that the source string is pointed by DS : SI. In other words, for the source string, the segment base address must be provided in DS, and the offset in SI. So also ES : DI must point to the destination address for the string.

Therefore, to facilitate the string movement we load SI with 1231H and DI with 3451H. Then the MOVSB instruction is executed. This results in the movement of character 'M' to offset location 3451H in the Extra segment. Then automatically, SI and DI are incremented to 1232H and 3452H respectively, so that next time through the loop, the MOVSB instruction moves the next character.

But, this automatic increment of SI and DI takes place only if the D *(direction)* flag is reset to 0 by executing the instruction CLD. If the D flag is set to 1 by executing the instruction STD, the SI and DI registers are automatically decremented after the execution of MOVSB instruction.

Note. The same string movement can be achieved by executing MOVSB in a loop 9 times, by initially loading SI with 1239H and DI with 3459, and setting D flag to 1.

1.7.12 A19-16/S6-3

As discussed early pins A19-16 provide the most significant (MS) 4 bits of memory address during the first clock cycle, T1, of a machine cycle. In subsequent clock cycles these 9 pins send out S6-3 status information, as explained below.

S4 and S3 indicate the segment register to be used for accessing data as shown in Table 7 below :

Table 7

S4	S3	Segment Register
0	0	ES
0	1	SS
1	0	CS (or none in case of I/O port access)
1	1	DS

S5 indicates the value of the Interrupt enable flag bit. S6 will be always in the 0 state and is not used.

1.8 THE SEGMENT REGISTERS

In 8086/8088 microprocessor there are 20-bit address lines, of 16-bits in length. Therefore to obtain 20-bit addresses from the available 16-bit registers, all 8086/8088 memory addresses are computed by summing the contents of a segments register and a effective memory address. The effective memory address is computed via a variety of addressing modes. The process of adding to obtain 20-bit address can be analysed as under.

First selected segment register contents are shifted-left four bits *(i.e., the contents are multiplied by 16 decimal),* and then added to the effective memory address to generate the actual physical address output. Table 8 below, depicts 16-pit of segment registers CS, DS, ES or SS which are displaced by 4-bits to left.

Table 8

Segment Register value (CS, DS, ES or SS)	x x x x x x x x x x x x x x x x H
Effective Memory Address	y y y y y y y y y y y y y y y y H
Physical Address	wwwwwwwwwwwwwwwwwwwwwwwww H

Effective address is calculated depending on the type of addressing mode, shown as "yyyyyyyyyyyyyyyy". The 20-bit physical address "wwwwwwwwwwwwwwwwwwwww" is obtained later adding the segment register value and effective address. *(Physical address is 20-bits wide).*

To know how the segmentation is used, it demands to know the memory structure of the 8086/8088 microprocessor.

Memory

The memory in an 8086/8088 system is a sequence of up to 2^{20} = one million bytes. A word is any two consecutive bytes in memory *(word alignment is not required)*, Words are stored in memory with the most significant byte at the higher memory address. These bytes are stored sequentially from byte 00000 to byte FFFFF hex.

A Program view memory space as a group of segments which is defined by the application. A segment is a logical unit of memory that may be up to 64K bytes long. Each segment is made up of contiguous memory locations and is an independent, separately addressable unit. Each segment is assigned a base address, which is its starting location in the memory space. All segments starts on 16-bit memory boundaries. Segments may be adjacent, disjoint, partially overlapped, or fully overlapped, is as shown in Fig. 1.8 below.

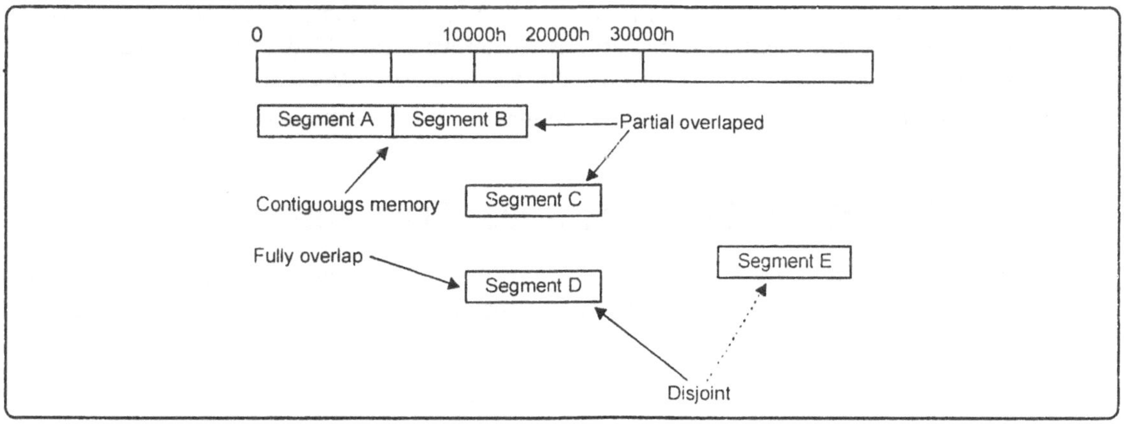

Fig. 1.8

Segment registers point to the four immediately addressable segments.

(*a*) Code Segment register [points to the instruction opcode]

(*b*) Data Segment register [points to the data memory]

(*c*) Stack Segment register [point to the Stack memory], and

(*d*) Extra Segment register [points to the data memory]

Now the Fig. 1.9 depicts the segment registers which are pointing to the various memory segments as below.

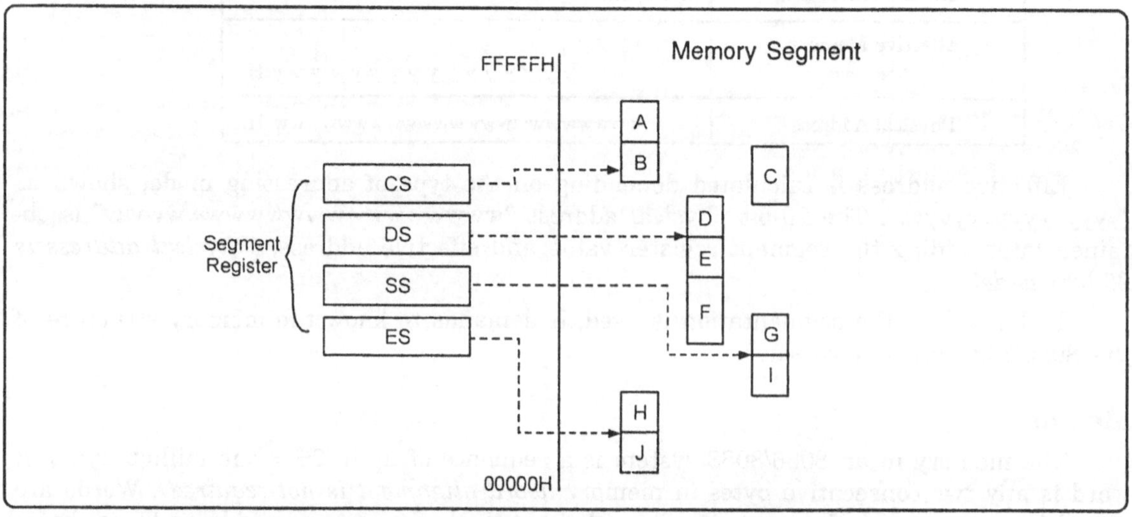

Fig. 1.9 Segment registers pointing memory segments

The logical address are of 16-bits wide, therefore 64k bytes or 65536 bytes can be addressed in a given segment.

Every time the CPU has to generate a memory address, one of the segment registers is automatically chosen and its contents added to a logical address. For an instruction fetch, the code segment register is automatically added to the logical address *(in this case, the contents of the instruction pointer)* to compute the value of the instruction address. Next for stack referencing the stack segment register is automatically added to the logical address *(the SP or BP register contents)* to compute the value of the stack address.

The segmentation makes it possible to re-locate and reentrant the programs.

1.9 COMPARISON BETWEEN 8086 AND 8088

These are many equalities between 8086 and 8088 microprocessors. However the following Table 9 depicts few important comparison between 8086 and 8088 microprocessors.

Table 9

S.No.	Microprocessor 8086	Microprocessor 8088
1.	It has sixteen data lines. Therefore it has AD_0 - AD_{15} signals	It has only eight data lines. Therefore, it has AD_0 - AD_7 and A_8 - A_{15} signals
2.	It has $\overline{BHE}$ signal to access higher byte	As data bus is 8-bit wide, it does not have $\overline{BHE}$ signal
3.	It has 6 byte instruction queue	It has 4 byte instruction queue. Due to 8-bit data bus, instruction fetching is slow and 4 bytes are sufficient for queue
4.	Its pin number 34 is $\overline{BHE}$ /S7. During T_1 (first clock cycle) $\overline{BHE}$ should be used to enable data on to the most significant byte of the data bus. During T_3, T_4 and T_4 status of this pin is logic 0. In maximum mode, 8087 monitors this pin to identify the CPU as a 8088 or a 8086 and accordingly sets its own queue length to 4 or 6 bytes	Its pin number 34 is $\overline{SSO}$. It acts as $\overline{S_0}$ in the minimum mode. In maximum mode, $\overline{SSO}$, pin is always high
5.	In minimum mode its pin 28 is assigned to signal $\overline{M/O}$	In minimum mode its pin 28 is assigned to signal IO/M

Note. The differences between 8086 and 8088 are only in their BIU and not in EU, but the execution unit (EU) is the same for both. As EU is same, the programming instructions are exactly the same for each. Programs written for the 8086 can be run on the 8088 without any changes.

1.10 PROBLEMS

1.10.1 What will be the physical address of the datum, given EA of a datum as 2359 H and DS as 49030H ?

Sol.

Segment Reg.		**Hex values**
DS	–	490B0
EA	–	2359

Therefore the physical address will be (490B0 + 2359)

That is = **4B 409 H**

1.10.2 The physical branch address is 5A23OH when (CS) = 5200H. What will be the physical address if (CS) is changed to 7800 H ?

Sol. The physical address for Cs = 5200 H is

Cs – 5200

Offset – xxxx

Physical adds 5A230H (given)

The offset = Physical adds – *(segment adds in 4-bits)*
= 5A230 – 5200
= 8230H

Now if CS is changed to 7800 H physically adds is
7800 + 8230
= 80230 H

REVIEW QUESTIONS

1. Briefly explain the history of events which lead to Microprocessor development.
2. Describe with neat diagram the architecture of 8086 Microprocessor.
3. Draw the functional pin diagram of 8086 and explain briefly.
4. Briefly explain the action of 8086 when NMI and INTR pins an activated.
5. Describe with examples how a 20 bit physical address of an instruction in generated in 8086 microprocessor ?
6. Briefly explain the utility of introduction queue in 8086.
7. Explain various flags of 8086.
8. Write a note on segment registers.
9. Compair 8086 and 8088 microprocessors.

OBJECTIVE TYPE QUESTIONS

1. *World's first microprocessor was developed in 1971 by ....*
2. *A binary bit stores a .... or a ....*
3. *In a computer 1K is equal to .... bytes.*
4. *In a computer 1M is equal to .... K bytes.*
5. *In a computer 1G is equal to .... M bytes.*
6. *The 8086 microprocessor addresses .... bytes of memory.*
7. *In a microprocessor .... circuits are not included.*
   (i) Memory (ii) Logic (iii) Computational (iv) All
8. *In a microprocessor the fetch/execute cycle is used to get and carry out ....*
   (i) ALU (ii) Logic (iii) Instruction (iv) Arithmetic

2 HARDWARE FEATURES OF 8086/8088

2.0 OBJECTIVE

In this chapter the details of hardware features of 8086/8088 microprocessor will be discussed. The details of pin-outs of the processor is explained. The functional pins and internal block diagram of 8284A is discussed. The Bus buffering and bus timing diagram are also discussed. Finally the Minimum and Maximum mode configuration is explained.

2.1 FUNCTIONAL PIN DIAGRAM OF 8086 / 8088

In this section we will observe the pin function of both 8086 and 8088 microprocessor. Later the explanation of the function of each of the microprocessors pin, along with DC characteristics will be discussed. Now Fig. 2.1 depicts the pin outs of 8086 and 8088 microprocessors.

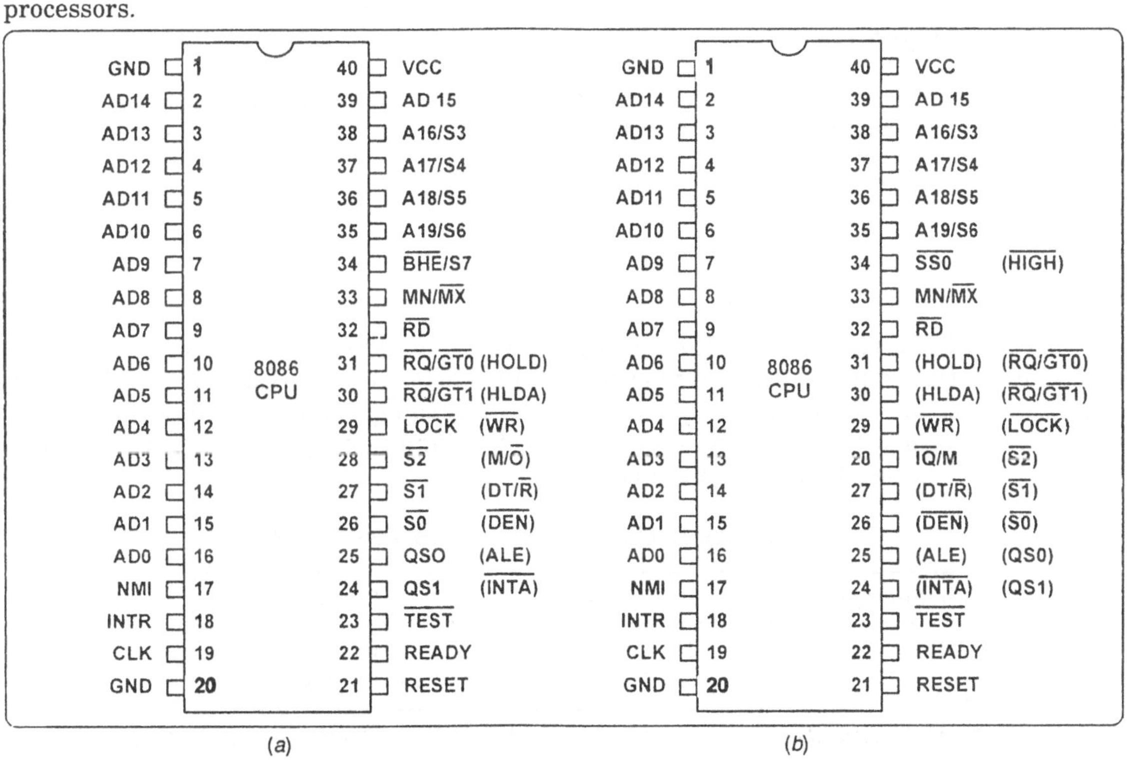

Fig. 2.1 Functional pins or pin out of (a) 8086 and (b) 8088 Microprocessors

From the observation of Fig. 2.1 (*a*) and 2.1 (*b*) it is clear that virtually there is no difference between 8086 and 8088 microprocessors. But both processors are packaged in 40-pin, DIPS dual-in-line packages.

8086 is a 16-bit microprocessor with 16-bit data bus, and the 8088 is a 16-bit microprocessor with an 8-bit data bus. *(As the pin-outs show, the 8086 has pin connections* $AD_0 - AD_{15}$, *and the 8088 has pin connections* $AD_0 - AD_7$*).* Therefore data bus width is the only major difference between these microprocessors. However, a minor difference in one of the control signals. That is the 8086 has an $M/\overline{IO}$ pin, and the 8088 has an $IO/\overline{M}$ pin. The only other hardware difference appears on Pin 34 of both chips : on the 8088, it is an $\overline{SSO}$ pin, while on the 8086, it is a $\overline{BHE}/S_7$ pin.

2.1.1 Power Supply for 8086/8088

Both the microprocessors require + 5.0 V with a supply voltage tolerance of ± 10 percent. 8086 draws a maximum supply current of 360 mA, and the 8088 draws a maximum of 340 mA. Both microprocessors operate in ambient temperatures ranges between 32°F and about 180°F. Which is not wide enough to be used outdoors in the winter or even in the summer, but extended temperature-range versions of the 8086 and 8088 microprocessors are available. There is also a CMOS version, requires very low supply current and also has an extended temperature range, *(80C88 and 80C86 are CMOS versions that require only 10 mA of power supply current and function in temperature extremes of –40°F through +225°F).*

2.1.2 DC Characteristics

It is not possible to connect anything to the pins of the microprocessor without knowing the input current requirement for an input pin and the output current drive capability for an output pin. This knowledge allows hardware designer to select the proper interface components for use with the microprocessor without the fear of damaging anything. Therefore we have to know :

(*a*) Input Characteristics, and

(*b*) Output Characteristics.

(*a*) Input Characteristics. Input characteristics of these microprocessors are compatible with all the standard logic components available today. The Table 1 depicts input voltage level and the input current requirements for any input pin on either microprocessor. The input current levels are very small because the inputs are the gates connections of MOSFETs and represent only leakage currents.

Table 1. Input Characteristics of 8086/8088

Logic level	Voltage	Current
0	0.8 V maximum	± 10 μA maximum
1	2.s0 V minimum	± 10 μA maximum

(*b*) Output Characteristics. The Table 2 illustrates output characteristics of all the output pins of these microprocessors. The logic 1 voltage level of the 8086/8088 is compatible with that of most standard logic families, but the logic 0 level is not. Standard logic circuits have a maximum logic 0 voltage of 0.4V, and the 8086/8088 microprocessor have a maximum of 0.45V, that mean there is a difference of 0.05V.

Table 2. Output Characteristics of 8086/8088

Logic level	Voltage	Current
0	0.45 V maximum	2.0 μA maximum
1	2.4 V minimum	–400 μA maximum

This difference of 0.05 V reduces noise immunity from a standard level of 400 mV (0.8V – 0.45 V) to 350 mV. This reduced noise immunity may result in problems with long wire connections or too many loads. Therefore it is recommended that not more than 10 loads of any type or combination be connected to an output pin without buffers. When this loading is exceeded, noise will begin to take its toll in timing problems, of processor.

Now Table 3 lists some of the more common logic families and the recommended fan-out from the 8086/8088. The best choice of component types for the connection to an 8086/8088 output pin is an LS, 74ALS, or 74HC logic component. Note that even though some of the current calculates to more than 10 loads, it is recommended that if a fan-out of more than 10 loads is required, the system should be buffered.

Table 3. Prescribed Pin-out (fan-out) from any 8086/8088 Pin Connections

Family	Sink Current	Source Current	Fan-out
TTL (74)	–1.6 mA	40 μA	1
TTL (74LS)	–0.4 mA	20 μA	5
TTL (74S)	–2.0 mA	50 μA	1
TTL (74ALS)	–0.1 mA	20 μA	10
TTL (74AS)	–0.5 mA	25 μA	10
TTL (74F)	–0.5 mA	10 μA	10
CMOS (74HC)	–10 μA	10 μA	10
CMOS (CD4)	–10 μA	10 μA	10
NMOS	–10 μ	10 μA	10

2.1.3 Pin Connections of 8086/8088

In this sub section we analyse the pin connections details of 8086/8088 microprocessor, with reference to Fig. 2.1 there are general pin outs and

(*a*) Minimum Mode pins and

(*b*) Maximum Mode pins. Now details of general pin-out of 8086/8088, are :

Pins	**Details**
■ AD_7–AD_0	8088 **address/data** bus lines compose the multiplexed address data bus of the 8088 and contain the rightmost 8-bits of the memory address of I/O port number whenever ALE is active *(logic 1)* or data whenever ALE is active *(logic 10).* These pins are at their high-impedance state during a hold acknowledge.

- A_{15}–A_8 — 8088 **address bus** provides upper-half memory address bit that **are present** throughout a bus cycle. These address connections go to their high-impedance state during a hold acknowledge.
- AD_{15}–AD_8 — The 8086 **address/data bus** lines compose upper multiplexed address/data bus on the 8086. These lines contain address bits A_{15} –A_8 whenever ALE is a logic 1, and data bus connections D_{15}–D_8. These pins enter a high-impedance state whenever a hold connections D_{15}–D_8. These pins enter a high-impedance state whenever a hold acknowledge occurs.
- A_{19}/S_6–A_{16}/S_3 — The **address/status** bus bits are multiplexed to provide address signals A_{19}–A_{16} and also status bits S_6–S_3. These pins also attain a high-impedance state during the hold acknowledge.

  The status bit S_6 always remains a logic 0, bit S_5 indicates the condition of the IF flag bits, and S_4 and S_3 show which segment is accessed during the current bus cycle. The truth table of S_4 and S_3 is as below in Table 4.

Table 4. Functional of Status bit S_3 and S_4

S_4	S_3	Function
0	0	Extra segment
0	1	Stack segment
1	0	Code or no segment
1	1	Data segment

These two status bits could be used to address four separate 1M byte memory banks by decoding them as A_{21} and A_{20}.

- $\overline{RD}$ — Whenever the **read signal** is a logic 0, the data bus is respective **to data** from the memory or I/O devices connected to the system. This pin floats to its high-impedance state during a hold acknowledge.
- **READY** — This input is controlled to insert wait states into the timing of the microprocessor. When the READY pin is placed at a logic 0 level, the microprocessor enters into wait states and remains idle, and when the READY pin is placed at a logic 1 level, it has no effect on the operation of the microprocessor.
- **INTR** — The Interrupt request is used to request a hardware interrupt. For INTR held high when IF = 1, the 8086/8088 enters an interrupt acknowledge cycle *(INTA becomes active)* after the current instruction has completed execution.
- $\overline{TEST}$ — **Test** pin is an input that is tested by the WAIT instruction. If $\overline{TEST}$ is a logic 0, the WAIT instruction functions as a NOP. If $\overline{TEST}$ is a logic 1, then the WAIT instruction waits for $\overline{TEST}$ to become a logic 0. This pin is most often connected to the 8087 numeric coprocessor.

- **NMI** — The **non-maskable interrupt** input is just similar to INTR except that the NMI interrupt does not check to see if the IF flag bit is a logic 1. When NMI is activated, this interrupt input uses interrupt vector 2.
- **RESET** — The reset input makes the microprocessor to reset itself if this pin is held high for a minimum of four clocking periods. If 8086 or 8088 is reset, it begins executing instructions at memory location FFFF0H and disables future interrupts by clearing the IF flag bit.
- **CLK** — The **clock pin** will provides the basic timing signal to the microprocessor. The clock signal must have a duty cycle of 33% *(high for one-third of the clocking period and low for two-thirds)* to provide proper internal timing for the 8086/8088.
- **V_{CC}** — This **power supply** input provides a +5.0 V, ± 10% signal to the microprocessor, to energy.
- **GND** — The **ground** connection is the return for the power supply. *(Note that the 8086/8088 microprocessors have two pins labelled GND i.e., both must be connected to ground for proper operation).*
- **$MN/\overline{MX}$** — The **minimum/maximum mode** pin selects either minimum mode or maximum mode operation for the microprocessor. When minimum mode is selected, the $MN/\overline{MX}$ pin must be connected directly to +5.0 V.
- **$\overline{BHE}/S_7$** — The bus high enable pin will help the 8086 to enable the most significant data bus bits (D_{15}–D_8) during a read or a write operation. The state of S_7 is always a logic 1.

(a) Minimum Mode Pins. The Minimum mode operation of the 8086/8088 is obtained by connecting the $MN/\overline{MX}$ pin directly to +5.0 V. Do not connect this pin to +5V through a pull-up register or it will not function correctly.

- **$IO/\overline{M}$ or $M/\overline{IO}$** — $IO/\overline{M}$ (8088) or the $M/\overline{IO}$ (8086) pin selects memory or I/O, which indicates that the microprocessor address bus contains either a memory address or an I/O port address. This pin is at its high-impedance state during a hold acknowledge.
- **$\overline{WR}$** — The **write line** is a strobe that indicates that 8086/8088 is outputting data to a memory or I/O device. At the time when the $\overline{WR}$ is a logic 0, the data bus contains valid data for memory or I/O. This pin floats to a high-impedance during a hold acknowledge.
- **$\overline{INTA}$** — The **interrupt acknowledge** signal is a response to the INTR input pin. The $\overline{INTA}$ pin is normally used to gate the interrupt vector number onto the data bus in response to an interrupt request.
- **ALE** — **Address latch enable** shows that the 8086/8088 address/data bus contains address information, whose address can be a memory

address or an I/O port number. *(Note that the ALE signal does not float during a hold acknowledge)*

- **DT / $\overline{R}$** The **data transmit/receive** signal shows that the microprocessor data bus its transmitting (DT / $\overline{R}$ = 1) or receiving (DT / $\overline{R}$ = 0) data. This signal is used to enable external data bus buffers.
- **DEN** **Data bus enable** activates external data bus buffers.
- **HOLD** The **hold** input requests a direct memory access (DMA). If the Hold signal is a logic 1, the microprocessor stops executing software and places its address, data and control bus at the high-impedance state. If the HOLD pin is a logic 0, the microprocessor executes software normally.
- **HLDA** The **Hold acknowledge** indicates that 8086/8088 microprocessors have entered the hold state.
- **$\overline{SSO}$** $\overline{SSO}$ status line is equivalent to the S_0 pin in maximum mode operation of the microprocessor. This signal is combined with IO/R and DT/R to decode the function of the current bus cycle as can be observed below in Table 5.

Table 5. The Bus Cycle Status of 8088 With $\overline{SSO}$

IO / $\overline{M}$	DT / $\overline{R}$	$\overline{SSO}$	Function
0	0	0	Interrupt acknowledge
0	0	1	Memory read
0	1	0	Memory write
0	1	1	Halt
1	0	0	Opcode fetch
1	0	1	I/O read
1	1	0	I/O write
1	1	1	Passive

(*b*) Maximum Mode Pins. To achieve maximum mode for use with external coprocessors, the MN / MX pin must be connected to ground.

- **$\overline{S2}$, $\overline{S1}$ and $\overline{S0}$** **Status bits** indicate the function of the current bus cycle, such signals are normally decoded by the 8288 bus controller described later in this chapter. The Table 6 gives the function of these three status bits in the maximum mode.

Table 6 : The Bus Control Functions Generated by the Bus Controller (8288) using $\overline{S2}$, $\overline{S1}$ and $\overline{S0}$

$\overline{S2}$	$\overline{S1}$	$\overline{S0}$	Function
0	0	0	Interrupt acknowledge
0	0	1	I/O read
0	1	0	I/O write
0	1	1	Halt
1	0	0	Opcode fetch
1	0	1	Memory read
1	1	0	Memory write
1	1	1	Passive

- $\overline{RO}/\overline{GT1}$ and
- $\overline{RO}/\overline{GT0}$ — The **request/grant** pins request direct memory accesses (DMA) during maximum mode operation. Both of there lines are bi-directional and are used to request and grant a DMA operation.
- $\overline{LOCK}$ — The **lock output** is used to lock peripherals off the system. This pin is activated by using the LOCK : prefix on any instruction.
- QS_1 and QS_0 — The queue status bits show the status of the internal instruction queue. These pins are provided for access by the numeric coprocessor (8087). Table 7 shows the operation of the queue status bits.

Table 7. The Queue Status Bits

QS_1	QS_0	Function
0	0	Queue is idle (No operation)
0	1	First byte of opcode
1	0	Queue is empty
1	1	Subsequent byte of opcode

2.2 MEMORY ADDRESSING IN 8086

In 8086 the memory address space can be viewed as a sequence of one million bytes in which any byte may contain an 8-bit data element and any two consecutive bytes may contain a 16-bit data element, and there will not be constraint on byte or word address boundaries. The address space is physically connected to a 16-bit data bus by dividing the address space into two 8-bit banks of up to 512K bytes each, as depicted in Fig. 2.2.

As observed from Fig. 2.2 one bank is connected to the lower half of the 16-bit data bus (D0-D7) and contains even address bytes. *i.e.*, when A0 bit is low, the bank is selected. The other bank is connected to the upper half of the data bus (D8-D15) and contains odd address bytes. That means when A0 is high and $\overline{BHE}$ *(Bus High Enable)* is low, the odd bank is selected. *(A specific byte within each bank is selected by address lines A1-A19).*

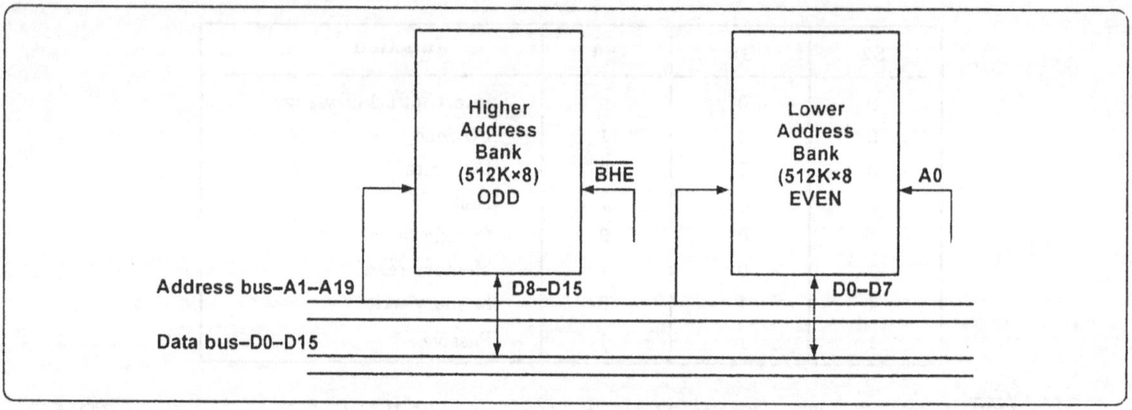

Fig. 2.2 Address space in 8086.

There are four possible ways to access the data from the memory. Such four ways are as listed below.

1. 8-bit data from Lower (Even) address Bank
2. 8-bit data from Higher (Odd) address Bank
3. 16-bit data starting from Even Address, and
4. 16-bit data starting from Odd Address.

Now let us observe arrangement for the above four methods of assessing data from memory.

2.2.1 The 8-bit Data from Lower/Even Address Bank

In this case to access memory bytes from Even address, information is transferred over the lower half of the data bus (D0-D7), where A0 is output LOW and $\overline{BHE}$ is output HIGH enabling only the even address bank. The Fig. 2.3 depicts the arrangement for this method as below :

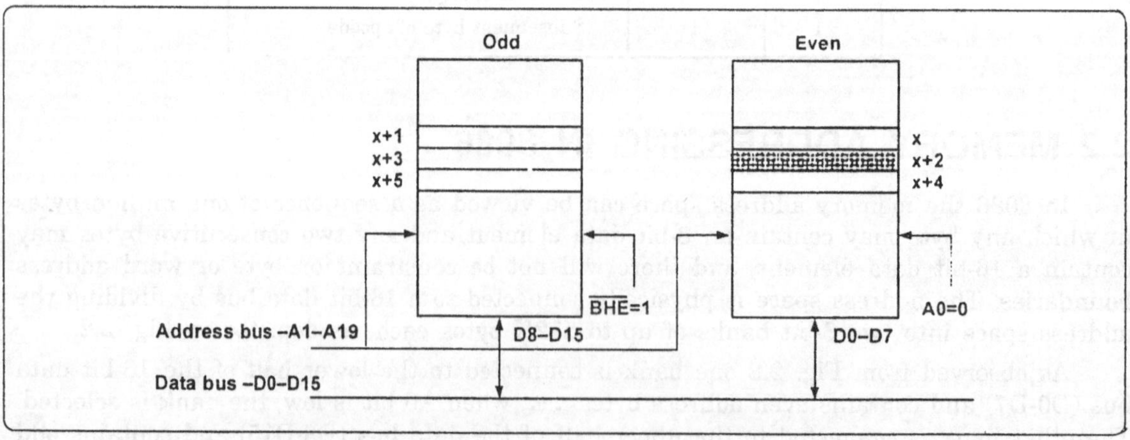

Fig. 2.3 8-bit Data Access from Lower/Even Address

Now let us consider an example for loading a byte of data into CH register *(i.e., higher order 8-bits of CX register)* from the memory location with an even address. Here the data will be accessed from the even bank via the (D0-D7) DATA BUS. Eventhough this data is transferred into the 8086 over lower 8-bit lines, the 8086 processor will automatically redirects the data to the higher 8-bits of its internal 16-bit data path and hence to the CH-register. Such capability permits bytes input-output transfer via the AL register to access I/O device connected to either the upper half of the data bus or the lower half of the 16-bit data bus, as observed in Fig. 2.3.

2.2.2 8-bit Data from Higher/Odd Address Bank

Here to access memory byte from an odd address information, is transferred over the higher half of the data bus (D8–D15). The BHE output low enables the upper memory bank, and whereas A0 is output high to disable lower memory bank. The arrangement is as depicted in Fig. 2.4 below.

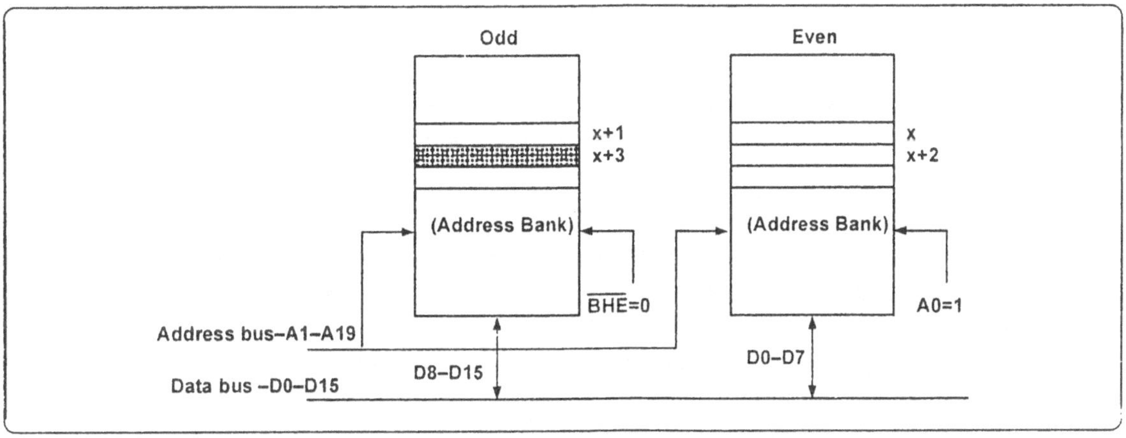

Fig. 2.4 8-bit Data from Odd Address Bank

2.2.3 The 16-bit Data Starting from Even-Address

Here, 16-bit data from an even address is accessed in a single bus cycle. The Address lines A1–A19 select the appropriate byte within each bank. A0 low and BHE low enables both banks simultaneously. The arrangement is as shown in Fig. 2.5 below.

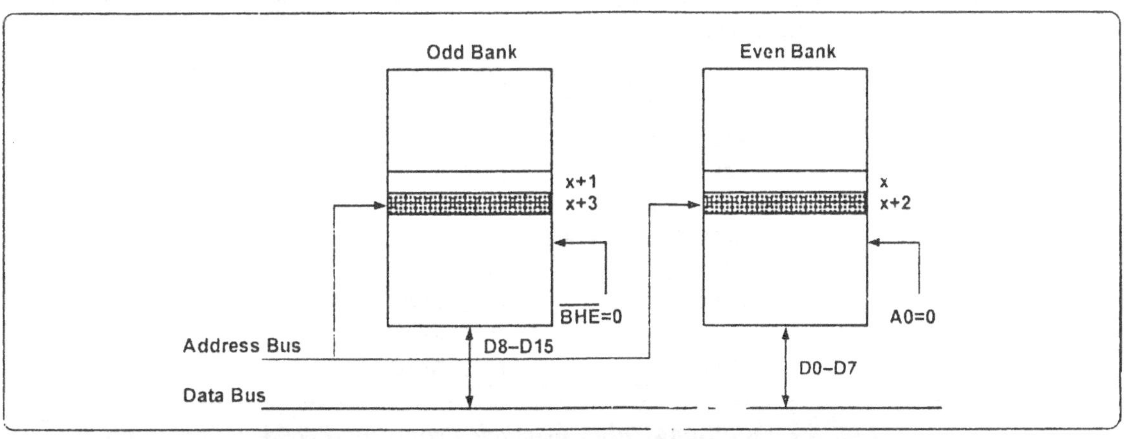

Fig. 2.5 16-bit Data Access[illegible] from Even Address

2.2.4 The 16-bit Data Access starting from Odd Address

Here a 16-bits word located at an odd address *(two consecutive bytes with the least significant byte at an odd byte address)* is accessed using two bus cycles. For first bus cycle, lower byte will be accessed. The arrangement with odd address as 0005 is as shown in Fig. 2.6 (*a*).

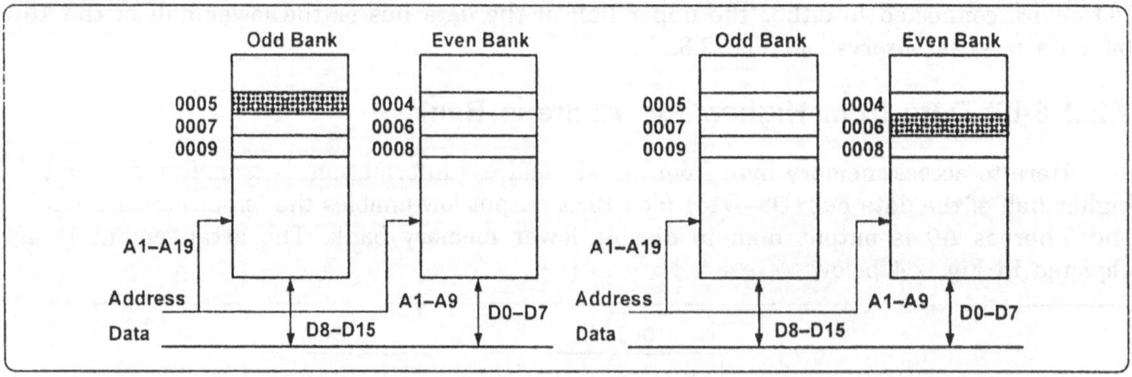

(*a*) First Access from Odd Address **(*b*) Next Access from Even Address**

Fig. 2.6 The 16-bit Data Access starting from Odd Address

Now during second bus cycle, the upper byte *[with the even address 0006H as observed in Fig. 2.6 (b)]* will be accessed. During first bus cycle, A1–A19 address bus specifies the address and A0 as 1 and $\overline{BHE}$ is low. That means the even memory bank will be disabled and odd memory bank is enabled and next, during second bus cycle, the address is incremented. That means, A0 is zero and $\overline{BHE}$ is made high, and even memory bank is enabled and the odd memory bank is disabled, as observed in Fig. 2.6.

2.3 CLOCK GENERATOR 8284

In this section we will study the details of the clock generator 8284A. Apart we will observe the RESET signal and introduction of READY signal to 8086 microprocessor. The 8284A is an important ancillary component to the 8086/8088 microprocessors. Without clock generator, it required many additional circuits to generate the clock in an 8086/8088-based system. 8284A provides following basic functions or signals : clock generation, RESET synchronization, READY synchronization, and a TTL level peripheral clock signal which are very important essential signal to microprocessors. Now Fig. 2.7 depicts the details of pin-out or functional pins of the clock generator 8284A.

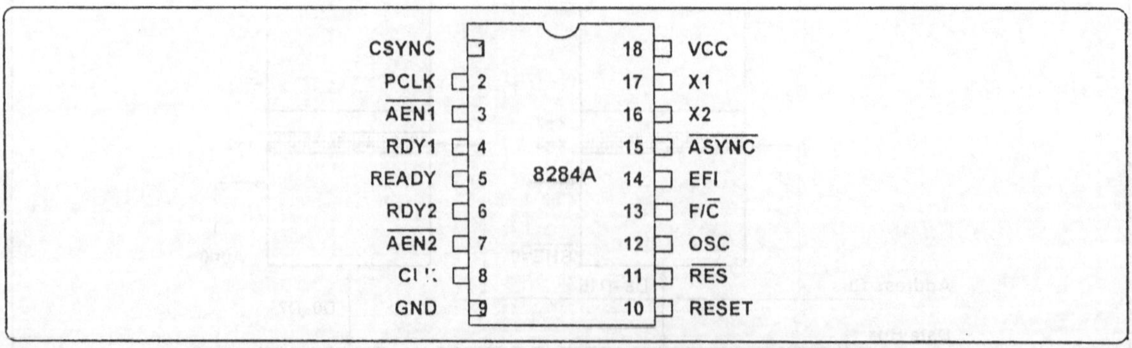

Fig. 2.7 Functional pins of clock generator 8284A

2.3.1 Pin Functions of 8284A

It is an 18-pin IC, which is specially designed to use with 8086/8088 microprocessors. Now let us observe the function of each pin of 8284A clock generator.

- $\overline{\text{AEN1}}$ **and** $\overline{\text{AEN2}}$ The **address enable** pins are provided to qualify the bus ready signals, RDY1 and RDY2, respectively, which are used to cause wait states, along with the RDY1 and RDY2 inputs. The Wait states are generates by READY pin of the 8086/8088 microprocessors, which is controlled by these two inputs.
- **RDY1 and RDY2** The bus ready inputs are provided in conjunction with the $\overline{\text{AEN1}}$ and $\overline{\text{AEN2}}$ pin to cause wait states in an 8086/8088 based μP system.
- $\overline{\text{ASYNC}}$ The **ready synchronization** selection input selects either one or two stages of synchronization for the RDY1 and RDY2 inputs.
- **READY** The Ready is an output pin which connects to the 8086/8088 READY input. This signal is synchronized with the RDY1 and RDY2 inputs.
- **X1 and X2** The **crystal oscillator** pins connect to an external crystal used as the timing source for the clock generator and all its functions.
- $\text{F}/\overline{\text{C}}$ The **frequency/crystal select** input results the clocking source for the 8284A. If this pin is held high, an external clock is provided to the EFI input pin, and if it is held low, the internal crystal oscillator provides the timing signal.
- **EFI** The **external frequency** input is used when the F/C is pulled high. EFI supplies the timing whenever the $\text{F}/\overline{\text{C}}$ pin is high.
- **CLK** The **clock** output pin provides CLK input signal to the 8086/8088 microprocessors *(and other components in the system).* The CLK pin has an output signal that is one-third of the crystal or EFI input frequency and has a 33 percent duty cycle, which is required by the 8086/8088.
- **PCLK** The **peripheral clock** signal is one-sixth the crystal or EFI input frequency and has a 50 percent duty cycle. The PCLK output provides a clock signal to the peripheral equipment in the system.
- **OSC** The **oscillator** output is a TTL level signal that is at the same frequency as the crystal or EFI input. *(The OSC output provides an EFI input to other 8284A clock generators in some multiple-processor systems).*
- $\overline{\text{RES}}$ The **reset** input is an active-low input to the 8284A. The $\overline{\text{RES}}$ pin is often connected to an RC network that provides power-on resetting.
- **RESET** The reset output is connected to the 8086/8088 RESET input pin.

- **CSYNC** The **clock synchronization** pin is used whenever the EFI input provides synchronization in systems with multiple processors. When the internal crystal the internal crystal oscillator is used, this pin must be grounded.
- **GND** The ground pin connects to ground
- **V_{CC}** The power supply pin connects to +5.0 V with a tolerance of ± 10%.

2.3.2 The Operation of Clock Generator 8284A

The clock generator 8284A is a simple device, its internal logic diagram is as shown in Fig. 2.8.

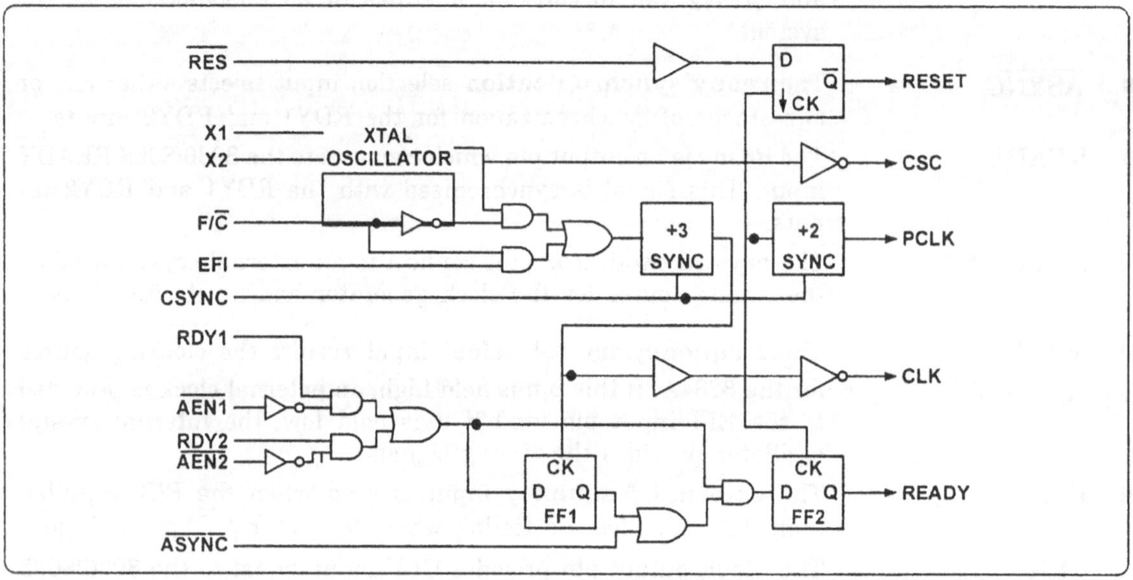

Fig. 2.8 Internal block diagram of 8284A

Now the two sections of operation those are of interested in 8284A clock generator are :

(*a*) Operation of the clock section, and

(*b*) Operation of the Reset section.

(*a*) Operation of the Clock Section. The top half of the logic diagram in Fig. 2.8 represents the clock and reset synchronization section of the 8284A clock generator. As observed, the crystal oscillator has two inputs : X1 and X2 when a crystal is attached to X1 and X2, oscillator generates a square-wave signal at the same frequency as the crystal. The square-wave so generated signal is fed to an AND gate and also to an inverting buffer that provides the OSC output signal. *(The OSC signal is sometimes used as an EFI input to other 8284A circuits in a system).* The inspection of AND gate reveals that when $F / \overline{C}$ is a logic 0, the oscillator output is steered through to the divide-by-3 counter. If $F / \overline{C}$ is a logic 1. then EFI is steered through to the counter.

The output of the divide-by-3 counter generates timing for ready synchronization, a signal for another counter *(divide-by-2)*, and the CLK signal to the 8086/8088 microprocessors. The CLK signal is also buffered before it leaves the clock generator. Observe that the output of the first counter feeds the second. These two cascaded counters provide the divide-by-6 output at PCLK, the peripheral clock output, the Fig. 2.9 below; shows the connection, arrangement of 8284A to 8086/8088 microprocessor.

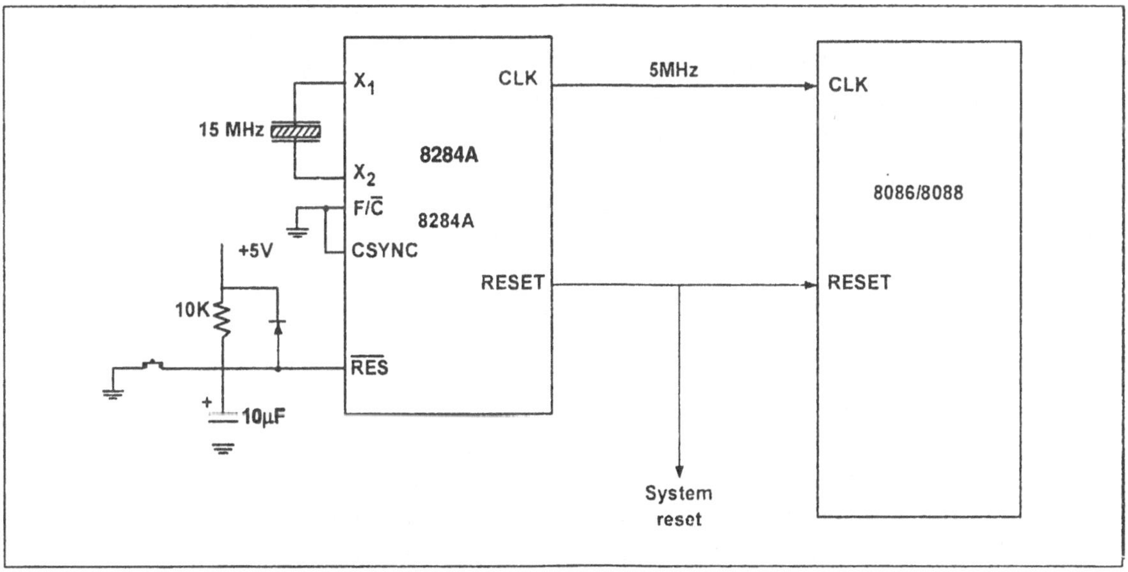

Fig. 2.9 Clock generator (8284A) and the 8086/8088 microprocessor illustrating the connection for the clock and reset signals. (A 15 MHz crystal provides the 5 MHz clock for the microprocessor)

From Fig. 2.9 two points to be observed ; *i.e.*,

1. That $F/\overline{C}$ and CSYNC are grounded to select the crystal oscillator and
2. That a 15 MHz crystal provides the normal 5 MHz clock signal to the 8086/8088 as well as a 2.5 MHz peripheral clock signal.

(*b*) Operation of the Reset Section. The reset section of the 8284A is simple, and it consists of a Schmitt trigger buffer and a single D-type flip-flop circuit. The D-type flip-flop ensures that the timing requirements of the 8086/8088 RESET input are fulfilled. This circuit applies the RESET signal to the microprocessor on the negative edge *(1 - to - 0 transition)* of each clock. The 8086/8088 microprocessors sample RESET at the positive edge *(0 - to - 1 transition)* of the clocks; therefore *(this circuit meets the timing requirements of the 8086/8088)*.

From Fig. 2.9 it can be observed that an RC circuit provides a logic 0 to the $\overline{\text{RES}}$ input pin when power is first applied to the system. After a short time, the $\overline{\text{RES}}$ input becomes a logic 1 because the capacitor charges toward +5.0 V through the resistor. A push-button switch allows the microprocessor to be reset by the operator. Exact rest timing requires the RESET input to become a logic 1 no later than four clocks after system power is applied and to be held high for at least 50 μs. *(The flip-flop makes certain that RESET goes high in four clocks, and the RC time constant ensures that it stays high for at least 50 μs).*

2.4 THE BUS BUFFERING AND LATCHING

To make use of 8086/8088 microprocessors with memory or I/O interfaces, the multiflexed buses of the μP has to be demultiflexed. Therefore we have to discuss about demultiplexing the buses, and mention how buses are buffered in the case of large system, depending on fan-out of the components. Now let us discuss about :

(1) Demultiplexing the Buses of 8086/8088, and

(2) Buffered system for 8086/8088.

2.4.1 Demultiplexing the Buses of 8086/8088

The address/data bus on the 8086/8088 is multiplexed shared so as to reduce the number of pins required for the 8086/8088 integrated circuit. But this burdens the hardware designer with the task of extracting or demultiplexing information from these multiplexed pins. So can we not leave buses multiplexed ?

The memory and I/O require that address remains valid and stable throughout a read or write cycle. When the buses are multiplexed, the address changes at the memory and I/O, which causes them to read or write data in the wrong locations.

In all computer systems they have three buses : (1) an address bus that provides the memory and I/O with the memory address or the I/O port number, (2) a data bus that transfers data between the microprocessor and the memory and I/O in the system, and (3) a control bus that provides control signals to the memory and I/O. (These buses must be present in order to interface to memory and I/Q).

Now let us observe the demultiplexing arrangement for (*a*) 8086 and (*b*) 8088 microprocessor.

(*a*) Demultiplexing the 8086. The 8086 system requires separate address, data and control buses, which differs primarily in the number of multiplexed pins. *(In the 8088, only $AD_7 - AD_0$ and $A_{19}/S_6 - A_{16}/S_3$ will be multiplexed).* In 8086, the multiplexed pins include $AD_{15} - AD_0$, $A_{19}/S_6 - A_{16}/S_3$, and $\overline{BHE}/S_7$. All of these signals must be demultiplex.

The Fig. 2.10 depicts a demultiplexed 8086 with all three buses : address ($A_{19} - A_0$ and $\overline{BHE}$), data ($D_{15} - D_0$), and control $(M/\overline{IO}, \overline{RD} \text{ and } \overline{WR})$, as below.

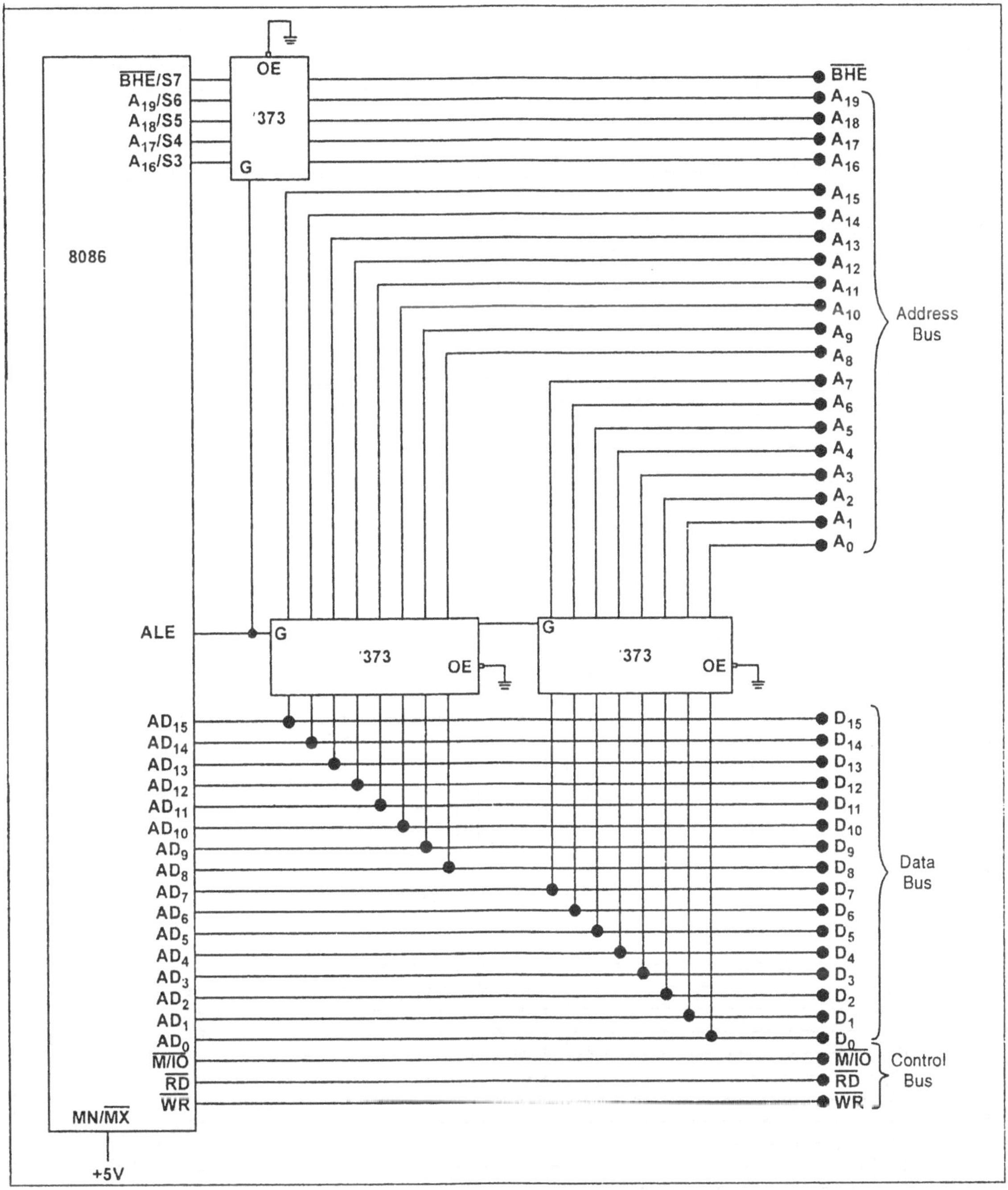

Fig. 2.10. 8086 microprocessor shown with a demultiplexed address bus (model is used to build many 8086-based systems)

In the circuit shown in Fig. 2.10 an additional 74LS373 latch has been added to demultiplex the address/data bus pins $AD_{15} - AD_8$ and a $\overline{BHE} / S_7$ input has been added to the top 74LS273 to select the high-order memory bank in the 16-bit memory system of the 8086 if compared with that for 8088. The memory and I/O system observe the 8086 as a device with a 20-bit address bus ($A_{19} - A_0$), a 16-bit data bus ($D_{15} - D_0$), and a 3 line control bus $\left(M / \overline{IO}, \overline{RD} \text{ and } \overline{WR}\right)$.

***(b)* Demultiplexing the 8088.** The Fig. 2.11 shows the 8088 microprocessor and the components required to demultiplex its buses. Here, two 74LS373 transparent latches are used to demultiplex the address/data bus connections $AD_7 - AD_0$ and the multiplexed address/status connections $A_{19}/S_6 - A_{16}/S_3$ as shown :

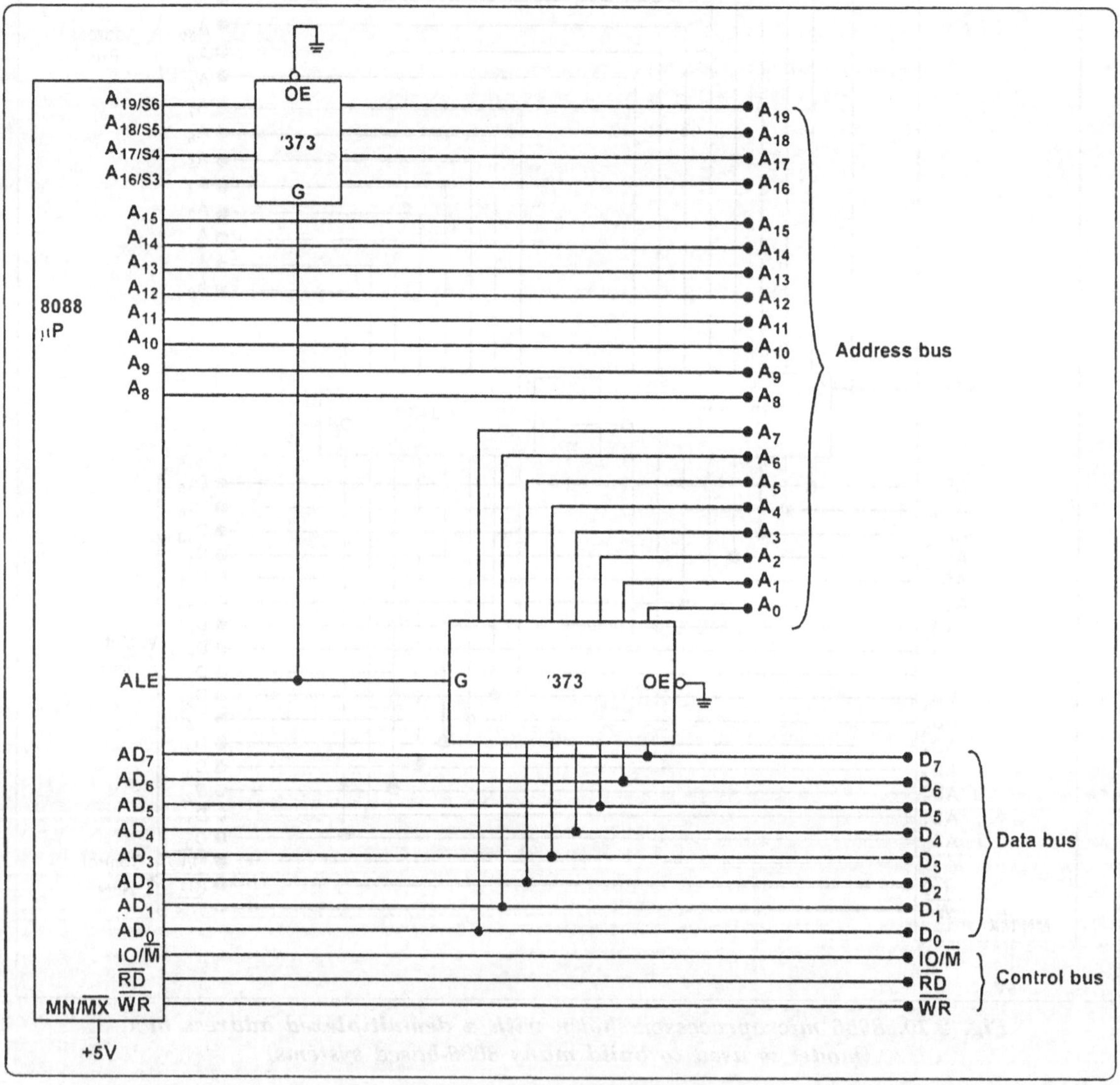

Fig. 2.11. 8088 microprocessor shown with a demultiplexed address bus (model is used to build many 8088-based systems)

The transparent latches, which are looking like wires whenever the address latch enable pin (ALE) becomes a logic 1 they pass the inputs to the outputs. After a small duration ALE returns to its logic 0 condition, which causes the latches to remember the inputs at the time of the change to a logic 0. In this case, $A_7 - A_0$ are stored in the bottom latch and $A_{19} - A_{16}$ in the top latch. This yields a separate address bus with connection $A_{19} - A_0$. These address connections allow the 8088 to address 1M byte of memory space. *(The fact that the data bus is separate allows it to be connected to any 8-bit peripheral device or memory component).*

2.4.2 Buffered System for 8086/8088

When more than 10 unit loads are attached to any bus pin, the entire 8086 or 8088 system must be buffered. The demultiplexed pins are already buffered by the 74LS373 latches, which have been designed to drive the high-capacitance buses encountered in microcomputer systems. The buffer's output current have been increased so that more TTL unit loads may be driven: a logic 0 output provides up to 32 mA of sink current, and a logic 1 output provides up to 5.2 mA of source current. That means a fully buffered signal will introduce a timing delay to the system. This causes no difficulty unless memory or I/O devices are used, which function at near the maximum speed of the bus.

Now let us observe the arrangement for fully buffered arrangement for

(*a*) 8086 and

(*b*) 8088 microprocessors.

(*a*) Fully Buffered 8086. The Fig. 2.12 shows a fully buffered 8086 microprocessor. It address pins are already buffered by the 74LS373 address latches; its data bus employs two 74LS245 octal bi-directional bus buffers; and the control bus signals, and IO / M, $M / \overline{IO}$, $\overline{RD}$ and $\overline{WR}$ use a 74LS244 buffer.

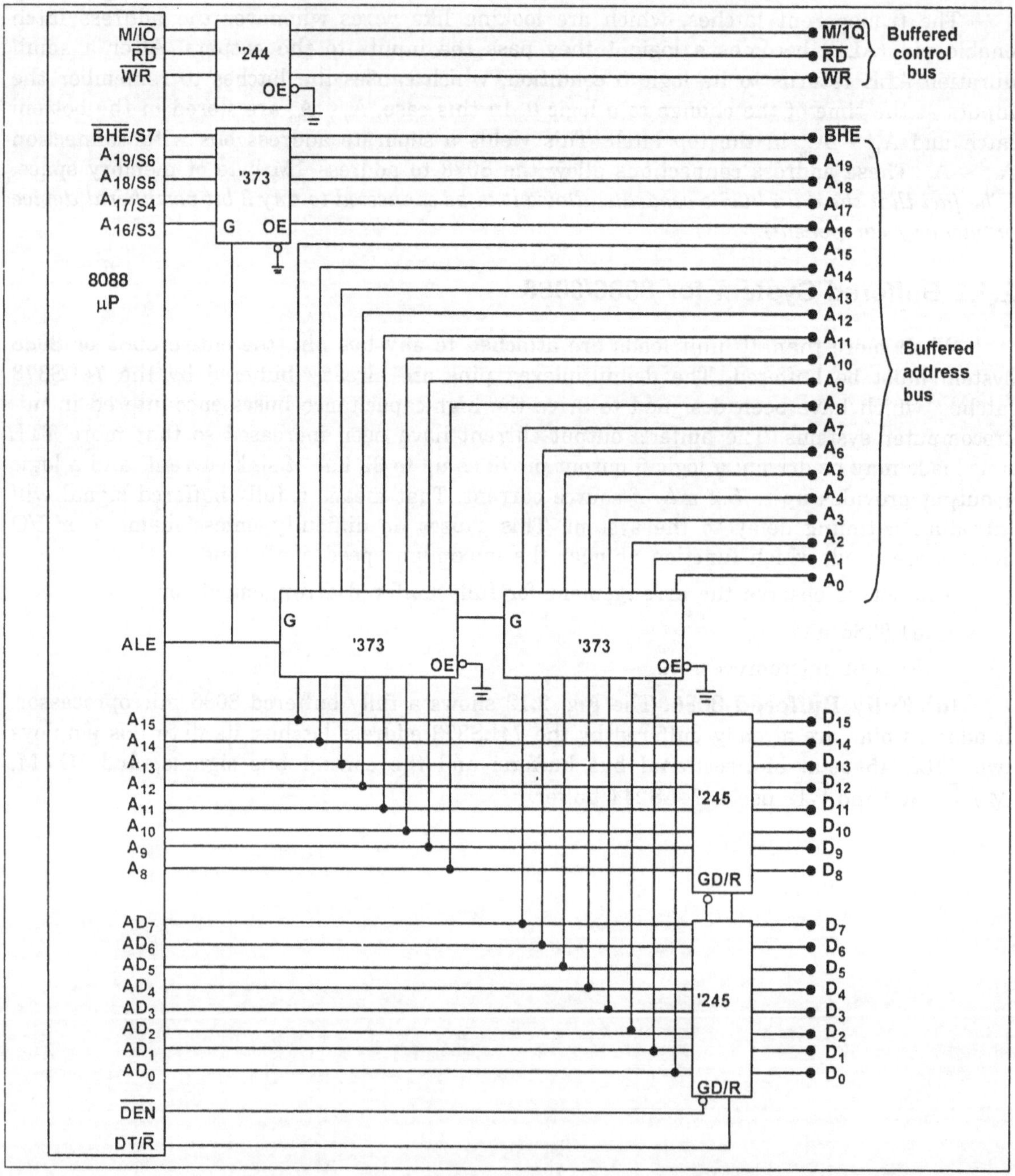

Fig. 2.12 Fully buffered 8086 microprocessor

Fully buffered 8086 system requires one 74LS244, two 74LS245s, and three 74LS373s.

Note. 8086 requires one more buffer than the 8088 because of the extra eight data bus connections, $D_{15} - D_8$. It also has a signal that is buffered for memory-ban selection.

(b) Fully Buffered 8088. The Figure 2.13 shows a fully buffered 8088 microprocessor. Observe that the remaining eight address pins, $A_{15} - A_8$, use a 74LS244 octal buffer; the eight data bus pins, $D_7 - D_0$, use a 74LS245 octal bi-directional bus buffer, and the control bus signals, IO / M, $\overline{RD}$ and $\overline{WR}$, use a 74LS244 buffer.

As seen in Fig. 2.13 a fully buffered 8088 system requires two 74LS244s, one 74LS245, and two 74LS373s. The direction of the 74LS245 is controlled by the DT / $\overline{R}$ signal, and is enabled and disabled by the $\overline{DEN}$ signal.

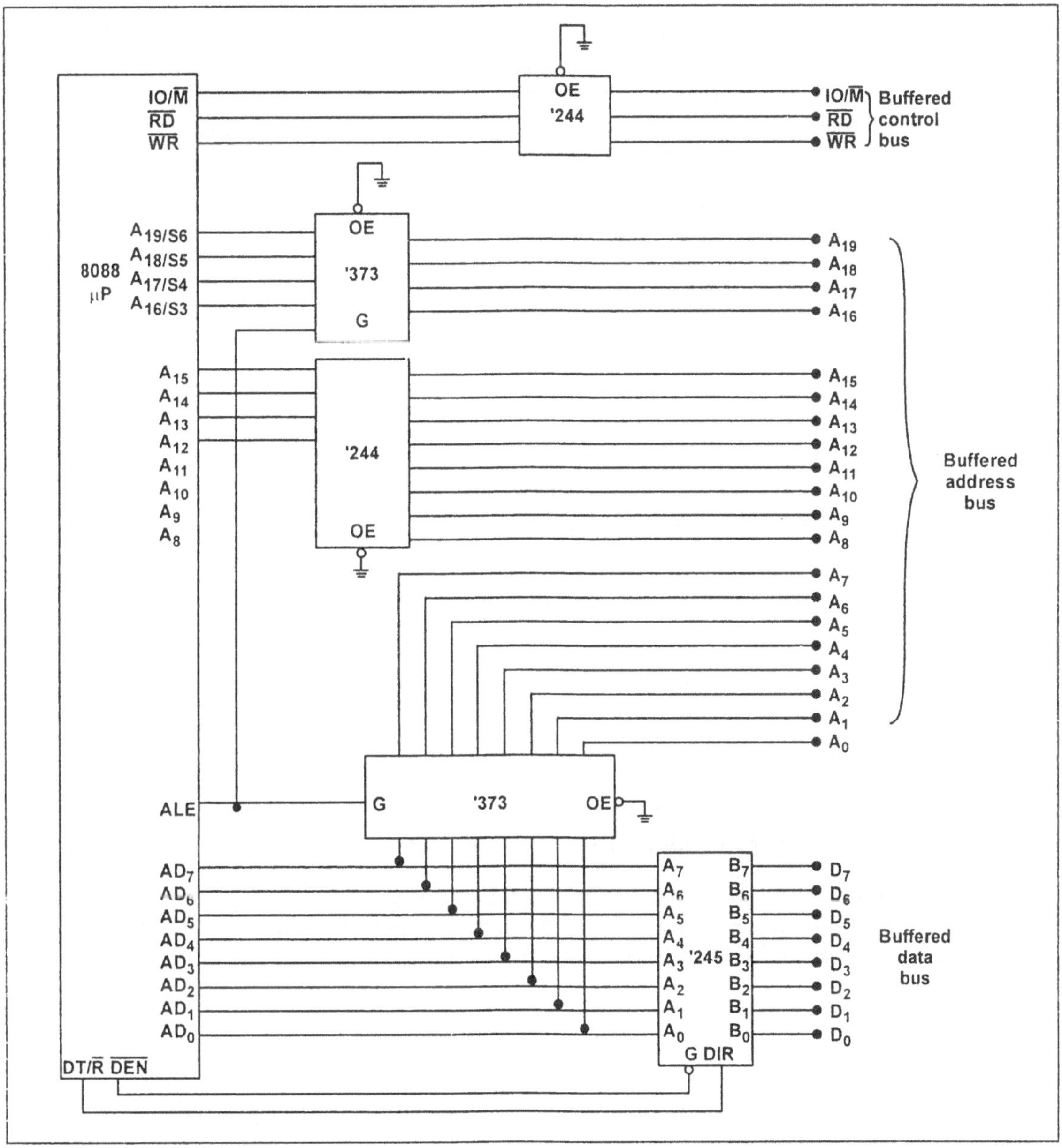

Fig. 2.13. Fully buffered 8088 microprocessor.

2.5 BUS TIMING DIAGRAMS

It is required to know system bus timing before choosing a memory or I/O device fcr interfacing to the 8086 or 8088 microprocessors. Now, we discuss the bus signals and the basic read and write timing of the 8086/8088. *(We discuss only the times that affect memory and I/O interfacing).*

2.5.1 The Basic Bus Operation

In 8086/8088 microprocessor, the three buses, address, data and control, will function similar to those of other microprocessors. For instant, when data are written to memory, the microprocessor outputs the memory address on the address bus, outputs the data to be written into memory on the data bus, and issues a write $(\overline{WR})$ to memory and $IO/\overline{M}$ = 0 for the 8088 and $M/\overline{IQ}$ = 1 for the 8086. The simple write bus cycle for 8086/8088 is as shown in Fig. 2.14.

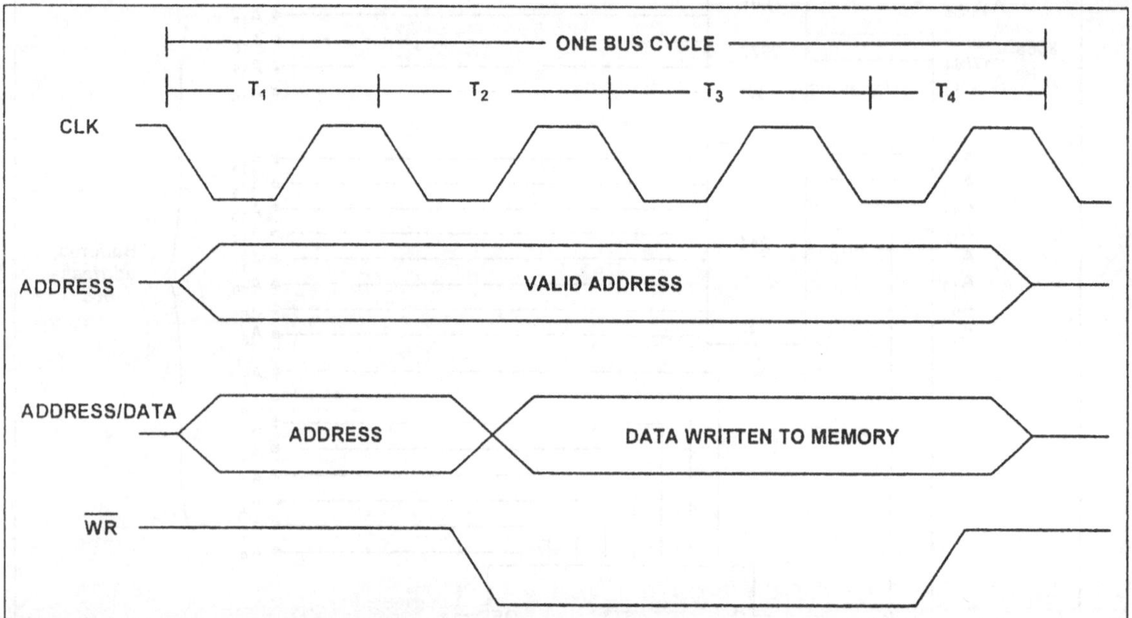

Fig. 2.14. Simplified 8086/8088 write bus cycle.

Now if data are read from the memory the microprocessor outputs the memory address on the address bus, issues a read memory signal, and accepts the data via the data bus. The simplified read bus cycle for 8086/8088 is as shown in Fig. 2.15.

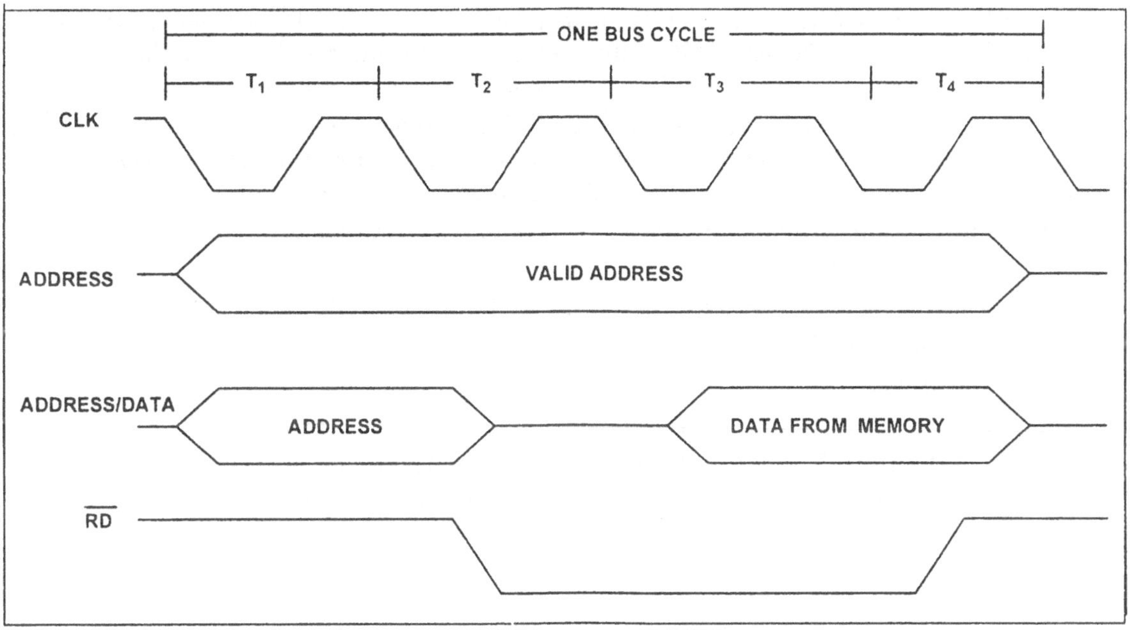

Fig. 2.15 Simplified 8086/8088 read bus cycle

2.5.2 Details of Timing in General

8086/8088 microprocessors use the memory and I/O in periods of time called bus cycles, each bus cycle equals four system-clocking periods *(T states)*. But some new microprocessors divide the bus cycle into as few as two clocking periods. When the clock is operated at 5 MHz *(the basic operating frequency for these two microprocessors)*, then on 8086/8088 a bus cycle will complete in 800 ns. That is microprocessor reads or writes data between itself and memory or I/O at a maximum rate of 1.25 million times per second. *(Because of the internal queue, the 8086/8088 can execute 2.5 million instructions per second (MIPS) in bursts)*.

During first clocking period in a bus cycle, T1, many things takes place. The address of the memory or I/O location is sent out via the address bus and the address/data bus connections. *(The address/data bus is multiplexed and sometimes contains memory addressing information, sometimes data)*. Also output during T1 are the control signals ALE, DT / $\overline{R}$ and IO / $\overline{M}$ (8088) or M / $\overline{O}$ (8086). The IO / $\overline{M}$ or M / $\overline{O}$ signal indicates whether the address bus contains a memory address or an I/O device *(port)* number.

Next during T2, the 8086/8088 microprocessor issue the $\overline{RD}$ or $\overline{WR}$ signal, $\overline{DEN}$, and, in the case of a write, the data to be written appears on the data bus. Such events cause the memory or I/O device to begin to perform a read or write. The $\overline{DEN}$ signal turns on the data bus buffers, if they are present in the system, so the memory or I/O can receive data to be written or so the microprocessor can accept the data read from the memory or I/O for a read operation. When this happens to be a write bus cycle, then the data are sent out to the memory or I/O through the data bus.

READY is sampled at the end of T2, if it is low at this time, T3 becomes a wait state (T_W). This clocking period helps to allow the memory time to access data. For the bus cycle happens to be a read bus cycle, the data bus is sampled at the end of T3.

Finally in T4, all bus signals are deactivated in preparation for the next bus cycle, which is also the time when the 8086/8088 microprocessors sample the data bus connections for data that are read from memory or I/O. Apart at this point, the trailing edge of the $\overline{WR}$ signal transfers data to the memory or I/O, which activate and write when the $\overline{WR}$ signal returns to a logic 1 level.

Now to let us observe the four T states, *i.e.*, T, Tr, T3 and T4 in

(*a*) Read timing, and

(*b*) Write timing

for minimum mode timing operation.

(*a*) Read Timing. The Fig. 2.16 depicts read timing for the 8088 microprocessors in detail when compaired to Fig. 2.15. The 8086 read timing is identical except that the 8086 has 16 rather then 8 data bus bits. A close look at this timing diagram should allow to identify all the main events described for each T state.

Most important item contained in the read timing diagram is the amount of time allowed the memory or I/O to read the data. The Memory is chosen by its access time, which is the fixed amount of time the microprocessor allows it to access data for the read operation. Therefore, it is important to note that the memory we choose comply with the limitations of the system.

Note. The microprocessor timing diagram does not, however, provide a straightforward memory access time. Instead, it is necessary to combine several times to arrive at the access time. To find memory access time in this diagram, we must first locate the point in T3 when data are sampled. When we examine the timing diagram closely. We will observe a line that extends from the end of T3 down to the data bus. The end of T3 is where the microprocessor samples the data bus.

The Memory access time starts when the address appears on the memory address bus and continues until the microprocessor samples the memory data at T3. Approximately three T states elapse between these times. The address does not appear until T_{CLAV} time *(110 ns if the clock is 5 MHz)* after the start of T1. This means that T_{CLAV} time must be subtracted from the three clocking states (600 ns) that separate the appearance of the address (T1) and the sampling of the data (T3). One other time must also be subtracted : the data setup time (T_{DVCL}), which occurs before T3. Memory access time is thus three clocking states minus the sum of T_{CLAV} and T_{DVCL}. Because T_{DVCL} is 30ns with a 5 MHz clock, the allowed memory access time is only 460 ns (access time = 600 ns – 110 ns – 30 ns).

Note. In specific the memory devices chosen for connection to the 8086/8088 operated at 5 MHz must be able to access data in less than 460 ns because of the time delay introduced by the address decoders and buffers in the system. At least a 30 or 40 ns margin should exist for the operation of these circuits. That means, memory speed will not be slower than about 420 ns to operate correctly with the 8086/8088 microprocessors.

Note. Other timing factor that may affect memory operation is the width of the $\overline{RD}$ strobe. On the timing diagram, the read strobe width is given as T_{RLRH}, the time for which is 325 ns *(5 MHz clock rate),* which is wide enough for almost all memory devices manufactured with an access time of 400 ns or less.

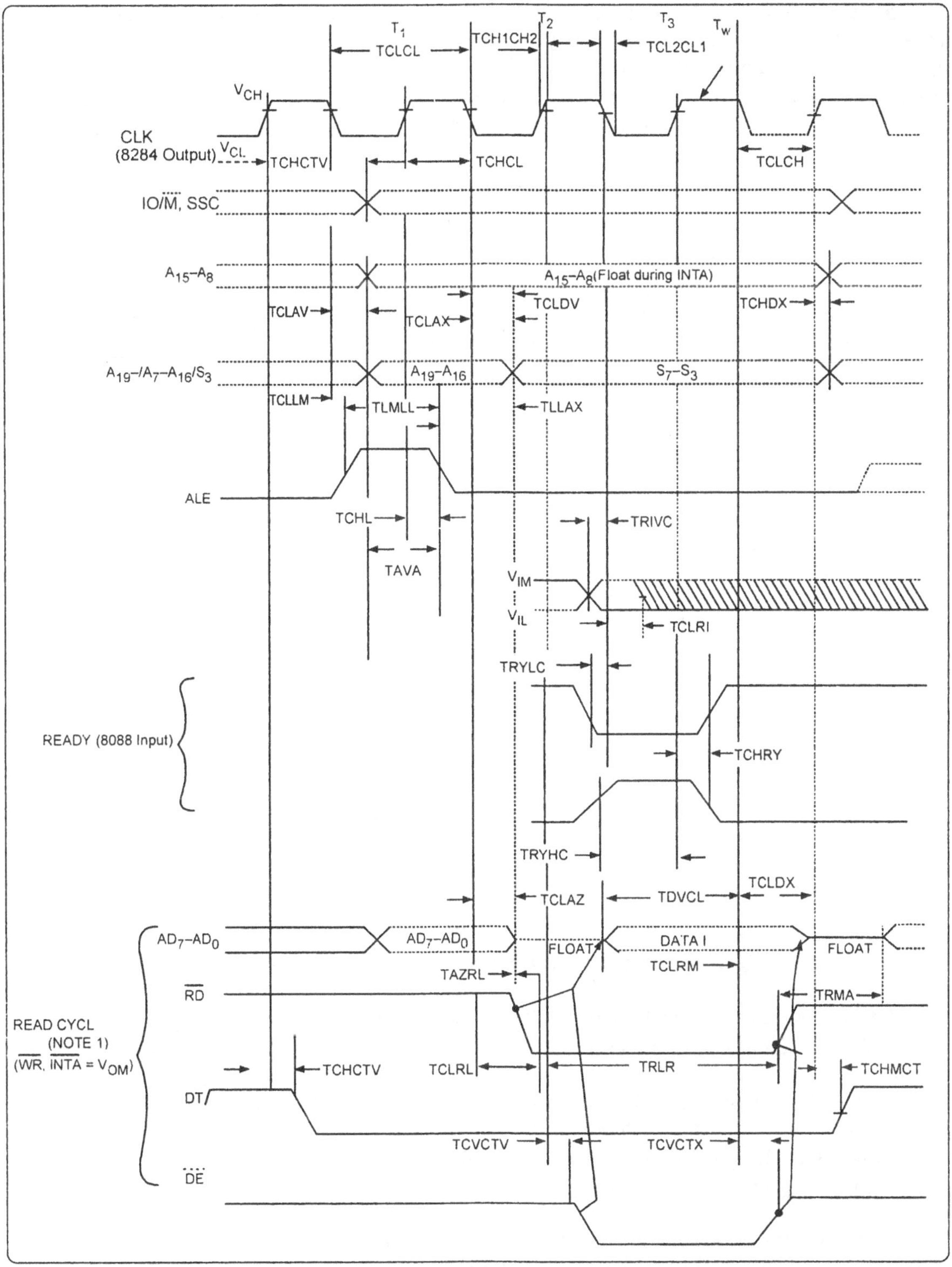

Fig. 2.16 Bus timing for a read operation (for 8088 under minimum mode)

***(b)* Write Timing.** The Fig. 2.17 depicts the write timing diagram for the 8088 microprocessor. Again, the 8086 is so nearly identical that it need not be presented here in a separate timing diagram. Main differences between read and write timing are minimal. The strobe is replaced by the $\overline{WR}$ strobe, the data bus contains information for the memory rather than information from the memory, and $DT/\overline{R}$ remains a logic 1 instead of a logic 0 throughout the bus cycle.

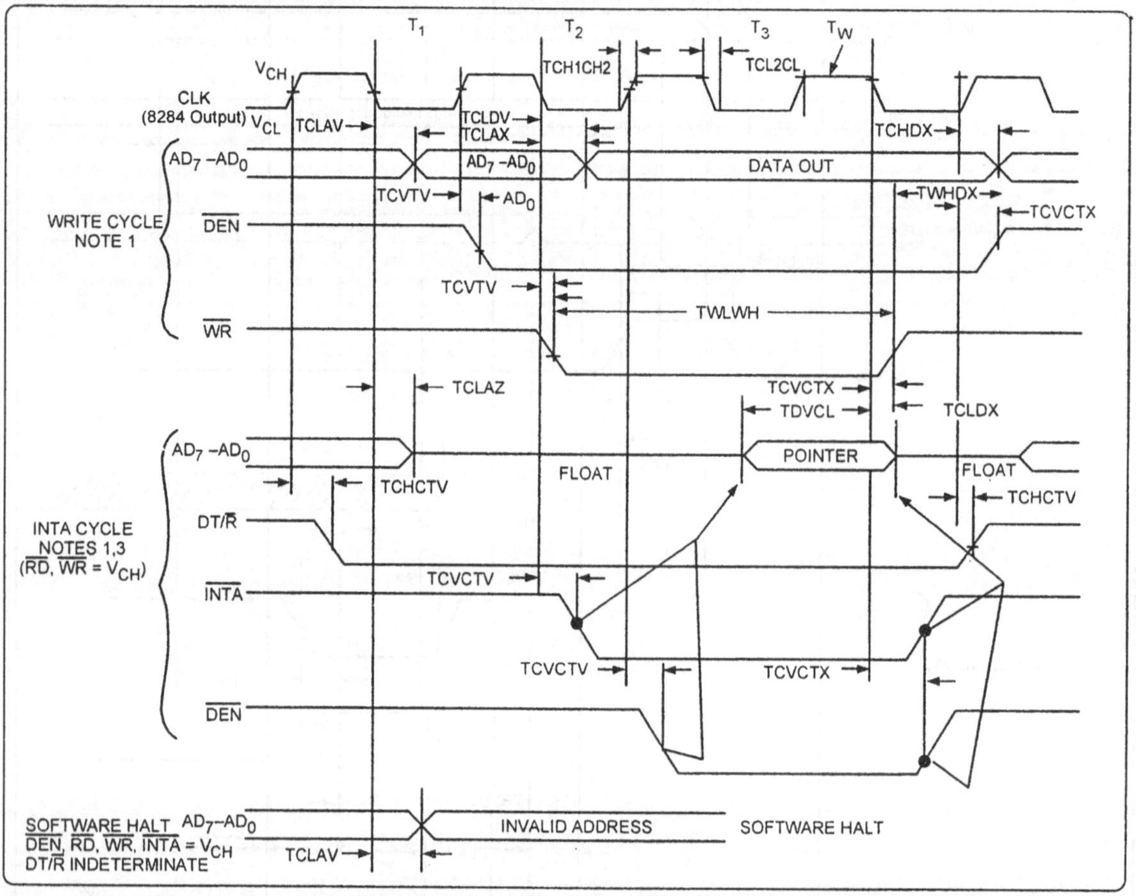

Fig. 2.17. Write bus timing of 8088 in minimum mode

Notes. 1. All signals switch between VOH and VOl unless otherwise specified.

2. RDY is sampled near the end of T_2, T_3, T_W to determine if Tw machines state are to be inserted.

3. Two INTA cycles run back-to-back. The 8088 local ADDR/Data bus is floating during both INTA cycles. Control signals are shown for the second INTA cycle.

4. Signals at 8284 are shown for reference only.

5. All Timing Measurements are made at 1.5V unless otherwise noted.

While interfacing some memory devices, timing may be especially critical between the point at which $\overline{WR}$ becomes a logic1 and the time when data are removed from the data bus. It is case because, as we can recall, memory data are written at the trailing edge of the $\overline{WR}$ strobe. As per the timing diagram, this critical period is T_{WHDX} or 88 ns when the 8088 processor is operated with a 5 MHz clock. The Hold time is often much less than this, and is in fact often 0 ns for memory devices. The width of the $\overline{WR}$ strobe is T_{WELWH} or 3240 ns at a 5 MHz clock rate. This rate is also compatible with most memory devices that have an access time of 400 ns or less.

2.5.3 Instruction Execution Timing

The execution time for instructions varies and depend on the type of instruction and addressing modes. This basic instruction time can be determined by multiplying the number of clock cycles needed to execute the instruction by the clock's period. The total instruction execution time is the sum of the basic time plus the time required to calculate the effective address if a memory operand is accessed. The basic instruction time assumes that the instruction to be executed has already been prefetched and stored in the instruction queue. If instruction is not in the instruction queue then additional clock cycles must be added to find the total execution time.

Total instruction execution time will depend on

(*a*) **Number of memory references required by the instruction.** If operand is a single byte or word with an even address only one memory reference is required. However, if operand is a word with an odd address two memory references are required. Each memory reference requires four clock cycles, hence, each odd address word reference requires four extra clock cycles.

(*b*) **Addressing Mode.** In case of two operand instructions, register-to-register addressing gives the faster operation than register-to-memory and memory-to-register operations. Between memory-to-register and register-to-memory operations memory to register is faster than the register-to-memory, (because register-to-memory operation requires extra time to store the result into the memory).

(*c*) **Variable Basic Execution Times.** Some instructions are data dependent and requires variable basic execution times. These are multiply, divide, shift, and rotate instructions.

(*d*) **Conditional Branch Instructions.** For conditional branch instructions, two different timings are specified. The smaller time should be considered when the condition is not met, and a branch is not taken. The larger time should be considered when the condition is met and the branch is taken. In the latter case, extra cycles are required to fetch the next instruction due to flushing of instruction queue.

2.6 THE READY AND WAIT STATES

The READY input causes wait states for slower memory and I/O components. A wait state (Tw) is an extra clocking period, inserted between T2 and T3, that lengthens the bus cycle. When one wait state is inserted, then the memory access time, normally 460 ns with a 5 MHz clock, is lengthened by one clocking period (200 ns) to 660 ns. Now let us generator and show how to insert one or more wait states selectively into the bus cycle, and examine the READY input and the synchronization times it requires.

2.6.1 READY Input

READY input is sampled at the end of T2 and again, if applicable, in the middle of T2. If READY is a logic 0 at the end of T2, then T3 is delayed and Tw is inserted between T2 and T3. READY is next sampled at the middle of Tw to determine if the next state is Tw or T3. It is tested for a logic 0 on the 1-to-0 transition of the clock at the end of T2 and for a 1 on the 0-to-1 transition of the clock in the middle of Tw. READY input 8086/8088 has some stringent timing requirements. The timing diagram in Fig. 2.18 depicts READY causing one wait state (Tw) along with the required setup and hold times from the system clock.

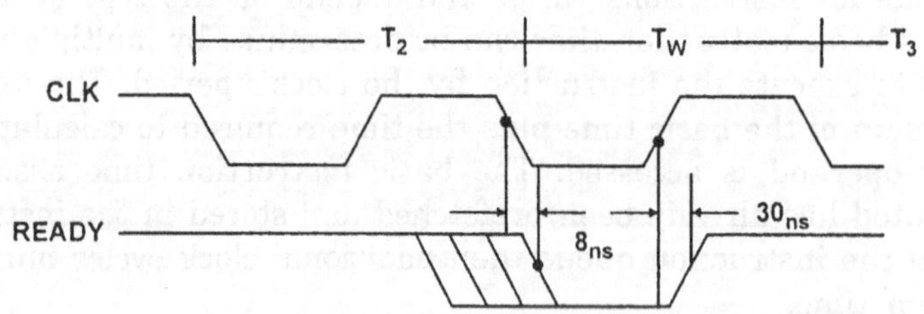

Fig. 2.18. READY input timing in 8086/8088.

The timing requirement for such operations is fulfilled by the internal READY synchronization circuitry of 8284A clock generator. When 82824A is used for READY, the RDY *(read input to the 8284A)* input occurs at the end of each T state.

2.6.2 RDY and 8284A

RDY is the synchronized ready input to the 8284A clock generator. Timing diagram for this input is as shown in Fig. 2.19. Eventhough it differs from the timing for the READY input to be 8086/8088, the internal 8284A circuitry guarantees the accuracy of the READY synchronization provided to the 8086/8088.

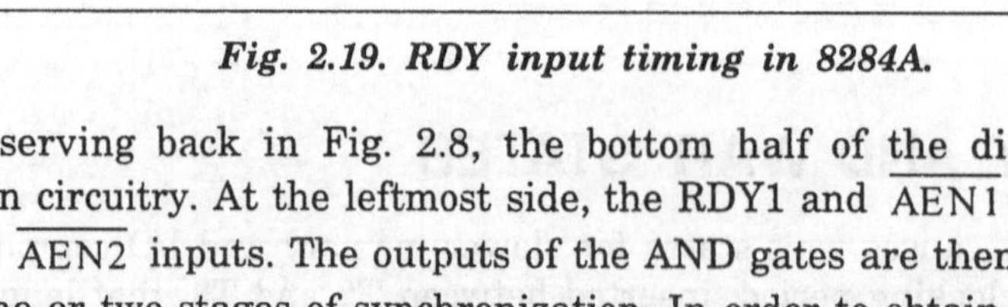

Fig. 2.19. RDY input timing in 8284A.

While observing back in Fig. 2.8, the bottom half of the diagram is the READY synchronization circuitry. At the leftmost side, the RDY1 and $\overline{AEN1}$ inputs are ANDed, as the RDY2 and $\overline{AEN2}$ inputs. The outputs of the AND gates are then ORed to generate the input to the one or two stages of synchronization. In order to obtain a logic 1 at the inputs to the flip-flops, RDY1 ANDed with $\overline{AEN1}$ must be active or RDY2 ANDed with $\overline{AEN2}$ must be active.

Next $\overline{ASYNC}$ input selects one stage of synchronization when it is a logic 1 and two stages when a logic 0. When one stage is selected, the RDY signal is kept from reaching the 8086/8088 READY pin until the next negative edge of the clock, and when two stages are selected, the first positive edge of the clock captures TDY in the first flip-flop. The output of this flip-flop is fed to the second flip-flop so that second flip-flop. The output of this flip-flop is fed to the second flip-flop so that the second flip-flop captures RDY on the next negative edge of the clock.

2.6.3 Wait State

The Fig. 2.20 depicts a circuit used to introduce any number of wait states for 8086/ 8088 microprocessors, where an 8-bit serial shift register (74LS164) shifts a logic 0 for one or more clock periods from one if its Q outputs through to the RDY1 input of the 8284A. With suitable strapping, this circuit can provide various numbers of wait states.

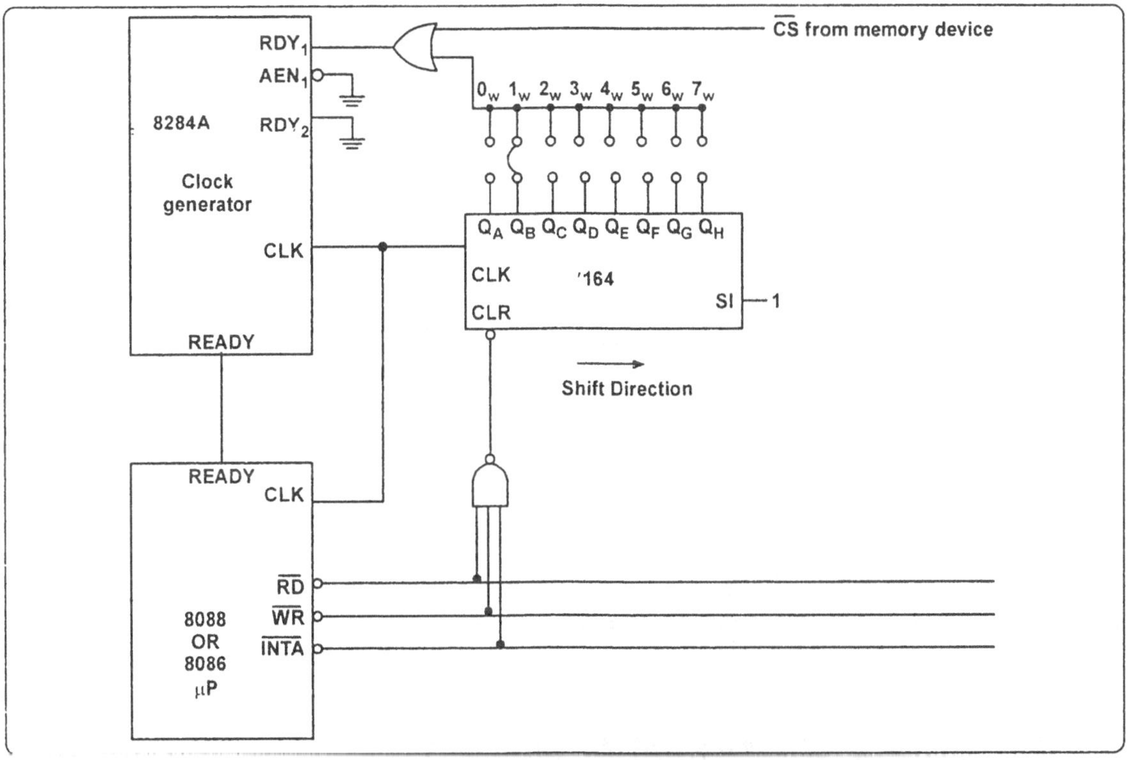

Fig. 2.20 A circuit that can cause between 0 and 7 wait states.

It can also be observed that how, the shift register is cleared back point. The output of the register is forced high when the second flip-flop captures $\overline{RD}$, $\overline{WR}$ and $\overline{INTA}$ pins are all logic l's. These three signals are high until state T2, so the shift register shifts for the first time when positive edge of the T2 arrives. If one wait is desired, then output Q_B is connected to the OR gate. If two waits are desired, output Q_C is connected, and so forth.

In Fig. 2.20 it can observed that this circuit does not always generate wait states, but is enabled from the memory only for memory devices that require the insertion of waits. If the selection signal from a memory device is a logic 0, the device is selected; then this circuit will generate a wait state. Now Fig. 2.21 depicts the timing diagram for this shift register wait state generator when it is wired to insert for one wait state.

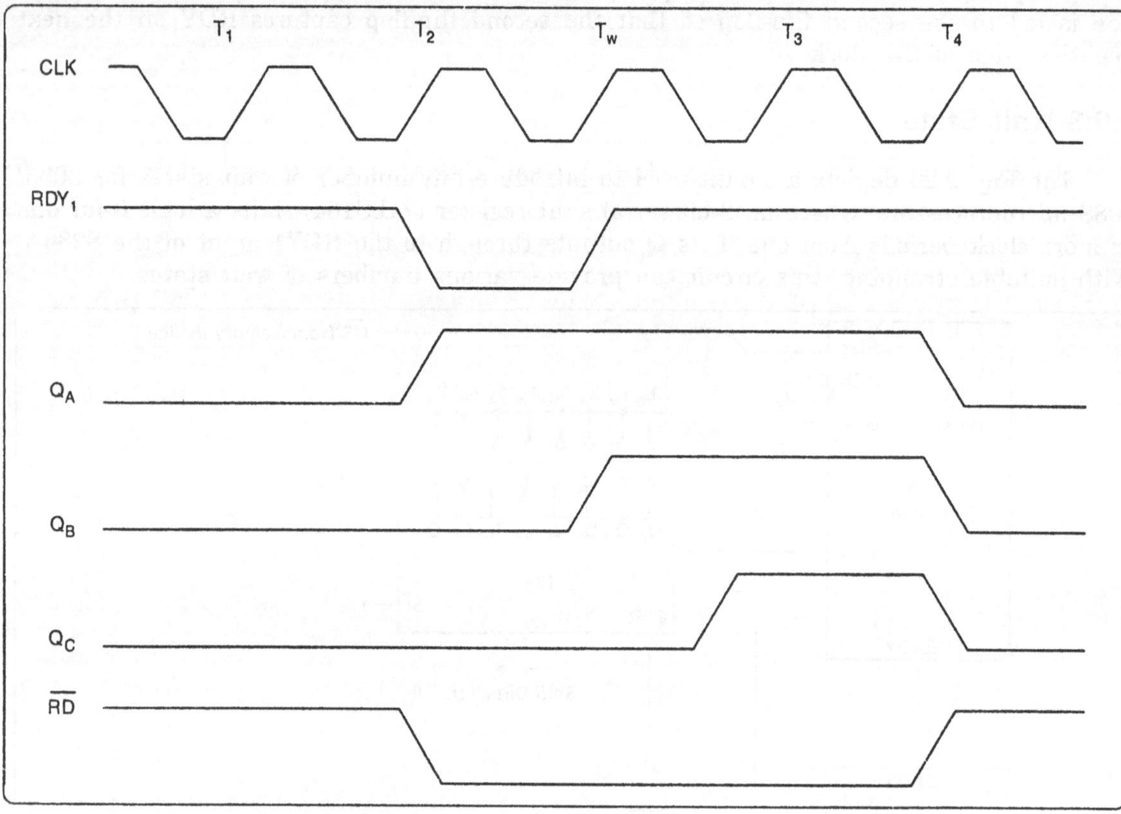

Fig. 2.21. Wait state generation timing of the circuit of Fig. 2.20.

The timing diagrams of Fig. 2.21 also illustrates the internal contents of the shift register's flip-flops to present a more detailed view of its operation. *(here one wait state is generated).*

2.7 MINIMUM AND MAXIMUM MODES

There are two available modes of operation for the 8086/8088 microprocessors : (1) Minimum mode and (2) Maximum mode. Minimum mode operation is obtained by connecting the mode selection pin MN / $\overline{MX}$ to +5.0 V, and maximum mode is selected by grounding this pin. Both modes enable different control structures for the 8086/8088 microprocessors.

2.7.1 Minimum Mode

In this mode the pin definition from 24 to 31 are to be observed, which is the main difference with respect to maximum mode, (refer Fig. 2.1). Now let us observe the pin definitions in minimum mode.

Pin 24; **(Interrupt Acknowledge) Output.** This indicates recognition or an interrupt request which consists of two negative going pulses in two consecutive bus cycles. The first pulse informs the interface that its request has been recognized and upon receipt of the second pulse, the interface is to send the interrupt type of the processor over the data bus.

Pin 25; **ALE (Address Latch Enable) Output.** This signal is provided by 8086 to demultiplex $AD_0 - AD_{15}$ into $AD_0 - AD_{15}$ and $D_0 - D_{15}$ using external latches.

Pin 26; **(Data Enable) Output.** This signal indicates the transceivers that the CPU is ready to send or receive data.

Pin 27; $DT/\overline{R}$ **(Data transmit/Receive) Output.** This signal is helps to control data flow direction. High on this pin indicates that the 8086 is transmitting the data and low indicates that the 8086 is receiving the data.

Pin 28; $M/\overline{O}$ **Output.** It is used to distinguish memory data transfer, ($M/\overline{IQ}$ = HIGH) and I/O data transfer ($M/\overline{IQ}$ = LOW).

Pin 29; $\overline{WR}$ **: Write Output.** $\overline{WR}$ is low whenever the 8086 is writing data into the memory or an I/O device.

Pin 30 and 31; **HOLD input, HLDA Output.** A HIGH on HOLD pin indicates that another master (DMA) is requesting to take over the system bus. On receiving HOLD signal processor outputs HLDA signal HIGH as an acknowledgement. At the same time, processor tristates the system bus. A low on HOLD gives the system bus control back to the processor. *(Processor then outputs low signal on HLDA).*

2.7.1 (*a*) The Minimum Mode Configuration

A typical minimum mode configuration is as illustrated in Fig. 2.22. As observed in the Fig. 2.22 $AD_0 - AD_{15}$, $A_{16}/S_3 - A_{19}/S_6$ and $\overline{BHE}/S_7$, singals are multiplexed. These signals are demultiplexed by external latches and ALE signal generated by the processor, which accomplished by using three latch ICs *(Intel 8282/8283)*, two of them are required for a 16-bit address and three are needed if a full 20-bit address is used.

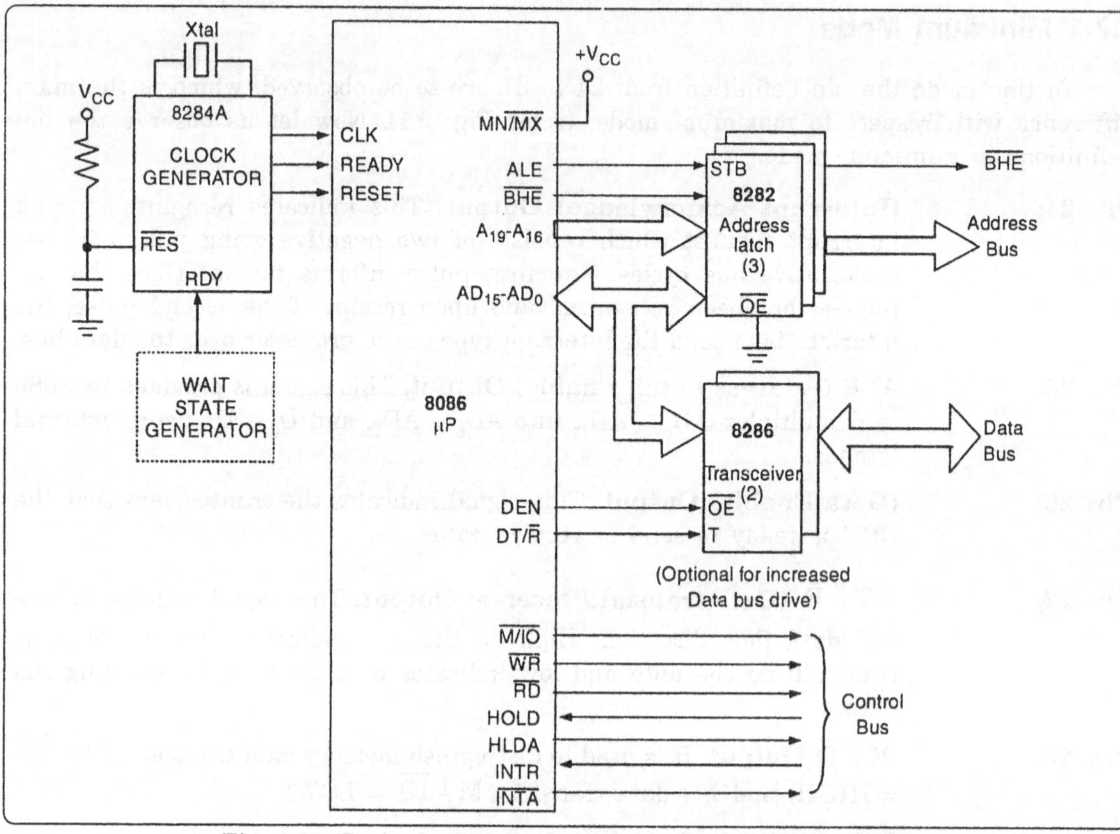

Fig. 2.22 Typical minimum mode configuration in 8086

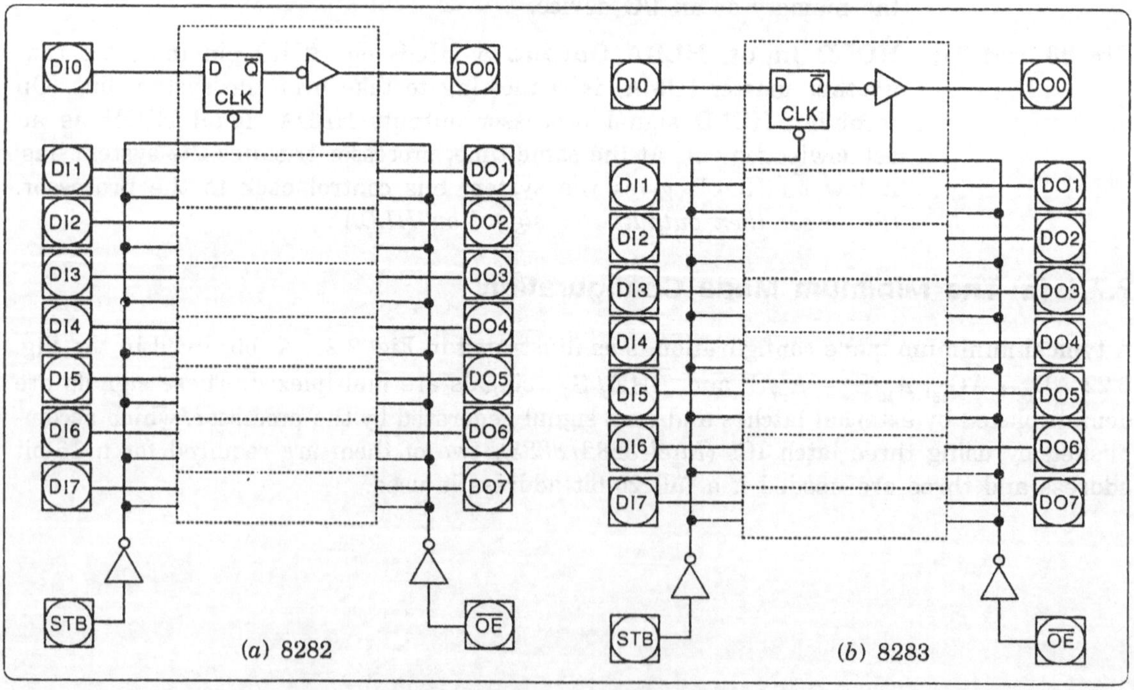

Fig. 2.23 Internal diagrams

Now Fig. 2.23 depicts the internal block of 8282/8283 latches, where 8282 provides noninverting outputs while the 8283 version inverts the input data. Apart from their demultiplexing function, these chips also buffer the address lines, providing increased output driving capability. The output low level is specified as 0.45 V maximum with a sink current of 32 mA maximum. The high level is specified as 2.4 V minimum while supplying a 5 mA maximum high level load current.

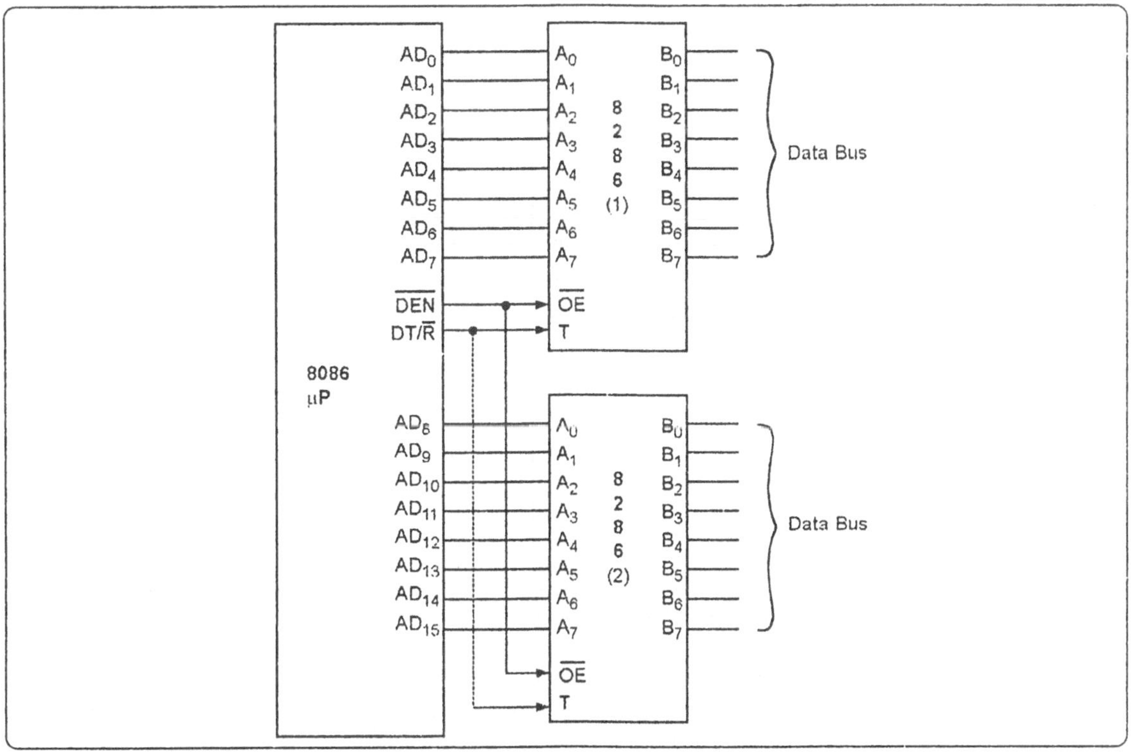

Fig. 2.24 The connection details of two 8286 with 8086

When a system includes several interfaces then to increase current sourcing/sinking capacities it is necessary to use drivers and receivers (transceiver) for data bus also. The Intel 8286 device is used to implement the transceiver block shown in Fig. 2.22 which contains 16 tristate elements, eight receivers, and eight drivers. That means two 8286s are required to service 16 data lines of 8086. The Fig. 2.24 shows the detailed connections of 8286.

The $DT/\overline{R}$ signal is connected to the T input, which controls the direction of the data flow. When this signal is low, receivers are enabled, so that 8086 can read data from memory or input device. To write data into memory or input device, the 8086's $DT/\overline{R}$ signal goes high. Since this drivers are enabled to transfer data from 8086 to the memory or the output device, at the time of data transfer, to enable output of transceiver its $\overline{OE}$ should be low. This accomplished by connecting $\overline{DEN}$ signal of 8086 to the $\overline{OE}$ pin of 8286, since $\overline{DEN}$ signal goes low when CPU is ready to send or receive data.

Then third component, other than the processor that appears in Fig. 2.22 is an 8284 clock generator. In addition to supplying a train of pulses at a constant frequency it synchronizes ready (RDY) signal, which indicates an interface is ready to complete a transfer, and reset $(\overline{RES})$ signal, which initializes the system, with the clock pulses. As observed in Fig. 2.25, there are two input pins for frequency source which are :

(*a*) From a pulse generator which is connected to the external frequency input (EFI)

(*b*) From an oscillator which is connected across X_1 and X_2.

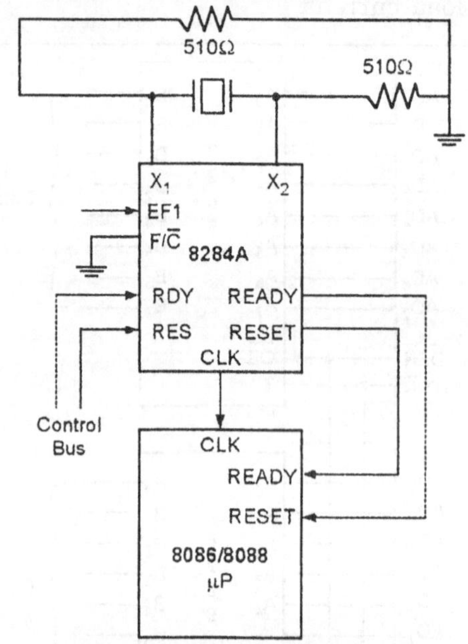

Fig. 2.25. The clock generator 8284.

As observed in fig. 2.25 the $F/\overline{C}$ signal selects one of the inputs. If $F/\overline{C}$ input is 1, the EFI input determines the frequency; otherwise oscillator determines the frequency. *(In both the cases the output clock frequency is one third of the input frequency).* Next the status on the $M/\overline{IO}$, $\overline{RD}$ and $\overline{WR}$ control bus lines will decide the type of data transfer. The Table 8 shows the details of data transfers as below.

Table 8

$M/\overline{IO}$	$\overline{RD}$	$\overline{WR}$	**Operation**
0	0	1	I/O read
0	1	0	I/O write
1	0	1	Memory read
1	1	0	Memory write

The HOLD and HLDA signals helps to interface other bus masters like DMA controller. Interrupt request (INTR) and interrupt acknowledge $(\overline{INTA})$ are used to extend the interrupt handling capacity of the 8086 with the help of interrupt controller.

2.7.1 (*b*) The Bus Timings for Minimum Mode

The bus timing diagrams for Input and Output transfers for 8086 microprocessor in minimum mode operation are as shown in Fig. 2.26 and Fig. 2.27 on next page.

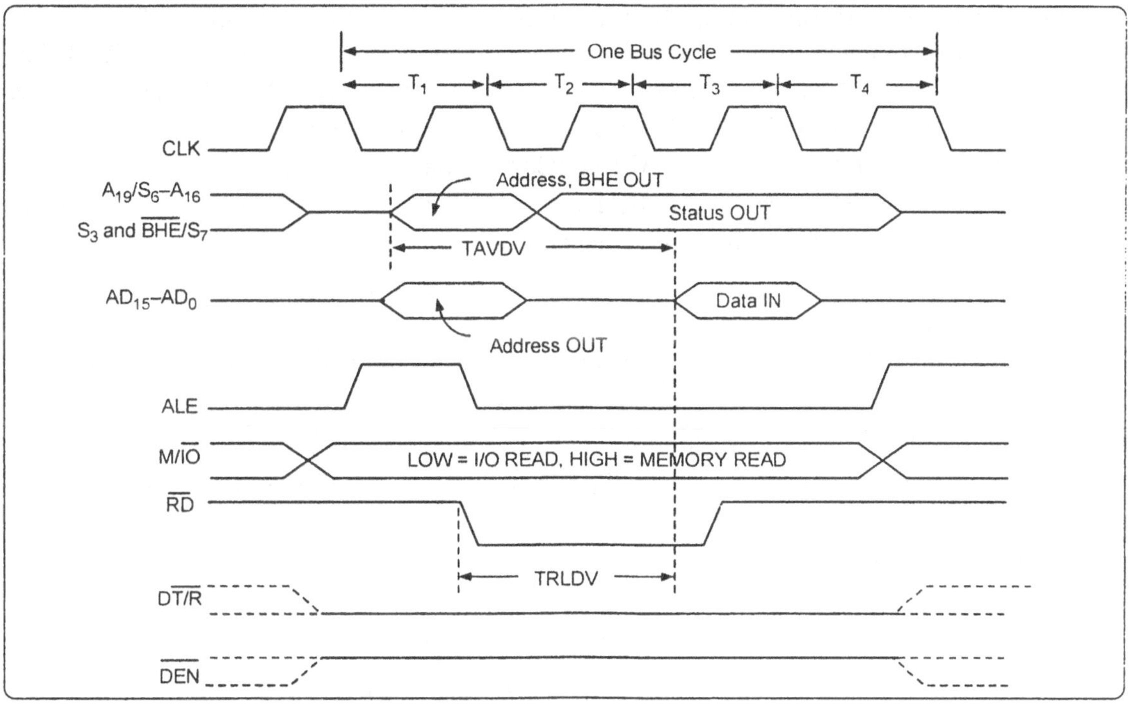

Fig. 2.26. Timing diagram for Input.

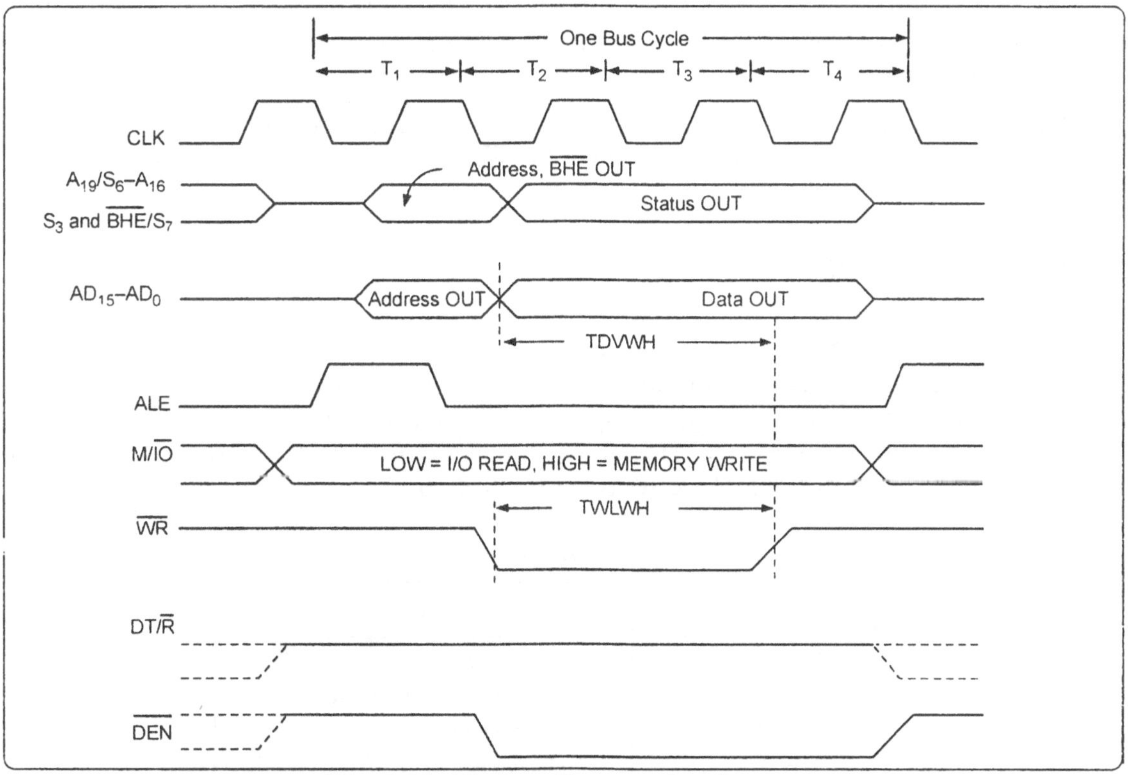

Fig. 2.27. Timing diagram for Output.

Now let us analyse the timing diagram of Fig. 2.26 and Fig. 2.27 in brief.

- If processor is ready to initiate the bus cycle, it applies a pulse to ALE during T1. Before the falling edge of ALE, the address, $\overline{BHE}$, $M/\overline{IO}$, $\overline{DEN}$ and $DT/\overline{R}$ must be stable *i.e.*, DEN = high and $DT/\overline{R}$ = 0 for output or $DT/\overline{R}$ = 1 for output.
- At trailing edge of ALE, ICs 72LS373 or 8282 latches the address.
- During T2 the address signals are disabled and $S_3 - S_7$ are available on $AD_{16}/S_3 - A_{19}/S_6$ and $\overline{BHE}/S_7$. Also $\overline{DEN}$ is lowered to enable transceiver.
- For input operation, $\overline{RD}$ is activated during T_2 and AD_0 to AD_{15} go in high impedance preparing for input.
- If memory or I/O interface can perform the transfer immediately; there are no wait states and data is output on the bus during T_3.
- After the data is accepted by the processor, $\overline{RD}$ is raised high at the beginning of T_4.
- Once detecting this transition during T_4, the memory or I/O device will disable its data signals.
- For an output operation, processor applies $\overline{WR}$ = 0 and then the data on the data bus signals.
- In T_4, $\overline{WR}$ is raised high and data signals are disabled.
- For either input or output operations, $\overline{DEN}$ is raised during T_4 to disable the transceiver and $M/\overline{IO}$ is set according to the next transfer at this or during next T_1 state. Thus length of bus cycle I 8086 is four clock cycle. When the bus is to be inactive after completion of bus cycle, then the gap between the successive cycles is filled by ideal stack clock cycles.

Note. If the memory or I/O device is not able to respond quickly during transfer, wait states (T_W) are inserted between T_3 and T_4, by disabling the READY input of the 8086. The bus activity during wait state is same as during T_3.

2.7.2 Maximum Mode

In this mode, the pin $MN/\overline{MX}$ is grounded. Now to observe the condition of the processors in this mode the pin definition for 24 to 31 is explained briefly.

Pin 24 and 25 **QS1 QS0 (Output).** These two output signals reflect status of the instruction queue. This status indicates the activity in the queue during the previous clock cycle, as in Table 9.

Table 9

QS_1	QS_0	Status
0	0	No operation (queue is idle)
0	1	First byte of an opcode
1	0	Queue is empty
1	1	Subsequent byte of an opcode

Pin 26, 27, 28 **(Output).** These three status signals indicate the type of transfer to be take place during the current bus cycle as shown in Table 10.

Table 10

$\bar{S}_1$	$\bar{S}_2$	$\bar{S}_0$	**Machine Cycle**
0	0	0	Interrupt Acknowledge
0	0	1	I/O Read
0	1	0	I/O Write
0	1	1	Halt
1	0	0	Instruction fetch
1	0	1	Memory read
1	1	0	Memory write
1	1	1	Inactive-Passive

Pin 29 $\overline{\text{LOCK}}$. This signal indicates that an instruction with a LOCK prefix is being executed and bus is not to be used by another potential master.

Pin 30 and 31 $\overline{\text{RQ}}/\overline{\text{GT}_1}$ **and** $\overline{\text{RQ}}/\overline{\text{GT}_0}$. In maximum mode, HOLD and HLDA pins are replaced by $\overline{\text{RQ}}$ (Bus request)/ $\overline{\text{GT}_0}$ (Bus Grant), and $\overline{\text{RQ}}/\overline{\text{GT}_1}$ signals. By using bus request signal one more master can request for the system bus and processor communicate that the request is granted to the requesting master by using bus grant signal. However both signals are similar except the $\overline{\text{RQ}}/\overline{\text{GT}_0}$ has higher priority than $\overline{\text{RQ}}/\overline{\text{GT}_1}$.

2.7.2 (*a*) The Maximum Mode Configuration

The Fig. 2.28 illustrates the typical maximum mode configuration, where additional circuitry is required to translate the control signals. The additional circuitry converts the status signals $(\bar{S}_2 - \bar{S}_0)$ into the I/O and memory transfer signals. It also generates the control signals required to direct the data flow and for controlling 8282 latches and 8286 transceiver. The Intel 8288 bus controller is used to implement this control circuitry.

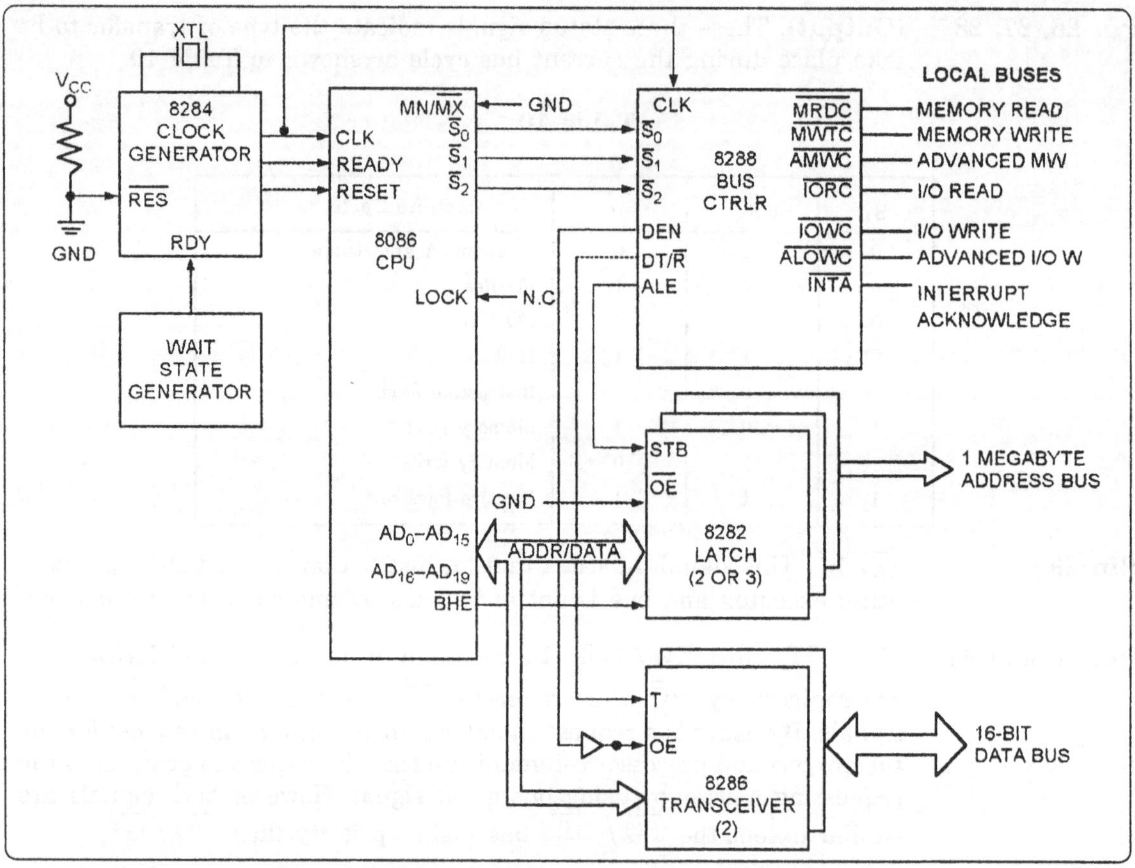

Fig. 2.28 Typical maximum mode configuration

Now Fig. 2.29 shows that 8288 is able to originate address latch enable signal to the 8282's, the enable and direction signals to the 8286 transceivers, and the interrupt acknowledge signal to the interrupt controller. Apart it also decodes the $\overline{S}_2 - \overline{S}_0$ signals to generate $\overline{MRDC}$, $\overline{MWTC}$, $\overline{IORC}$, $\overline{IOWC}$, MCE / $\overline{PDEN}$, $\overline{AEN}$, IOB, CEN, ALOWC and AMWC signals.

The commands are briefed as below :

- $\overline{\text{MRDC}}$ **(Memory Read Command).** It instructs memory to put the contents of the addressed location on the data bus.
- $\overline{\text{MWTC}}$ **(Memory Write Command).** It instructs memory to accept the data on the data bus and load the data into the addressed memory location.
- $\overline{\text{IORC}}$ **(I/O Read Command).** It instructs an I/O device to put the data contained in the addressed port on data bus.
- $\overline{\text{IOWC}}$ **(I/O Write Command).** It instructs an I/O device to accept the data on the data bus and load the data into the addressed port.

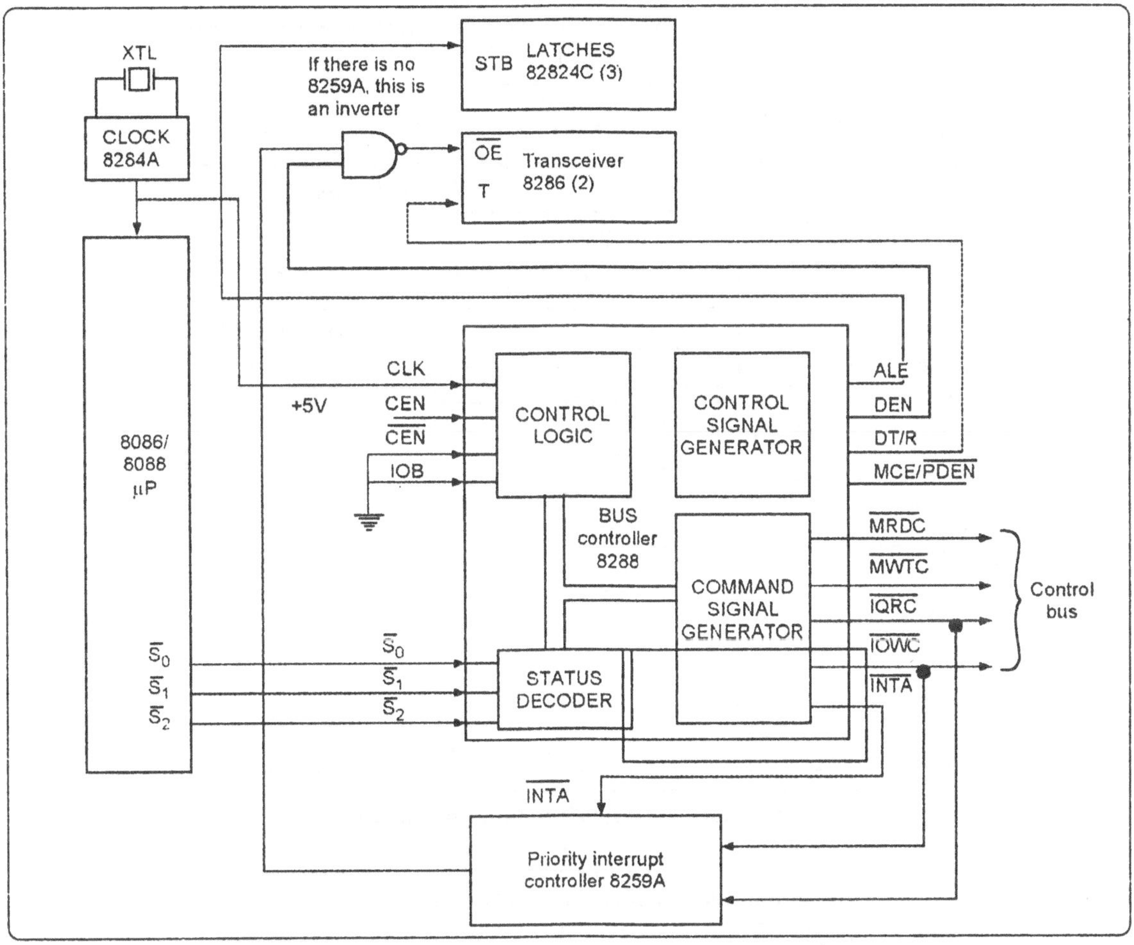

Fig. 2.29 The 8288 bus controller

- **$MCE/\overline{PDEN}$ (Master Cascade Enable/Peripheral Data Enable).** It controls the mode of operation of 8259.
- **$\overline{AEN}$ IOB and CEN.** These pins helps in multiprocessor system, with a single processor in the system, $\overline{AEN}$ and IOB are grounded and CEN in tied high.
- **$\overline{AIOWC}/\overline{AMWC}$ (Advance I/O Write Command/Advance Memory Write Command).** These signals are just similar to IOWC and MWTC except that they are activated one clock pulse earlier. This gives slow interfaces an extra clock cycle to prepare to input data.

2.7.2 (*b*) The Bus Timings for Maximum Mode

The Fig. 2.30 and Fig. 2.31 shows the timing diagrams for Input and Output transfers respectively as below :

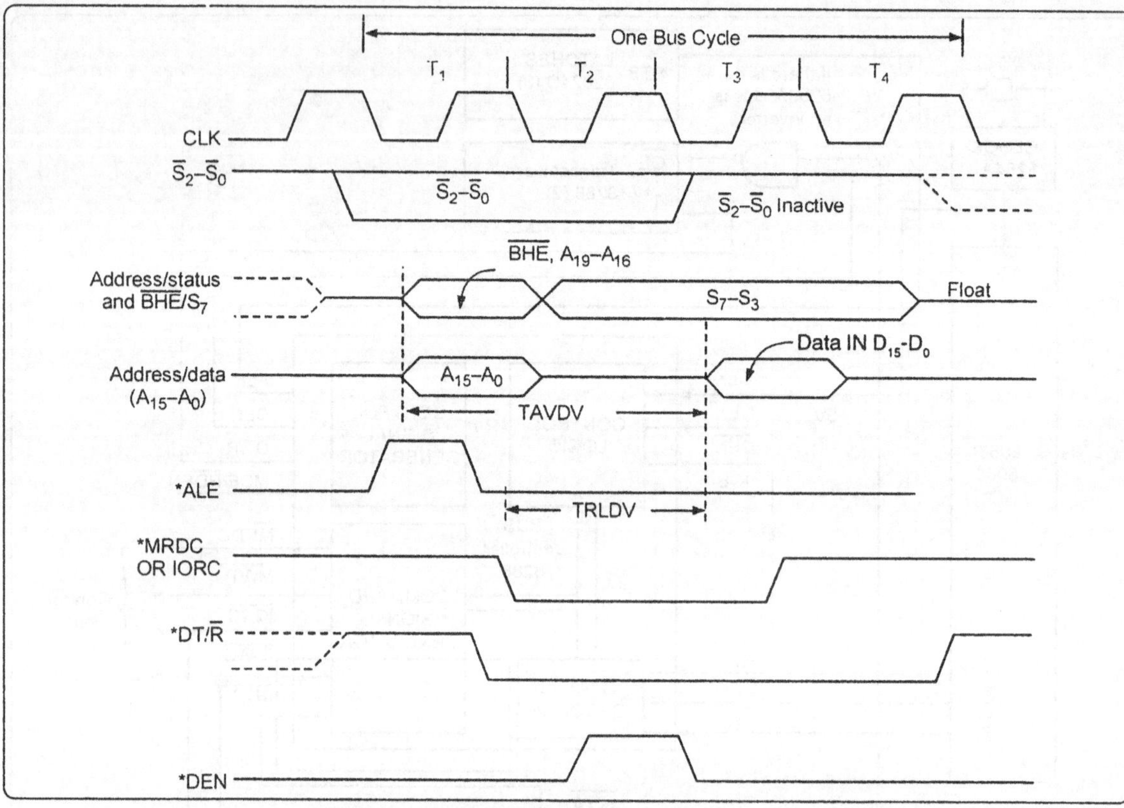

Fig. 2.30 Timing diagram for Input

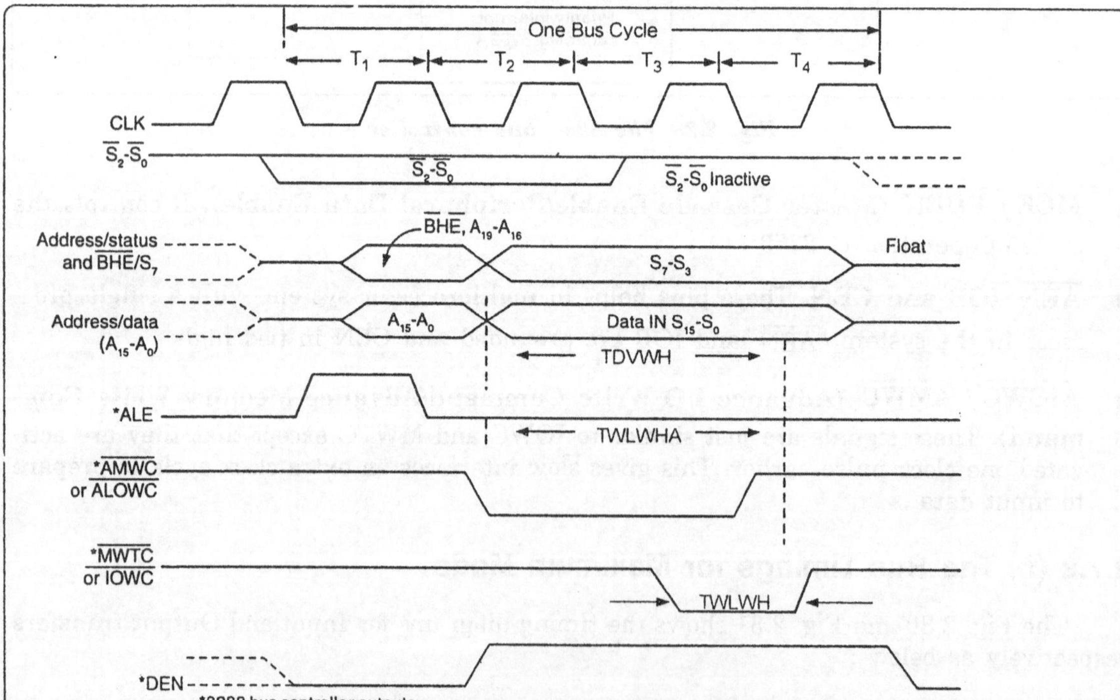

Fig. 2.31 Timing diagram for Output

The timing diagram for Input and Output are briefed as below :

- $\overline{S_0}$, $\overline{S_1}$, $\overline{S_2}$ are set at the beginning of bus cycle, once detecting the change on passive state $\overline{S_0}$, $\overline{S_1}$, $\overline{S_2}$ = 1, the 8288 bus controller will output a pulse on its ALE and apply a required signal to its DT / $\overline{R}$ pin during T_1.
- In T_2, 8288 will set DEN = 1 which enabling transceiver. For an input, 8288 it will activates $\overline{MRDC}$ or $\overline{IORC}$. These signals are activated until T_4. For an output, the $\overline{AWWC}$ or $\overline{AIOWC}$ is activated from T_2 to T_4 and $\overline{MWTC}$ or $\overline{IOWC}$ is activated from T_3 to T_4.
- The status bits $\overline{S_0}$ to $\overline{S_2}$ remain active until T_3, and become passive during T_3 and T_4.
- If ready input is not activated before T_3, wait state will be inserted between T_3 and T_4.

REVIEW QUESTIONS

1. List out the differences between 8086 and 8088 microprocessors.
2. Explain wether 8086/8088 are TTL compatible.
3. What is the purpose of status bits S3 and S4 ?
4. What happens when HOLD input to the 8086/8088 is placed at its logic 1 level ?
5. What condition does a logic 0 an 8086/8088 $\overline{RD}$ pin indicate ?
6. Briefly explain the operation of $\overline{LOCK}$ pin.
7. With the help of block diagrams explain the operation of 8284 clock generation.
8. Which bus connection on 8086 microprocessor are typically demultiplexed ?
9. Which TTL integrated circuit is often used to demultiplex the buses as 8086/8088 ?
10. Why buffers are often needed on the address data and control buses in a microprocessor system?
11. Draw the block diagram showing 8086 in maximum mode configuration and explain.
12. Mention the signals which are different when 8086 in minimum mode and is maximum mode.
13. Explain the function of following pins in 8086 μP :

    (*a*) NMI (*b*) $\overline{TEST}$ (*c*) DEN (*d*) $QS_0 - QS_1$ (*e*) $\overline{BHE}$

OBJECTIVE TYPE QUESTIONS

1. *The PCLK output of 8284A is .... MHz, if crystal oscillator is operating at 14 MHz.*
2. *A bus cycle is equal to clocking .... periods.*
3. *The INTR input is .... sensitive.*
4. *8086/8088 use multiplexed address and data bus because .... ?*
   - *(i) Multiplexing is supported by 8085 μP*
   - *(ii) 40 pins is a good size for IC*
   - *(iii) Multiplexing reduces number of liens between μP and auxiliary ICs.*
   - *(iv) All.*
5. *The SSO of 8088 μP indicate ....*
6. *If READY pin is grounded, it will introduce .... states into the bus cycle of 8086/8088 μP.*

3 MEMORY INTERFACING

3.0 OBJECTIVE

After completing this chapter, the reader will be familiar with :

- Memory Device
- Read/write Memory cycle
- Memory Interfacing concepts
- Address Decoding
- Basic 8/16-bit Memory Interfacing
- The DRAM controller.

3.1 INTRODUCTION

Every microprocessor-based system has a memory system, which may be simple or complex. Most of the systems contains mainly two types of memory : ROM, *i.e.*, Read Only Memory and RAM, *i.e.*, Random access memory or read/write memory. Interfacing of memory device with μP will be discussed in this chapter. The memory interfacing circuits are used to access memory often to read instruction code and data stored in memory, such read/write operations are monitored by control signals. The microprocessor activates such signals when required to read from and write into memory.

3.2 OPERATION AND TERMINOLOGY OF MEMORY

The Memories are made up of registers, each register in the memory is one storage location and each location is identified by an address. The number of storage locations can vary from a few in some memories to hundreds of thousand in others. Each location can accommodate one or more bits. Normally, the total number of bits that a memory can store is its capacity. Most of the types the capacity is specified in terms of bytes. Each register consists of storage elements like flip-flop or capacitors in semiconductor memories and magnetic domain in magnetic storage, each of which stores one bit of data. A storage element is called a cell. Fig. 3.1 depicts simple Read/write operation is memory cells.

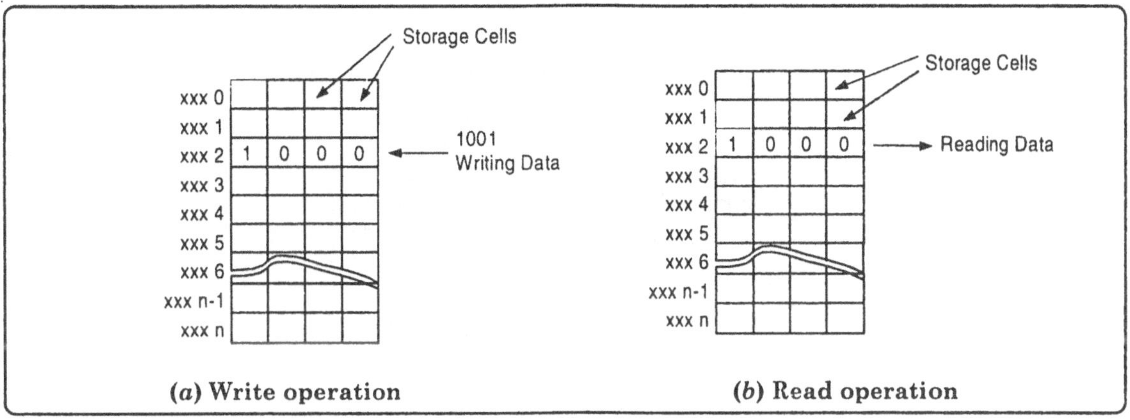

(*a*) **Write operation** (*b*) **Read operation**

Fig. 3.1 Generalized Memory operations.

As observed from Fig. 3.1, the data stored in memory by a process called writing and are retrieved from the memory by a process called reading.

3.3 MEMORY DEVICES

Before trying to interface memory to the microprocessor, it is suitable to understand the operation of memory components. Now let us explain the function of the four common types of memory : **read-only memory** (ROM), **flash memory** (EEPROM), **static random access memory** (SRAM), and **dynamic random access memory** (DRAM). The classification of semiconductor memory is as shown belows.

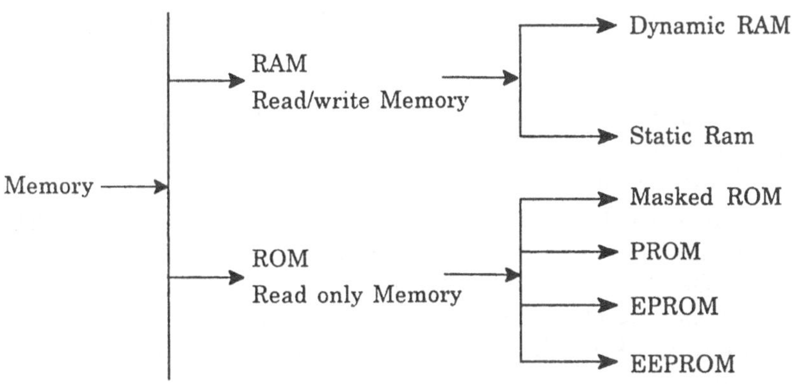

3.3.1 Memory Pin Connections

The Pin connections are common to all memory devices are the address inputs, data outputs or input/outputs, some type of selection input, and at least one control input used to select a read or write operation. Now Fig. 3.2 illustrates for ROM and RAM generic-memory devices.

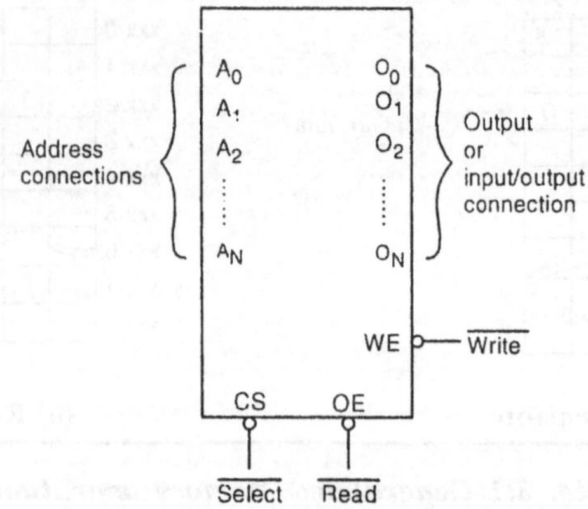

Fig. 3.2 Pseudo memory component illustrating the address, data, and control connection (for ROM and RAM generic-memory devices).

Before discussing different memory devices let us analyze the different pin connections for different purposes. The important connection such as :

(*a*) Address Connections

(*b*) Data Connections

(*c*) Selection Connections, and

(*d*) Control Connections

(*a*) Address Connections

All memory devices have address inputs that select a memory location within the memory device. The address inputs are almost always labelled from A_0, the least significant address input, to A_n, where subscript $_n$ can be any value but is always labelled as one less that the total number of address pins. For instant a memory device that has 10 address pins has its address pins labelled from A_0 to A_9. Number of address pins found on a memory device is determined by the number of memory locations found within it.

The more common memory devices have between 1K (1,024) to 16M (16,777,216) memory locations, with 256M memory location devices on the horizon. A 1K memory device has 10 address pins (A_0 – A_9); therefore, 10 address inputs are required to select any of its 1,024 memory locations. It takes a 10-bit binary number *(1,024 different combinations)* to select any single location on a 1,024-location device. When a memory device has 11 address connections (A_0 – A_{10}), it has 2,048 (2K) internal memory locations. The number of memory locations can thus be extrapolated from the number of address pins. For instant, a 4K memory device has 12 address connections, an 8K device has 13, and so forth. A device that contains 1M locations requires a 20-bit address (A_0 – A_{19}).

A 400H represents a 1K-byte section of the memory system, when a memory device is decoded to begin at memory address 10000H, and it is a 1K device, its last location is at address 103FFH – one location less than 400H. One more important hexadecimal number to remember is a 1000H, because 1000H is 4K. A memory device containing a starting address of 14000H that is 4K bytes in size ends at location 14FFFH – one locations less than 1000H. A third number is 64K or 10000H. A memory that starts at location 3000H and ends at location 3FFFFH is a 64K byte memory. At the end since 1M of memory is not uncommon, a 1M memory contains 100000H memory locations.

(*b*) Data Connections

All memory devices have a set of data outputs or input/outputs. The device illustrated in Fig. 3.1 has a common set of input/output (I/O) connections. Now a days memory devices have bi-directional common I/O pins.

The data connections are the points at which data are entered for storage or extracted for reading. Data pins on memory devices are almost always labelled D_0 through D_7 for an 8-bit-wide memory device. In this sample memory device, there are eight I/O connections, which means that the memory device stores 8-bits of data in each of its memory locations. An 8-bit-wide memory device is often called a **byte-wide** memory. Although most devices are currently 8-bits wide, not all memory devices are 8-bits wide. Some devices are 16-bits, 4-bits, or just 1-bit wide.

Catalogue listings of memory devices often refer to memory locations times bits per location. For instant, a memory device with 1K memory locations and 8-bits in each location is often listed as a 1K × 8 by the manufacturer. A 16K × 1 is a memory device containing 16K 1-bit memory locations. The Memory devices are also often classified according to total bit capacity. For instant, a 1K × 8 bit memory device is sometimes listed as an 8K memory device, or a 64K × 4 memory device is listed as a 256K device. *(These variations occur from one manufacturer to an other).*

(*c*) Selection Connections

All memory device has an input — sometimes more than one — that selects or enables the memory device. Such kind of input is most often called a **chip select** $(\overline{CS})$, **chip enable** $(\overline{CE})$, or simply **select** $(\overline{S})$ input. RAM memory generally has at least one $\overline{CS}$ or $\overline{S}$ input, and ROM at least one $\overline{CE}$. If the $\overline{CE}$, $\overline{CS}$ or $\overline{S}$ input active *(a logic 0 in this case, because of the over-bar)*, the memory device performs a read or a write; if it is inactive *(a logic 1 in this case)*, the memory device cannot do a read or a write because it is turned off or disabled. *(When more than one* $\overline{CS}$ *connection is present, all must be activated to read or write data).*

(*d*) Control Connections

Each memory devices have some form of control input or inputs. A ROM normally has only one control input, while a RAM often has one or two control inputs.

The control input, while a ROM is the **output enable** $(\overline{OE})$ or gate $\overline{G}$ connection, which allows data to flow out of the output data pins of the ROM. If $\overline{OE}$ and the selection input are both active, then the output is enabled; if $\overline{OE}$ is inactive, the output is disabled at its high-impedance state. The $\overline{OE}$ connection enables and disables a set of three-state buffers located within the memory device and must be active to read data.

RAM memory device has either one or two control inputs, when there is one control input, it is often called $R/\overline{W}$. This pin selects a read operation or a write operation only if the device is selected by the selection input $\left(\overline{CS}\right)$. If the RAM has two control inputs, they are usually labelled $\overline{WE}$ (or $\overline{W}$) and $\overline{OE}$ (or $\overline{G}$). Here $\overline{WE}$ **(write enable)** must be active to perform a memory write operation, and OE must be active to perform a memory read operation. If these two controls ($\overline{WE}$ and $\overline{OE}$) are present, they must never both be active at the same time. When both control inputs are inactive (logic 1's), then data are neither written nor read and the data connections are at their high-impedance state.

3.4 READ ONLY MEMORY (ROM)

The read-only will permanently store programs and data that are resident to the system and must not change when power is disconnected. Since ROM is permanently programmed so data are always present, even when power is disconnected. This type of memory is often called nonvolatile memory, available in many forms today. A device we call a ROM is purchased in mass quantities from a manufacturer and programmed during its fabrication at the factory itself. The EPROM ***(erasable programmable read-only memory),*** a type of ROM, is more commonly used when software must be changed often or when too limited a number are in demand to make the ROM economical. For a ROM to be practical, we usually must purchase at least 10,000 devices. An EPROM is programmed in the field on a device called an EPROM programmer. The EPROM is also erasable is exposed to high-intensity ultraviolet light for about 20 minutes or less, depending on the type of EPROM.

The PROM memory devices are also available, but they are not as common today. The PROM ***(programmable read-only memory)*** is also programmed in the field by burning open tiny Nichrome or silicon oxide fuses, but once programmed it cannot be erased.

One more, newer type of **read-mostly memory** (RMM) is called the flash memory. The flash memory is also often called an EEPROM *(electrically erasable programmable ROM), EAROM (electrically alterable ROM),* or a NOVRAM (nonvolatile RAM). These memory devices are electrically erasable in the system, but require more time to erase than a normal RAM. The flash memory device is used to store setup information for systems such as the video card in the computer. It may also soon replace the EPROM in the computer for the BIOS memory. Some systems contains a password stored in the flash memory device.

3.4.1 Architecture of ROM

Fig. 3.3 illustrate simple ROM for four byte diode ROM, which consists of only diodes and a decoder.

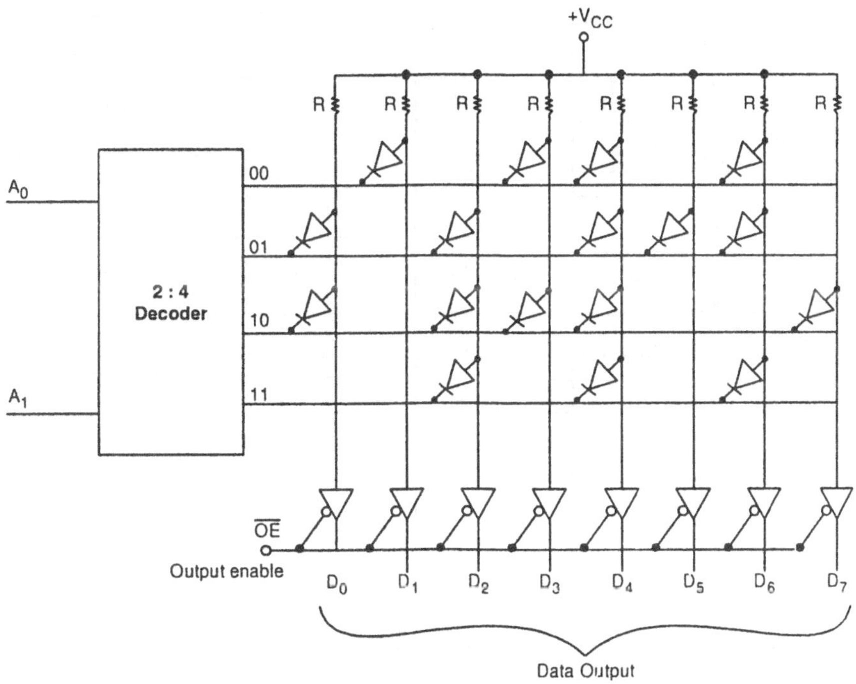

Fig. 3.3 Four byte diode ROM.

As observed in the Fig. 3.3 address lines A_0 and A_1 are decoded by 2 : 4 decoder and used to select one of the four rows. As decoder output is active low, it places a logic 0 on the selected row. Each output data line goes to logic 0 if a diode connects the output data column to the selected row. The data is available on the output data lines only when output enable $(\overline{OE})$ signal is low. Now Table 1 shows the contents of ROM at four locations, as below.

Table 1. Contents of ROM

Address in Binary	The Binary Data								Data in Hex
	D_0	D_1	D_2	D_3	D_4	D_5	D_6	D_7	
00	1	0	1	0	0	1	0	1	A5
01	0	1	0	1	0	0	0	1	51
10	0	1	0	0	0	1	1	0	46
11	1	1	0	1	0	1	0	1	D5

ROMs use MOS technology instead of diode in present day. Fig. 3.4 depicts four nibble *(half-byte)* ROM using MOS transistors, where diodes and pull up resistors are replaced by MOS transistors.

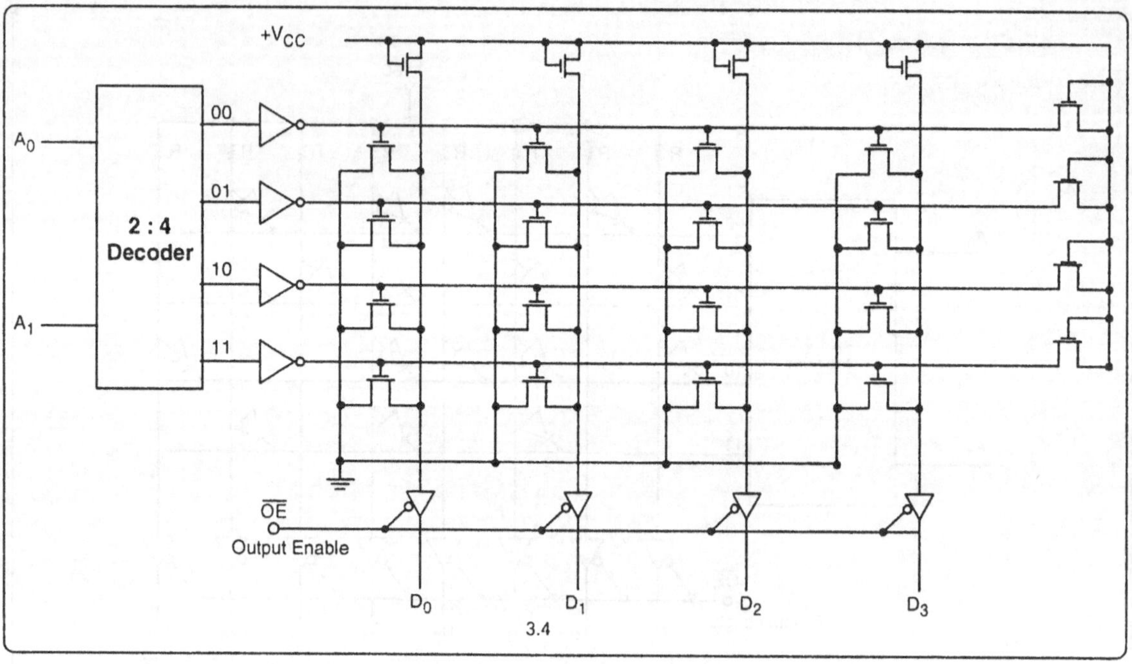

Fig. 3.4. Four half-byte ROM.

Address on the address lines *(A_0 and A_1)* is decoded by 2 : 4 decoder, which selects one of the four rows making it logic 0. The inverter connected at the output of decoder inverts the state of selected row *(i.e., logic 1)*. Hence each output data line goes to logic 0 if a gate of MOS transistor is connected to row select lines. When gate of the MOS transistor is connected to the selected row, MOS transistor is turned on, which pulls the corresponding column data line to logic 0.

Note. In integrated circuits, a thin metallized layer connects the gates of some transistors to the row select lines. The gate connections of MOS transistors depend on the data to be stored in the ROM. That means, according to the user truth table, manufacturer can deposit thin layer of metal to connect gates of the transistors. Once the pattern/mask is decided, it is possible to make thousands of such ROMs. Such ROMs are called Mask-programmed ROMs. The Masked ROMs are used in microprocessor based toys, TV games, home computers and other such high volume consumer products etc.

3.4.2 Programmable Read Only Memory (PROM)

The Fig. 3.5 depicts a four byte PROM, which has diodes in every bit position; hence, the output is initially all 0s, and each diode, however has a fusible link in series with it.

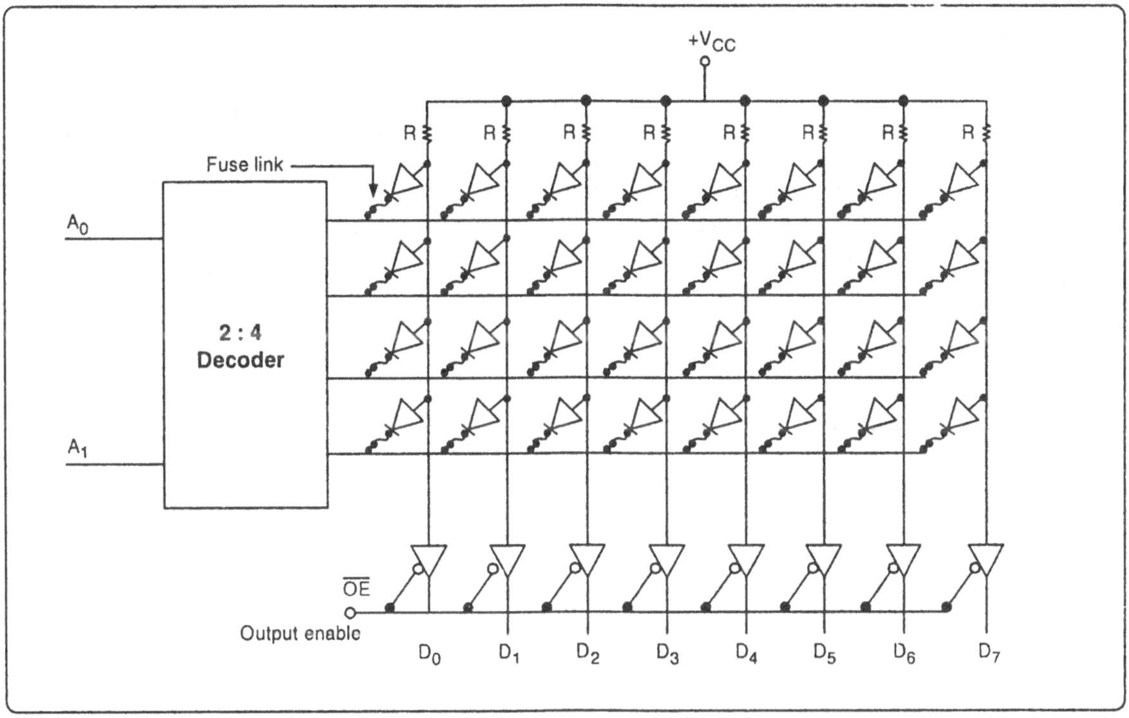

Fig. 3.5 Simple four byte PROM.

By addressing bit and applying proper current pulse at the corresponding output, we can blow out the fuse, storing logic 1 at that bit position. The fuse uses material like nichrome and polycrystalline. For blowing the fuse it is necessary to pass around 20 to 50 mA of current for period 5 to 20 ms. Blowing of fuses according to the truth table is called programming of ROM. The user can program PROMs with special PROM programmer. PROM programmer selectively burns the fuses according to pattern to be stored. This process is also known as burning of PROM, which are one time programmable. Hence once programmed, the information stored is permanent.

3.4.3 Erasable Programmable Read Only Memory (EPROM)

The Erasable programmable ROMs use MOS circuitry, they store 1s and 0s as a packet of charge in a buried layer of the IC chip. There EPROMs can be programmed by the user with a special EPROM programmer. We can erase the stored data in the EPROMs by exposing the chip to ultraviolet light through its quartz window for 15 to 20 minutes. The Fig. 3.6 depicts the 2716 EPROM, is representative of most EPROMs.

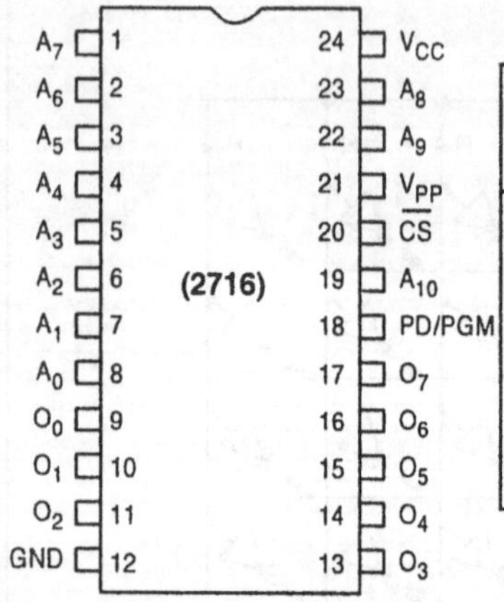

(*a*) **Pin Configuration**

MODE \ PINS	PD/PGM (18)	$\overline{CS}$ (20)	V_{PP} (21)	V_{CC} (21)	Outputs (9-11, 13-17)
Read	V_{IL}	V_{IL}	+5	+5	D_{OUT}
Deselect	Don't care	V_{IH}	+5	+5	High Z
Power Down	V_{IH}	Don't care	+5	+5	High Z
Program	Pulsed V_{IL} to V_{IH}	V_{IH}	+25	+5	D_{IN}
Program Verify	V_{IL}	V_{IL}	+25	+5	D_{OUT}
Program Inhibit	V_{IL}	V_{IH}	+25	+5	High Z

(*b*) **The Mode Selection**

$A_0 - A_{10}$	ADDRESSES
PD/PGM	POWER DOWN/PROGRAM
DS	CHIP SELECT
$O_0 - O_7$	OUTPUTS

(*c*) **Pin Names**

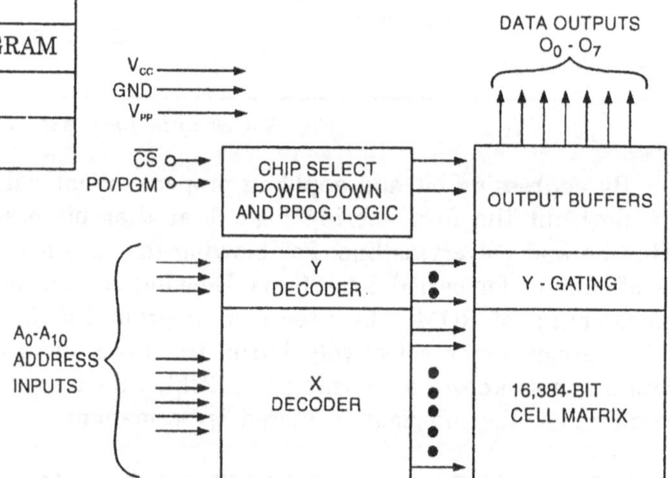

(*d*) **Block Diagram**

Fig. 3.6 Pin-out of the 2716, 2K × 8 EPROM.

The device shown in Fig. 3.6 contains 11 address input and 8 data outputs. The 2716 is a 2K × 8 memory device. The 27XXX series of the EPROMs contains the following part numbers : 2704 (512 × 8), 2708 (1K × 8), 2716 (2K × 8), 2732 (4K × 8), 2764 (8K × 8), 27128 (16K × 8), 27256 (32K × 8), 27512 (64K × 8), and 271024 (128K × 8). Each of these parts contains address pins, eight data connections, one or more chip selection inputs $\left(\overline{CE}\right)$ and an output enable pin $\left(\overline{OE}\right)$. The Fig. 3.7 illustrates the timing diagram for the 2716 EPROM.

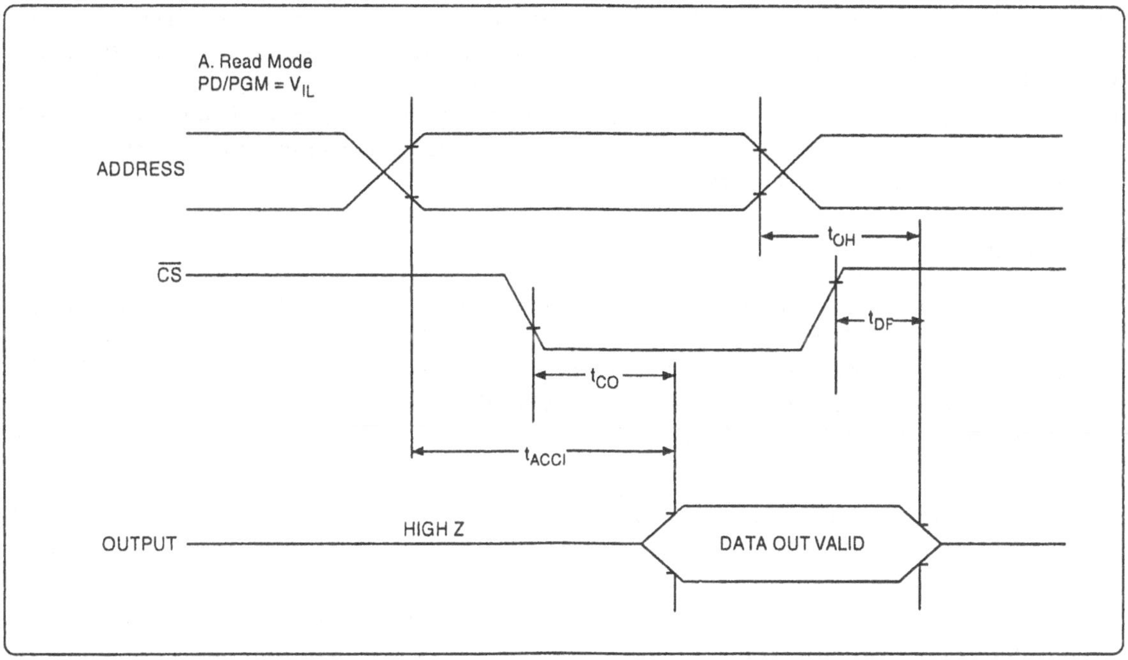

Fig. 3.7. Timing diagram of AC characteristics of the 2716 EPROM.

The Data only appear on output connections after a logic 0 is placed on both the $\overline{CE}$ and $\overline{OE}$ pin connections. If $\overline{CE}$ and $\overline{OE}$ are not both logic 0's the data output connections remain at their high-impedance or off states. Note that the V_{PP} pin must be placed at a logic 1 for data to be read from the EPROM. In some cases, the V_{PP} pin is in the same position as the $\overline{WE}$ pin on the SRAM. This can allow a single socket to hold either an EPROM or an SRAM. Examples are the 2716 EPROM and the 6116 SRAM, both 2K × 8 devices that have the same pin-out, except for V_{PP} on the EPROM and $\overline{WE}$ on the SRAM.

Note. Important piece of information provided by the timing diagram and data sheet is the memory access time–the time that it takes the memory to read information. As Fig. 3.7 illustrates, memory access time *(T_{ACC})* is measured from the appearance of the address at the address inputs until the appearance of the data at the output connections. Which is based on the assumption that the $\overline{CE}$ input goes low at the same time that the address inputs become stable. Also, $\overline{OE}$ must be a logic 0 for the output connections to become active. Basic speed of the EPROM is 450 ns. Such type of memory component requires wait states to operate properly with the 8086/8088 microprocessors because of its rather long access time. If wait states are not desired, higher speed versions of the EPROM are available at an additional cost. *(Today, EPROM memory is available with access times of as little as 100 ns).*

3.4.4 Electrical Erasable Programmable Read Only Memory (EEPROM)

Electrically erasable programmable ROMs also use MOS circuitry which are similar to that of EPROM. Data is stored as charge or no charge on an insulated layer or an insulated floating gate in the device. The insulating layer is made very thin (< 200 Å). Hence a voltage as low as 20 to 25V can be used to move charges across the thin barrier in either direction for programming or erasing. The EEPROM allows selective erasing at the register level rather than erasing all the information since the information can be changed by using electrical signals. EEPROM memory also has a special chip erase mode by which entire chip can be erased in 10 ms, which is quite small as compared to time required to erase EPROM, and it can be erased and reprogrammed with device right in the circuit. *(EEPROMs are most expensive and the least dense ROMs).*

3.4.5 NVRAM

It is non-volatile RAM created by the combination of a static RAM and an electrically erasable PROM (EEPROM) in the same chip. The EEPROM and static RAM have exact mirror memories structure of each other, when powered it functions as a normal RAM, for power failure, it saves the stored data in the EEPROM. The entire contents of the RAM can be transferred in parallel to the EEPROM in less than 4 ms. When repowered it is possible to get entire contents of EEPROM back into the RAM. Hence EEPROM is used for temporary storage data during power failure. The $\overline{\text{STORE}}$ and $\overline{\text{RECALL}}$ signals are used to store data in EEPROM and transfer data back into RAM, respectively *(main advantage of NVRAM is that a battery backup is not required to save working data in case of power failure).*

The information such as time, data, monitor setting and other computer configurations are normally stored in the CMOS static RAM or NVRAM of a computer. Few standard NVRAMs and their capacities are mentioned here in Table 2.

Table 2

S. No.	NVRAM	Capacity
1.	M48Z2M1	16 MB × 8
2.	M48Z512A	4 MB × 8
3.	M48Z129	1 MB × 8
4.	M48Z02/12	16 KB × 8

3.4.6 Flash Memory

The Flash memory *(sometimes called "flash RAM")* is a type of constantly-powered non-volatile memory which can be erased and reprogrammed in units of memory called blocks, which is a variation of electrically erasable programmable read-only memory (EEPROM) which, unlike flash memory, is erased and rewritten at the byte level, which is slower than flash memory updating. Flash memory is frequently used to hold control code such as the Basic Input/Output System (BIOS) in a personal computer. If BIOS needs to be changed *(rewritten)*, the flash memory can be written to in block *(rather than byte)* sizes, making it easy to update. *[flash memory is not useful as random access memory (RAM) because RAM needs to be addressable at the byte (not the block) level)].*

Flash memory is named because the microchip is organized so that a section of memory cells are erased in a single action or *"flash"*. The erasure is caused by Fowler-Nordheim tunneling in which electrons pierce through a thin dielectric material to remove an electronic charge from a floating gate associated with each memory cell. *(Intel offers a form of flash memory that holds two bits (rather than one) in each memory cell, thus doubling the capacity of memory without a corresponding increase in price).*

The Flash memory is used in digital cellular phones, digital cameras, LAN switches, PC Cards for notebook computers, digital set-up boxes, embedded controllers, and other devices.

Features of flash memory

S. No.	Features	Explanation
1.	High Speed	Flash memories operate at higher speeds
2.	Low voltage	They operate at lower voltages thus save power and consume less energy.
3.	Durability	Flash memories are able to tolerate severe shock and vibration without losing data

3.5 STATIC RAM DEVICES (SRAM)

The Static RAM memory devices retain data for as long as DC power is applied, since no special action *(except power)* is required to retain stored data, these devices are called static memory. They are also called volatile memory since they will not retain data without power. The main difference between a ROM and RAM is written under normal operation, while a **ROM** is programmed outside the computer and is only normally read. **The SRAM stores temporary data** and is used when the size of the read/write memory is relatively small. Today, a small memory is one that is less than 1M byte. The Fig. 3.8 illustrates 4016, which is a 2K × 8 read/write memory. This devices has 11 address inputs and 8 data input/output connections, which is representative of all SRAM devices.

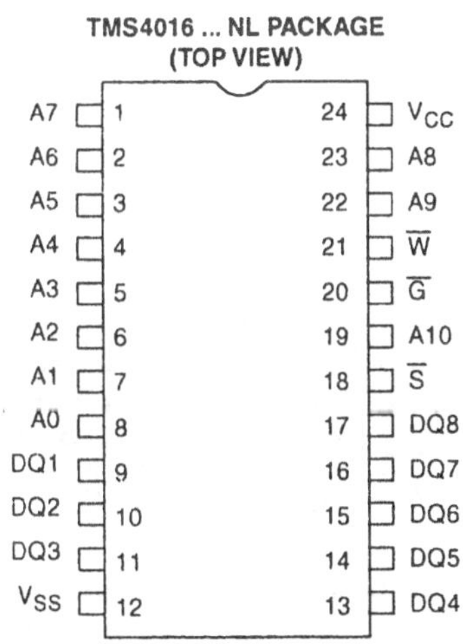

PIN NOMENCLATURE	
A0 – A10	Addresses
DQ_1 – DQ8	Data In/Data Out
$\overline{G}$	Output Enable
$\overline{S}$	Chip Select
V_{CC}	+5-V Supply
V_{SS}	Ground
$\overline{W}$	Write Enable

Fig. 3.8 Pin-out of the TMS4016, 2K 8 static RAM (SRAM)

Note. Control inputs of this RAM are slightly different from those presented earlier. The pin is labelled $\overline{G}$ the $\overline{CS}$ pin $\overline{S}$ and the $\overline{WE}$ pin $\overline{W}$. Despite the altered designations, the control pins function exactly the same as those outlined previously. *(Other manufacturers make this popular SRAM under the part numbers 2016 and 6116).*

3.5.1 TTL RAM CELL

The Fig. 3.9 illustrated a simplified schematic of a bipolar memory cell, which is implemented using TTL *(Transistor-Transistor-Logic)* multiple emitter technology. It stores 1 bit of information and is nothing but a flip-flop, which can store either 0 or 1 as long as power is applied, and it can set or reset to store either 1 or 0, respectively.

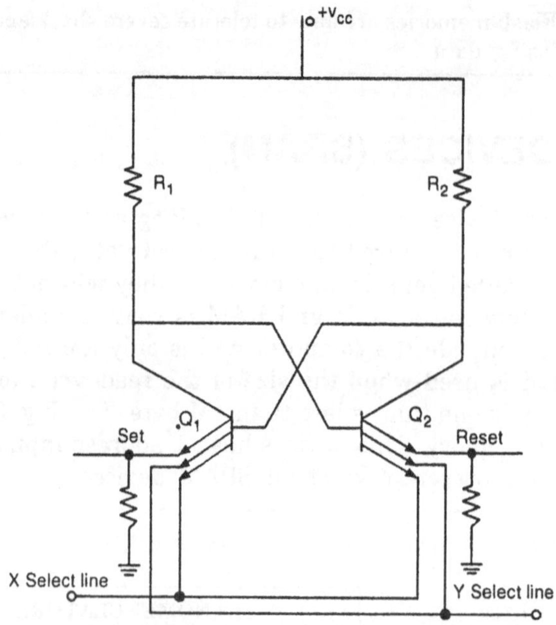

Fig. 3.9 TTL RAM cell

In Fig. 3.9 X select and Y select input lines select a cell from matrix and the Q_1 and Q_2 are cross coupled inverters, hence one is always OFF while the other is ON. A "1" is stored in the cell if Q_1 is conducting and Q_2 is OFF. A "0" is stored in the cell if Q_2 is conducting and Q_1 is OFF. The state of the cell is changed to a "0" by pulsing a HIGH on the Q_1 (SET) emitter. This turns OFF Q_1. When Q_1 is turned OFF, Q_2 is turned ON. As long as Q_2 is ON, its collector is LOW and Q_1 is held OFF.

A "1" can be rewritten by pulsing the Q_3 (reset) emitter high. Large number of these memory cells are organized on a row and column basis to form a memory chip. Fig. 3.10 illustrates the row and column organization of a 8192 bit memory chip as below.

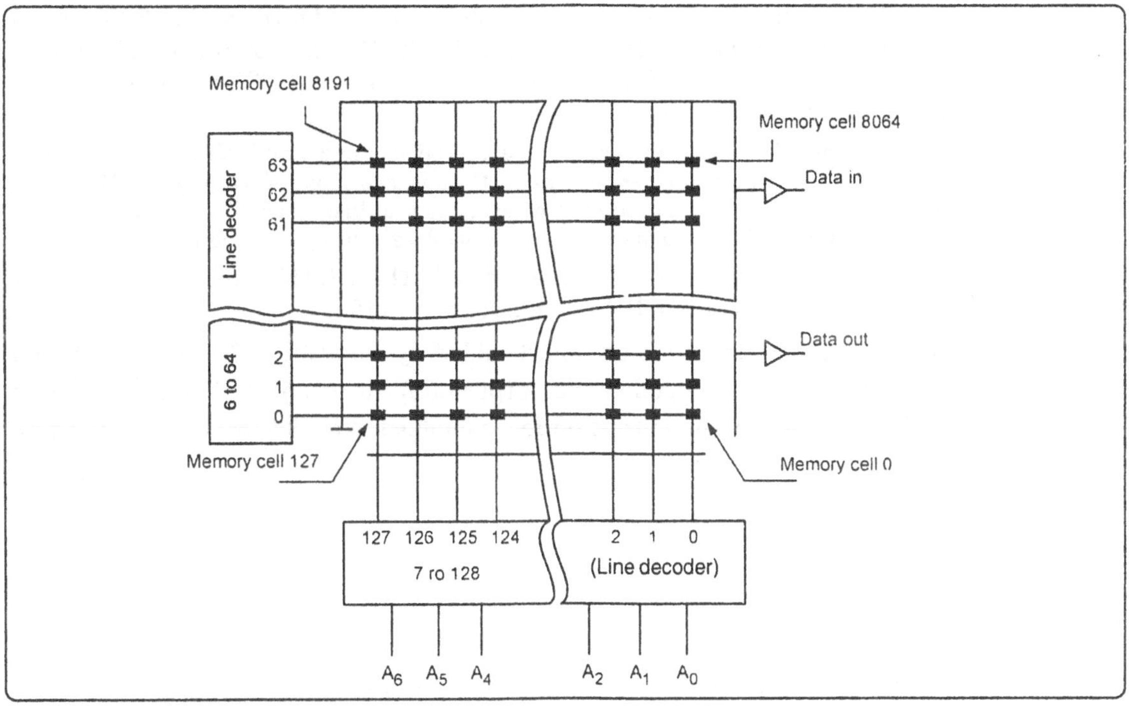

Fig. 3.10 Row and column organization (for 8192 static RAM).

This chip has 13 address lines, out of which the first seven address lines are connected to the column decoder to indicate one of the 128 columns, remaining 6 address lines are connected to the row decoder to indicate one of 64 rows. Where the decoded row and column cross, they select the desired individual memory cell *(Simple arithmetic shows that there are 64 × 128 = 8192, crossing, i.e., this memory has 8192 memory cells).*

3.5.2 Dynamic RAM Memory (DRAM)

About the largest static RAM available today is a 128K × 8. The Dynamic RAMs, on the other hand, are available in much larger sizes : upto 16M × 1. In all other respects, DRAM is essentially the same as SRAM, except that it retains data for only 2 or 4 ms on an integrated capacitor. After 2 or 4 ms, the contents of the DRAM must be completely rewritten (refreshed) because the capacitors, which store a logic 1 or logic 0, lose their charges. Fig. 3.11 depicts the DRAM cell.

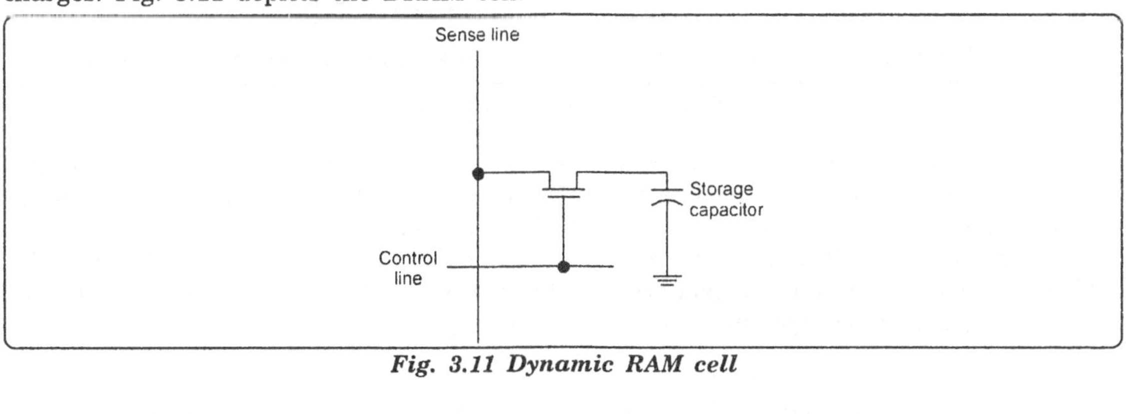

Fig. 3.11 Dynamic RAM cell

A dynamic RAM contains thousands of such memory cells shown in Fig. 3.11. When COLUMN *(Sence)* and ROW *(Control)* lines go high, the MOSEFET conducts and charges the capacitor, thus the COLUMN and ROW lines go low, the MOSFET opens and the capacitor retains its charge., thus it stores 1 bit.

Note. Since only a single MOSFET and capacitor are needed, the dynamic RAM contains more memory cells as compared to static RAM per unit area, the disadvantage of dynamic RAM is that is requires refreshing of charge on the capacitor after every few milliseconds. This complicates the system design, since it requires the extra hardware to control refreshing of dynamic RAMs.

The transistor acts as a switch in such type of cell. The DRAM consists of three tri-state buffers *i.e.*, Input buffer, Output buffer and refresh buffer.

Input and output buffers and enabled and disabled by controlling $R/\overline{W}$ line. When $R/\overline{W}$ line is LOW, input buffer is enabled and output buffer is disabled. When $R/\overline{W}$ line is HIGH, input buffer is disabled and output buffer is enabled. Now let us observe the three operations

(*a*) Write Operation,

(*b*) Read Operation, and

(*c*) Refresh Operation.

(*a*) Write Operation

To enable write operation line is made LOW which enables input buffer and disables output buffers, as shown in Fig. 3.12 as below :

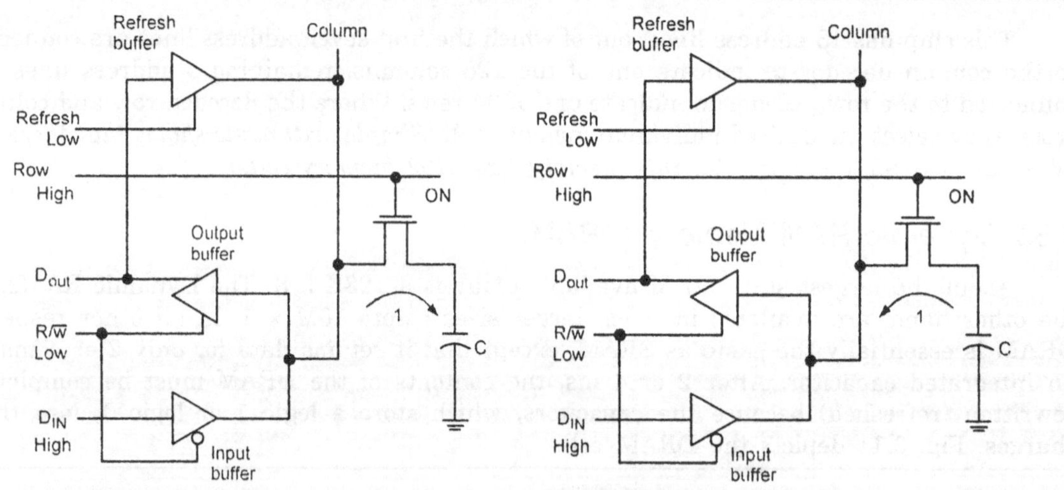

(*a*) Writing "1" into memory cell **(*b*) Writing "0" into memory cell**

Fig. 3.12 Write operation

As observed in Fig. 3.12 to write a 1 into the cell, the D_{IN} line is HIGH, and the transistor is turned ON by a HIGH on the ROW line, which allows the capacitor to charge to a positive voltage. When 0 is to be stored, a LOW is applied to the D_{IN} line. The capacitor remains uncharged, or if it is storing a 1, it discharges as indicated in Fig. 3.12 (*b*). If the ROW line is made LOW, the transistor turns off and disconnects the capacitor from the data line, thus storing the charge *(either 1 or 0)* on the capacitor.

(b) Read Operation

To read data from the cell, the $R/\overline{W}$ line is made HIGH, which enables output buffer and disables input buffer. Then ROW line is made HIGH. It turns transistor ON and connects the capacitor to the D_{OUT} line through output buffer. The Fig. 3.13 depicts the Read operation *i.e.*, reading "1" from the memory cell.

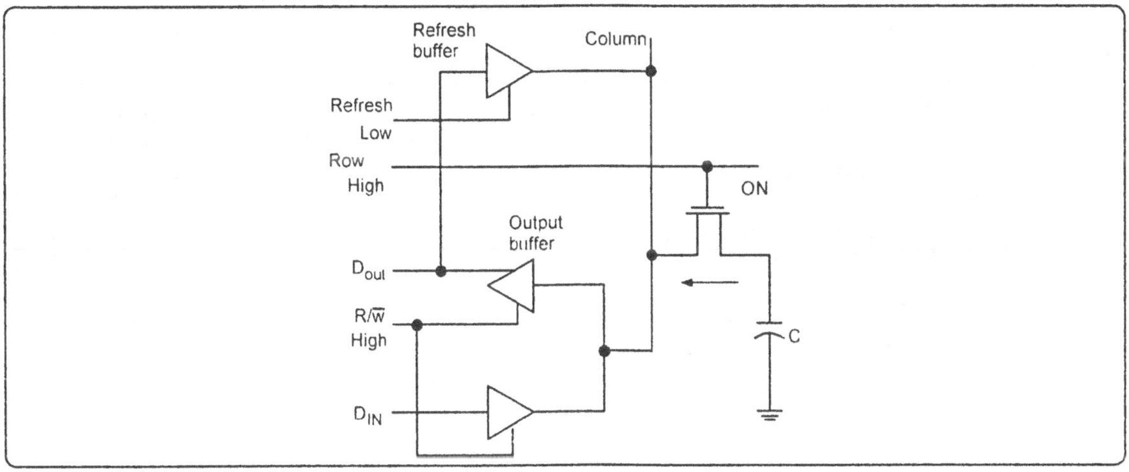

Fig. 3.13 Read operation.

(c) Refresh Operation

Now to enable refresh operation $R/\overline{W}$ line, ROW line and REFRESH line are made HIGH, which turns ON transistor and connects capacitor to COLUMN line. As $R/\overline{W}$ is HIGH, output buffer is enabled and the stored data bit is applied to the input of refresh buffer. The enabled refresh buffer then produces a voltage on COLUMN line corresponding to the stored bit and thus replenishing the capacitor which is illustrated in Fig. 3.14 below.

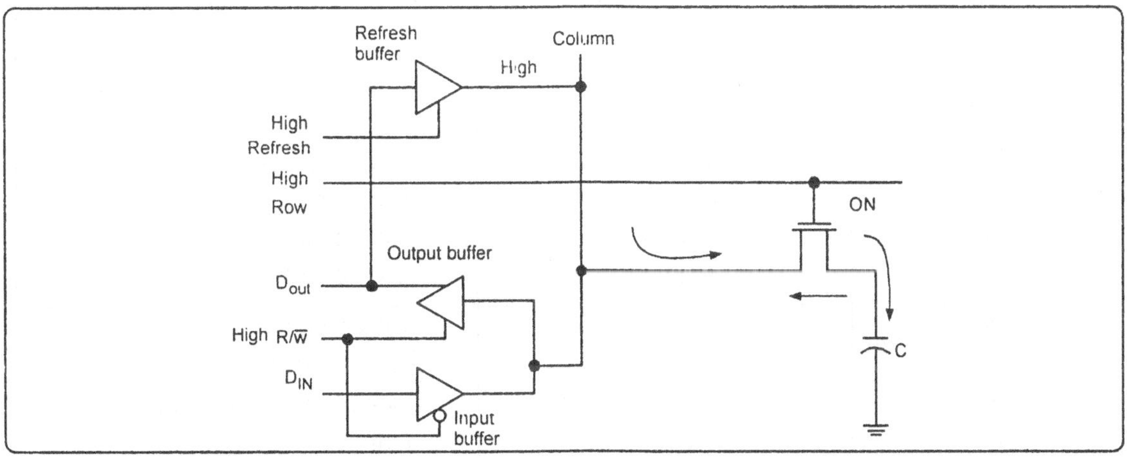

Fig. 3.14 Refreshing operation (for stored '1').

3.5.3 Comparison between SRAM and DRAM

There are few points which provides comparisons between SRAM and DRAM, they are listed in Table 3.

Table 3. Comparison between SRAM and DRAM

S. No.	SRAM	DRAM
1.	Static RAM contains less Dynamic.	RAM contains more memory cells as compared memory cells per unit area to static RAM per unit area.
2.	It has less access time hence faster memories.	Its access time is greater than static RAMs.
3.	Static RAM consists of number of Flip-flops. *(Each flip-flop stores one bit).*	Dynamic RAM stores the data as a charge on the capacitor. It consists of MOSFET and the capacitor for each cell.
4.	Refreshing circuitry is not needed.	Refreshing circuitry is needed to maintain the charge on the capacitors after every few milliseconds. Extra hardware is required to control refreshing, which makes system design complicated.
5.	Cost is more.	Cost is less.

3.5.4 Memory Chips and their Capacity

In market then are many memory chips are available. The user has to select the memory chips as required with respect to their memory capacity. Now the Table 4 provides the list of stand memory chip with their capacities.

Table 4. Available Memory Chips

S. No.	Memory Chip	Memory Capacity
1.	2114	1K × 4 Bit RAM
2.	6116	2K × 8 Bit RAM
3.	65116	2K × 8 Bit RAM
4.	6264	8K × 8 Bit RAM
5.	63256	32K × 8 Bit RAM
6.	84256	32K × 8 Bit RAM
7.	55257	32K × 8 Bit RAM
8.	4164	64K × 1 Bit Dynamic RAM
9.	41256	256K × 1 Bit Dynamic RAM
10.	6256	256K × 1 Bit Dynamic RAM
11.	511000	1 MB Dynamic RAM

12.	2316	2K × 8 Bit ROM
13.	2332	3K × 8 Bit ROM
14.	2364	8K × 8 Bit ROM
15.	65256	32K × 8 Bit ROM
16.	23512	64K × 8 Bit ROM
17.	521000	128K × 8 Bit ROM
18.	2708	1K × 8 Bit EPROM
19.	2716	2K × 8 Bit EPROM
20.	2732	4K × 8 Bit EPROM
21.	2764	8K × 8 Bit EPROM
22.	2816	2K × 8 Bit EPROM
23.	27128	16K × 8 Bit EPROM
24.	27256	32K × 8 Bit EPROM
25.	27512	64K × 8 Bit EPROM
26.	27F64	8K × 8 Bit Flash Memory
27.	28F256	32K × 8 Bit Flash Memory
28.	571000	128K × 8 Bit EPROM

3.6 READ AND WRITE MEMORY CYCLES

Now let us analyze the read and write memory cycles along with their timing parameters.

3.6.1 Read Cycle

The read cycle for static RAM is as shown in Fig. 3.15 which is drawn as the basis of different parameters of timing.

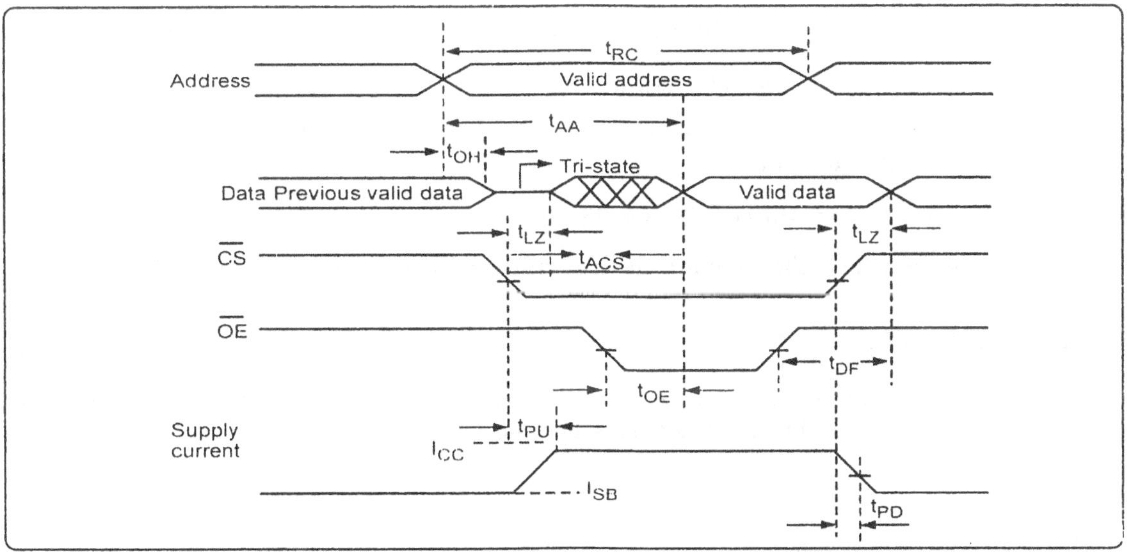

Fig. 3.15. Timing waveform for read cycle.

The different timing parameters are :

- **Read Cycle Time**

  Is the minimum time for which an address must be held stable on the address bus, in read cycle.

- **Address Access Time**

  Is the maximum specified time within which a valid new data is put on the data bus after an address is applied.

- **Output Hold from Address Change**

  The time for which a data from previously read address will be present on the output after a new address is applied.

- **Chip Selection to Output in Tri-state**

  The minimum time for which the data output will remain in tri-state after goes low.

- **Chip Select Access Time**

  Is the minimum time required for memory device to put valid data on the data bus after $\overline{CS}$ goes low.

- **Chip Deselection to Output in Tri-state**

  Is the minimum within which output returns to its high impedance state (tri-state) after $\overline{CS}$ goes high.

- **Output Enable to Output Delay**

  Is the minimum time required for the memory device to put valid data on the data bus after $\overline{CS}$ goes low.

- **Output Enable High to Output Float**

  Is the time for which data will remain valid on the data bus after $\overline{CS}$ goes high.

- **Chip Selection to Power Up Time**

  Is the time within which the memory device is powered upto its normal operating current after $\overline{OE}$ goes low.

- **Chip Deselection to Power Down Time**

  Is the time for which the memory device will remain powered up after $\overline{CS}$ goes high.

3.6.2 Write Cycle

The write cycle for static RAM is an shown in Fig. 3.16 where the timing diagram is drawn on the basis of different parameters of time.

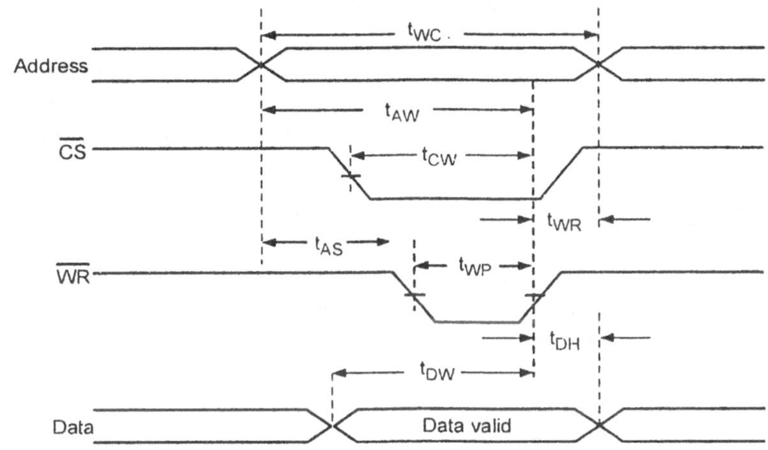

Fig. 3.16. Timing waveforms for write cycle.

The different timing parameters are :

- **Write Cycle Time**

  Is the minimum time for which an address must be held stable on the address bus, in write cycle.

- **Address Valid to End of Write**

  Is the time at which address must be applied on the address bus before $\overline{WR}$ goes high.

- **Write Recovery Time**

  Is the time for which address will remain on address bus after $\overline{WR}$ goes high.

- **Address Setup Time**

  If address is applied, it is the time after which $\overline{WR}$ can be made low.

- **Chip Selection to the End of Write**

  Is the time at which the $\overline{CE}$ must be made low to select the device before $\overline{WR}$ goes high.

- **Write Pulse Width**

  Is the time for which $\overline{WR}$ goes low.

- **Data Valid to the End of Write**

  Is the minimum time for which data must be valid on the data bus before $\overline{WR}$ goes high.

- **Data Hold Time**

  Is the time for which data must be held valid after $\overline{WR}$ goes high.

3.7 MEMORY INTERFACING CONCEPTS

The memory interfacing for 8088 (and 80188) with their 8-bit data buses and 8086 (and 80188) with their 16-bit data buses are to be discussed. However for 32-bit and 64-bit memory interfacing is not discussed here. While interfacing memory initially it needs to : select chip, identify the register and enable suitable buffer.

The Microprocessor system includes memory devices and I/O devices, and it is to be noted that microprocessor can communicate *(read/write)* with only one device at a time, since the data, address and control buses are common for all the devices. In order to communicate with memory or I/O devices, it is necessary to decode the address from the microprocessor. Due to this each device *(memory or I/O)* can be accessed independently.

3.7.1 The Memory Addressing

In most of the memory, ICs are byte oriented *i.e.*, each memory location can store only one byte of data, where 8086 is a 16-bit microprocessor, which can transfer 16-bit data. That means in addition to byte, word (16-bit) has to be stored in the memory. This is stored by using two consecutive memory locations, one for least significant byte and other for most significant byte. The address of word is the address of least significant byte. To implement this, the entire memory is divided into two memory banks : bank 0 and bank 1. Fig. 3.17 shows interfacing diagram to these memory banks. The Bank 0 is selected only when A0 is zero and Bank 1 is selected when $\overline{BHE}$ is zero, and A0 is zero for all even addresses.

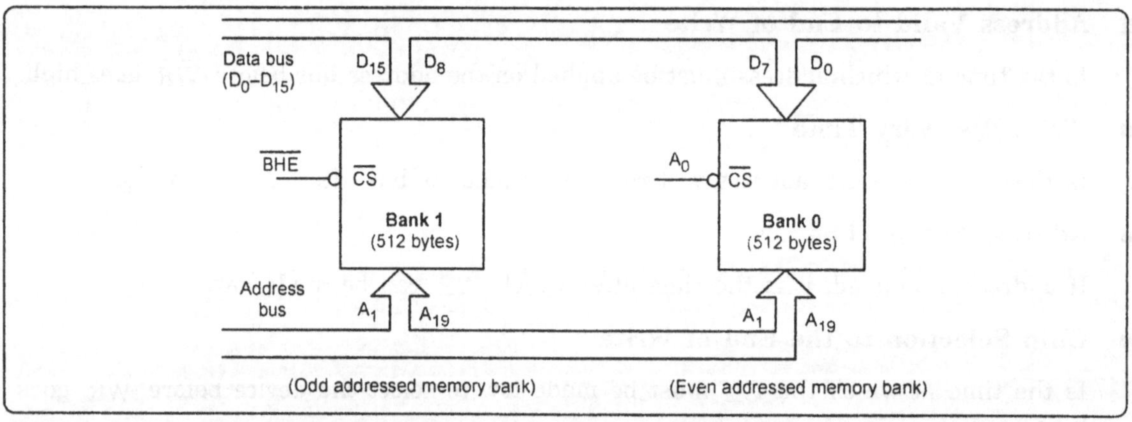

Fig. 3.17. Memory interfacing diagram.

That means Bank is normally referred as even addressed memory bank. $\overline{BHE}$ is used to access higher order memory bank, referred to as odd addressed memory bank. Now $\overline{BHE}$ and A0 will tell the interface how data appears on the bus. The four possible combination are as shown in Table 5 below :

Table 5. Four Possible Combination of $\overline{BHE}$ and A0

S. No.	The Operation	$\overline{BHE}$	A0	Data Lines Used
1.	Read/Write a byte at an even address	1	0	$D_7 - D_0$
2.	Read/Write a byte at an odd address	0	1	$D_{15} - D_8$
3.	Read/Write a word at an even address	0	0	$D_{15} - D_0$
4.	Read/Write a word at an odd address	0	1	$D_{15} - D_0$ in first operation byte from odd bank is transferred
		1	0	$D_7 - D_0$ in second operation byte from even bank is transferred

All most all microprocessor based system has a memory system, which contain two basic types of memory, read-only memory (ROM) and random access memory (RAM) or read/write memory. Read only memory contains system software and permanent system data such as lookup tables, while Random Access Memory contains temporary data and application software. ROMs/PROMs/EPROMs are mapped to cover the CPU's reset address, since these are non-volatile. If 8086 is reset, the next instruction is fetched from memory location FFFF0H. So in the 8086 systems, the location FFFF0H must be ROM location. 8086 microprocessor provides 20-bit memory address that allows up to 1M byte main memory. In any memory interface one block of memory must not be allowed to overlap another memory block. That means in order to attack a memory device to the microprocessor, it is necessary to decode the address from the microprocessor to access each memory IC independently.

3.7.2 Address Decoding

There are few techniques for address decoding which is important and required in memory interfacing. The three address decoding techniques are :

(*a*) Absolute decoding,

(*b*) Linear decoding and

(*c*) Block decoding.

(*a*) Absolute Decoding

In this technique memory chip is selected only for the specified logic level on the address lines; no other logic levels can select the chip. Fig. 3.18 shows the memory interface with absolute decoding. Two 8K EPROMs (2764) are used to provide even and odd memory banks, here.

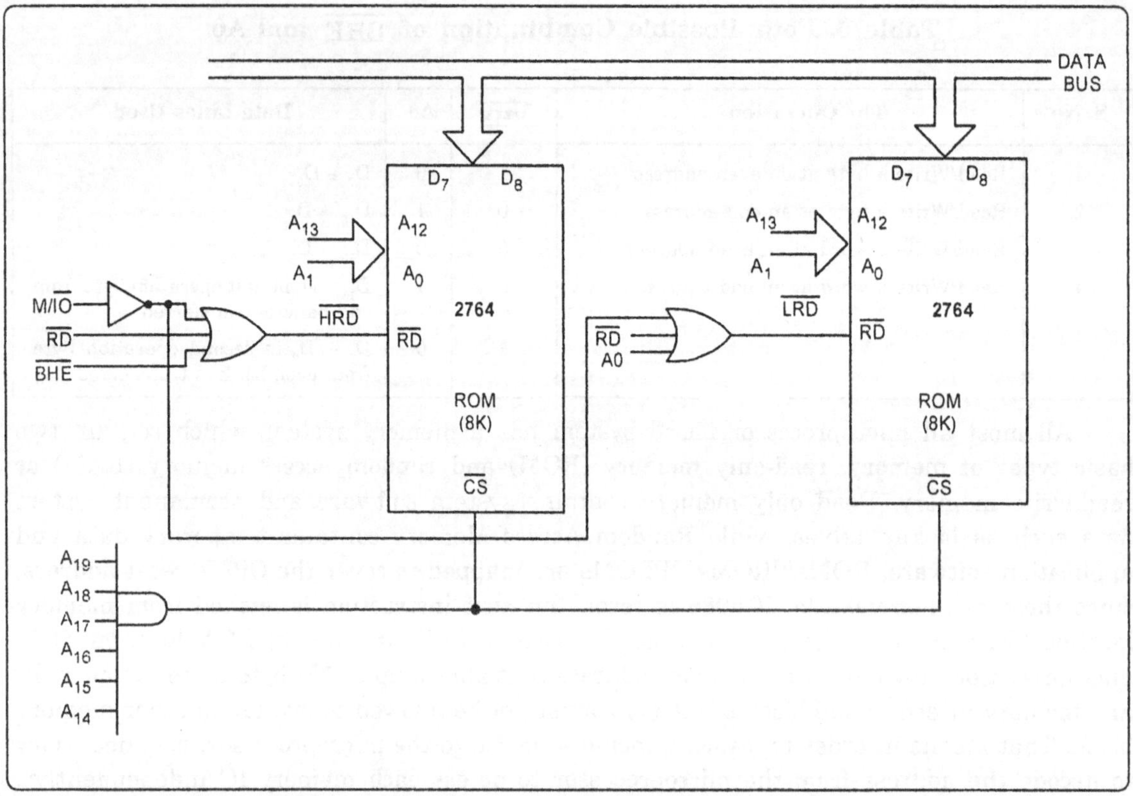

Fig. 3.18 Absolute decoding.

In Fig. 3.18 control signals $\overline{BHE}$ and A0 are used to enable outputs of odd and even memory banks respectively. As each memory chip has 8K memory locations, thirteen address lines are required to address each locations, independently. The remaining address lines are uses to generate an unique chip select signal. *(This addressing technique is normally used in large memory systems).*

(*b*) Linear Decoding

In small systems, hardware for the decoding logic can be eliminated by using only required number of addressing lines (not all). Other lines are simply ignored. This technique is referred as linear decoding or partial decoding. Fig. 3.19 shows the addressing of 16K RAM (6264) with linear decoding.

Control signals $\overline{BHE}$ and A_0 are used to enable odd and even memory banks, respectively. The address line A_{19} is used to select the RAM chips. When A_{19} is low, chip is selected, otherwise it is disabled. Address lines A_{11} to A_{18} chips are selected, which gives multiple addresses *(shadow addresses). (This technique reduces the most cost of decoding circuit, but it has drawback of multiple addresses).*

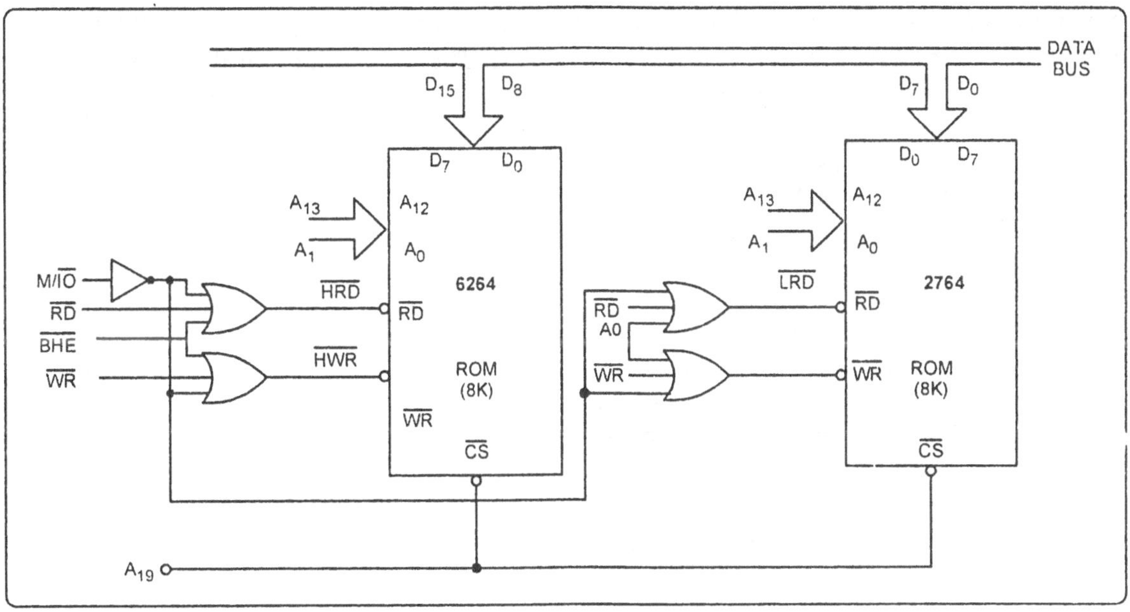

Fig. 3.19. Linear decoding.

(c) Block Decoding

The Fig. 3.20 illustrate the block decoding technique using 74138, 3 : 8 decoder, on next page :

In the case of partial microcomputer the memory array often consists of several blocks of memory chips. Where each block of memory decoding circuit. To avoid separate decoding for each memory block special decoder IC is used to generate chip select signal for each block.

3.8 8/16 BIT MEMORY INTERFACING

The Memory interfacing for 8-bit and 16-bit will be discussed with 8088 and 8086 microprocessors respectively. Memory interfacing for 8088 with its 8-bit data bus and for 8086 with its 16-bit data bus will be explained in this section. The memory interfacing for RAM and ROM are examined and explained.

3.8.1 Basic 8-bit Memory Interfacing (8088)

The 8088 microprocessors has 8-bit data bus, which makes it ideal to connect to the common 8-bit memory devices available today. The 8-bit memory size makes the 8088 ideal as a simple controller. For the 8088 to function correctly with the memory, however, the memory system must decode the address to select a memory component and it must use the RD, WR and IO/M control signals provided by the 8088 to control the memory system.

Here the minimum mode configuration is used in the section and is essentially the same as the maximum code system for memory interface. The main difference is that, in maximum mode, $IO/\overline{M}$ is combined with $\overline{RD}$ to generate an $\overline{MRDC}$ signal, and $IO/\overline{M}$ is combined with $\overline{WR}$ to generate an $\overline{MWTC}$ signal. These maximum mode control signals

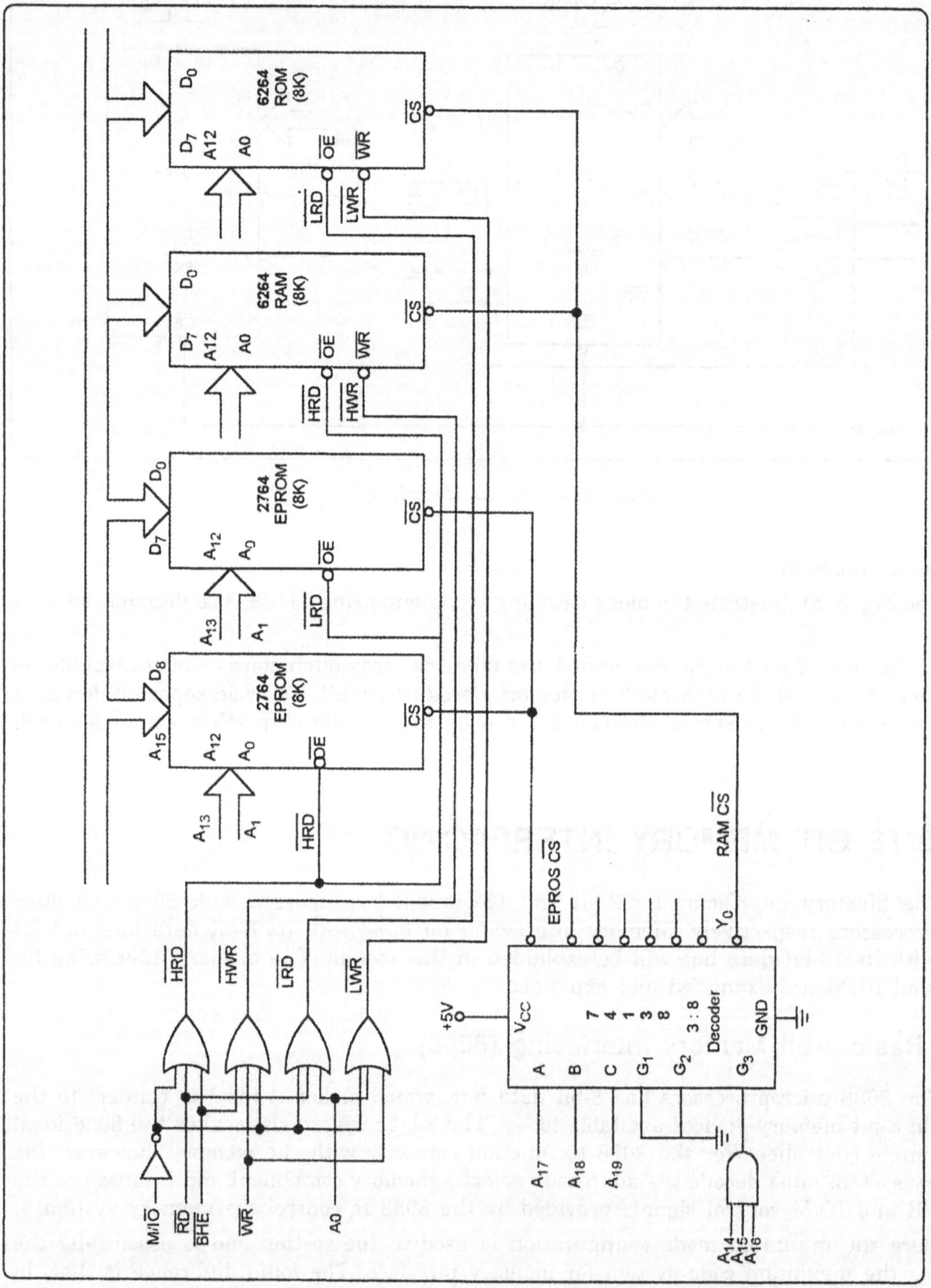

Fig. 3.20 Block decoding

are developed inside the 8288 bus controller. Whereas in minimum mode, the memory sees the 8088 as a device with 20 address connections (A_{19} – A_0), 8 data bus connections (AD_7 – AD_0) and the control signals $IO/\overline{M}$, $\overline{RD}$ and $\overline{WR}$.

(*a*) Interfacing EPROM to 8088

Here the wait states and the use of $IO/\overline{M}$ signal to enable decoder will be discussed. The Fig. 3.21 depicts an 8088 μP connected to eight 2732 EPROMs 4K 8 memory devices that are commonly used.

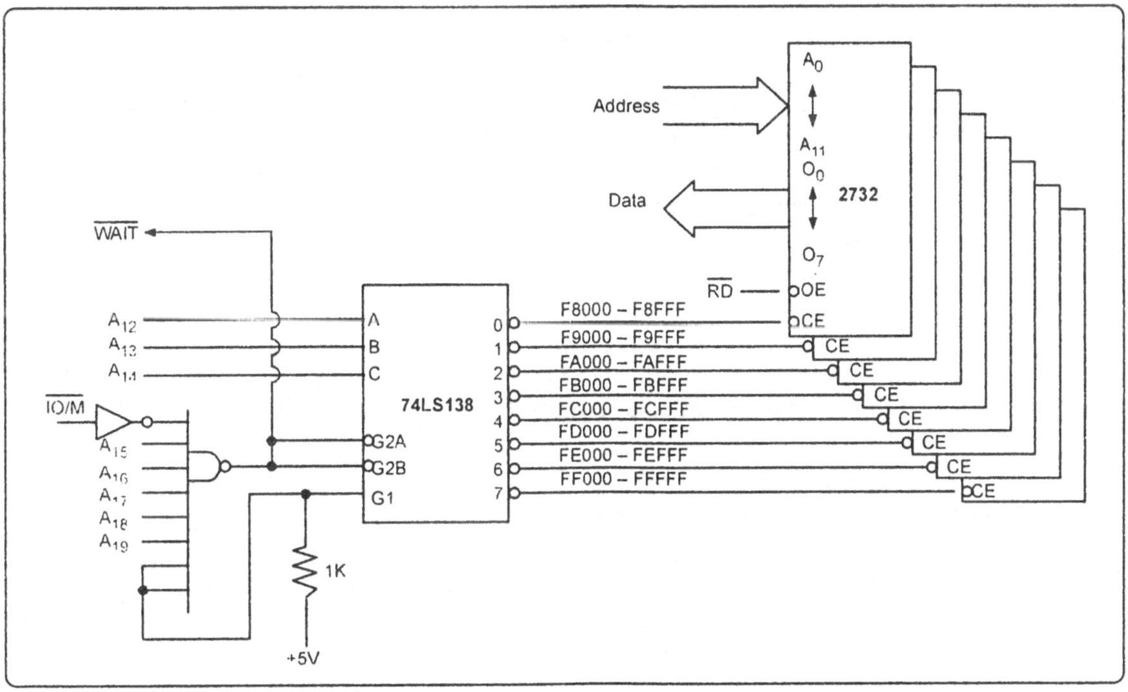

Fig. 3.21 ***Eight 2732 EPROMs interfaced to the 8088 microprocessor (the output of the NAND gate is used to cause a wait state whenever this section of the memory is selected).***

In Fig. 3.21 the 2732 has one more address input (A_{11}) than the 2716, and twice the memory. The device in this illustration decodes eight 4K × 8 blocks of memory, for a total of 32K × 8 bits of the physical address space for the 8088.

The decoder (74LS138) is connected a little differently than might be expected because the slower version of this type of EPROM has a memory access time of 450 ns. We know that when the 8088 is operated with a 5 MHz clock it allows 460 ns for the memory to access data. Since the decoder's added time delay (12 ns), it is impossible for this memory to function within 460 ns. In order to correct this problem, we must add a NAND gate to generate a signal to enable the decoder and a signal for the wait state generator. With a wait state inserted every time this section of the memory is accessed, the 8088 will allow 660 ns for the EPROM to access data. It is known that an extra wait state adds 200 ns (1 clock) to the access time. The 660 ns is sample time for a 450 ns memory component to access data, even with the delays introduced by the decoder and any buffers added to the data bus.

Note. The decoder is selected for a memory address range that begins at location F8000H and continues through location FFFFFH – the upper 32K bytes of memory. This section of memory is an EPROM because FFFF0H is where the 8088 starts to execute instructions after a hardware reset. We call location FFFF0H the cold-start locations. Software stored in this section of memory would contain a JMP instruction at location FFFF0H that jumps to location F8000H so that the remainder of the program can execute.

(*b*) Interfacing RAM to the 8088

The RAM is little easier to interface when compaired with EPROM because most RAM memory components do not require wait states. An ideal section of the memory for the RAM is the very bottom, which contains vectors for interrupts. Interrupt vectors are often modified by software packages, so it is rather important to encode this section of the memory with RAM. The Fig. 3.22 shows sixteen 62256 32K × 8 static RAMS are interfaced to the 8088, beginning at memory location 00000H, as on next page.

In Fig. 3.22 this circuit board uses two decoders to select the sixteen different RAM memory component and a third to select the other decoders for the appropriate memory sections. Sixteen 32K RAMs fill memory from location 00000H through location 7FFFFH, for 512K bytes of memory.

The first decoder (U4) in this circuit selects the other decoders Address beginning with 00 selects decoder U3, and an address that begins with 01 selects decoder U9. Observe that extra pins remains at the output of decoder U4 for future expansion, which allow more 256K × 8 blocks of RAM, for a total of 1M × 8, simply by adding the RAM and the additional secondary decoders.

It can also be noticed from the circuit in Fig. 3.22 that all the address inputs to this section of memory are buffered, as are the data bus connections and control signals $\overline{RD}$ and $\overline{WR}$, which is important when many devices appear on a single board or in a single system. Suppose that three other boards like this are plugged into a system. Then without buffers on each board, the load on the system address, data, and control buses would be enough to prevent proper operation. *(Excessive loading causes the logic 0 output to rise above the 0.8 V maximum allowed in a system).* The Buffers are normally used if the memory will contain additions at some future date, and when the memory will never grow, then buffers may not be required.

3.8.2 Basic 16-bit Memory Interfacing (8086)

The 8086 μP difference from 8088 μP in three ways :

(1) the data bus is 16-bits wide instead of 8-bits wide as on the 8088

(2) the $IO/\overline{M}$ pin of the 8088 replaced with an $M/\overline{IO}$ pin, and

(3) there is a new control signal called bus high enable $(\overline{BHE})$. *(The address bit A_0 or $\overline{ALE}$ is also used differently).*

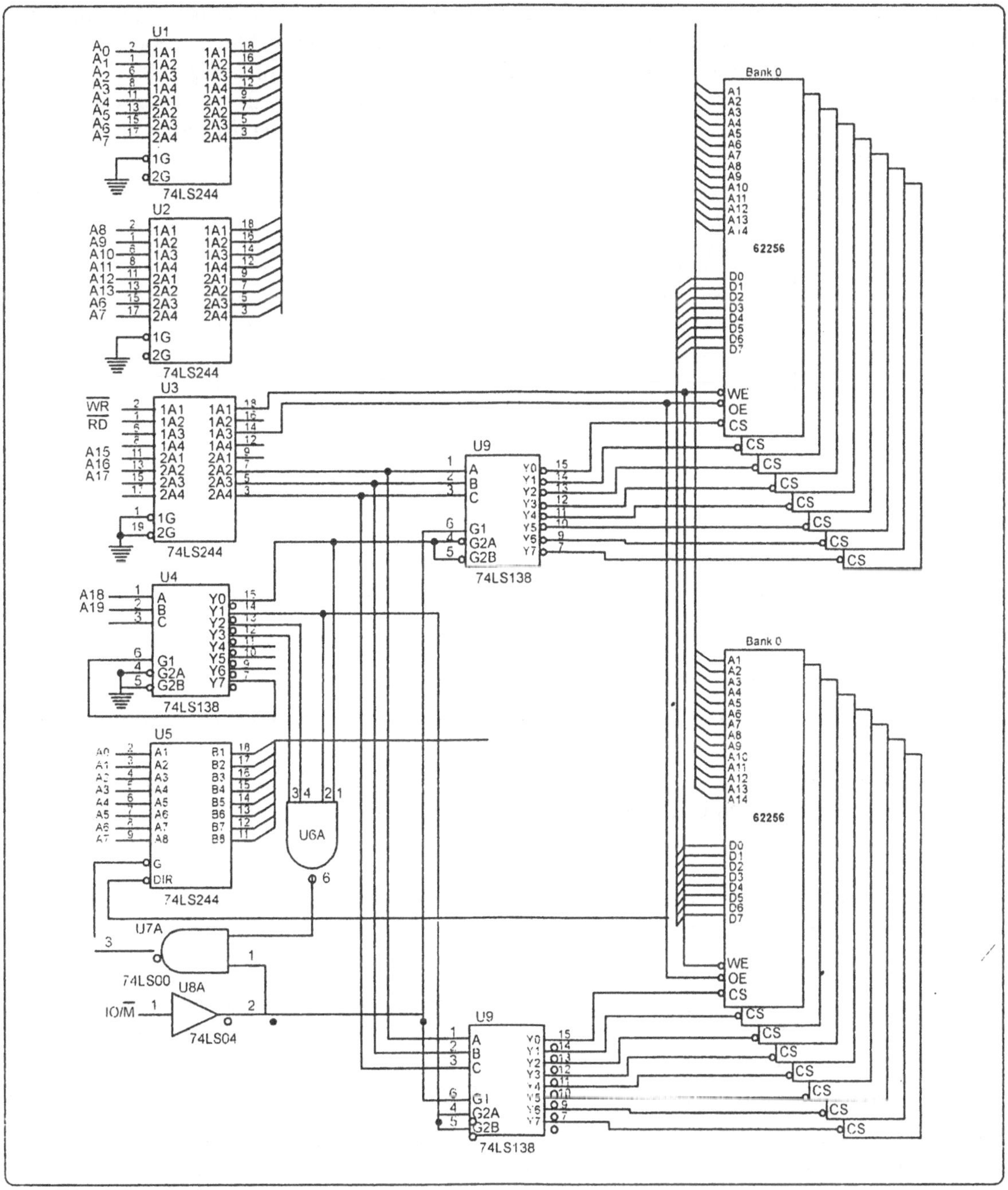

Fig. 3.22 A 512K byte static memory system using 16 62255 SRAMS

The 16-bit Bus Control

Data bus of the 8086, is twice as wide as the bus for the 8088. This wider data bus presents us with a unique set of problems that have not been encountered before. The 8086, must be able to write data to any 16-bit location-or any 8-bit location. This means that the 16-bit data bus must be divided into two separate sections (banks) that are 8-bits wide so the microprocessor can write to either half (8-bit) or both halves (16-bit). Fig. 3.23 illustrates the two banks of the memory. Where one bank *(low bank)* holds all the even-numbered memory locations, and the other bank *(high bank)* holds all the odd-numbered memory locations, as below :

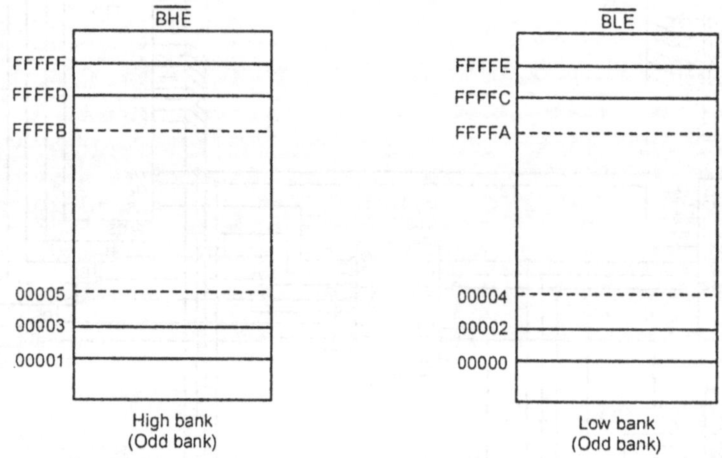

Fig. 3.23 The high (odd) and low (even) 8-bit memory banks of the 8086 microprocessors.

The 8086, use the $\overline{BHE}$ signal (high bank) and the A_0 address bit or $\overline{BLE}$ *(bus low enable)* to select one or both banks of memory used for the data transfer. The Table 6 depicts the logic levels on these two pins and the bank or banks selected below :

Table 6. Memory Bank Selection Using BHE and BLE (A_0)

S. No.	$\overline{BHE}$	$\overline{BLE}$ (A_0)	Function
1	0	0	Both banks enabled for a 16-bit transfer
2	0	1	High bank enabled for an 8-bit transfer
3	1	0	Low bank enabled for an 8-bit transfer
4	1	1	No banks enabled

Bank selection is accomplished in two ways : (1) a separate write signal is developed to select a write to each bank of the memory or (2) separate decoders are used for each bank.

Note. As a careful comparison reveals, the first technique is by far the least costly approach to memory interface for the 8086 microprocessors.

The Separate Bank Decoders

The use of separate bank decoders is often the least-effective way to decode memory address for the 8086, such method is sometimes used, but it is difficult to understand why

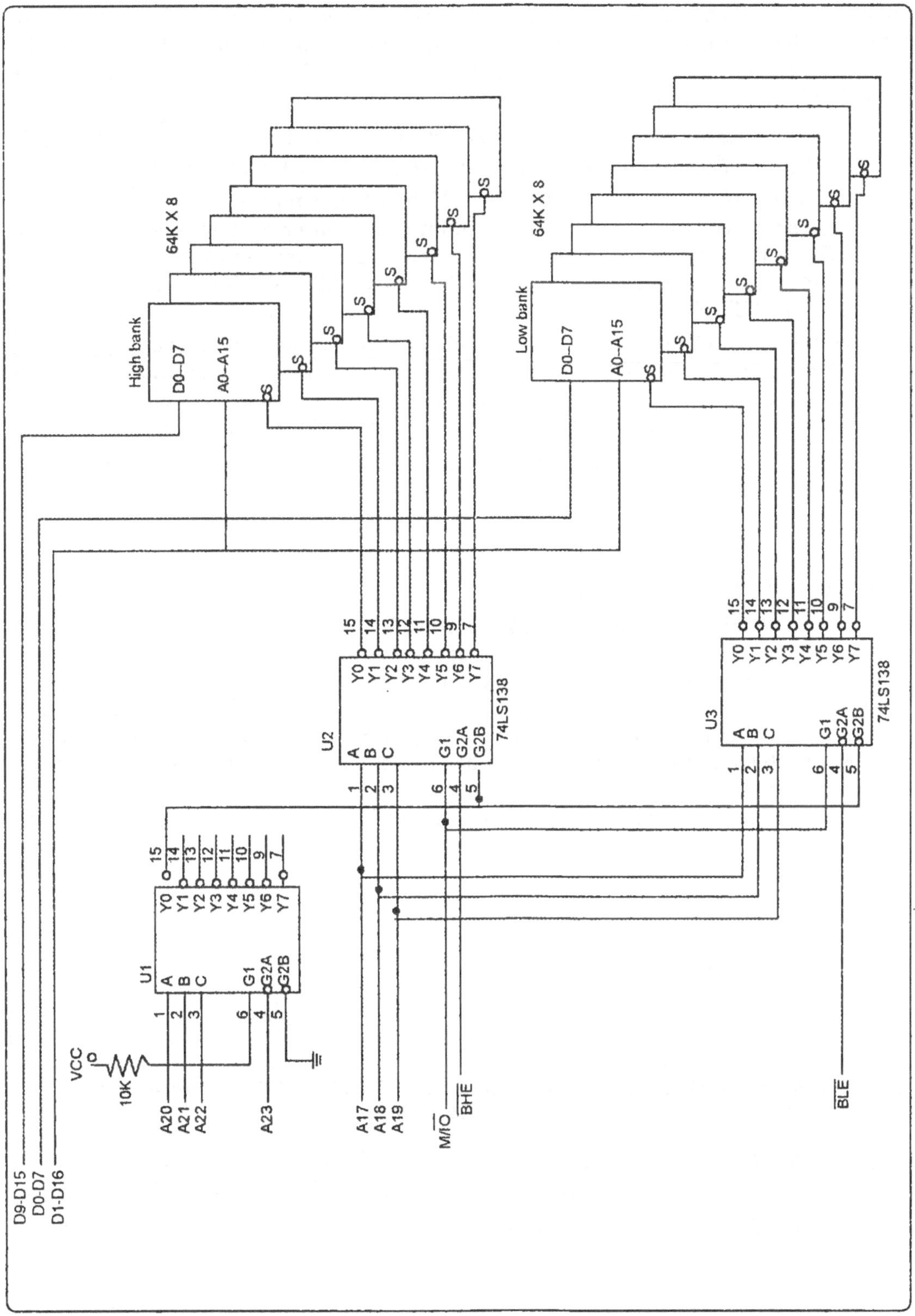

Fig. 3.24 Separate bank decoders.

in most cases. One reason may be to conserve energy, because only the bank or banks selected are enabled. This is not always the case with separate bank read and write signals. The Fig. 3.24 illustrates two 74LS138 decoders used to select 64K RAM memory components for the 80386sX microprocessor (24-bit address).

Her decoder U2 has the $\overline{BLE}$ pin (A_0) attached to $\overline{G2A}$, and decoder U3 has the $\overline{BHE}$ signal attached to its $\overline{G2A}$ input. Since the decoder will not activate until all its enables inputs are active, decoder U2 only activates for a 16-bit operation or an 8-bit operation from the low bank, and decoder U3 activates for a 16-bit operation or an 8-bit operation to the high bank. These two decoders and the sixteen 64K RAMs they control represent a 1M range of the 80386SX memory system. Decoder U1 enables U2 and U3 for memory address range 000000H-0FFFFFH.

Observe from Fig. 3.24 that the A_0 address pin does not connect to the memory; in fact, it does not even exist on the 80386SX microprocessor. Also observe that address bus bit position A_1 is connected to memory address input A_0, A_2 is connected to A_1, and so forth. The reason is that A_0 from the 8086/80186 is already connected to decoder U2 and does not need to be connected again to the memory. When A_0 or $\overline{BLE}$ is attached to the A_0 address pin of memory, every other memory location in each bank of memory would be used. *(This means that half of the memory is wasted if A_0 or $\overline{BLE}$ is connected to A_0).*

The Separate Bank Write Strobes

The most effective way to handle bank selection is to develop a separate write strobe for each memory bank, which requires only one decoder to select a 16-bit wide memory. This often saves money and reduces the number of components in a system. To generate separate read strobes for each memory bank is normally unnecessary, because the 8086 and 80386SX microprocessors read only the byte of data that they need at any given time from half of the data bus. If 16-bit sections of data are always presented to the data bus during a read, the microprocessor ignores the 8-bit section it doesn't need without any conflicts or special problems. The Fig. 3.25 depicts the generation of separate 8086 write strobes for the memory.

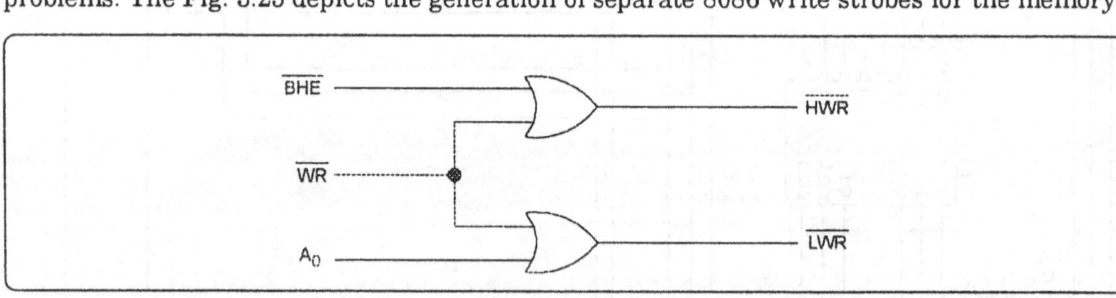

***Fig. 3.25** Memory bank write selection inputs signals: HWR (high bank write) and LWR (low bank write).*

In Fig. 3.25 a 74LS32 OR gate combines A_0 with $\overline{WR}$ for the low bank selection signal $(\overline{LWR})$ and $\overline{BHE}$ combines with $\overline{WR}$ for the high bank selection signal $(\overline{HWR})$. (Write strobes for the 80286/80386SX are generated using the $\overline{MWTC}$ signal instead of $\overline{WR}$).

Note. A memory system that uses separate write strobes is constructed differently from either the 8-bit system (8088) or the system using separate memory banks. Memory in a system that uses separate write strobes is decoded as 16-bit wide memory. For instant, suppose that a memory system will contain 64K bytes of SRAM memory. Such memory requires two 32K byte memory devices (62256) so that a 16-bit wide memory can be constructed. Since memory is 16-bit wide, and another circuit generates the bank write signals, address bit A_0 becomes a don't care.

Example 1. It is required to design an 8088 based system with the following specifications :

(i) 8088 in minimum mode (ii) 8K byte EPROM (iii) 8K byte RAM

And draw complete schematic of the design indicating all the address selected decoders used, etc.

Sol. Since RAM and EPROM are 8K byte, 13 address lines are required from A_0 to A_{12}. The remaining address lines are used for address decoding circuitry. The address lines A_0 to A_7 and A_{16} to A_{19} are latched using IC 74373 (octal latch) and ALE signal. Fig. 3.26 shows the interface between 8088 and two memory chips and Table 7 the Memory Map.

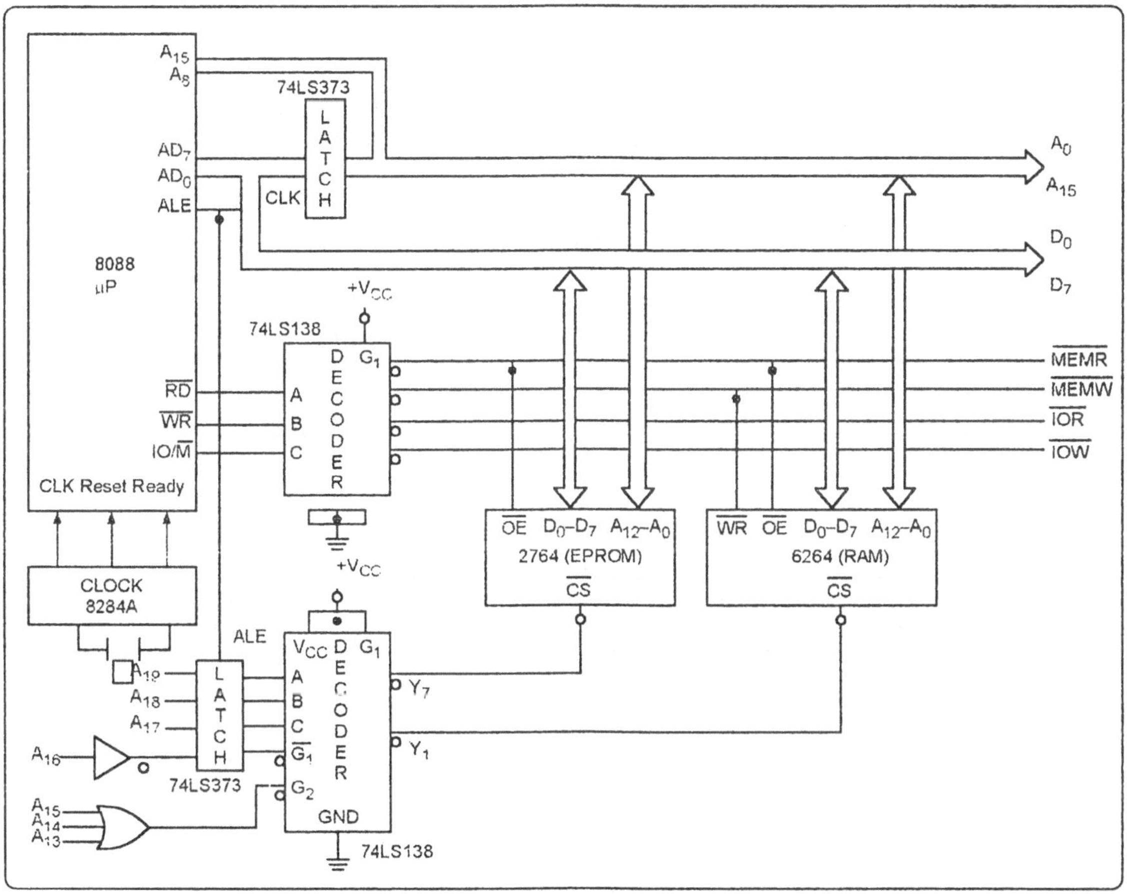

Fig. 3.26 Interfacing 8K RAM and 8K EPROM with 8088 in minimum mode (for example 1).

Table 7. Memory Map

A_{19}	A_{18}	A_{17}	A_{16}	A_{15}	A_{14}	A_{13}	A_{12}	A_{11}	A_{10}	A_9	A_8	A_7	A_6	A_5	A_4	A_3	A_2	A_1	A_0	Address	Memory
1	1	1	1	1	1	1	0	0	0	0	0	0	0	0	0	0	0	0	0	FE00H	Start EPROM
1	1	1	1	1	1	1	1	1	1	1	1	1	1	1	1	1	1	1	1	FFFFFH	End EPROM
0	0	0	1	1	1	1	0	0	0	0	0	0	0	0	0	0	0	0	0	1E000H	Start RAM
0	0	0	1	1	1	1	1	1	1	1	1	1	1	1	1	1	1	1	1	1FFFFH	End RAM

Example 2. It is required to Design an 8086 based system with the following specifications :

(i) 8086 in maximum mode (ii) 64K Byte EPROM (iii) 64K Byte RAM

Then draw the complete schematic of the design indicating address map.

Sol. Since 8086 is a 16 bit microprocessor, it can access 16-bit data simultaneously. For interfacing memory module to 8086 CPU, it is required to have odd and even memory banks which can be achieved by using two 32K byte EPROMs and two 32K byte RAMs, one for odd bank and another for even bank.

In the maximum mode, memory and I/O read/write, address latch enabled (ALE), Data Enable (DEN), Data transmit/receive $\left(DT / \overline{R}\right)$ signals must be decoded externally using bus controller 8288. Fig. 3.27 shows the memory interface with 8086 in the maximum mode and the Table 8 provides Memory Map.

Table 8. Memory Map

$\overline{BHE}$	A_{19}	A_{18}	A_{17}	A_{16}	A_{15}	A_{14}	A_{13}	A_{12}	A_{11}	A_{10}	A_9	A_8	A_7	A_6	A_5	A_4	A_3	A_2	A_1	A_0	Address	Address
1	1	1	1	1	0	0	0	0	0	0	0	0	0	0	0	0	0	0	0	0	F0000H	Even
1	1	1	1	1	1	1	1	1	1	1	1	1	1	1	1	1	1	1	1	0	FFFFEH	EPROM1
0	1	1	1	1	0	0	0	0	0	0	0	0	0	0	0	0	0	0	0	1	F0001H	Odd
0	1	1	1	1	1	1	1	1	1	1	1	1	1	1	1	1	1	1	1	1	FFFFFH	EPROM2
1	0	0	1	1	0	0	0	0	0	0	0	0	0	0	0	0	0	0	0	0	30000H	Even
1	0	0	1	1	1	1	1	1	1	1	1	1	1	1	1	1	1	1	1	0	3FFFEH	RAM1
0	0	0	1	1	0	0	0	0	0	0	0	0	0	0	0	0	0	0	0	1	30000H	Odd
0	0	0	1	1	1	1	1	1	1	1	1	1	1	1	1	1	1	1	1	1	3FFFFH	RAM2

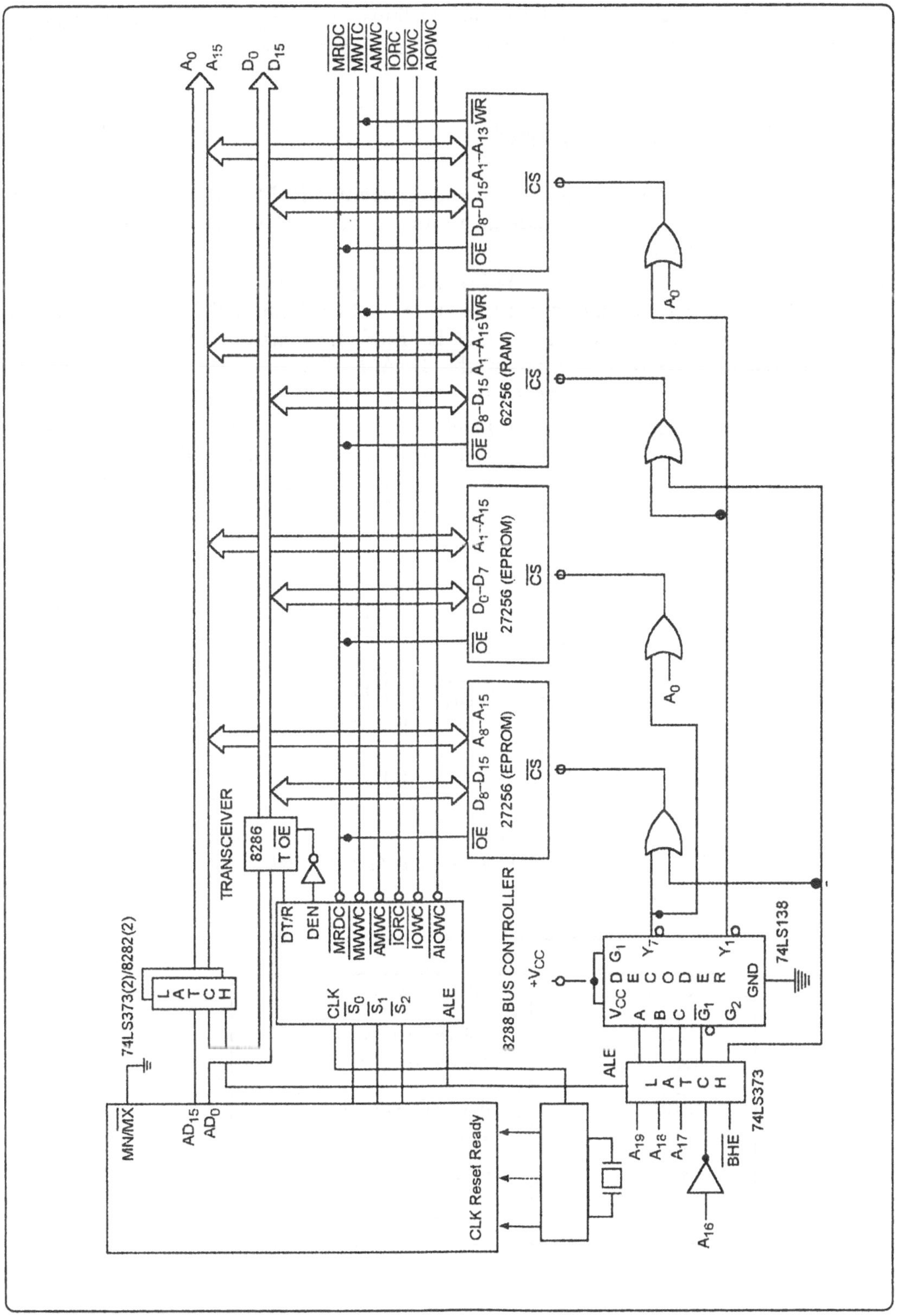

Fig. 3.27 Interfacing 64K Ram and 64K EPROM with 8086 in maximum mode (for example 2)

Since 32K byte RAM and EPROM need 15 address lines, A_1 to A_{15} lines are used. A_0 and $\overline{BHE}$ are used to select even and odd memory banks respectively as shown in Fig. 3.27.

3.9 DRAM MEMORY SYSTEM

Since RAM memory is often very large, it needs many SRAM devices at a great cost or just a few DRAMs at reduced cost. A DRAM retains data for only 2 to 4 ms and requires multiplexing of address inputs.

A DRAM must be refreshed periodically because it stores data internally on capacitors that lose their charge in a short period of time. In order to refresh a DRAM, the contents of a section of the memory must periodically be read or written. Any read or write automatically refreshes an entire section of the DRAM. *(The number of bits refreshed depends on the size of the memory components and its internal organization).*

Therefore, DRAM contains more memory cells as compared to static RAM per unit area and they are available in much larger sizes : upto 16M × 1. Here we see the interfacing of DRAM to the processor, *i.e.*, DRAM memory system. Fig. 3.28 depicts the pin diagram and Table 9 provides pin definition for a typical 64 × 4 DRAM – TMS4464.

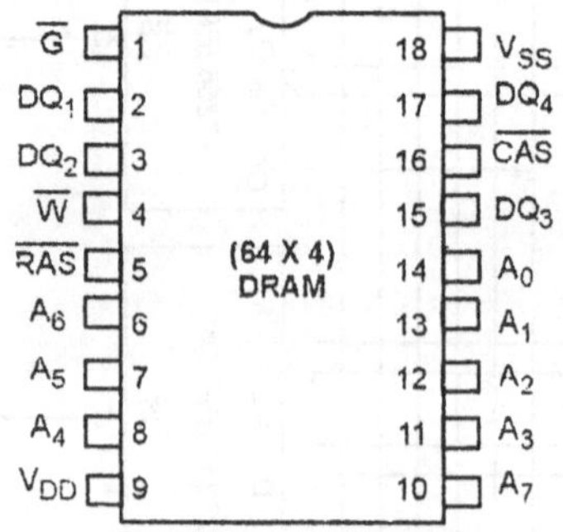

Fig. 3.28 Pin diagram of DRAM-TMS4464.

Table 9. Pin Definitions

$A_0 - A_7$	Address lines
$\overline{CAS}$	Column address strobe
$DQ_1 - DQ_4$	Data lines
$\overline{RAS}$	Row address strobe
V_{DD}	+5V supply
$\overline{W}$	Write enable
V_{SS}	Ground

It is known that to address 64K memory registers we require 16 address lines. But in DRAM to reduce total number of address lines, address is given in two phases, which allows DRAM to multiplex lower address and higher address. DRAM TMS4464 (64K × 4) has eight address lines ($A_0 - A_7$) as shown in Fig. 3.28. If **column address strobe** $\overline{CAS}$ is low these address lines give column address and when **row address strobe** $\overline{RAS}$ is low these address lines give row address. First, $A_0 - A_7$ are placed on the address bus and strobed into an internal row latch by asserting $\overline{RAS}$. Then, the address lines $A_8 - A_{15}$ are placed on the same eight address inputs and strobed into an internal column latch by asserting $\overline{CAS}$.

That means 16-bit address gets latched in two internal 8-bit latches. The internally latched 16-bit access the contents of one of the 4-bit memory locations. Now Fig. 3.29 shows the circuit for multiplexing $A_0 - A_7$ and $A_8 - A_{15}$ address lines.

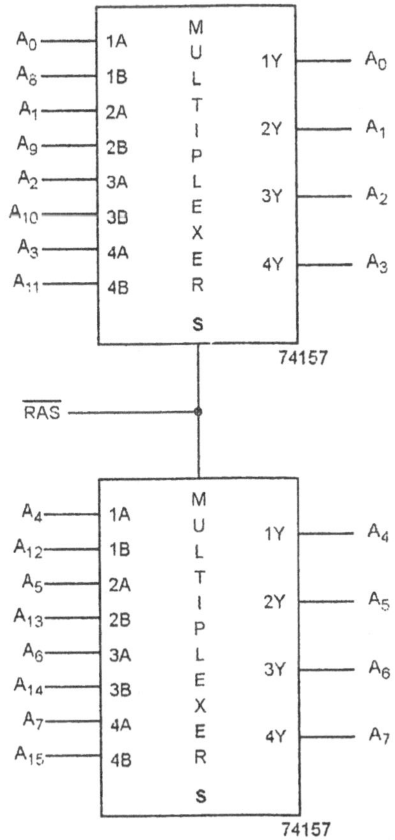

Fig. 3.29 Address multiplex for 64 × 4 DRAM – TMS4464.

From Fig. 3.29 it can be observed that here, $\overline{RAS}$ line is used to strobe row address into the DRAM. If $\overline{RAS}$ is low row address is strobed and when $\overline{RAS}$ is high column address is strobed into the DRAM.

3.9.1 The DRAM Controller

In many systems DRAM controller integrated circuit is used to do address multiplexing and the generation of the RAM control signals, let us the details of 82C08 *(DRAM controller)*. The 82C08 can control upto 1M byte of memory for 8086, which contains an address multiplexer that multiplexes an 18-bit address onto 9 address connections. Fig. 3.30 shows pin diagram of DRAM controller, where AL_0 – AL_{08} and AH_0 – AH_{08} are the address inputs for DRAM controller and AO_0 – AO_{08} are the address outputs from the DRAM controller.

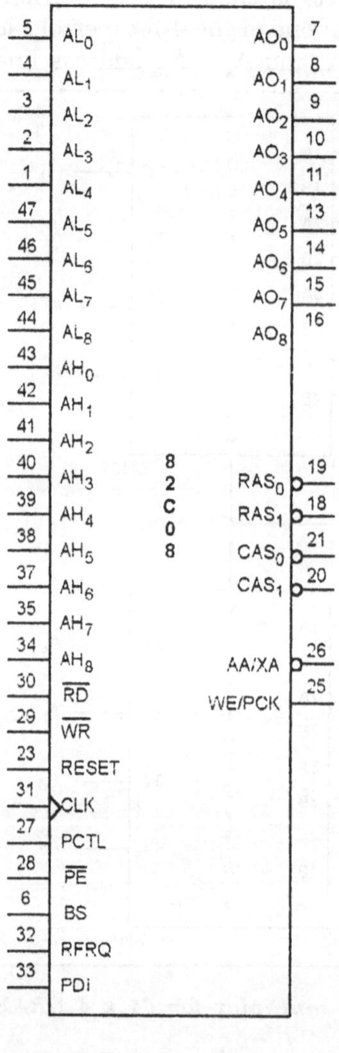

Fig. 3.30 Pin diagram of 82C08 DRAM controller (can control two banks of memory).

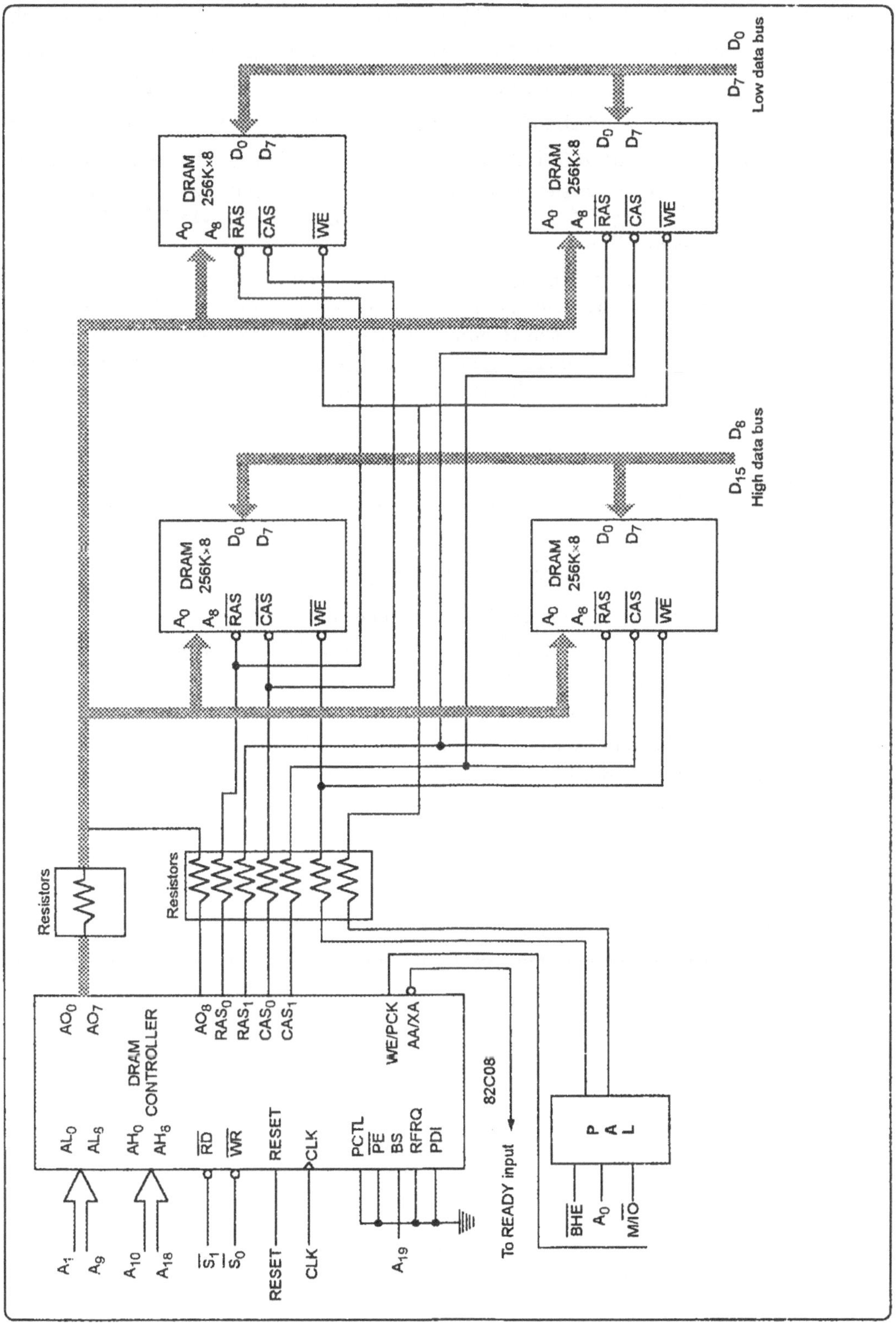

Fig. 3.31 Interface of 1M byte DRAM memory to 8086 µP

As observed in Fig. 3.30 the 82C08, DRAM controller has internal circuitry which generates the $\overline{CAS}$ and $\overline{RAS}$ signals for the DRAM. The 82C08 generates these signals internally using the CLK, $\overline{S_1}$ and $\overline{S_0}$ signals. The processor provides $\overline{S_1}$ and $\overline{S_0}$ signals as $\overline{RD}$ and $\overline{WR}$ inputs to the 82C08 DRAM controller. The $\overline{AACK} / \overline{XACK}$ signal of 80C08 is an acknowledge output that is used to indicate the ready condition for the microprocessor. This signal is normally connected to the READY input of the microprocessor or the clock generator READY input. Now Fig. 3.31 illustrates the interface of 1M byte DRAM memory for 8086 μP using DRAM controller.

Since DRAM controller accepts 18 address lines it can address upto 256 K memory locations. Therefore 1M byte DRAM memory is interfaced using a series of four 256 K × 8 bit DRAM modules. The PAL 16L8 is used to generate bank select signals and other control signals, as shown in the Fig. 3.31. The A_{19} signal selects the upper bank or the lower bank through the BS ***(Bank select)*** input to 82C08 DRAM controller.

REVIEW QUESTIONS

1. What are the common types of connections to all memory devices ?
2. What is the purpose of $\overline{CS}$ or $\overline{CE}$ pin on a memory components ?
3. How many bytes of storage do the following EPROM memory devices contain ?
   (*i*) 2708 (*ii*) 2716 (*iii*) 2764 (*iv*) 2732
4. Mention the classification of semiconductor memory.
5. With the help of a neat diagram explain the working of TTL RAM cell.
6. What is refreshing ? How it is done ?
7. Compair SRAMs and DRAMs.
8. Explain various addressing decoding techniques.
9. Explain the use of typical pin of DRAM.
10. Explain how address multiplexing is achieved for DRAM interface ?
11. Explain with diagram for eight 2732 EPROMs interfaced to 8088 μP in 8-bit memory interfacing.

4 ADDRESSING MODES AND INSTRUCTION SET OF 8086

4.0 OBJECTIVE

After completion of this chapter, you can be able to explain levels of programming, explain different addressing modes of 8086. Also can describe different instruction set of 8086 like, Data transfer instruction, Arithmetic Instructions, Logical Instructions, string Instructions, Program Execution Transfer Instructions and Processor Control Instructions, *(control, transfer instructions).*

4.1 LEVELS OF PROGRAMMING

In a computer the program has to be loaded into the main memory for execution. It is loaded into the memory in the form of binary codes. The computer can be programmed at three different levels based on the languages used for programming. They are,

1. Machine/Low Level Programming
2. Assembly/Intermediate Level Programming
3. High Level Programming

The relative comparison of different levels of programming with respect to program portability, speed of execution, flexibility in programming, etc., Which is illustrated in Fig. 4.1 below.

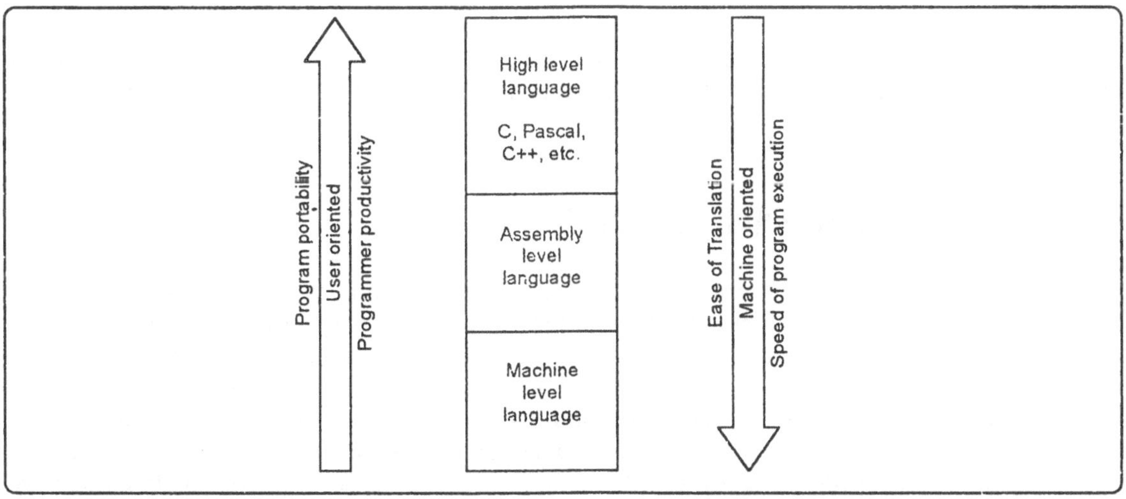

Fig. 4.1 Levels of programming.

4.1.1 Machine Level Programming *(Low Level Programming)*

The Machine level programming is lowest level of programming. It is written using the instructions which are represented by the binary codes known as the machine codes. The language in which instructions are represented in binary is known as the machine language. Machine language programs the programs that the computer can understand and execute directly. The binary instructions are different on different microprocessors for the same operation and hence, program written in binary language is machine dependent.

4.1.2 Assembly Level Programming (*Intermediate Level of Programming*)

Assembly level programming is an intermediate level of programming. It is written using instructions which are represented by two, three, or four english letters known as mnemonics. The mnemonics are shortened form of English words (*s*) for the operation performed by the instructions. For example, the mnemonic for addition is ADD, the mnemonic for branch is JMP, the mnemonic for data movement is MOV. Language in which instructions are represented by mnemonics is known as the assembly language. It is much easier to write programs in assembly language than in machine language. Assembly language instructions match machine language instructions more or less on one-to-one basis. The assembly language program is to be converted into machine language for execution. It is performed by the software known as assembler, *(assembly language is machine dependent as different as different microprocessors have different mnemonics for the same operation).*

4.1.3 High Level Programming

The language in which instructions resemble English-like statements is called as the High Level Language *(HLL)*. The HLL has a rich vocabulary of words, symbols, phrases and sentences.The programs written in this language are easy to understand and machine independent and hence they are known as portable programs. The HLL statements are converted into several instructions of the assembly language. A HLL program has to be converted into the machine language in order to executed. It is performed by the software known as the compiler *(Some of the HLLs are BASIC, COBAL, Pascal, C, C++, etc).*

Note. Machine level language is known as the first generation programming language. Assembly level language is known as the second generation programming language and High level language is known as the third generation programming language. In the recent past, high level programming languages have evolved known as the Fourth Generation Languages *(4 GLs)* and the Fifth Generation Languages *(5 GLs).*

4.2 ASSEMBLER DIRECTIVES

The Assembly language programs are composed of two types of statements, that is : first, the instructions, which are translated to machine code by the assembler and second, the directives, that directs the assembler during the assembly process for which no machine code is generated. The directive is statement which gives direction to the assembler.The directives are also called pseudo-instructions and are not translated into the machine language. There directives are necessary for the assembler to assemble efficiently and correctly. It is difficult to differentiate between the directives and instructions by looking at the assembly language program. The directives control the generation of machine code and organisation of program. The *'assembler'* supports directives for data definition, segment organisation, procedure, macro definition, etc.

Assembler directives are classified into following eleven categories based on the functions performed by them :

1. Data Definition and Storage Allocation Directives,
2. Program Organization Directives,
3. Alignment Directives,
4. Programs End Directives,
5. Value-Returning Attribute Directives,
6. Procedure Definition Directives,
7. Macro Definition Directives,
8. Data Control Directives,
9. Branch Displacement Directives,
10. Header File Inclusion Directives and
11. Target Machine Code Generation Control Directives.

4.3 ADDRESSING MODES OF 8086

In all microprocessors for the convenience of the programmer, various methods for accessing data needed in the execution of an instruction. These various methods of accessing data are called the *'Addressing Modes'*. In 8086 instructions which involve 2 operands, like 'MOV BX, CX' the first operand specifies the destination, and the second specifies the source, just like in 8085. That means instruction moves the contents of CX register to BX register.

The 8086 addressing modes can be classified into four major categories, as mentioned below :

1. Immediate addressing
2. Register addressing
3. Memory addressing modes
4. I/O port addressing

Now let us analyse each addressing modes of 8086 in brief.

4.3.1 Immediate Addressing Mode

The data to be moved to a register or a memory location in this mode, is provided as part of the instruction itself. The data may be 8 bits or 16 bits size. The immediate operand can only be the source operands.In the examples below, the quantities in parentheses, like *(DX)*, are to be read as *'contents of DX'*. Let us make clear with few examples below.

Example 1 : MOV DX,0034H

Here, the source is specified using Immediate addressing. It is 16 bit value 0034H. The destination is specified using Register addressing. The instruction moves the immediate 16 bit data 0034H to DX register. It is a 3 byte instruction.One byte is used to specify the code for 'MOV DX' and 2 bytes are used for the 16 bit data 0034H. Observe the content of 'DX' before addressing as below :

	Before	After
(DX)	AC12	0034

In assembly language, the above instruction can also be written as MOV DX, 34H. *(It still occupies 3 bytes only).*

Example 2 : MOV CH, 23H

In this example, the source is the 8 bit immediate data 23H. The destination is specified using register addressing. It moves 23H to CH. This is a 2 byte instruction. One byte is used to specify the code for 'MOV CH', and another byte is used for the 8 bit data 23H.

	Before	After
(CH)	12	23

Example 3 : MOV [BX],0AC24H

Here in this example, the source is specified using Immediate addressing. It is the 16 bit data AC24H. Observe that in assembly language AC24H should be written as 0AC24H. This is because the assembler treats anything which starts with an alphabet as a symbolic name. To overcome this, a 0 is written first to indicate that it is number. The destination is a memory word location specified using Register indirect addressing. It moves AC24H to a memory location whose 16 bit EA is provided in BX.

	Before	After
(BX)	A3D2	A3D2
(DS : A3D2)	1246	AC24

Note. In the above example, contents of byte location DS : A3D2 becomes 24H, and contents of byte location DS : A3D3 becomes ACH. Thus, in memory, 8086 stores the LS byte of a word at a Lower address, and the MS byte of the word at the next Higher address. This is similar to that in 8085 μP. It is a 4 byte instruction. This is because, to specify a memory location more number of this are needed. Two bytes are used to code *'MOV [BX]'*, and another 2 bytes specify the 16 bit data AC24H.

4.3.2 Register Addressing Mode

In register addressing mode the source and/or destination operand is specified in a register. The operand can be of 8 bit or 16 bit size. The 8 bit operand can be in one of AH, AL, BH, BL, CH, CL, DH or DL. The 16 bit operand can be in one of AX, BX, CX, DX, SI, DI, BP or SP. If a segment register is to be used as an operand in an instruction, the Segment register can be one of CS, DS, ES or SS. It is still made clear with following examples.

Example 1 : MOV CX, SI

Here, both the source and destination operands are specified in registers. It moves the contents of SI to CX register. This is a 2 byte instruction. In fact, the minimum size for a MOV instruction is 2 bytes, and this happens when both the operands are in registers, or we are moving a 8 bit immediate data to a 8 bit register like BL. That is ;

	Before	**After**
(CX)	124C	23A2
(SI)	23A2	23A2

Example 2 : MOV LOC,BL

The source is a 8 bit data specified using register addressing mode. The destination is a memory location, specified using Direct addressing. It moves the contents of BL register to the symbolic memory location LOC.

	Before	**After**
(DS : LOC)	D2	34
(BL)	34	34

That is a 4 byte instruction. This is because, to specify a memory location using direct addressing 2 extra bytes are needed.

Example 3 : MOV DX, AX

Here, both the source and destination operands are specified in registers. It moves the contents of AX to DS register. This is a 2 byte instruction, that is :

	Before	**After**
(DS)	1200	F2A2
(AX)	F2A2	F2A2

4.3.3 Memory Addressing Modes

To access data in memory, 8086 must generate a 20 bit address, which is called Physical Address or PA *(as explained earlier in Chapter 1).* This is done by adding a 16 bit value, called Effective Address or EA, to the starting address of a segment, which is given by the contents of a segment register with a 0H appended as the LS hex digit. EA represents the offset of the desired operand from the stating address of the segment. Generally, DS register is used as the segment register to generate the PA, while accessing data from memory, or for addressing the source string in string instructions. In string instructions, the ES register is generally used as the segment register to generate the PA of the destination string.

But, when EA is derived using BP register, the PA is generated using SS as the segment register. Of course, when the EA is provided in SP, the PA is generated using SS as the segment register. Segment override prefix can be used to generate PA using any other segment register. Memory addressing modes fall into two groups as under :

(*a*) Direct addressing mode, and

(*b*) Indirect addressing modes

(*a*) Direct Addressing Mode

In Direct addressing mode, the 16 bit EA is provided in the instruction itself. The 20 bit PA is generated using DS register in the direct addressing mode. Let us observe few examples to make it more and clear.

Example 1 : MOV BL,LOC

Here, the source operand is specified in Direct addressing mode. LOC is a symbolic memory location, specified using DB assembler directive in the data segment. This directive is discussed later. The assembler generates a 16 bit offset or EA in the instruction code, to specify this location. The destination is specified using register addressing. Suppose the offset value corresponding to LOC is 1B57H. Then, the contents of the byte at memory location DS : 1B57H is moved to BL.

	Before	After
(DS : LOC)	D2	D2
(BL)	34	D2

Note. It is a 4 byte instruction, because it has to specify a memory location using Direct addressing 2 extra bytes are needed.

Example 2 : MOV BX,DS : 34H

In this example the source operand is given in direct addressing mode. This instruction moves the word value at an offset of 0034H in the Data Segment to BX. 0034H is the EA. The PA is generated using the DS register. This method of specifying the direct address is less common than specifying it using symbolic location. If we omit specifying the Segment register, the assembler treates it as immediate addressing mode.

	Before	After
(DS : 0034)	A2D2	A2D2
(BX)	1234	A2D2

Note. It is a 4 byte instruction. This is because, to specify a memory location using Direct addressing 2 extra bytes are needed.

(*b*) Indirect Addressing Modes

In direct addressing, the EA is calculated from the contents of one or two registers, along with a displacement value, if any, provided in the instruction. Accordingly, Indirect addressing can be subdivided into five addressing modes, as mentioned below :

(*i*) Register indirect
(*ii*) Based addressing with displacement
(*iii*) Indexed addressing with displacement
(*iv*) Based Indexed addressing
(*v*) Based Indexed addressing with displacement.

Now let us analyse the five subdivided Indirect addressing modes in brief.

(*i*) Register Indirect Addressing. In this mode, the EA is provided in an Index register or a Base register. The Index register can be SI or DI. But the Base register can only BX. It cannot be BP.

Example 1 : MOV [DI],0AC24H

Here in this example, the source is specified using Immediate addressing. It is the 16 bit data AC24H. The destination is a memory location specified using Register indirect addressing. It moves AC24H to a memory location whose 16 bit EA is provided in DI. That is,

	Before	After
(DI)	A3D2	A3D2
(DS : A3D2)	1246	AC24

Note. It is a 4 byte instruction. This is because, to specify a 16 bit immediate data 2 extra bytes are needed.

(*ii*) Based Addressing with Displacement. If an operand is specified using based addressing with displacement, a part of the EA is specified as the contents of a Pointer register *(BP or BX),* and the other part is provided in the instruction itself. This part provided in the instruction is called as the Displacement. Here we can discuss few concept to analyse, based addressing, with displacement.

- **Sign Extension.** Let us degrees for a moment to understand the term *'Sign Extension'*. Inside a computer anything and everything must be represented in the form of 1s and 0s only. 8086, like most other processors, uses 2's complement notation to represent signed numbers.
- **Representation of positive numbers.** In 2's complement notation, a positive number starts with a 0 in the MS bit position, and the other bits provide its, magnitude. MS bit of a signed number is generally called the *'Sign bit'*. Therefore, +2 is represented using 8 bits as 0 000 0010 and same +2 using 16 bits is represented as **0000 0000 0** 000 0010.

Note. This has been achieved by simply putting copies of the sign bit in the MS byte. This process is known as Sign Extension.

- **Representation of negative numbers.** In 2's complement notation, a negative number is represented as the 2's complement of the same magnitude positive number. Thus, because +2 is represented as **0** 000 0010 using 8 bits, –2 is represented as 1 111 1110 or FEH, which is the 2's complement of +2. So also, as +2 is represented as 0000 0000 0000 0010 using 16 bits, –2 is represented using 16 bits as **1111 1111 1**111 1110 or FFFEH. Once again, observe that we have simply put copies of the sign bit of the 8 bit number FEH in the MS byte of the word.
- **Generation of EA.** The displacement is a signed number. Depending on the magnitude of the displacement, the assembler codes the displacement using one or two bytes. If the displacement is a small value in the range – 128 to 127, it is coded as a one byte value. This value sign extended to 16 bits is called the *'Effective displacement'*.

When the magnitude of the displacement is larger than the above mentioned range, the assembler codes it as a two byte value. In this case, the effective displacement is same as the actual displacement.

For instant, in the instruction *'MOV –2[BX],AC24H'* the displacement value is –2. The assembler codes this displacement as FEH. This FEH sign extended as FFFEH forms part of the EA. If BX contents is 4200H, the other part of the EA is taken as 4200H. So, the complete EA is taken as 41FEH, which is the sum of 4200H and FFFEH. In this addition any carry that is generated is ignored.

When the displacement is greater than 127, or less than –128, the assembler codes it as a 2 byte displacement. For instant, in the instruction *'MOV 300H[BX],AC24H'* as the displacement is greater than 127, the assembler codes the displacement as the 2 byte value 0300H. The sum of 0300H and the contents of BX provides the EA. The 20 bit PA is then generated using the appropriate Segment register. Now let us observe few examples.

Example 1 : MOV 2345[BX],0AC24H

In this example, the source is specified Immediate addressing. It is the 16 bit data AC24H. The destination is a memory location specified using Based addressing with 16 bit displacement. It moves AC24H to a memory location whose 16 bit EA is provided as the sum of the contents in BX, and the 16 bit displacement 2345H, observe the changes after.

	Before	After
(BX)	A3D2	A3D2
(DS : C717)	1246	AC24

In this example, C717 is nothing but the sum of A3D2H, which is the contents of BX, and 2345, which is the 16 bit displacement provided as part of the instruction. It is a 6 byte instruction. This is because, 2 bytes are to specify the displacement, and another 2 extra bytes are needed to provided the 16 bit immediate data. MOV[BX+2345H],0AC24H is an alternative method of writing the same instruction in the assembly language of 8086.

Now, MOV[BX-05H],0AC24H results in moving the immediate data AC24H to the memory location whose effective address is 05H less than the contents of BX register. This will be a 5 byte instruction in which the displacement is given by the byte value FBH, which is –05H in 2's complement notation. The Effective Displacement will be taken as FFFBH, which is the 16 bit sign extended value of FBH.

Example 2 : MOV 45H[BP],0AC24

Here in this example, the source is specified using Immediate addressing. Which is the 16 bit data AC24H. The destination is a memory location specified using Based addressing with 8 bit displacement. It moves AC24H to a memory location whose 16 bit EA is provided as the sum of the contents in BP and the 16 bit effective displacement 0045H.

	Before	After
(BP)	A3D2	A3D2
(SS : A417)	1246	AC24

Here in this example, A417 is the sum of A3D2H, which is the contents of BP, and 0045H, which is the 16 bit effective displacement. Note that SS register is used for generating the PA, as BP is used for obtaining the EA. It is a 5 byte instruction. This is because, in the instruction code only 1 byte is used for specifying the displacement.

Example 3 : MOV [BX+85H],0AC24H

Here, the source is specified using immediate addressing. It is the 16 bit data AC24H. The destination is a memory location specified using Based addressing with 16 bit displacement. It moves AC24H to memory location whose 16 bit EA is provided as the sum of the contents in BX, and the 16 bit displacement 0085H.

	Before	After
(BX)	A3D2	A3D2
(DS : A457)	1246	AC24

In this example, A457 is the sum of A3D2H, which is the contents of BX, and 0085H, which is the 16 bit displacement. It is a 6 byte instruction. This is because, 2 bytes *(0085H)* are used in the instruction code to specify the displacement.

***(iii)* Indexed Addressing with Displacement.** It is very similar to the Based addressing with displacement. If an operand is specified using Indexed addressing with displacement, a part of the EA is specified as the contents of an Index register *(SI or DI)*, and the other part is provided in the instruction itself. This part provided in the instruction is called as the Displacement. *(For explanation of the displacement part, see Based addressing with displacement).*

Note. The sum of the contents of part of the EA specified in an Index register, and the 16 bit effective displacement as described above, provides the complete 16 bit EA. The 20 bit PA is then generated using DS as the Segment register.

Now let us observe few examples.

Example 1 : MOV 2345H[DI],0AC24H

Here in this example, the source is specified using immediate addressing. It is the 16 bit data AC24H. The destination is a memory location specified using Based addressing with 16 bit displacement. It moves AC24H to a memory location whose 16 bit EA is provided as the sum of the contents in DI, and the 16 bit displacement 2345H.

	Before	After
(DI)	A3D2	A3D2
(DS : C717)	1246	AC24

Observe in this example that C717 is the sum of A3D2H, which is the contents of DI and 2345H, which is the 16 bit displacement provided as part of the instruction. It is a 6 byte instruction.

Example 2 : MOV [DI+45H],0AC24H

Here the source is specified using Immediate addressing, which is a 16 bit data AC24H. The destination is a memory location specified using Based addressing with 8 bit displacement. It moves AC24H to a memory location whose 16 bit EA is provided as the sum of the contents in DI and the 16 bit effective displacement 0045H.

	Before	After
(DI)	A3D2	A3D2
(DS : A417)	1246	AC24

Now, in this example, A417 is the sum of A3D2H, which is the contents of DI and 0045H, which is the 16 bit effective displacement. It is a 5 byte instruction. This is because, in the instruction code only 1 byte is used for specifying the displacement.

Example 3 : MOV 85H[DI],0AC24H

For this example, source is specified using Immediate addressing which is 16 bit data AC24H. The destination is a memory location specified using Based addressing with 16 bit displacement. It moves AC24H to a memory location whose 16 bit EA is provided as the sum of the contents if DI, and the 16 bit displacement 0085H :

	Before	After
(DI)	A3D2	A3D2
(DS : A457)	1246	AC24

Here, A457 is the sum of A3D2H, which is the contents of DI, and 0085H, which is the 16 bit displacement. It is a 6 byte instruction. This is because, 2 bytes (0085H) are used in the instruction code to specify the displacement.

(iv)* Based Indexed Addressing.** In this addressing mode, the EA is provided partly in a Base register *(BX or BP)*, and partly in an Index register *(SI or DI)*. Their sum provides the EA. In this mode, displacement is not provided as a part of the instruction. The final 20 bit PA is generated using the appropriate ***'Segment Register'. Let us analyse by few examples :

Example 1 : MOV [DI] [BX],0AC24H

In this example, the source is specified using. Immediate addressing which is 16 bit data AC24H. The destination is a memory location specified using Based Indexed addressing which moves AC24H to a memory location whose 16 bit EA is provided as the sum of the contents in DI and BX, that is,

	Before	After
(DI)	A3D2	A3D2
(BX)	7234	7234
(DS : 1606)	1246	AC24

Here, 1606 is the sum of A3D2H, which is the contents of DI, and 7234H which is the contents of BX. The carry generated in this addition is ignored. It is a 4 byte instruction. This is because, there is not displacement in the instruction. MOV [BX+DI],0AC24H is an alternative way of writing the same instruction.

Example 2 : MOV [DI+BP],0AC24H

In this example, the source is specified using Immediate addressing. Which is 16 bit data AC24H. The destination is a memory location specified using Based Indexed addressing It moves AC24H to a memory location whose 16 bit EA is provided as the sum of the contents in DI and BP.

If DI contents is A3D2H and BP contents is 1234H, the EA is their sum, which is B606H. This is also a 4 byte instruction. Note that SS is used for generating PA ; that is,

	Before	After
(DI)	A3D2	A3D2
(BP)	1234	1234
(SS : B606)	1246	AC24

(v) Based Indexed Addressing with Displacement. It is same as ***'Based Indexed addressing'***, with displacement of 8 bits or 16 bits, also provided as part of the instruction. Now analysis by few examples.

Example 1 : MOV 37H[DI] [BX],0AC24H

In this example, the source is specified using Immediate addressing. Which is 16 bit data AC24H. The destination is a memory location specified using Based Indexed addressing with 8 bit displacement. It moves AC24H to memory location whose 16 bit EA is provided as the sum of the contents in DI, BX, and the 16 bit effective offset of 0037H.

	Before	**After**
(DI)	A3D2	A3D2
(BX)	7234	7234
(DS : 163D)	1246	AC24

Here, 163D is the sum of A3D2H, *(which is the contents of DI)*, 7234H *(which is the contents of BX)*, and 0037H which is the effective displacement. The carry generated in this addition is ignored. It is a 5 byte instruction. It is because, one extra byte is used for specifying the displacement in the instruction. MOV [BX+DI+37H],0AC24H is alternative way of writing the same instruction.

Example 2 : MOV [DI+BP+85H],0AC24H

In this example, the source is specified using 'Immediate addressing' which is 16 bit data AC24H. The destination is a memory location specified using Based Indexed addressing with 16 bit displacement. It moves AC24H to memory location whose 16 bit EA is provided as the sum of the contents in DI, BP, and the 16 bit displacement of 0085H.

	Before	**After**
(DI)	A3D2	A3D2
(BP)	1234	1234
(SS : B68B)	1246	AC24

Here, B68B is the sum of A3D2H, *(which is the contents of DI)*, 1234H *(which is the contents of BP)*, and 0085H which is the 16 bit displacement. Which is a 6 byte instruction. It is because, the instruction code uses 2 bytes to specify the displacement of 0085H. *(Note that SS is used for generating the PA).*

Example 3 : MOV [DI+BX+123H],0AC24H

In this example, the source is specified using 'Immediate addressing' which 16 bit data ACX24H. The destination is a memory location specified using 'Based Indexed addressing' with 16 bit displacement. It moves AC24H to a memory location whose 16 bit EA is provided as the sum of the contents in DI, BX, and 16 bit offset of 1234H.

	Before	**After**
(DI)	A3D2	A3D2
(BX)	7234	7234
(DS : 283A)	1246	AC24

Here, 283A is the sum of A3D2H, which is the contents of DI, 7234H which is the contents of BX, and 1234H which is the displacement. The carry generated in this addition is ignored. It is a 6 byte instruction. This is because two extra bytes are used for specifying the displacement in the instruction.

Memory Modes as Derivatives of Base Indexed Addressing with Displacement

The 6 memory addressing modes can be treated as the special cases of **Based Indexed addressing** with displacement. Thus, Based Indexed addressing with Displacement can be treated as the **General Memory Addressing mode.**

I. If there is only displacement, but no Base or Index register, the General addressing mode transforms to Direct Addressing mode. Of course, the displacement has to be always a 16 bit quantity, and this provides the EA.

II. If only a Base register or an Index register is specified, without any displacement, the General addressing mode transforms to Register Indirect addressing. But, BP is not allowed to be used in Register Indirect addressing. In case, the programmer uses an instruction like **'MOV [BP], 1234H'**, the assembler treats it as **'MOV [BP+0], 1234H'**, and generates the code.

III. If a Base register is specified along with displacement, but without an Index register, the General memory addressing mode transforms to Based addressing with displacement.

IV. If an Index register is specified along with displacement, but without a Base register, the General memory addressing mode transforms to Indexed addressing with displacement.

V. If both Base and Index registers are specified without a displacement, the General memory addressing mode transforms to Based Indexed addressing mode.

VI. At the end, if Base and Index registers are used along with displacement, the General memory addressing mode is the same as Based Indexed addressing with displacement.

4.3.4 I/O Port Addressing

The 8086 will not directly communicate with input/output (I/O) devices, as the I/O devices are electromechanical and so are very slow compared to the speed of operation of the μP. So a μP will communicate with I/O ports which are semiconductor chips working at electronic speeds, which in turn will communicate with I/O devices. An I/O port is like a 8 bit register, which will be outside the μP, with 8 pins for communication with the fast μP and another 8 pins for communication with the slow I/O device. There are two types of addressing I/O ports,

(*a*) Fixed port addressing, and

(*b*) Variable port addressing

(*a*) **Fixed Port Addressing.** In fixed port addressing, a one byte port address will be provided, after the opcode byte. In this type of addressing a port, the program cannot change the port address as the execution of the program proceeds. For instant, IN AL,83H instruction moves the contents of input port whose address is 83H to AL.

	Before	After
(AL)	34	78
(i/p port 83)	78	78

One more another example IN AX,83H instruction moves the contents of input port whose address is 83H to AL, and the contents of input port with address 84H to AH. That is :

	Before	After
(AX)	1234	5678
(i/p port 83)	78	78
(i/p port 84)	56	56

Note. The OUT 83H, AL instruction moves the contents of AL to the output port whose address is 83H. Similarly, OUT 83H,AX instruction moves the contents of AL to output port whose address is 83H, and the contents of AH to output port with address 84H.

(*b*) Variable Port Addressing. In variable port addressing, port address will not be provided explicitly in the instruction. The address of the port is implied to be the contents of DX register always. This method of addressing a I/O ports is known as variable port addressing. Here in this type of addressing a port, the program can change the port address as the execution of the program proceeds.

For instant IN AL,DX instruction moves the contents of input port, whose 16 bit port address is gives in DX, to A1 that is :

	Before	After
(AL)	34	78
(DX)	ACBC	ACBC
(i/p port ACBC)	78	78

One more example, IN AX,DX instruction moves to AL the contents of input port whose 16 bit port address is given in DX, and to AH from the port with the next higher address.

	Before	After
(AX)	1234	5678
(DX)	ACBC	ACBC
(i/p port ACBC)	78	78
(i/p port ACBD)	56	56

Note. The OUT DX,AL instruction moves the contents of AL to output port, whose 16 bit port address is given in DX. Similarly, OUT DX,AX instruction moves the contents of AL to the output port whose 16 bit port address is given in DX, and contents of AH to the port with the next higher address.

4.4 INSTRUCTION SET OF 8086

The 8086 instruction set includes equivalents of the 8085 instructions plus few more new ones. New instructions contain operations such as signed/unsigned multiplication and division, bit manipulation instructions, string instructions, and interrupt instructions. The 8086 has approximately 117 different instructions with about 300 opcodes, 8086 instruction set contains no operand, single operand, and two operand instructions. Except for string instructions which involve array operations, the 8086 instructions do not permit memory to memory operations. This chapter consists a dictionary of all the 8086 instructions. Now we can list the instruction directionary as below where complete instruction set of 8086 is divided into sin major groups.

1. Data Transfer Instructions,
2. Arithmetic Instructions,
3. Bit Manipulation or Logical Instructions,
4. String Instruction,
5. Program Execution Transfer Instructions and
6. Processor Control Instructions.

Now each type of instruction set of 8086 can be analysed for applications to write programs.

4.4.1 Data Transfer Instructions

In this group of there are large number of instructions which perform data transfer of the following types :

(*a*) General purpose Byte or Word transfer Instructions,
(*b*) Special Address Transfer Instructions,
(*c*) Flag Transfer Instructions and
(*d*) Simple Input/Output Transfer Instructions.

(*a*) General Purpose Byte or Word Transfer Instructions

Here also there are different transfer instructions let us observe with few examples.

■ **MOV Instruction.** MOV destination, source.

MOV instruction copies a word or a byte of data from some source to a destination where destination can be a register or a memory location. The source can be a register, a memory location, or an immediate number. The source and destination in an instruction can't both be memory locations. The source and destination in a MOV instruction must be of same type *i.e.,* either both must be byte or word.

Note. MOV instruction does not affect any flags.

Examples :

Instructions	Explanation
■ MOV BX, 592FH	; Loads the immediate number 592FH in BX
■ MOV CL, [357AH]	; Copy the contents of memory location, at a
	; displacement of 357AH from data segment base, into
	; the CL register.2

■ MOV [734AH], BX	; Copy the contents of BX register to two memory ; locations in the data segment. Copy the contents of BL ; register to memory location at a displacement of ; 734AH and BH register to memory location at a ; displacement of 734BH
■ MOV DS, CX	; Copy word from CX register to data segment register
■ MOV TOTAL [BP], AX	; Copy AX to two memory locations. AL to first ; location, AH to second. Effective address, EA, ; is the sum of displacement represented by TOTAL ; and contents of BP. Physical address = EA + SS
■ MOV CS : TOTAL [BP], AX	; Same as above instruction, but physical address ; = EA+CS. Because the segment override prefix is CS.

■ **PUSH Instruction :** Push source

The PUSH instruction decrements stack pointer by two and copies a word from some source to the location in the stack wherever the stack pointer point. Here the source must be word *(16 bit)*. The source of the word can be a general purpose register, a segment register or memory.

Examples :

Instruction	Explanation
■ PUSH CX	; Decrement SP by 2, copy CX to stack
■ PUSH DS	; Decrement SP by 2, copy DS to stack
■ PUSH NEXT [BX]	; Decrement SP by 2, copy a word from memory in DS ; (*i.e.*, PA = EA + DS) to stack wtih EA = NEXT + [BX]

■ **POP Instruction :** POP destination

The POP instruction copies a word from the stack location pointed by the stack pointer to a destination. Where the destination can be a general purpose register, a segment register, or a memory location. After the word is copied to the specified destination, the stack pointer is automatically incremented by 2.

Note. POP CS is illegal.

Examples :

Instruction	Explanation
■ POP CX	; Copy a word from top of stack ; to CX and increment SP by 2
■ POP DS	; Copy a word from top of stack ; to DS and increment SP by 2
■ POP NEXT [BX]	; Copy a word from top of stack to memory in DS ; (*i.e.*, PA = EA + DS) with EA = NEXT + [BX]

■ **XCHG Instruction :** XCHG destination, source

XCHG instruction exchanges the contents of a register with the contents of another register or the contents of a register with the contents of a memory location*(s)*. The instruction cannot exchange the contents of two memory locations. The source and destination both must be words or bytes. The segment registers can't be used in these instructions.

Examples :

Instruction	**Explanation**
■ XCHG BX, CX	; Exchange word in BX with word in CX
■ XCHG AL, CL	; Exchange byte in AL with byte in CL
■ XCHG AL, SUM [BX]	; Exchange byte in AL with byte in memory at
	; EA = SUM + [BX]. PA = EA + DS

■ **XLAT Instruction :** Translate byte in AL

XLAT instruction replaces a byte in the AL register with a byte from a lookup table in memory. The BX register stores the offset of the starting address of the lookup table and AL register stores the offset of the starting address of the lookup table and AL register stores the byte number from the lookup table. This instruction copies byte from address pointed by [BX + AL] back into AL. One more equivalent mnemonic is XLATB which stands for translate byte.

(*b*) Special Address Transfer Instructions

■ **LEA Instruction :** Load Effective Address : LEA register, source

The LEA instruction determines the offset of the variable or memory location named as the source and loads this address in the specified 16-bit register. The flags are not affected by LEA instruction.

Examples :

Instruction	**Explanation**
■ LEA CX, TOTAL	; Load CX with offset of TOTAL in DS
■ LEA BP, SS : STACK_TOP	; Load BP with offset of STACK_TOP in SS
■ LEA AX, [BX] [DI]	; Load AX with EA = [BX] + [DI]

■ **LDS Instruction :** Load register and DS with words from memory. LDS register, memory address of first word.

This instruction copies a word from two memory locations into the register specified in the instruction, then it copies a word from the next-two memory locations into the DS register.

Examples :

Instruction	**Explanation**
■ LDS CX, [391AH]	; Copy contents of memory at displacement of 391AH
	; 321BH to CX. Then copy contents at displacement of
	; 391CH and 391DH in DS

■ **LES Instruction :** Load register and ES with words from memory. LES register, memory address of first word.

This instruction loads new values into the specified register and into the ES register from four successive memory locations. The word from the first two memory location is copied into the specified register and the word from the next two memory locations is copied into the ES register.

Examples :

Instruction	Explanation
■ LES CX, [3483H]	; Copy contents of memory at displacement of 3483H in ; DS to CL, contents of 348H in DS to CH and copy ; the contents of memory at displacement of 3485H and ; 3486H in DS to ES register

(*c*) Flag Transfer Instruction

The different instructions in this category are as under :

- **LAHF Instruction :** Load lower byte of flag register in AH.

  This instruction copies the contents of lower byte of 8086 flag register to AH register.
- **SAHF Instruction :** Copy AH register to low byte of flag register

  The contents of the AH register are copied into the lower byte of the 8086 flag register.
- **PUSHF Instruction :** Push flag register on the stack

  This instruction decrements the stack pointer by 2 and copies the word in the flag register to the memory location(s) pointed by the stack pointer.
- **POPF Instruction :** POP word from top of stack to flag register

  This instruction copies a word from the two memory locations at the top of stack to the flag register and increments stack pointer by 2.

There are many instructions with PUSH and POP. The eight following instructions can be obseved as listed below :

1. Push contents of a register above the stack top
2. Push contents of a memory location above the stack top
3. Push contents of a segment register above the stack top
4. Push contents of Flags register above the stack top
5. Pop to a register from the stack top
6. Pop to a memory location from the stack top
7. Pop to a segment register from the stack top and
8. Pop to Flags register from the stack top.

1. Push contents of a register above the stack top

Example : Code for PUSH SI

It is a 1 byte instruction, with the code 01010110 = 56H. In this code SI is specified by the bits 110.

Note. Stack is LIFO *(Last in, First out)* data structure implemented in the RAM area.

2. Push contents of a memory location above the stack top

Example : Code for PUSH 83H[SI]

This instruction pushes above the top of the stack, the contents of the word location whose EA is the sum of the contents of SI and the 16 bit offset 0083H. The MS byte of the word is stored at a higher address and the LS byte at a lower address. It is a 4 byte instruction, with the code as depicted below.

opcode	MOD	opcode	R/M	16 bit disp	
1111 1111	10	110	100	00 83H	= FF B4 00 83H

It is stored in memory as FF B4 83 00H.

3. Push contents of a segment register above the stack top

Example : Code for PUSH DS

It is a 1 byte instruction, with the code 000 11 110 = 1EH. In this code DS is specified by the bits 11.

4. Push contents of Flags register above the stack top

The mnemonic for this instruction is PUSHF and the template for this instruction is

1	0	0	1	1	1	0	0	= 9CH

	Before	After
(FLAGS)	1234	1234
(SP)	5678	5676
(SS : 5676)	C5	34
(SS : 5677)	B2	12

5. Pop to a register from the stack top

Example : Code for POP SI

It is a 1 byte instruction, with the code 01011 110 = 5EH. In this code SI is specified by the bits 110.

6. Pop to a memory location from the stack top

Example : Code for POP 83H[SI]

This instruction pops the top of the stack information to the word location whose EA is the sum of the contents of SI and the 16 bit offset 0083H. The LS byte of the word is loaded from the top of the stack and the MS byte is loaded from the next higher address in the stack. It is a 4 byte instruction, with the code as mentioned below.

opcode	MOD	opcode	R/M	16 bit disp	
1000 1111	10	000	100	00 83H	= 8F 84 00 83H

It is stored in memory as 8F 84 83 00H. Suppose if in the memory we have following instruction code.

opcode	MOD	opcode	R/M	
1000 1111	11	000	110	= 8F C6H

Then there is another possible code for the instruction POP SI. *(But, this is 2 bytes compared to the 1 byte code of 5EH for the same instruction. So the assembler always translates POP SI as 5EH instead of 8F C6H).*

7. Pop to a segment register from the stack top

Template for this instruction is

0	0	0	S	R	1	1	1

Since there are 2 bits in the template for identifying the segment register, there are 4 instructions in this category.

Example : Code for POP DS

It is a 1 byte instruction, with the code 000 11 111 = 1FH. In this code DS is specified by the bits 11.

	Contents Before	Contents After
(DS)	1234	B2C5
(SP)	5678	567A
(SS : 5678)	C5	C5
(SS : 5679)	B2	B2

8. Pop to Flags register from the stack top

The mnemonic for this instruction is POPF and the template for this instruction is as under :

1	0	0	1	1	1	0	1	= 9DH

	Contents Before	Contents After
(FLAGS)	1234	B2C5
(SP)	5678	567A
(SS : 5678)	C5	C5
(SS : 5679)	B2	B2

This instruction is used for changing the 16 bit contents of Flags register to the desired value.

(*d*) Simple Input and Output Port Transfer Instructions

There are mainly IN and OUT instruction here, which can be analysed.

- **In Instruction :** Input a byte or word from port.

The IN instruction will copy data from a port to the accumulator. If an 8-bit port is read, the data will go to AL and if an 16-bit port is read the data will go to AX.

The IN instruction can be executed mode 8-bit address of the port is a part of the instruction.

1. Direct: In direct mode 8-bit address of the port is a part of the instruction.

Examples:

Instruction	Explanation
IN AL, 0F8H	; Copy a byte form port 0F8H to AL
IN AX, 95H	; Copy a word from port 95h to AX

2. Indirect: In indirect addressing, the address of the port is referred from DX register. Since DX is a 16-bit register, the port address can be any number between 0000H to FFFFH. Therefore, it is possible address to upto 65,536 ports in this mode.

Examples:

Instruction	Explanation
MOV DX, 30F8H	; Load 16-bit address of the port in DX
IN AL, DX	; Copy a byte from 8-bit port 30F8H to AL
IN AX, DX	; Copy a word from 16-bit port 30F8H to AX

■ **OUT Instruction :** Send a byte or word to a port

The OUT instruction copies a byte from AL or a word from AX to the specified port, it can be executed in two different addressing modes, as Direct and Indirect.

1. Direct : In direct addressing mode 8-bit address of the port is a part of the instruction.

Examples:

Instruction	Explanation
OUT 0F8H, AL	; Copy contents of AL to 8 bit port 0F8H
OUT 0FBH, AX	; Copy contents of AX to 16-bit port 0FBH

2. Indirect: In indirect addressing, the address of the port is referred from DX register. It has advantage of accessing 215 *i.e.,* 65536 ports as mentioned earlier.

Examples:

Instruction	Explanation
MOV DX, 30F8H	; Load 16-bit address of the port in DX
MOV DX, AL	; Copy the contents of AL to port 30F8H
MOV DX, AX	; Copy the contents of AX to port 30F8H

4.4.2 Arithmetic Instructions

The Arithmetic Instruction are grouped into four sub-groups as mentioned below :

(*a*) Addition Instructions,

(*b*) Subtraction Instructions,

(*c*) Multiplication Instructions and

(*d*) Division Instruction.

(a) Addition Instruction : Here, we have ADD and ADC.

(i) ADD/ADC Instruction : ADD destination, source/ADC destination, source.

These instructions add a number from source to a number from destination and put the result in the destination. The ADC, instruction also adds the status of carry flag into the result. The source may be an immediate number, a register, or a number location. The source and the destination in an instruction cannot both be memory locations. The source and destination both must be a word or byte. If we want to add a byte to a word, we must copy the byte to a word location and fill the upper byte of the zeroes before adding.

Flags affected are ; AF, CF, OF, PF, SF, ZF.

Examples :

Instruction	**Explanation**
ADD AL, 0F0H	; Add immediate number 0F0H to contents of AL
ADC DL, CL	; Add contents of CL to contents of DL with carry and
	; store result in DL *i.e.*, DL ← DL + CL + CY
ADC DX, BX	; Add contents of BX to contents of DX with carry and
	; store result in DX *i.e.*, DX ← DX + BX + CY
ADD CL, TOTAL [BX]	; Add byte from effective address
	; Total [BX] to contents of CL
ADD CX, TOTAL [BX]	; Add word from effective address
	; TOTAL [BX] to contents of CX

(i) ADD Destination, Source

The sum of the two operands, which may be bytes or words, replaces the destination operand. Both operands may be signed or unsigned binary numbers. *(Here ADD updates AF, CF, OF, PF, SP and ZF).* Now let us observe different additions for ADD.

■ **Register Addition**

ADD Reg1, Reg2; Any general purpose register can be used for the addition

Reg1 = Reg1 + Reg2

Example 1 : *ADD CX, DX; CX = CX + DX*

Contents Before : CX = 1000H, DX = 2000H, CF = 0 AF = 0 OF = 0 PF = 0 ZF = 0 SF = 0

Contents After : CX = 3000H, DX = 2000H, CF = 0 AF = 0 OF = 0 PF = 1 ZF = 0 SF = 0

■ **Register, Memory Addition**

For ADD Reg, Memory ; Contents of a general purpose Register is added to Memory and the result is stored back into the register.

Example 2 : *ADD DI, [BX], Alpha*

Contents Before : DI = 1200H, BX = 2000, DS = F000H, Alpha = 0200H
M [F2200] = 310H, CF = 0, AF = 0, OF = 0, PF = 0, ZF = 0, SF = 0

Contents After : *DI = 15FFh, BX = 2000, DS = F000H, Alpha = 0200H*
M [F2200] = 190FH, CF = 0, AF = 1, OF = 0, PF = 0, ZF = 0, SF = 0

■ **ADD Memory, Reg**

The contents of Memory are added to a general-purpose register and the result is stored into Memory.

Example 3 : *ADD TEMP, CL*
Contents Before : *DS = E000H, TEMP = 3200H, M [E3200] = 69H, CL = 97H*
CF = 0 AF = 0 OF = 0 PF = 0 ZF = 0 SF = 0
Contents After : *DS = E000H, TEMP = 3200H, M [E3200] = 00H, CL = 97H*
CF = 1 AF = 1, OF = 1, PF = 1, ZF = 1, SF =0

■ **Immediate data**

It adds immediate data with the general purpose register/memory and stores the result back into register/memory.

Example 4 : *ADD SI, 1234H*
Contents Before : *SI = 4000H, CF = 0 AF = 0 OF = 0 PF = 0 ZF = 0 SF = 0*
Contents After : *SI = 5234H, CF = 0 AF = 0 OF = 0 PF = 1 ZF = 0 SF = 0*

Example 5 : *ADD ALPHA, 2220H*
Contents Before : *ALPHA = 3000H, M [DS*16 + 3000] = 400H*
CF = 0 AF = 0 OF = 0 PF = 0 ZF = 0 SF = 0
Contents After : *ALPHA = 3000H, M[DS*16 + 3000] = 2620H*
CF = 0 AF = 0 OF = 0 PF = 1 ZF = 0 SF = 0

(The First operand cannot be immediate data. The assembler will produce an error message if the first operand is specified as immediate data).

■ **Encoding will be further categories here**

For Memory or Register Operand with Register Operand :

000000 dw	mod reg r / m

If d = 1 then LSRC = REG, RSRC = EA, DEST = REG
else LSRC = EA, RSRC = REG, DEST = EA

For Immediate Operand to Memory or Register Operand :

100000 sw	mod 000 r / m	data	data if s : w = 01

LSRC = EA, RSRC = data, DEST = EA

For Immediate operand to Accumulator :

00000 10 w	data	data if w = 01

If w = 0 then LSRC = AL, RSRC = data, DEST = AL
else LSRC = AX, RSRC = data, DEST = AX

(*ii*) ADC Destination, Source

The ADC *(Add with Carry)* sums the operands, which may be bytes or words, adds one if CF is set and replaces the destination operand with the result. Both operands may be signed or unsigned binary numbers. *(ADC updates AF, CF, OF, PF, SF and ZF).* As ADC incorporates a carry from a previous operation, it can be used to write routines to add numbers longer than 16-bits.

Example 1 : *ADC AX, SI; AX = AX + SI + CF*

Contents Before : *AX = 1800H, SI = 2300H, CF = 1 AF = 0 OF = 0 PF = 0 ZF = 0 SF = 0*

Contents After : *AX = 3B01H, SI = 2300H, CF = 0 AF = 0 OF = 0 PF = 0 ZF = 0 SF = 0*

Note. All the addressing modes supported by the Add instruction can be used here also.

■ Encoding will be

For memory or Register Operand with Register Operand :

000100 *dw*	mod *reg r / m*

If d = 1 then LSRC = REG, RSRC = EA, DEST = REG

else LSRC = EA, RSRC = REG, DEST = EA

For Immediate Operand to Memory or Register Operand :

10 000 sw	mod 010 r / m	data	data if s : w = 01

LSRC = EA, RSRC = data, DEST = EA

For Immediate Operand to Accumulator :

0001010 w	data	data if w = 01

If w = 0 then LSRC = AL, RSRC = data, DEST = AL

else LSRC = AX, RSRC = data, DEST = AX

■ INC (Instruction Destination)

The INC *(Increment)* adds one to the destination operand. The operand may be a byte or a word and is treated as an unsigned binary number. *(INC updates AF, OF, PF, SF and ZF; it does not affect CF).*

Example 1 : *INC CX ; CX = CX + 1 ;*

Contents Before : *CX = 1234 ; CF = 1 AF = 0 OF = 0 PF = 0 ZF = 0 SF = 0*

Contents After : *CX = 1235 ; CF = 1 AF = 0 OF = 0 PF = 1 ZF = 1 SF = 0*

Example 2 : *INC [DI + BX] ; M [DS * 16 + DI + BX] = 1 + M [DS * 16 + DI + BX]*

Contents Before : *DI = 1000, BX = 0000, DS = 1000, M [11000] = FFFFH*

CF = 0 AF = 0 OF = 0 PF = 0 ZF = 0 SF = 0

Contents After : *DI = 1000, BX = 0000, DS = 1000, M [11000] = 0000H*
CF = 0 AF = 1 OF = 1 PF = 1 ZF = 1 SF = 0

- **Encoding will be**

For memory or Register Operand :

1111111 w	mod 000 r / m

DEST = EA

16-Bit Register Operand :

01000 reg

DEST = REG

- **AAA Instruction**

AAA *(ASCII Adjust for Addition)* changes the contents of register AL to a valid unpacked decimal number; the higher-order half-byte is zeroed. AAA updates AF and CF; the contents of OF, PF, SF and ZF are undefined following execution of AAA. The AAA adjustment is performed as mentioned below :

1. If the low-order 4 bits of the AL register are between 0 and 9 and the AF is 0 then go to step 3.
2. If the low-order 4-bits of the AL register are between A and F or the AF = 1, then add 06 to the AL register, and add 1 to AH register, and set the AF to 1.
3. Clear the higher-order four bits of the AL register.

Example : AAA; ASCII Adjust Accumulator

(1) *Contents Before :* *AX = 3421H, CF = 0 AF = 1 OF = 1 PF = 1 ZF = 1 SF = 0*
Contents After : *AX = 3507H, CF = 1 AF = 1 OF = 1 PF = 1 ZF = 1 SF = 0*

(2) *Contents Before :* *AX = 3421H, CF = 0 AF = 0 OF = 1 PF = 1 ZF = 1 SF = 0*
Contents After : *AX = 3401H, CF = 0 AF = 0 OF = 1 PF = 1 ZF = 1 SF = 0*

- **Encoding (for AAA) will be**

00110111

- **DAA Instruction (Decimal Adjust for Addition)**

The DAA instruction corrects the result of previously added two valid packed decimal operands. *(The destination operand must be register AL).* DAA changes the contents of AL to pair of valid packed decimal digits. It updates AF, CF, PF, SF and ZF; the contents of OF are undefined following the execution of DAA. The algorithm for the conversion is DAA is as under :

1. If AF is 1 or the lower-order four bits of the AL register is A through F, then add 06 to the AL register and set AF to 1.
2. If CF is 1 or the higher-order bits of the AL register are greater than 9, then add 60H to the AL register and set CF to 1.

Example : MOV AX, 3421H; AX = 3421H, CF = 0 AF = 1 OF = 1 PF = 1 ZF = 1 SF = 0

ADD AL, 92H; AX = 34B3H, CF = 0 AF = 0 OF = 0 PF = 0 ZF = 0 SF = 0

DAA ; AX = 3423H, CF = 1 AF = 0 OF = 0 PF = 0 ZF = 0 SF = 0

Here the AL register content is changed after performing DAA operation.

- **Encoding (for DAA) will be**

00100111

(*b*) Subtraction Instructions

In subtraction instruction again we have

(*i*) SUB – SUB destination, source

(*ii*) SBB – SBB destination, source

These instructions subtract the number in the source from the number in the destination and put result in the destination. The SBB, instruction also subtracts the status of carry flag from the result. The source may be an immediate number, a register, or a memory location. The destination may be a register or a memory location. The source and the destination both cannot be memory locations. The source and destination both must be word or byte. If we want to subtract a byte from a word, we must copy the byte to a word location and fill the upper byte of the word with zeroes before subtracting.

The Flags affected are : AF, CF, OF, PF, SF and ZF.

Examples :

Instructions	**Explanation**
■ SUB AL, 0F0H	; Subtract immediate number 0F0H
	; from contents of AL store result in AL
■ SBB DL, CL	; Subtract contents of CL and status of carry flag
	; from the contents of DL and store result in DL
	; *i.e.*, DL ← DL – CL – CY
■ SBB DX, BX	; Subtract contents of BX and status of carry
	; flag from the DX and store result in DX
	; *i.e.*, DX ← BX – BX – CY
■ SUB CL, TOTAL [BX]	; Subtract byte from effective address TOTAL [BX]
	; from the contents of CL and store result in CL
■ SYB CX, TOTAL [BX]	; Subtract word from effective address TOTAL [BX]
	; from the contents of CX and store result in CX

(i) SUB Destination, Source

The source operand is subtracted from the destination operand, and the result replaces the destination operand. The operand may be bytes or words. Both operand may be signed or unsigned binary numbers. *(SUB updates AF, CF, OF, PF, SF and ZF).* Here there are different types of subtraction that can be performed.

■ Register Subtraction

For SUB Reg1, Reg2; Any general purpose register can be consider for the subtraction.
Reg1 = Reg1 – Reg2.

Example *SUB CX, DX; CX = CX – DX*
Contents Before : *CX = 2000H, DX = 1000H*
CF = 0 AF = 0 OF = 0 PF = 0 ZF = 0 SF = 0
Contents After : *CX = 1000H, DX = 1000H*
CF = 0 AF = 0 OF = 0 PF = 1 ZF = 0 SF = 0

■ Register Memory Subtraction

For, SUB Reg, Memory; Contents of general purpose Register is subtracted from Memory and the result is store back into register.

Example 1 : *SUB DI, [BX] Alpha*
Contents Before : *DI = 2100H, BX = 2000, DS = F000H, Alpha = 200H*
M [F2200] = 310H, CF = 0 AF = 0 OF = 0 PF = 0 ZF = 0 SF = 0
Contents After : *DI = 1DF0H, BX = 2000, DS = F000H, Alpha = 200H*
M [F2200] = 190FH, CF = 0 AF = 0 OF = 0 PF = 1 ZF = 0 SF = 0

For SUB Memory, Reg; Contents of Memory is subtracted from the General purpose register and the result is stored into Memory.

Example 2 : *SUB TEMP, CL*
Contents Before : *DS = E000H, TEMP = 3200H, M[E3200] = 69H, CL = 97H*
CF = 0 AF = 0 OF = 0 PF = 0 ZF = 0 SF = 0
Contents After : *DS = E000H, TEMP = 3200H, M[E3200] = D2H, CL = 00H*
CF = 1 AF = 0 OF = 1 PF = 1 ZF = 0 SF = 0

■ Immediate Data

For SUB Reg, Immediate data/SUB Mem, Immediate data. It Subtract the immediate data from the general-purpose register/memory and store the result back into register/memory.

Example 1 : *SUB SI, 1234H*
Contents Before : *SI = 4000H, CF = 0 AF = 0 OF = 0 PF = 0 ZF = 0 SF = 0*
Contents After : *SI = 2ABBH, CF = 0 AF = 0 OF = 0 PF = 1 ZF = 0 SF = 0*
Example 2 : *SUB ALPHA, 2220H*
Contents Before : *ALPHA = 3000H, M [DS*16 + 3000] = 400H;*
CF = 0 AF = 0 OF = 0 PF = 0 ZF = 0 SF = 0
Contents After : *ALPHA = 3000H, M [DS*16 + 3000] = D1D0H*
CF = 1 AF = 0 OF = 0 PF = 0 ZF = 0 SF = 1

(First operand cannot be immediate data and the assembler will give an error message if the first operand is specified as immediate data).

■ **Encoding will be**

For Memory or Register Operand and Register Operand :

001010 dw	mod reg r / m

If d = 1 then LSRC = REG, RSRC = EA, DEST = REG

else LSRC = EA, RSEC = REG, DEST = EA

For Immediate Operand from Memory or Register Operand :

10 0000 sw	mod 101 r / m	data	data if s : w = 01

LSRC = EA, RSRC = data, DEST = EA

For, Immediate Operand to Accumulator :

00101 10 w	data	data if w = 1

If w = 0 then LSRC = AL, RSRC = data, DEST = AL

else LSRC = AX, RSRC = data, DEST = AX

(ii) SBB Destination, Source (Subtract with Borrow)

For, SBB it subtracts the source from the destination and subtracts one if CF is set, and returns the result to the destination operand. Both operands may be signed or unsigned binary numbers. SBB updata AF, CF, OF, PF, SF and ZF. It incorporates a borrow from a previous operation, SBB instruction be used to write routines that subtract numbers longer than 16-bits.

Example : *SBB AX, SI; AX = AX – SI – CF*

Contents Before : AX = 1800H, SI = 2300H,

CF = 1 AF0 OF = 0 PF = 0 ZF = 0 SF = 0

Contents After : AX = F500H, SI = 2300H,

CF = 1 AF = 1 OF = 0 PF = 1 ZF = 0 SF = 1

(All the addressing modes used in SUB instruction can also be used here).

■ **Encoding will be**

For Memory or Register Operand and Register Operand :

000110 dw	mod reg r / m

If d = 1 then LSRC = REG, RSRC = EA, DEST = REG

else LSRC = EA, RSRC = REG, DEST = EA

For Immediate Operand from Memory or Register Operand :

10 0000 sw	mod 001 r / m	data	data if s : w = 01

LSRC = EA, RSRC = data, DEST = EA

For Immediate Operand to Accumulator :

00011 10 w	data	data if w = 1

If w = 0 then LSRC = AL, RSRC = data, DEST = AL
else LSRC = AX, RSRC = data, DEST = AX

■ **DEC Instruction :** Decrement destination

DEC instruction subtract 1 from the specified destination. The destination may be register or memory location. The AF, OF, PF, SF and ZF flags are affected here.

Examples :

Instruction	**Explanation**
DEC AL	; Subtracts 1 from the contents of AL
DEC BX	; Subtracts 1 from the contents of BX

(They carry flag CF is not affected).

If the contents of 8-bit register are 00H and 16-bit register are 0000H, after DEC instruction contents of registers will be FFH and FFFFH respectively without affecting carry flag.

Instruction	**Explanation**
DEC BYTE PTR [BX]	; Decrement byte at offset of BX in DS
	; BYTE PTR directive indicates to the assembler
	; that the **byte from memory** is to be decremented
DEC WORD PTR [BX]	; Decrement word at offset of BX in DS
	; WORD PTR directive indicates to the assembler
	; that the **word from memory** is to be decremented

Example 1 : *DEC CX; CX = CX – 1;*
Contents Before : *CX = 1234; CF = 1 AF = 0 OF = 0 PF = 0 ZF = 0 SF = 0*
Contents After : *CX = 1233; CF = 1 AF = 0 OF = 0 PF = 1 ZF = 0 SF = 0*

Example 2 : *DEC [DI+BX]; M [DS * 16 + DI + BX] = 1-M [DS * 16 + DI + BX]*
Contents Before : *DI = 1000, BX = 0000, DS = 1000, M [11000] = FFFFH*
CF = 0 AF = 0 OF = 0 PF = 0 ZF = 0 SF = 0
Contents After : *DI = 1000, BX = 0000, DS = 1000, M [11000] = FFFEH*
CF = 0 AF = 1 OF = 0 PF = 1 ZF = 0 SF = 1

■ **Encoding will be**

For Memory or Register Operand :

1111111 dw	mod 001 r / m

DEST = EA

For 16-Bit Register Operand .

01001 reg

DEST = REG

NEG Instruction : ***(Form 2's complement)***

This instruction replaces the number in a destination with the 2's complement of that number, and the destination can be a register or a memory location. This instruction can be implemented by inverting each bit and adding 1 to it.

Note. The negate instruction updates the AF, CF, SF, PF, ZF and OF flags.

Examples :

Instruction	**Explanation**
	; AL = 0011 0101 35H
NEG AL	; Replace number in AL with its 2's complement
	; AL = 1100 1011 = CBH

Example 1 : *NEG SI*
Contents Before : *SI = 3200H, CF = 1 AF = 0 OF = 0 PF = 0 ZF = 0 SF = 0*
Contents After : *SI = CEFF, CF = 1 AF = 0 OF = 0 PF = 0 ZF = 0 SF = 0*

Example 2 : *NEG BL*
Contents Before : *BL = FF, CF = 0 AF = 0 OF = 0 PF = 0 ZF = 0 ZF = 0 SF = 0*
Contents After : *BL = 01, CF = 1 AF = 1 OF = 0 PF = 0 ZF = 0 ZF = 0 SF = 0*

Example 3 : *NEG CL*
Contents Before : *CL = 80H, CF = 0 AF = 0 OF = 0 PF = 0 ZF = 0 SF = 0*
Contents After : *CL = 80H, CF = 0 AF = 0 OF = 0 PF = 1 ZF = 0 SF = 1*

■ **Encoding for NEG**

1111011 dw	mod 011 r/m

If w = 0 then SRC = FFH
else SRC = FFFFH

■ **CMP Instruction :** CMP destination, source

This instruction compares a byte/word from the specified source with a byte/word from the specified destination, where the source and destination both must be byte or word. The source may be an immediate number, a register, or a memory location. The destination may be a register or a memory location. But the source and destination both can't be memory locations. The comparison is done by subtracting the source byte or word from the destination byte or word. But the result is not stored in the destination, *(source and destination remain unchanged, only flags are updated).*

Note. The AF, OF, SF, ZF and CF flags are updated by the CMP instruction.

Examples :

Instruction	**Explanation**
CMP BL, 01H	; Compare immediate number 01H with byte in BL
CMP CX, BX	; Compare word in BX with word in CX
CMP CX, TOTAL	; Compare word at displacement
	; TOTAL is DS with word in CX.

The comparison instruction reflect on the three important condition flags. They are CF, SF and ZF, as under.

	CF	ZF	ZF
Destination = Source	0	1	0
Destination > Source	0	0	0
Destination < Source	1	0	1

Example 1 : *CMP CX, BX*

Contents Before : *CX = 1000, BX = 1000*
F = 0 AF = 0 OF = 0 PF = 0 ZF = 0 SF = 0

Contents After : *CX = 1000, BX = 1000*
CX = 0 AF = 0 OF = 0 PF = 1 ZF = 1 SF = 0

Example 2 : *CMP DH, ALPHA*

Contents Before : *DH = 3FH, ALPHA = 1000, M [DS*16 + 1000] = 42H*
CF = 0 AF = 0 OF = 0 PF = 0 ZF = 0 SF = 0

Contents After : *DH = 3FH, ALPHA = 1000, M [DS*16 + 1000] = 42H*
CF = 1 AF = 0 OF = 0 PF = 0 ZF = 0 SF = 1

Example 3 : *CMP BL, 02H*

Contents Before : *BL = 30H, CF = 0 AF = 0 OF = 0 PF = 0 ZF = 0 SF = 0*

Contents After : *BL = 30H, CF = 0 AF = 1 OF = 0 PF = 1 ZF = 1 SF = 0*

- **Encoding will be**

For Memory or Register Operand with Register Operand :

001110 dw	mod reg r / m

If d = 1 then LSRC = REG, RSRCH = EA,
else LSRC = EA, RSRC = REG

Fog Immediate Operand with Memory or Register Operand :

10 0000 sw	mod 111 r / m	data	data if s : w = 01

LSRC = EA, RSRC = data, DEST = EA

For Immediate Operand with Accumulator :

00111 10 w	data	data if w = 1

If w = 0 then LSRC = AL, RSRC = DATA
else LSRC = AX, RSRC = data

■ **AAS Instruction :** ASCII Adjust after subtraction

Here the number 0-9 are represented as 30-39 in ASCII code. When we want to subtract two decimal digits which are represented in ASCII code, it is necessary to mask upper nibble (3) from the code before instruction. The 8086 allows to subtract the ASCII codes for two decimal digits without marking off the "3" in the upper nibble of each digit. The AAA instruction can be used subtraction to get the current result in unpacked BCD form.

Examples :

Instruction	**Explanation**
	; AL = 0011 1000 ASCII 8
	; CL = 0011 0010 ASCII 2
SUB AL, CL	; AL = 0000 0110 BCD 06
	; CF = 0
AAS	; AL = 0000 0010 = BCD 06
	; CF = 0 no borrow required
	; AL = 0011 0010 ASCII 2
	; CL = 0011 1000 ASCII 8
SUB AL, CL	; AL = 1111 1010 = FAH
	; CF = 1
AAS	; AL = 0000 0110 = BCD 6
	; CF = 1 borrow needed means (–6)

By the instruction AAS, adjustment is performed as mentioned below :

1. If the lower-order 4 bits of the AL register are between 0 and 9 and the AF is 0 then go to step 3.
2. If the lower-order bits of the AL register are between A and F or the AF is 1, then subtract 6 from AL register, subtract 1 from the AH register, and set the AF flag to 1.
3. Clear the higher order four bits of the AL, register.
4. Set, the CF to the value of the AF.

Note. The AAS instruction updates AF and CF; the contents of OF, PF, SF and ZF are undefined following the execution of AAS.

Example 1 : *If AX = 0438 ;*
SUB AL, 35h
Then, *AAS ; AX = 04 03 ; AF = 0 CF = 0*

Example 2 : *If AX = 0F92 ;*
SUB AL, 94
Then, *AAS ; AX = 0E08 ; AF = 1 CF = 1*

■ **Encoding will be**

00111111

■ **DAS Instruction :** Decimal Adjust After Subtraction

This instruction is used subtracting two packed BCD numbers to make sure the result is correct packed BCD. The Instruction words by two steps as :

1. If the value of the low-order four bits *(D_3–D_0)* in the AL is greater than 9 or if AF is set ; the instruction subtracts 6 *(06)* from the low-order four bits.
2. If the value of the high-order four bits *(D_7–D_4)* in the AL is greater than 9 or if carry flag is set, the instruction subtracts *6 (60)* from the high-order four bits.

Examples :

Instruction	**Explanation**
	; AL = 0011 0010 = 32 BCD
	; CL = 0001 0111 = 17 BCD
SUB AL, CL	; AL = 0001 1011 = 1BH
	; Subtract 0110 Because 1011 > 9
	; AL = 0001 0101 = 15 BCD
	; AL = 0010 0011 = 23 BCD
	; CL = 0101 1000 = 58 BCD
SUB AL, CL	; AL = 1100 1011 = CBH CF = 1
	; Subtract 0110 (6) because 1011 > 9
	; AL = 1100 0101 = C5H
	; Subtract 0110 0000 because 1100 > 9
	; AL = 0110 0101 = 65 BCD CF = 1, CF = 1 Borrow
	; is needed means number is negative (–65)

The DAS instruction updates the AF. CF, PF and ZF, and OF flag is undefined after DAS instruction.

(DAS instruction only works for AL)

Example 1 : *If AL = 86 BCD* *; BH = 57 BCD*
SUB AL, BH *; AL = 2FH, CF = 0*
Then, *DAS* *, Subtract 06H because lower nibble > 9*
AL = 29 BCD

Example 2 : *If AL = 49 BCD*
BH = 72 BCD
SUB AL, BH *; AL = D7H, CF = 1*

Then,	DAS	; Subtract 60H, Because the upper nibble is > 9
	AL = 77 BCD	; CF = 1 means borrow was needed

■ **Encoding will be**

0010111

(c) Multiplication Instructions

Here there are three instructions to be discussed;

(i) MUL (ii) IMUL and (iii) AAM instruction.

(i) MUL Instruction : MUL source

This instruction multiplies an unsigned byte from source and unsigned byte in AL register or unsigned word from source and unsigned word in AX register. The source can be a register or unsigned word from source and unsigned word in AX register. The source can be a register or a memory location. If the byte is multiplied by the contents of AL, the result is stored in AX. The most significant byte is multiplied by the contents of AL, the result is stored in AX. The most significant byte is stored in AH and least significant byte is stored in AL. When a word is multiplied by the contents of AX, the most significant word of result is stored in DX and least significant word of result is stored in AX.

For MUL instruction flags affected are AF, PF, SF and ZF.

Examples :

Instruction	**Explanation**
MUL BL	; AL × BH, result in AX
MUL BX	; AX × BX, result high word in DX low word in AX
MUL WORD PTR [BX]	; AX times word in DS pointed by [BX]
	; result high word in DX low word in AX

Example 1 : If AL = 32H, AH = 23H, CL = 06H

MUL CL ; The result is in AX register with AH having the higher order 8-bits

AX = 09C4H

The flags affected here are CF and OF ; Values are CF = 1, OF = 1

Example 2 : If AX = 0023H, DX = 0089H, BX = 0022H ;

MUL AX, BX ; (After the multiplication the result is in DX and AX registers)

AX = 04C9, DX = 0000, CF = 0 OF = 0

■ **Encoding will be**

1111011 w	mod 100 r / m

If w = 0 then LSRC = AL, RSRC = EA, DEST = AX, EXT = AH

else LSRC = AX, RSRC = EA, DEST = DX ; AX, EXT = DX

(ii) IMUL Instruction : *(IMUL Source)*

This instruction multiplies a signed byte from some source and a signed byte in AL, or a signed word from some source and a signed word in AX. The source can be register or memory location. If a signed byte is multiplied by AL a signed result will be put in AX. When a signed word is multiplied by AX, the high-order word of the signed result is put in DX and the low-order word of the signed result is put in AX, register.

When the upper byte of a 16-bit result or the upper word of 32-bit result contains only copied of the sign bit *(all 0's or all 1's)*, then the CF and the OF flags will both be 0's. The AF, PF, SF and ZF flags are undefined after IMUL Instruction.

For multiplying a signed byte by a signed word it is necessary to move the byte into a word location and fill the upper byte of the word with copies of the sign bit. This can be done using CBW instruction.

Examples :

Instruction	Explanation
IMUL BL	; AL × BL, result in AX
IMUL CX	; AX × CX, high-order word of result in DX and
	; low-order word of result in AX

Example 1 : *If AL = 69H, BL = 14H*
Then *IMUL BL ; AX = +0834H, Flags : SF = 0, CF = 1, OF = 1*

Example 2 : *If AX = –28H, BL = 59H*
Then *IMUL BL ; AX = F218H Flags : SF = 1, CF = 1, OF = 1*
; In place of IMUL if MUL instruction is used then AX = 4B18H

- **Encoding will be**

1111011 w	mod 101 r / m

If w = 0 then LSRC = AL, RSRC = EA, DEST = AH, EXT = AH, LOW = AL
else LSRC = AX, RSRC = EA, DEST = DX : AX, EXT = DX, LOW = AX

(iii) AAM Instruction : BCD Adjust After Multiply

After the two unpacked BCD digits are multiplied, the AAM instruction is used to adjust the product to two unpacked BCD digits in AX.

Examples :

Instruction	Explanation
	; AL = 0000 0100 = Unpacked BCD 4
	; CL = 0000 0110 = Unpacked BCD 6
MUL CL	; AL × CL Result in AX
	; AX = 0000 0000 0001 1000 = 0018H
AAM	; AX = 0000 0010 0000 0100 = 0204H
	; Which is unpacked BCD for 24

Example : *If AL = 05 (Unpacked BCD), BH = 09 (Unpacked BCD)*
MUL BH ; The result is in AX and is equal to 002DH
Then for AAM ; The unpacked result in obtained in AX and is equal to 04 05H
ZF = 0, SF = 0, PF = 0

■ **Encoding will be**

11010100	00001010

(*d*) Division Instructions : (DIV, IDIV, CBW and CWD)

■ **DIV Instruction :** DIV source

DIV instruction is used to divide an unsigned word a byte or to divide an unsigned double word by a word while dividing a word by a byte, the word must be in AX register. After the division AL will contain an 8-bit quotient and AH will contain an 8-bit remainder. If an attempt is made to divide by 0 or the quotient is too large to fit in AL *(greater than FFH),* the 8086 will automatically execute a type 0 interrupt.

While double word is divided by a word, the most significant word of the double word must be in DX and the least-significant word must be in AX. After the division AX will contain a 16-bit quotient and DX will contain a 16-bit remainder. Next, if an attempt is made to divide by 0 or the quotient is too large to fit in AX register *(greater than FFFFH),* the 8086 will do a type 0 interrupt. For DIV instruction source may be a register or memory location.

To divide a byte by a byte, it is required to put the divided byte in AL and fill AH with all 0's. So also to divide a word by a word, it is necessary to put the divided word in AX and full DX with all 0's.

Note. All flags are undefined after a DIV instruction, *i.e.*, contents of AF, CF, OF, PF, SF and ZF.

Examples :

Instruction	**Explanation**
DIV CL	; Word in AX/byte in CL,
	; Quotient in AL, remainder in AH
DIV CX	; Double word in DX and AX/word in CX,
	; Quotient in AX, remainder in DX

Example 1 : DIV BH ;
Contents Before : AX = 37D7H, BH = 97H
Contents After : Quotient in AX = 0083H, Remainder in DX = 05E5

■ **Encoding will be**

1111011 w	mod 110 r / m

If w = 0 then NUMR = AX, DIVR = EA, QUO = AL, REM = AH, MAX = FFH
else NUMR = DS : AX, DIVR = EA, QUO = AX, REM = DX, MAX = FFFFH

■ **IDIV Instruction :** IDIV Source

The IDIV *(Integer Divide)* instruction performs a signed division of the accumulator *(and its extension)* by the source operand. The source operand may be a register or the contents of memory *(any memory addressing mode)*. When the source operand is a byte, it is divided into the double-length dividend assumed to be in register AH and AL. The single-length quotient is stored in AL, and the single-length remainder is stored in AH. To a byte integer division, the maximum positive quotient is +127 *(7F)* and the minimum negative quotient is –128 *(81FH)*. When the source operand is a word, it is divided into the double length dividend in register DX and AX *(the high-order 16-bits are in DX and the low order 16-bits in AX)* and the single-length remainder is stored in DX. For word integer division, the maximum positive quotient is +32767 *(7FFFH)* and the minimum negative quotient is –32, 768 *(8001H)*. If the quotient is positive and exceeds the maximum, or is negative and is less than the minimum, the quotient and the remainder are undefined, and a type-0 interrupt is generated. A division by zero interrupt results in the action as mentioned below :

1. Push the flag register onto the stack
2. Clear IF and TF *(Interrupt Flag and Trap Flag)*
3. Push the CS register onto the stack
4. Load the word at the location 00002H into the CS
5. Push the IP onto the stack
6. Load the word at memory location 00000H into the IP.

Note. The value of OF, SF, ZF, AF, PF and CF are undetermined for this operation. IDIV instruction is used to divide a signed word by a signed byte, or to divide a signed double word *(32 bits)* by a signed word. Rest all is similar to DIV instruction.

Example : *IDIV BL;*
Contents Before : *AX = 03ABH ; BL = D3H = –2DH*
Contents After : *Quotient in AL = ECH ; Remainder in AH = 27H*

■ **Encoding will be**

1111011 w	mod 111 r / m

If w = 0 then
NUMR = AX, DIVR = EA, QUO = AL, REM = AH, MAX = 7FH
else NUMR = DS : AX, DIVR = EA, QUO = AX, REM = DX, MAX = 7FFFH

■ **AAD Instruction :** Binary Adjust before division. *(ASCII Adjust before Division)*

AAD instruction converts two unpacked BCD digits in AH and AL to the equivalent binary number in AL. This adjustment must be made before dividing the two unpacked BCD digits in AX by an unpacked BCD byte. After division AL, will contain the unpacked BCD quotient and AH will contain the unpacked BCD remainder. The PF, SF and ZF flags are updated, and AF, F and OF are undefined after AAF instruction.

Examples :

Instruction	Explanation
AAD	; AX = 0403 unpacked BCD for 43 decimal, CL = 07H
	; Adjust to binary before division,
	; AX = 002BH = 2BH = 43 decimal
DIV CL	; Divide AX by unpacked BCD in CL
	; AL = quotient = 06 unpacked BCD
	; AH = remainder = 01 unpacked BCD

The algorithm for AAD instruction is as mentioned below :

1. Multiply the contents of the AH register by 0A.
2. Add AH to AL.
3. Store 00 into AH register.
4. Setting flags will be in the following manner.
   CF, OF, AF : Undefined
   PF, ZF : based on the AL register contents
   SF : based on the MSB bit of the AL register

Example : *AAD ; AX = 0043H = 67 decimal*
DIV CH ;

Contents Before : AX = 0607 ; Unpacked BCD FOR 67 decimal ; CX = 09

Contents After : AL = Quotient = 07 unpacked BCD
AH = Reminder = 04 Unpacked BCD ; PF = 0, SF = 0, ZF = 0

■ **Encoding will be**

11010101	0000 1010

■ **CBW Instruction :**

This instruction copies the sign of a byte in AL to all the bits in AH, AH is then said to be sign extension of AL.

Examples :

Instruction	Explanation
	; AX = 0000 0000 1001 1000 = –152 decimal
CBW	; Convert signed byte in AL to signed word in AX
	; Result in AX = 1111 1111 1001 1000 = –152 decimal

Contents Before : AX = –155 decimal = 0000000100011011 binary
Contents After : AX = –155 decimal = 1111111100011011 binary

■ **Encoding will be**

10011000

■ **CWD Instruction :** *(Convert Word to Double Word)*

This instruction copies the sign bit of a word in AX to all the bits of the DX register. DX is then said to be sign extension of AX.

Note. CWD affects no flag.

Examples :

Instruction	Explanation
	; DX = 0000 0000 0000 0000
	; AX = 1111 0000 1100 0001 = – 28865 decimal
CWD	; Convert signed word in AX to signed double
	; word in DX : AX
	; DX = 1111 1111 1111 1111
	; AX = 1111 0000 1100 0001 = – 28865 decimal

Contents Before : DX = 00000000 00000000 = 0000 H;
AX = 11110000 11000111 = F0C7 H

Contents After : DX = 11111111 11111111 = FFFF H;
AX = 11110000 11000111 = F0C7 H.

■ **Encoding will be**

10011000

4.4.3 Bit Manipulation Instruction (Logical Instructions)

The Bit Manipulation Instruction are divided/grouped into three subgroups as mentioned below :

(*a*) Logical Instructions
(*b*) Shift Instructions, and
(*c*) Rotate Instructions.

(*a*) Logical Instructions : (NOT, AND, OR, XOR and TEST)

■ **NOT Instruction** : NOT destination

The NOT instruction with invert each bit of a byte or a word. The destination can be register or a memory location. This instruction will not affect any flags.

Examples :

Instruction	Explanation
	; AL = 0110 1100
NOT AL	; AL = 1001 0011
	; CX = 1010 1111 0010 0110
NOT CX	; CX = 0101 0000 1101 1001

Example 1 : *NOT AX*
Contents Before : *AX = 1234H*
Contents After : *AX = EDCBH*

Example 2 : *NOT [SI]*
Contents Before : *SI = 1000H DS = 2000 M [21000] = 00FFH*
Contents After : *SI = 1000H DS = 2000 M [21000] = 00FFH*

■ **Encoding will be**

1111011 w	mod 011 r / m

If w = 0 then SRC = FFH
else SRC = FFFFH

■ **AND Instruction :** AND destination, source

AND instruction logically ANDs each bit of the source byte or word with the corresponding bit in the destination and stores result in the destination. The source may be an immediate number, a register or a memory location. The destination may be a register or a memory location. The source and destination both cannot be memory locations in the same instruction. The CF and OF are both 0 after AND. The PF, SF and ZF are affected, and AF is undefined.

Examples :

Instruction	**Explanation**
	; AL = 1001 0011 = 93H
	; BL = 0111 0101 = 75H
AND BL, AL	; AND byte in AL with byte in BL
	; BL = 0001 0001 = 11H
	; CX = 0110 1011 1001 1110
AND CX, 00F0H	; CX = 0000 0000 1001 0000

Example 1 : *AND AX, BX*
Contents Before : *AX = 2314H, BX = F090H, CF = 0, AF = 0, PF = 0, SF = 0, ZF = 0*
Contents After : *AX = 2010H, BX = F090H, CF = 0, AF = 0, PF = 0, SF = 0, ZF = 0*

Example 2 : *AND SI, FF00H*
Contents Before : *SI = 3030 H, CF = 0, AF = 0, PF = 0, SF = 0, ZF = 0*
Contents After : *SI = 3000H, CF = 0, AF = 0, PF = 1, SF = 0, ZF = 0*

■ **Encoding will be**

For Memory or Register Operand with Register Operand :

001000 dw	mod reg r / m

If d = 1 then LSRC = REG, RSRC = EA, DEST = REG
else LSRC = EA, RSRC = REG, DEST = EA

For Immediate Operand to Memory or Register Operand :

100000 0w	mod 100 r / m	data	data if w = 1

LSRC = EA, RSRC = data, DEST = EA

For Immediate Operand to Accumulator :

0010010 w	data	data if w = 1

If w = 0 then LSRC = AL, RSRC = data, DEST = AL
else LSRC = AX, RSRC = data, dest = AX

■ **OR Instruction** : OR destination, source

This instruction logically ORs each bit of the source byte or word with the corresponding bit in the destination and stores result in the destination. The source may be an immediate number, a register or a memory location. The destination may be a register or a memory location. The source and destination both cannot be memory locations in the same instruction. The CF and OF are both 0 after OR. The PF, SF and ZF are affected, and AF flag is undefined.

Examples :

Instruction	**Explanation**
	; AL = 1001 0011 = 93H
	; BL = 0111 0101 = 75H
OR BL, AL	; OR byte in AL with byte in BL
	; BL = 1111 0111 = F7H
	; CX = 0110 1011 1001 1110
OR CX, 00F0H	; CX = 0110 1011 1111 1110

Example 1 : *OR AX, BX*
Contents Before : *AX = 2314H, BX = F090H, CF = 0, AF = 0, PF = 0, SF = 0, ZF = 0*
Contents After : *AX = F394H, BX = F090H, CF = 0, AF = 0, PF = 0, SF = 1, ZF = 0*

Example 2 : *OR SI, FF00H*
Contents Before : *SI = 3030H, CF = 0, AF = 0, PF = 0, SF = 0, ZF = 0*
Contents After : *SI = FF30H, CF = 0, AF = 0, PF = 1, SF = 0, ZF = 0*

■ **Encoding will be**

For Memory or Register Operand with Register Operand :

000010 dw	mod reg r / m

If d = 1 then LSRC = REG, RSRC = EA, DEST = REG
else LSRC = EA, RSRC = REG, DEST = EA

For Immediate Operand with Memory or Register Operand :

100000 0w	mod 001 r / m	data	data if w = 1

LSRC = EA, RSRC = data, DEST = EA

For Immediate Operand with Accumulator :

0000110 w	data	data if w = 1

If w = 0 then LSRC = AL, RSRC = data, DEST = AL

else LSRCH = AX, RSRC = data, DEST = AX

■ **XOR Instruction :** XOR destination, source

This instruction logically XORs each bit of the source byte or word with the corresponding bit in the destination and stores result in the destination. The source may be an immediate number, a register or a memory location, and destination may be a register or a memory location. The source and destination both cannot be memory locations in the same instruction. The CF and OF are both 0 after XOR. The PF, SF and ZF are affected and AF flag is undefined.

Example 1 : *XOR AX, BX*

Contents Before : AX = 2314H, BX = F090H, CF = 0, AF = 0, PF = 0, SF = 0, ZF = 0

Contents After : AX = D384H, BX = F090H, CF = 0, AF = 0, PF = 0, SF = 0, ZF = 0

Example 2 : *XOR SI, FF00H*

Contents Before : SI = 3030H CF = 0, AF = 0, PF = 0, SF = 0, ZF = 0

Contents After : SI = CF30H CF = 0, AF = 0, PF = 1, SF = 0, ZF = 0

■ **Encoding will be**

For Memory or Register Operand with Register Operand :

001100 dw	mod reg r / m

If d = 1 then LSRC = REG, RSRC = EA, DEST = REG

else LSRC = EA, RSRC = REG, DEST = EA

For Immediate Operand to Memory with Register Operand :

100000 0 w	mod 110 r / m	data	data if w = 1

LSRC = EA, RSRC = data, DEST = EA

For Immediate Operand with Accumulator :

0011010 w	data	data if w = 1

If w = 0 then LSRC = AL, RSRC = data, DEST = AL

else LSRC = AX, RSRC = data, DEST = AX

■ **TEST Instruction :** TEST destination, source

This instruction ANDs the contents of a source byte or word with the contents of the specified destination word. Flags are affected. But neither operand is changed.

Examples :

Instruction	**Explanation**
TEST AL, CL	; AND CL with AL
	; Update flags, result is not stored
TEST BX, CX	; AND CX with BX
	; Update flags, result is not stored

Note. The flags CF, OF, PF, SF and ZF are affected and AF is undefined. Normally OF and CF will be cleared.

Example 1 : *TEST AX, BX*
Contents Before : *AX = 2314H, BX = F090H, CF = 0, AF = 0, PF = 0, SF = 0, ZF = 0*
Contents After : *AX = 2314H, BX = F090H, CF = 0, AF = 0, PF = 1, SF = 0, ZF = 0*

Example 2 : *TEST SI, FF0H*
Contents Before : *SI = 3030H, CF = 0, AF = 0, PF = 0, SF = 0, ZF = 0*
Contents After : *SI = 3000H, CF = 0, AF = 0, PF = 1, SF = 0, ZF = 0*

- **Encoding will be**

For Memory or Register Operand with Register Operand :

1000011 w	mod reg r / m

SRC = EA, DEST = REG

For Register Operand with Accumulator :

10010 reg

SRC = REG, DEST = AX

(*b*) Shift Instruction (SAL/SHL, SHR, and SAR)

Here the bits in bytes and words may be shifted arithmetically or logically. Up to 255 shifts can be performed according to the value of the count operand coded in the instruction. The count may be specified as the constant *(only one shift if a constant is specified)*, or as register CL, allowing the shift count to be a variable supplied at execution time.

The Arithmetic shifts can be used to multiply and divide binary numbers by powers of two, whereas logical shifts can be used to isolate bits in bytes or words. Shift instructions affect the flags as mentioned below.

Flag	**Changes**
AF	; is undefined
PF, SF, ZF	; are updated normally, as in logical instructions
CF	; always contains the value of the last bit shifted out of the destination
OF	; Undefined for multi-bit shift

Note. The sign-bit is set if the value of the high-order (sign) bit was changed by the operation; if the sign bit retains its original value, OF will be cleared.

■ SAL/SHL Instruction : *(Shift Arithmetic Left and Shift Logical Left)* SA/SHL destination, count

SAL and SH are two mnemonics for the same instruction, and they shifts each bit in the specified destination to the left and 0 is stored at LSB position. The MSB is shifted into the carry flag. The destination can be a byte or a word. It can be in a register or in a memory location. The number of shifts are indicated by count. But if the number of shifts required is one, you can place 1 in the count position. If number of shifts are greater than 1 then shift count must be loaded in CL register and CL must be placed in the count position.

The Diagram shows SAL instruction for byte operation.

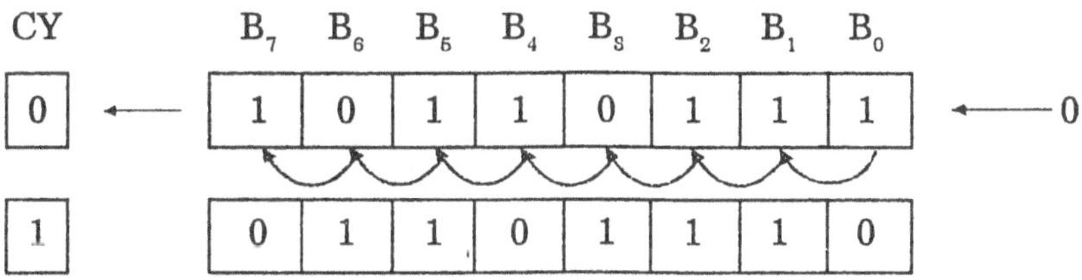

Flags : All flags are affected, as mentioned earlier.

Examples :

Instruction	**Explanation**
SAL CX, 1	; Shift word in CX 1 bit position left, 0 in LSB
MOV CL, 05H	; Shift word in AX left 5 time
SAL AX, CL	; Shift word in AX left 5 time
	; 0's in 5 least-significant bits

Example 1 : *SAL AH, 01;*

Contents Before : AH = 32H, CF = 0, AF = 0, PF = 0, SF = 0, ZF = 0

Contents After : AH = 64H, CF = 0, AF = 0, PF = 0, SF = 0, ZF = 0

Example 2 : *SAL DI, CL*

Contents Before : DI = 0032H, CL = 04, CF = 0, AF = 0, PF = 0, SF = 0, ZF = 0

Contents After : DI = 0320H, CL = 00, CF = 0, AF = 0, PF = 0, SF = 0, ZF = 0

■ Encoding will be

11010 0 vw	mod 100 r / m

If v = 0 then COUNT = 1; else COUNT = (CL)

■ SHR Instruction : *(Shift Logical Right)* SHR destination, count

SHR instruction shifts each bit in the specified destination to the right and 0 is stored at MSB position. The LSB is shifted into the carry flag. The destination can be a byte or a word. It can be in a register or in a memory location. The number of shifts are indicated by count. If number of shifts required is one, you can place 1 in the count position. But if the number of shifts are greater than 1 then count must be loaded in CL register and CL must be placed in the count position.

The Diagram shows SHR instruction for byte operation, as below :

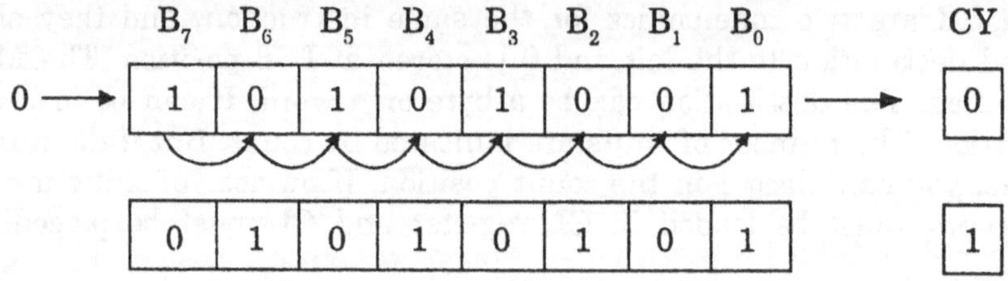

Flags : Here all flags are affected.

Examples :

Instruction	**Explanation**
SHR CX, 1	; Shift word in CX 1 bit position right, 0 in MSB
MOV CL, 05H	; Shift word in AX right 5 time
SHR AX, CL	; Shift word in AX right 5 time
	; 0 ; s in 5 most significant bits.

Example 1 : *SHR SI, 01*
Contents Before : *SI = 1246H*
Contents After : *SI = 0923H*

Example 2 : *SHR IN_BY [SI], CL*
Contents Before : *SI = 1000, CL = 02, M[DS*16 + 1000 + IN_BY] = 2000H*
Contents After : *SI = 1000, CL = 02, M[DS*16 + 1000 + IN_BY] = 0800H*

■ **Encoding will be**

11010 0 vw	mod 101 r / m

If v = 0 then COUNT = 1
else COUNT = (CL)

■ **SAR Instruction :** *(Shift Arithmetic Right)* SAR destination, count.

SAR instruction shifts each bit in the specified destination some number of bit positions to the right. As a bit is shifted out of the MSB position, a company of the old MSB is put in the MSB is put in the MSB position. The LSB will be shifted into CF. For the case of multiple shifts, CF will be contain the bit most recently shifted in from the LSB. Bits shifted into CF previously will be lost.

The destination can be a byte or a word. It can be in a register or in a memory location. The number of shifts are indicated by count. If number of shifts required is one, you can place I in the count position. When number of shifts are greater than 1 then shifts count must be loaded in CL register CL must be placed in the count position.

The diagram shows SAR instruction for byte operation, as below :

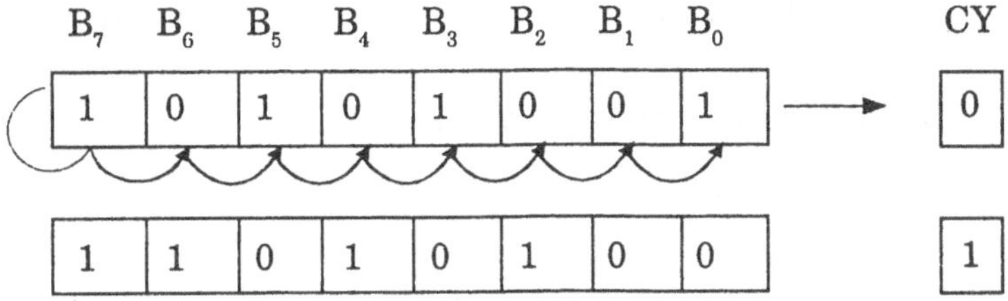

Flag : All flags are affected here ;

Examples :

Instruction	**Explanation**
SAR BL, 1	; Shift byte in BL one bit position right
MOV CL, 04H	; Load desired number of shifts in CL
SAR DX, CL	; Shift word stored in DX 4 bit positions right

> **Note.** SAR does not produce the same result as the divide of an equivalent IDIV instruction if the destination operand is negative and 1-bit shifted out.

Example : Shifting-5 right by 1 yields-3, while integer division of –5 by 2 yields –2. (The difference in the instructions is that IDIV truncates all numbers towards zero, while SAR truncates positive numbers towards zero and negative numbers towards infinity).

1. SAR DX, 01

Contents Before : DX = 2345H, Contents After : 11A2H

2. SAR DI, CL

Contents Before : DI = 3000H, CL = 04H, Contents After : DI = 0300H, CL = 00

■ **Encoding will be**

11010 0 vw	mod 111 r / m

If v = 0 then COUNT = 1
else COUNT = (CL)

(*c*) Rotate Instructions (ROL, ROR, RCL and RCR)

From rotate instructions bits in bytes and words may also be rotated. The bits rotated out of an operand are not lost as in a shift, but are circled back into the other end of the operand. As in the case of shift instructions, the number of bits to be rotated is taken from the count operand, which may specify either a constant 1, or the CL register. The carry flag may act as an extension of the operand in the rotate instructions, allowing a bit to be isolated in CF and then tested by a JC or JNC instruction.

Rotate instruction affect the carry and overflow flags. The CF flag always contains the value of the last bit rotated instruction out. In multi-bit rotate, the value of OF is always undefined. In single-bit rotate, OF is set is the operation changes the high-order *(sign)* bit of the destination operand. *(If the sign bit retains it original value, OF is cleared).*

■ **ROL Instruction :** *(Rotate Left)* ROL destination, count

This instruction rotates all bits in a specified byte or word to the left some number of bit positions. MSB is placed as a new LSB and a new CF.

The diagram shows ROL instruction for byte rotation as below :

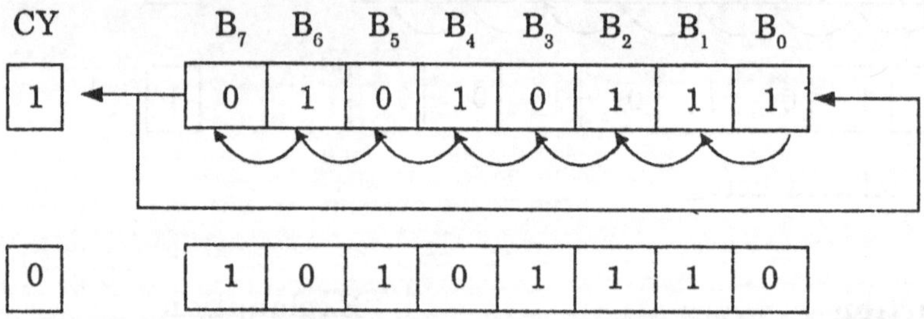

The destination can be of byte or a word, which can be in a register or in a memory location. The number of shifts are indicated by count. If number of shifts required is one you can place 1 in the count position. When number of shifts are greater than 1 then shift count must be loaded in CL register and CL must be placed in the count position.

Examples :

Instruction	**Explanation**
ROL CX, 1	; Word in CX one bit position left, MSB to LSB and CF
MOV CL, 03H	; Load desired number of bits to rotate in CL
ROL BL, CL	; Rotate BL three positions

Example 1 : *ROL BX, 01*

Contents Before : BX = 8200H, CY = 0, *Contents After : BX = 0401H, CY = 1*

Example 2 : *ROL DI, CL*

Contents Before : DI = 9233H, CL = 4, CY = 0 ;

Contents After : DI = 2339H, CL = 4 CY = 1

■ **Encoding will be**

11010 0 vw	mod 000 r / m

If v = 0 then COUNT = 1

else COUNT = (CL)

■ **ROR Instruction :** *(Rotate Right)* ROR destination, count

ROR instruction rotates all bits in a specified byte or word to the left some number of bit positions. LSB is placed as a new MSB and a new CF.

The destination may be a byte or a word, which can be in a register or in a memory location. The number of shifts are indicated by count. If number of shifts required is one, you can place 1 in the count position. If number of shifts are greater than 1 then shift count must be loaded in CL register and CL must be placed in the count position.

The Diagram shows ROR instructiion for byte rotation, as under :

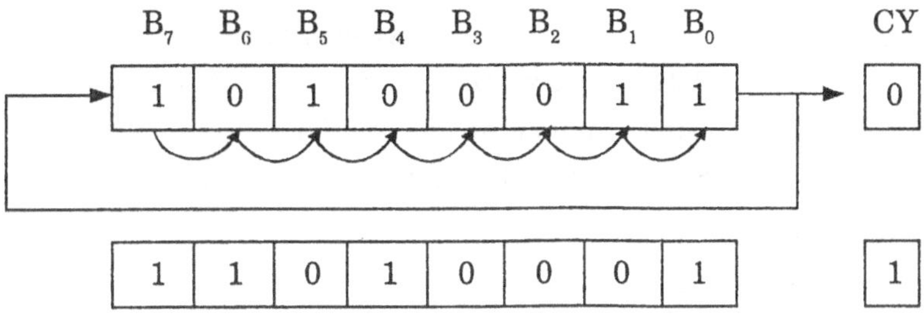

Examples :

Instruction	Explanation
ROR CX, 1	; Rotated word in CX one bit position left,
	; LSB to MSB and CF
MOV CL, 03H	; Load number of bits to rotate in CL
ROR BL, CL	; Rotate BL three positions

Example 1 : *ROR CX, 01*
Contents Before : CX = 3456H, CF = 0
Contents After : CX = 1A2BH, CF = 0

Example 2 : *ROR AL, CL*
Contents Before : AL = 56H, CF = 0 ; CL = 4 ;
Contents After : AL = 65, CF = 0 ; CL = 0 ;

■ **Encoding will be**

11010 0 vw	mod 001 r / m

If v = 0 then COUNT = 1
else COUNT = (CL)

■ **RCL Instruction :** *(Rotate through Carry)* RCL destination, count

RCL instruction rotates all of the bits in a specified word or byte some number of bit positions to the left along with the carry flag. MSB is placed as a new carry and previous carry is placed as a new LSB.

The destination may be a byte or a word which can be in a register or in a memory location. The number of shifts are indicated by count. When number of shifts required is one, you can place 1 in the count position. When the number of shifts are greater than 1 then shift count must be loaded in CL register and CL must be placed in the count position.

The Diagram shows RCL instruction for byte rotation, as under :

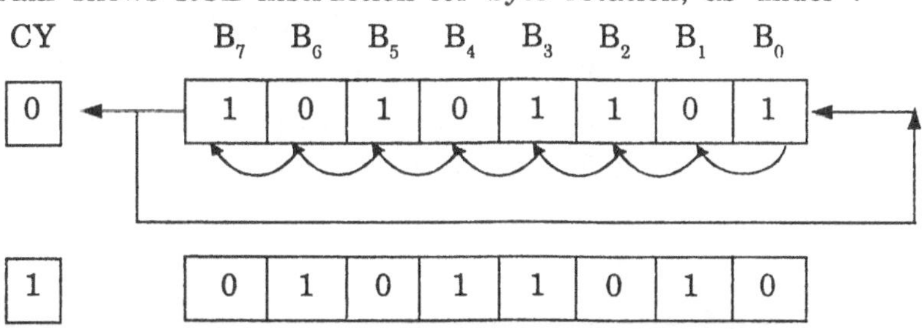

Examples :

Instruction	Explanation
RCL, CX, 1	; Rotated word in CX 1 bit left, MSB to CF, CF to LSB
MOV CL, 04H	; Load number of bit positions to rotate in CL
RCL AL, CL	; Rotate AL 4 bits left

Example 1 : *RCL CX, 01*
Contents Before : *CX = 9876H, CY = 0 ;*
Contents After : *CX = 30ECH, CY = 1*

Example 2 : *RCL AL, CL*
Contents Before : *AL = 34H, CL = 04H, CY = 0*
Contents After : *AL = 41H, CY = 1 ;*

- **Encoding will be**

11010 0 vw	mod 010 r/m

If v = 0 then COUNT = 1
else COUNT = (CL)

- **RCR Instruction :** *(Rotate through Carry Right)* RCR destination, count

RCR instruction rotates all of the bits in a specified word or byte some number of bit positions to the left along with the carry flag. LSB is placed as a new carry and previous carry is placed as a new MSB.

The destination may be a byte or a word, which can be in a register or in a memory location. The number of shifts are indicated by count. If number of shifts required is one you can place 1 in the count position. When the number of shifts are greater than 1 then shift count must be loaded in CL register and CL must be placed in the count position. The diagram shows RCR instruction for byte rotation, as under

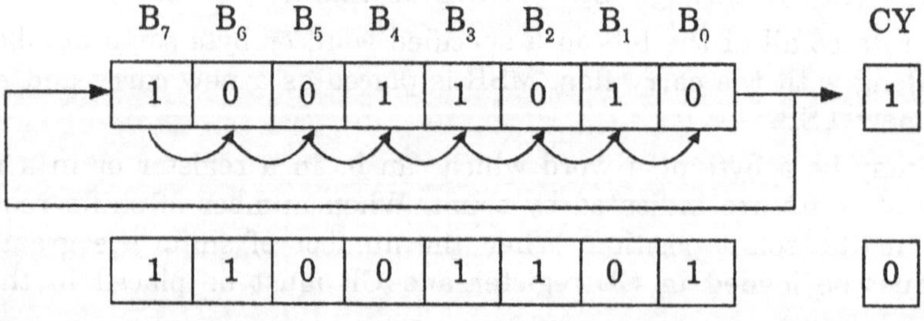

Examples :

Instruction	Explanation
RCR CX, 1	; Word in CX 1 bit right, LSB to CF, CF to MSB
MOV CL, 04H	; Load number of bit positions to rotate in CL
RCR AL, CL	; Rotate AL 4 bits right

Example 1 : *RCR BX, 01*
Contents Before : *BX = 3400H;*
Contents After : *BX = 1A00H;*

Example 2 : *RCR BL, CL*
Contents Before : *BL = 89H, CY = 0, CL = 4H;*
Contents After : *BL = 28H, CY = 1;*

- **Encoding will be**

11010 0 vw	mod 011 r / m

If v = 0 then COUNT = 1
else COUNT = (CL)

4.4.4 Program Execution Transfer Instructions

In this type there are mainly two types of instructions

(*a*) Unconditional transfer Instructions and

(*b*) Conditional Transfer Instructions.

(*a*) Unconditional Transfer Instructions (CALL, RET and JMP)

- **CALL Instruction**

CALL instruction is used to transfer execution to a subprogram or procedure. The two basic types of CALLs, are near and far. A near CALL is a cell to a procedure which is in the same code segment as the CALL instruction. If 8086 executes a near CALL instruction it decrements the stack pointer by two and copies the offset of the next instruction after the CALL on the stack. It loads IP with the offset of the first instruction of the procedure in same segment.

A far CALL is a call to a procedure which is in a different segment from that which contains the CALL instruction. It 8086 executes a far CALL it decrements the stack pointer by two and copies the contents of the CS register to the stack. It then decrements the stck painter by two again and copies the offset of the instruction after the CALL to the stack. At the end, it loads CS with the segment base of the segment which contains the procedure and IP with the offset of the first instruction of the procedure in that segment.

Examples :

Instruction	**Explanation**
Direct within segment (near)	; PRO is the name of the procedure
CALL PRO	; The assembler determines displacement of pro
	; from the instruction after the CALL and codes
	; this displacement in as part of the instruction
Indirect within-segment (near)	
CALL CX	; CX contains, the offset of the first instruction
	; of the procedure. Replaces contents of IP with
	; contents of register CX

Indirect to another segment (far);

```
CALL DWORD PTR [BX]          ; New values for CS and IP are fetched from four
                             ; memory locations in DS. The new value for CS
                             ; is fetched from [BX] and [BX + 1], the new IP
                             ; is fetched from [BX + 2] and [BX + 3]
```

■ RET Instruction

RET instruction will return execution from a procedure to the next instruction after the CALL instruction in the calling program. When the procedure is a near procedure *(in the same code segment as the CALL instruction)*, then the return will be done by replacing the instruction pointer with a word from the top of the stack.

If the procedure is a far procedure *(in a different code segment from the CALL instruction which calls it)*, then the instruction pointer will be replaced by the word at the top of the stack and stack pointer will then be incremented by two. The code segment register is then replaced with a word from the new top of the stack. After the code segment word is popped off the stack, the stack pointer is again incremented by two. Such words/word are the offset of the next instruction after the CALL. That means 8086 will fetch the next instruction after the CALL.

RET instruction can be followed by a number, for instant, RET 4. Here the stack pointer will be incremented by an additional four address after the IP or the IP and CS are popped off the stack. This form is used to increment the stack pointer up over parameters passed to the procedure on the stack.

Note. RET instruction will affects not any flags.

■ JMP Instruction

JMP instruction will always cause 8086 to fetch its next instruction from the location specified in the instruction rather than from the next location after the JMP instructions.

Two basic types of JMPs are near and far. A near JMP is a jump where destination location is in the same code segment. In this case only IP is changed. A far JMP is a jump where destination location is from a different segment. In this case both IP and CS are changed as specified in the destination.

The Near and Far jumps are further described as either direct or indirect. If the destination address for the jump is specified directly within the instruction, then the jump is described as direct. If the destination address for the jump is contained in a register or memory location, the jump is referred as indirect, because the 8086 has to access the specified register or memory location to get the required destination address.

Examples :

Instruction	**Explanation**
JMP NEXT	; Fetch next instruction from address at label NEXT
	; If label is in same segment, an offset coded as part of the
	; instruction will be added to the instruction pointer to produce

```
                        ; the new fetch address. If the label is in another segment then
                        ; IP and CS will be replaced with values codes in as part of
                        ; the instruction. This type of jump is referred to as direct
                        ; because the displacement of the destination or the destination
                        ; itself is specified directly in the instruction.
JMP BX                  ; Replace the contents of IP with the contents of BX
                        ; BX must first be loaded with the offset of the destination
                        ; instruction in CS. This is a near jump. It is also referred
                        ; to as an indirect jump because the new value for IP comes
                        ; from a register rather than from the instruction itself as in a
                        ; direct-type jump
JMP WORD PTR [BX]       ; Replace IP with a word from a memory location
                        ; pointed to by BX in DS. This is an indirect near jump
JMP DWORD PTR [SI]      ; Replace IP with word pointed to by SI in DS
                        ; Replace CS with word pointed to by SI + 2 in DS
                        ; This is an indirect far jump
```

As discussed earlier, a near type jump instruction can cause the next instruction to be fetched from anywhere in the current code segment. In order to produce the new instruction fetch adress, this instruction adds a 16-bit signed displacement contained in the instruction to the contents of the instruction pointer register. A 16 bit signed displacement means that the jump can be to a location anywhere from +32,767 to –32,768 bytes from the current instruction pointer location. A positive displacement usually means jump is *"ahead"* in the program, and a negative displacement usually jump is *"backward"* in the program.

Note. A special case of the direct near jump instruction is direct short jump. If the destination for the jump is within a displacement range of +127 to –128 bytes from the current instruction pointer location, the destination can be reached with just an 8 bit displacement.

(*b*) Conditional Transfer Instructions

These instruction will cause a jump to a label given in the instruction when the desired conditions(s) occurs in the program before the execution of the instruction. The destination must be in the range of –128 bytes to +127 bytes from the address of the instruction after the conditional transfer instruction. If the jump is not taken, execution simply goes on the next instruction. Following table depicts different conditional transfer instructions with description and condition for jump.

Table 1

Instruction Code	Description/Explanation	Condition for Jump
JA/JNBE	Jump if above/Jump if not below nor equal	CF = 0 and ZF = 0
JAE/JNB	Jump if above or equal/Jump if not below	CF = 0 and ZF = 1
JB/JNAE	Jump if below/Jump if not above nor equal	CF = 1 and ZF = 0
JBE/JNA	Jump if below or equal/Jump if not above	CF = 1 and ZF = 1
JC	Jump if carry flag = 1	CF = 1
JE/JZ	Jump if equal/Jump if zero flag	ZF = 1
JG/JNLE	Jump is greater/Jump if not less than nor equal	ZF = 0 and CF = OF
JGE/JNL	Jump if greater than or equal/Jump if not less than	SF = 0F
JL/JNGE	Jump if less than/Jump if not greater than nor equal	SF = 0F
JGE/JNG	Jump if less than or equal/Jump if not greater than	ZF = 1 or SF ≠ OF
JNC	Jump if no carry	CF = 0
JNE/JNZ	Jump if not equal/Jump if not zero	ZF = 0
JNO	Jump if no overflow	OF = 0
JNP/JPO	Jump if not parity/Jump if parity odd	PF = 0
JNS	Jump if not sign	SF = 0
JO	Jump if overflow flag = 1	OF = 1
JP/JPE	Jump if parity/Jump if parity even	PF = 1
JS	Jump if sign flag = 1	SF = 1
JCXZ	Jump if CX is zero	CX = 0

Note that the terms greater and less are used to refer to the relationship of two signed numbers.

4.4.5 Iteration Control Instructions

These instructions are used to execute a series of instructions for specific number of times. The number is specified in the CX register, which is automatically decremented by one, each time after execution of LOOP instruction. Until CX = 0, execution will jump to a destination specified by a label in the instruction.

The destination address for the jump must be in the range of –128 bytes to +127 bytes from the address of the instruction after the iteration control instruction. For LOOPE/LOOPZ and LOOPNE/LOOPNZ instructions there is one more condition for exit from loop, which is shown in the following table. If the loop is not taken, execution simply goes on to the next instruction after the iteration control instruction.

Table 2

Instruction Code	Description/Explanation	Condition for Exit
1. LOOP	Loop through a sequence of instructions	CX = 0
2. LOOPE/LOOPZ	Loop through a sequence of instructions	CX = 0 or ZF = 0
3. LOOPNE/LOOPNZ	Loop through a sequence of instructions	CX = 0 or ZF = 1

4.4.6 Processor Control Instructions

Different processor control instructions are briefly explained as below :

- **STC Instruction :**

  This instruction sets the carry flag, STC does not affect any other flag.

- **CLC Instruction :**

  This instruction resets the carry flag to zero. CLC does not affect any other flag.

- **CMC Instruction :**

  This instructions complements the carry flag. CMC does not affect any other flag.

- **STD Instruction :**

  This instruction is used to set the direction flag to one so that SI and/or DI can be decremented automatically after execution of string instructions. STD does not affect any other flag.

- **CLD Instruction :**

  This instruction is used to reset the direction flag to zero, so that SI and/or DI can be incremented automatically after execution of string instructions. CLD does not after any other flag.

- **STI Instruction :**

  This instruction sets the interrupt flag to one. This enables INTR interrupt of the 8086. STI does not affect any other flag.

- **CLI Instruction :**

  This instruction resets the interrupt flag to zero. Due to this 8086 will not respond to an interrupt signals on its INTR input. CLI does not affect any other flag.

4.5 EXTERNAL HARDWARE SYNCHRONIZATION INSTRUCTIONS

In this type there are five instructions to be discussed, viz :

(*i*) HLT Instruction,

(*ii*) WAIT Instruction,

(*iii*) ESC Instruction,

(*iv*) LOCK Instruction and

(*v*) NOP Instruction.

(*i*) HLT Instruction

HLT instruction will cause the 8086 to stop fetching and executing instructions. The 8086 will enter a halt state. The only ways to get the processor out of the halt state are with an interrupt signal on the INTR pin, an interrupt signal on the NMI pin, or a reset signal on the RESET input.

(*ii*) WAIT Instruction

When this instruction executes, 8086 processor enters an idle condition where it is doing no processing. Processor 8086 will stay in this idle state until a signal is asserted on the 8086 TEST input pin, or until a valid interrupt signal is received on the INTR or the NMI interrupt input pins. If a valid interrupt occurs while the 8086 is in this idle state, the 8086 will return to the idle state after the interrupt service procedure executes. The WAIT instruction affects no flags. The WAIT instructions is used to synchronize the 8086 with external hardware such as the 8087 math coprocessor.

(*iii*) ESC Instruction

ESC instruction is used to pass instructions to a coprocessor such as 8087 math coprocessor which shares the address and data bus with an 8086. Instructions for coprocessor are represented by a 6-bit code embedded in the escape instruction. When the 8086 fetches an ESC instruction, the coprocessor decodes the instruction and carries out the action specified by the 6-bit specified in the instruction. In most cases the 8086 treats the ESC instruction as a NOP. *(In some cases the 8086 will access a data item in memory for the coprocessor).*

(*iv*) LOCK Instruction

In a multiprocessor system each microprocessor will have its own local buses and memory. The individual microprocessors are connected together by a system bus so that each can access system resources such as disk drives or memory. Each microprocessors only takes control of the system bus when it needs to access some system resources. The LOCK prefix allows a microprocessor to make sure that another processor does not take control of the system bus while it is in the middle of a critical instruction. When an instruction with a LOCK prefix executes, the 8086 will assert its bus lock signal output. *(This signal is connected to an external bus controller device which then prevents any other processor from taking over the system bus. LOCK affects no flags).*

Examples :

Instruction	Explanation
LOCK XCHG SEMAPHORE, AL	; The XCHG instruction requires two bus accesses
	; The LOCK prefix prevents another processor from taking
	; control of the system bus between the two accesses.

(*v*) NOP Instruction

If NOP instruction is executed no operation is performed except fetch and decode. It takes three clock cycles to execute the instruction. NOP instruction does not affect any flag. This instruction is used to fill in time delays or to delete and insert instructions in the program while trouble shooting.

4.6 INTERRUPT INSTRUCTIONS : [INT, INTO, AND IRET]

(*i*) **INT Instruction :** INT Type

This instruction causes 8086 to call a far procedure. The term type in the instruction refers to a number between 0-255 which identifies the interrupt. The address of the procedure is taken from the memory whose address is four times the type number.

(*ii*) **INTO Instruction**

If the overflow flag is set then, this instruction will cause the 8086 to do an direct far call to a procedure you write to handle overflow condition. To do call 8086 will read a new value for IP from address 00010H and a new value of CS from address 00012H.

(*iii*) **IRET Instruction**

IRET instruction is used at the end of the interrupt service routine to return execution to the interrupted program. The 8086 copies return address from stack into IP and CS registers and the stored value of flags back to the flags register.

(The RET instruction does not copy the flags from the stack back to the flag register).

4.7 STRING INSTRUCTIONS

In this type there are six subdivision or types of instruction, which are listed as below :

(*i*) REP/REPE/REPZ/REPNE/REPNZ Instruction,

(*ii*) MOVS/MOVSB/MOVSW Instruction,

(*iii*) CMPS/CMPSB/CMPSW Instruction,

(*iv*) SCAS/SCASB/SCASW Instruction,

(*v*) LODS/LODSB/LODSW Instruction, and

(*vi*) STOS/STOSB/STOSW Instruction.

(*i*) **REP/REPE/REPZ/REPNE/REPNZ Instruction**

The REP instruction is a prefix which is written before one of the string instructions. These instructions repeat until specified conditions exists, as mentioned below :

Table 3

Instruction Code	Condition for Exit
REP	CX = 0
REPE/REPZ	CX = 0 OR ZF = 0
REPNE/REPNZ	CX = 0 OR ZF = 1

Examples :

Instruction	Explanation
REPZ CMP SB	; Compare string bytes until CX = 0
	; or until string bytes not equal

(ii) MOVS/MOVSB/MOVSW Instruction

This instruction copies a byte or word from a location in the data segment to a location in the extra segment. The offset of the source byte or word in the data segment must be in the SI register, and the offset of the destination in the extra segment must be contained in the DI register.

In the case of multiple byte or multiple word moves the number of elements to be moved in put in the CX register so that it can function as a counter. Later the byte or word is moved SI and DI are automatically adjusted to point to the next source and the next destination. When the direction flag is 0, then SI and DI will be incremented by 1 after a byte move and they will incremented by 2 after a word move. If the DF is a 1, then SI and DI will be decremented by a after a byte move and they will be decremented by 2 after a word move. MOVS affects no flags.

The way to say that the assembler whether to code the instruction for a byte or word move is to add a *"B"* or a *"W"* to the MOVS mnemonic. MOVSB, for instant says move a string as bytes. MOVSW says move a string as words.

Examples :

Instruction	Explanation
CLD	; Clear Direction Flag to autoincrement SI and DI
MOV AX, 00H	
MOV DS, AX	; Initialize data segment register to 0
MOV ES, AX	; Initialize extra segment register to 0
MOV SI, 2000H	; Load offset of start of source string into SI
MOV DI, 2400H	; Load offset of start of destination into DI
MOV CX, 04H	; Load length of string in CS as counter
REP MOVSB	; Decrement CX and MOVSB until CX will be 0

Note. After move SI will be one greater than offset of last byte in source string. DI will be one greater than offset of last byte of destination string. CX will be 0.

(iii) CMPS/CMPSB/CMPSW Instruction

A string is said to be a series of the same type of data items is sequential memory locations. The CAMPS instruction can be used to compare a byte in one string with a byte in another string or to compare a word in one string with a word in another string. SI is used to hold the offset of a byte or word in the source string and DI is used to hold the offset of a byte or a word in the other string. Comparison is done by subtracting the byte or word pointed to by DI from te byte or word pointed to by SI.

Note that the AF, CF, OF, PR, SF and ZF flags are affected by the comparison, but neither operand is affected.

Examples :

Instruction	Explanation
	; Point SI at source string, Point DI at destination string
MOV SI, OFFSET F_STRING	
MOV DI, OFFSET S_STRING	
CLD	; DF cleared so SI and DI will
	; autoincrement after compare
CMPS F_STRING, S_STRING	; The assembler uses names to determine whether
	; strings were declared as type byte or as type word
MOV CX, 100	; Put number of string elements in CX, Point SI at
	; source of string and DI at destination of string
MOV SI, OFFSET F_STRING	
MOV DI, OFFSET S_STRING	
STD	; DF set so SI and DI will autodecrement after compare
REPE CMPSB	; Repeat the comparison of string bytes until and
	; of string or until compared bytes are not equal

Note : After comparison SI and DI will be automatically incremented or decremented according to direction flag to point to the next element in the two string *(if DF = 0, SI and DI ↑ otherwise ↓)* CX functions as a counter which is decremented after each comparison. This will go on until CX = 0.

Examples :

Instruction	Explanation
	; Point SI at source string, Point DI destination string
MOV SI, OFFSET F_STRING	
MOV DI, OFFSET S_STRING	
CLD	; DF cleared so SI and DI will
	; autoincrement after compare
CMPS F_STRING, S_STRING	; The assembler uses names to determine
whether	
MOV CX, 100	; Put number of string elements in CX, Point SI at
	; source of string and DI at destination of string
MOV SI, OFFSET F_STRING	
MOV DI, OFFSET S_STRING	
STD	; DF set so SI and DI will autodecrement after compare
REPE CMPSB	; Repeat the comparison of string bytes until end
	; of string or until compared bytes are not equal

(iv) SCAS/SCASB/SCASW Instruction

The SCAS instruction compares a string byte with a byte in AL or a string word with word in AX. The instruction affects te flags, but it does not change either the operand in AL (AX) or the operand in the string. The string to be scanned must be in the extra segment and DI must contain the offset of the byte or the word to be compared.

Note. After comparison DI will be automatically incremented or decremented according to direction flag to point to the next element in the two strings (if DF = 0, SI and DI ↑ otherwise ↓) CX functions as a counter which is decremented after each comparison. This will (continue, till) CX = 0. SCAS affects the AF, CF, OF, PF, SF and ZF flags.

Examples :

Instruction	Explanation
	; Scan a text string of 80 characters
	; for a carriage return
MOV AL, 0DH	; Byte to be scanned for into AL
MOV DI, OFFSET TEST_STRING	; Offset of string to DI
MOV CX, 80	; CX used as element counter
CLD	; Clear DF, so DI autoincrements
REPNE SCAS TEXT_STRING	; Compare byte in string with byte in AL.

(v) LODS/LODSB/LODSW Instruction

This instruction copies a byte from a string location pointed to by SI and AL, or a word from a string location pointed to by SI to AX, LODS instruction does not affect any flags.

Examples :

Instruction	Explanation
CLD	; Clear direction flag so SI is autoincremented
MOV SI, OFFSET S_STRING	; Point SI at string
LODS S_STRING	

(vi) STOS/STOSB/STOSW Instruction

STOS instruction copies a byte from AL or a word from AX to a memory location in the extra segment. DI is used to hold the offset of the memory location in the extra segment. After the copy, DI is automatically incremented or decremented to point to the next string element in memory. When the direction flag, DF, is cleared, then DI will automatically be incremented by one for a byte string or incremented by two for a word string. When the direction flag is set, DI will be automatically decremented by one for a byte string or decremented by two for a word string. STOS instruction does not affect any flags.

Examples :

Instruction	Explanation
MOV DI, OFFSET D_STRING	; Point DI at destination string
STOS D_STRING	; Assembler uses string name to determine whether string
	; is of type byte or type wore. If byte string, then string

```
                           ; byte replaced with contents of AL. If word string
                           ; then string word replaced with contents of AX
MOV DI, OFFSET D_STRING    ; Point DI at destination string
STOSB                      ; "B" added to STOS mnemonic directly tells assembler
                           ; to replace byte in string with byte from AL. STOSW
                           ; would tell assembler directly to replace a word
                           ; in the string with a word from AX
```

REVIEW QUESTIONS

1. Briefly describe the need for addressing modes in a processor.
2. Compair the addressing modes of 8085 (8-bit) and 8086 (16-bit) microprocessors.
3. Using an example explain :
   (*a*) Immediate Addressing
   (*b*) Register Addressing, and
   (*c*) Direct Addressing Modes.
4. What are the type of Addressing modes for ?
   (*i*) MOV CX
   (*ii*) ROL AX, CL
   (*iii*) ADD AX, [BX] [SI]
   (*iv*) JMP 023 AH
   (*v*) INAL, DX
5. What is the difference in index addressing and base addressing ?
6. Explain the usefulness of the following instructions in 8086 processor.
   (*i*) TEST (*ii*) LES (*iii*) LOCK (*iv*) XLAT
7. Mention two instructions that clear AX register to zero.
8. Which segment register cannot be popped from the stack ?
9. Describe how XLAT instruction transforms the contents of AL register.
10. Write the difference between following instructions :
    (*a*) MOV BL, 437AH and MOV BL, DS : BYTE PTR [437AH]
    (*b*) MOVCX, 437AH and MOV CX, [437 AH]
11. With the help of an example explain the action performed by mP 8086 for the following instructions :
    (*i*) CMP SB (*ii*) AAM (*iii*) ROL (*iv*) IMUL
12. Explain the difference between a jump and a call instruction. What does the processor do in executing it ?
13. What is a iteration instruction ? Explain.

OBJECTIVE TYPE QUESTIONS

1. *If a JMP instruction is stored at memory location 100H within the current code segment, it cannot be a .... jump if it is jumping to memory to action 200H within the current code segment.*
2. *The instruction pointer length of 8088 is ....*
   (i) 8 bits (ii) 32 bits (iii) 4 bits (iv) 16 bits
3. *By a separate segment resister, which of the following is not addressed.*
   (i) Code (ii) Data (iii) Stack (iv) all have separate segment registers
4. *Separate ALU is used with .... register to perform memory location calculation.*
   (i) Segment (ii) Flag (iii) General purpose resistor
   (iv) Instruction pointer (v) all
5. *Segment override may not be used in .... and ....*
6. *A way to use two stacks and quickly switch between two stack pointer is to use .... instruction.*
   (i) POP (ii) MOV (iii) XCHG (iv) PUSH
7. *............. register cannot be changed by using MOV instruction.*
8. *Immediate data cannot be loaded into .... register.*
9. *8086 TEST instruction performs same function as .... instruction but without storing any result effecting condition flags only.*
   (i) OR (ii) XOR (iii) SUB (iv) AND

5 TECHNIQUES FOR ASSEMBLY LANGUAGE PROGRAMMING

5.0 OBJECTIVE

After completion of this chapter you will know program segments, application of assembly language and its development. You will understand what is meant by program development and tools. Apart you will be familiar with program formatting and Assembler directives.

5.1 INTRODUCTION TO ASSEMBLY LANGUAGE PROGRAMMING

The Assembly Language Programming continues to be critical in system programming the advancement in high level programming. The programs written in this language are compact and efficient, which has to be converted to machine code for execution and it is performed by the translator known as assembler. Assembly language programming needs good knowledge of machine architecture, operating system, and programming principles. Assembly language is case insensitive, therefore program can be coded either in uppercase, lowercase, or combination of lower/uppercase characters, which will be discussed.

The program segments, procedures, and macro programming has to be discussed at great length. The program development tools such as editor, assembler, linker, etc., are also to be discussed with the program development process.

Next it also covers software support available on the IBM-PCs with access methods and examples.

5.2 PROGRAM SEGMENTS

8086 microprocessor organizes main memory as segments of 64Kb size and also provides with four segment registers *(CS, DS, SS and ES)* to access these segments. The 8086 processor organizes the segments into four groups as mentioned below :

(*a*) Code Segment

(*b*) Data Segment

(*c*) Stack Segment

(*d*) Extra Segment

The segments code, data, stack, and extra are pointed to by the segment registers CS, DS, SS and ES respectively. Code segment is used to store the program code, the data, stack and extra segments are normally used to store data used in the program. The code segment can also be used to store the program instructions and data *i.e.*, the code and data segment register *(CS, DS)* points to the same segment.

5.2.1 Definition of Segment

The assembler directives SEGMENT and ENDS are made use to define a segment. The directive SEGMENT is made used to indicate the beginning of a logical segment and the directive ENDS informs the end of the segment to the assembler in its processes. The directives SEGMENT and ENDS must enclose the data or code of a segment and next both the SEGMENT and ENDS follows the name of the segment.

The General Form

```
        Segment Name     SEGMENT [WORD/PUBLIC]
        ..
        ' Storage definition, allocation,
        ; Program code, and alignment directives
        ..
Segment Name ENDS
```

where :

- Segment Name is the name of the user defined segment
- WORD/PUBLIC is the type specifiers

5.2.2 Multiple Data and Code Segments

The data, stack and extra segments are normally used to store data used in the program and the code segment is used to store instructions of a program. The program having data and code which cannot fit into a segment of 64Kb must be define in multiple segments. The branch to a procedure located in a different segment is made by using FAR call. The outline of a segment mentioned below indicates the arrangement of the two code segments.

```
_MYCODE1  SEGMENT
          ASSUME CS : _MYCODE1,     DS :_DATA
          ..
          CALL FACTORIAL
          ..
_MYCODE1  ENDS
_MYCODE2  SEGMENT
          FACTORIAL PROC FAR
```

```
          ASSUME CS : _MYCODE2
          ..
          RET
          FACTORIAL ENDP
_MYCODE2  ENDS
```

Note. In the case of multiple segments, the segment registers must be explicitly setup. The procedures defined in a different segment must be defined as the FAR procedure, so that the assembler recognizes a call to that procedure and codes the procedure call as FAR cell.

Number of code and data segments that can be referenced in a program is defined by the memory model. The data or procedure code located in the same segment is referenced using two byte address known as NEAR pointer. The data or a procedure code located in a different segment is referenced using four byte address *(explicit segment and offset value)* known as the FAR pointer. The memory model specifies the number of segments that can be used in a program. The available memory models which are shown in Table 1 as under.

Table 1. Program Memory Models

Memory Model	Description
Tiny	Both code and data must fit in a single 64Kb segment
	Code and data reference are NEAR
Small	Code must fit within a single 64Kb segment
	Data must fit within a single 64Kb segment
	Code and data reference are NEAR
Medium	Program code can be larger than 64Kb segment
	Data must fit within a single 64Kb segment
	Code reference is FAR and data reference is NEAR
Compact	Code must fit within a single 64KB segment
	Program data can be larger than 64Kb segment
	Code reference is NEAR and data reference is FAR
Large	Both program code and data may be larger than 64Kb segment
	Code reference is FAR and data reference is FAR
	Single data array cannot be larger than 64Kb
Huge	Both program code and data may be larger than 64Kb segment
	Code reference is FAR and data reference is FAR
	Pointers to elements of an array are also FAR
	and data array may exceed 64Kb in size

5.2.3 End of Program

Physical end of the program is indicated by the directive END. The directive END is generally followed by the label, which indicates the point at which execution of the program begins. Such concept is illustrated as :

```
_CODE SEGMENT
      ASSUME CS : _CODE
START :
   ..
_CODE ENDS
      END START
```

The directive END with label is used in the main program of a multiprogram module. *(Other than main program module, all other modules must end with the directive END without label).*

5.2.4 Program Termination

Directive END is not a command to terminate the execution of the program but it is just an assembler directive indicating the end of the source code in a program module. The program termination must be done explicitly using the program termination system call of the operating system which is explained with an example as :

Example 1 :

```
_CODE SEGMENT
      ASSUME CS : _CODE
START :
      ; Program code
      ..
      ; Program termination
      MOV  AH, 4CH             ; program terminate function
      MOV  AL, 004H            ; return code
      INT 21H                  ; call DOS service
_CODE ENDS
      END START
```

The instruction HLT may be used to stop execution of the program, however this does not transfer control back to the loader or the operating system shell for processing next command of the user. Instruction HLT is used as a program termination command in a microprocessor-kit or subsystem designed using 8086 processor.

Note. In case of the microcomputer *(IMB-PC),* the system does not hait after the execution of the program instead, it returns to the operating system shell which accepts the next command issued by the user.

Assembly language has very few restrictions rules ; *(nearly everything is left to the discretion of the programmer),* while high level language details various semantically and syntactical restrictions.

5.3 APPLICATION OF AN ASSEMBLY LANGUAGE

Usually programmers will not write large application program in assembly language, but they write short, specific routines, subroutines in assembly language and call them from a high level language program. *(Assembly language subroutines can be written to handle operations that are not available in the higher level language).*

For instant consider writing a business application program in COBOL for IBM-PC : There are no commands in the COBOL language definition to do :

- Check for the free space on the disk,
  create a sub directory,
  write protect a file, and
  create overlapping windows.

But all these can be done within the COBOL program by writing assembly language subroutines.

To develop an Assembly language, the process or the steps to follow as listed below :

(*a*) Specify the source code as per the assembly language definition

(*b*) Assemble the program to create the object code

(*c*) Link the object code to create the executable code

(*d*) Test and debug the program

The Fig. 5.1 depicts the process of assembly language programming in the form of flow chart, when each step will result in creating one or more files with extensions, as shown :

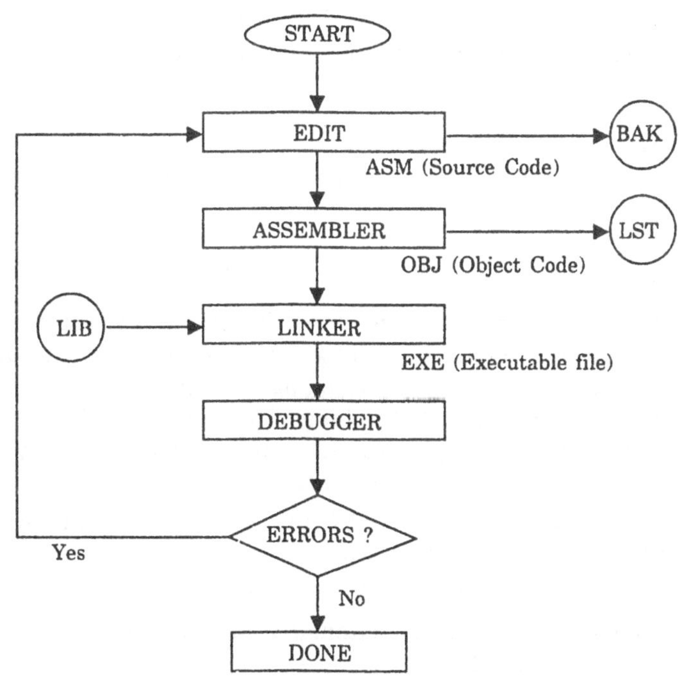

Fig. 5.1 Flow chart for Assembly language program.

Now let us briefly analyze each steps to develop assembly language, program *i.e.*,

(*i*) Creating Source Code

(*ii*) Assembling the Program

(*iii*) Listing File

(*iv*) Test and Debug

(*i*) Creating Source Code

Source code consists of 8086/8088 program memories, suitable pseudo opcodes and assembler directives. A file can be created with a text editor and is given an extension ASM. The text editor may be any Word Processor (like EDLIN, NE) which can produce standard ASCII *(unformatted text containing no control characters)* code.

(*ii*) Assembling the Program

For assembling the program assemblers are available for the IBM-PC,

(*a*) Microsoft Macro Assembler (MASM) and

(*b*) Borland Turbo Assembler (TASM)

A part doing tedious task of producing the binary codes for instruction statements, an assembler also allows to refer to data items by name rather than by numerical addresses, which makes the program much more readable. In addition to program instructions, the source program also contain directives to the assembler. The Pseudo instructions are assembler directives entered into the source code . ong with the assembly language.

Once the program is written completely, it can be assembled to obtain the OBJ file by executing MASM or TASM. The assembly language program file name should be mentioned along with the command, as shown below :

MASM <file name.ASM>; or TASM <file name.ASM>

The <file name.ASM> file contains the assembly language program is assembled and assembler generates error messages if there are any errors such as Syntax errors. Such errors are listed along with the line number and no errors then OBJ file is created. *(To obtain the EXE file the user has to LINK the OBJ file).*

LINK <file name>; or TLINK <file name>;

When the file is smaller than 64K bytes, it can be converted from an execution file to a command file *(.COM)*. Which is slightly different from an execution file (.EXE). In a command file the program must be originated at location 100H before it can execute. Which this means that the program must be no longer than *(64K-100H)* on length. The command file requires less space in memory than the equivalent execution file. The system loads. COM file off the disk into the computer memory more quickly than the execution file.

Not to create a .COM file from a .EXE file, we require EXE2BIn converter EXE2BIN converts .EXE file to .COM or binary file.

Example 1 : *EXE2BIN <filename> <filename.com>*

Here, the <filename> with an EXE extension in converted to <filename> with .COM extension with above command.

(*iii*) Listing File

In general, Turbo assembler produces only one file as the result of assembly : an object *(.OBJ)* file with the same as the source *(.ASM)* file. If required ask TURBO Assembler to produce a listing file with the extension. LST as well, simply required by typing two additional commas on the command line, as in following example :

Example 2 : TASM Hello,

Assembles Hello.ASM and produces the Object file Hello.OBJ

Example 3 : TASM Hello, ; Will Generate the listing file HELLO.LST along with the Hello.OBJ file. Similarly by using the command

TASM Hello, Or TASM/I

Hello generates listing file along with the .Obj file and so on.

(*iv*) Test and Debug

An executable program may be run under DOS or DEBUG with normal rule, run a program under DOS only when there is no error or it produces some not visible or audible result. When the program result is stored in registers or in memory, the result is visible. Therefore, it should be run using DEBUG or TD (Turbo Debugger) or code-view only. EXE files can be loaded into memory using DEBUG as :

Example 1 : DEBUG <filename.EXE>

By using DEBUG it is possible to find the bugs in the program and next the memory it is possible to check and correct the errors using different commands in DEBUG.

Now following Table 2 depicts few commands with format and action :

Table 2

S. No.	Command	Format and Action	
1	G – GO	**Format :**	G [offset], [offset]
		Action :	Executes a program starting at the current location. Offset values are temporary breakpoint instruction the processor stops and displays registers and flag contents
2	T – TRACE	**Format :**	T [Instruction Count]
		Action :	Executes one or more instructions and displays register and flag values for each of them
		Example :	T ; Executes only the next instruction
			T5 ; Executes the next 5 instructions.
3	P – PTRACE	**Format :**	P [Instruction Count]
		Action :	Same as Trace, but treats subroutine calls, interrupts, loop instructions, and repeat string instructions as a single instruction
4	Q – QUIT	**Format :**	Q
		Action :	Exists to DOS
5	D – DUMP	**Format :**	D seg : offset [offset] D seg : offset L byte count
		Action :	Displays the numeric contents of memory locations and their character equivalents (if any)

6	N – Name the Program	**Format :**	N <filename>
		Action :	Name the Program
7	W – Write the file to disk	**Format :**	W
		Action :	Bytes starting from the memory location whose address is provided by IP address and written as a COM file to the disk. The number of bytes that are to be stored is indicated by the contents of CX register. The name of the file is to be specified by means of the N command prior to executing the W command.
8	R – Register	**Format :**	R <register.name>
		Action :	The contents of the register are displayed additionally, the register content can replace by the value entered by the user. If no-register name is provided, the contents of all the registers are displayed.
9	A – Assemble	**Format :**	A <CS : offset>
		Action :	This command allows us to enter the assembler mnemonics directly
10	U – Unassemble	**Format**	U <CS : offset>
		Action :	This command lists a program from the memory. The memory start location is specified by CS : offset.
11	L – Load	**Format :**	L [address] [drive] [first sector] [number]
		Action :	Reads sectors from the disk into memory. The memory start address is provided in the command.
12	E – Enter	**Format :**	E <address> [list]
		Action :	It enables us to change the contents of the specified memory location. List is an optional data that has to be entered.

Now observe five more ideas/techniques to develope a program and debug :

(*a*) Carefully define the program to solve the problem in hand and work out with best algorithm.

(*b*) If the program have several parts, write test and debug each part individually and then add parts one at a time.

(*c*) When a program or program section does not work then first recheck the algorithm to make sure it really does what you want it to.

(*d*) If the algorithm seems correct, then check to make sure that you have used the correct instructions to implement the algorithm. Work out on paper the effect that a series of instructions will have on some sample data. *(Such predictions on paper can later be compared with the actual results produced when the program section runs).*

(*e*) If you don't find a problem in the algorithm or the program instruction use debugger to help you localize the problem. Use single step or trace for short program sections. In the case of longer programs use breakpoints, which is often a faster technique to narrow the source of a problem down to a small region.

5.4 THE PROGRAM DEVELOPMENT TOOLS

The Program development is a process of program design, development, implementation, translation, testing debugging, and maintenance. Availability of right tools makes the program development process easier. The program development tools are listed below :

(*a*) Text Editor

(*b*) Pre-Processor

(*c*) Assembler

(*d*) Compiler

(*e*) Linker

(*f*) Loader

(*g*) Debugger

(*h*) EXE to COM Converter

(*i*) Library Builder

Now let us observe the model of each tool, in brief.

(*a*) Text Editor. A Text Editor is defined as a software that allows the user to enter program and modify the program and the program created using such text editor is known as source program. The editor provides a set of commands for insertion, deletion, modification of a single letter, statement, or block of statements, etc. Text Editor model as is shown in Fig. 5.2.

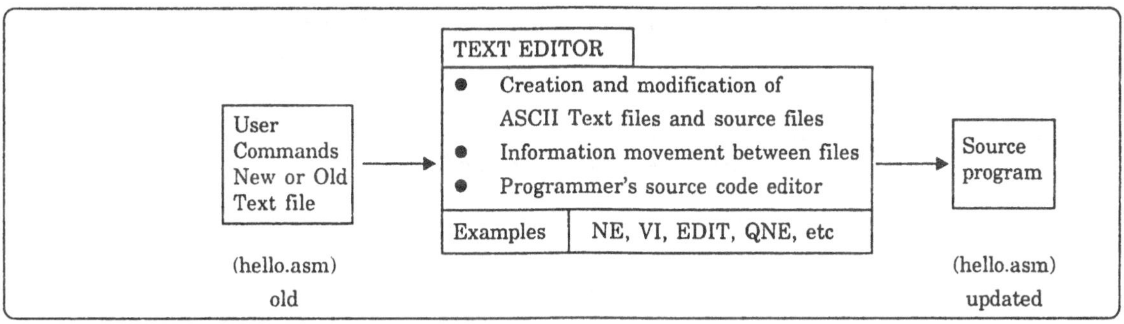

Fig. 5.2 Text Editor Model.

Examples of editors are Norton Editor (NE), DOS editor (EDIT), Non Document Mode of Wordstar (WS), Visual Editor (VI), etc.

(*b*) Pre-Processor. A Pre-Processor is a software that translates a source program with macros and include header files to the source file with the macros resolved and header files included in the source program. Pre-Processors output is then transferred to the next phase of the program translation. Normally, pre-processors are built into the assemblers or the compliers called as microprocessors. The model of a pre-processor is as shown in Fig. 5.3 below :

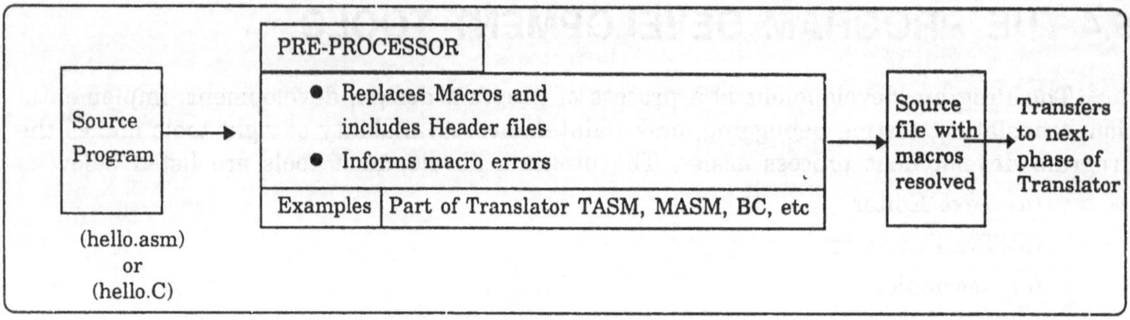

Fig. 5.3 Pre-processor Model.

Examples of pre-processors are built-in Borland's Turbo assembler (TASM), Microsoft's Macro assembler (MASM), High Level Language Compilers such as Borland C, Microsoft C, etc.

***(c)* Assembler.** An Assembler is a software that translates assembly language program to a machine language program and the input to the assembler is known as a source file. Assembler produces the output as relocatable object code and reports program errors during translation. The model of an assembler is as shown in Fig. 5.4.

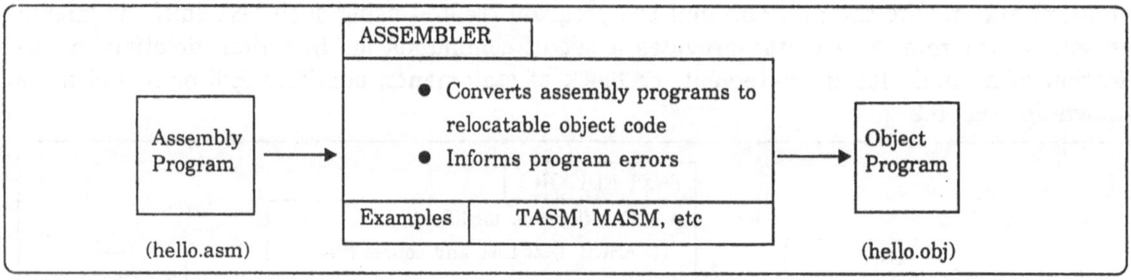

Fig. 5.4 Assembler Model.

Examples of assemblers are Borland's Turbo assembler (TASM), Microsoft's Macro assembler (MASM), etc.

***(d)* Compiler.** Compiler is a software that translates high level language program to an equivalent object code in a relocatable form and the input to the compiler is a source file. Compiler produces the output as a relocatable object code and reports program errors during translation. The model of a compiler is as shown in Fig. 5.5

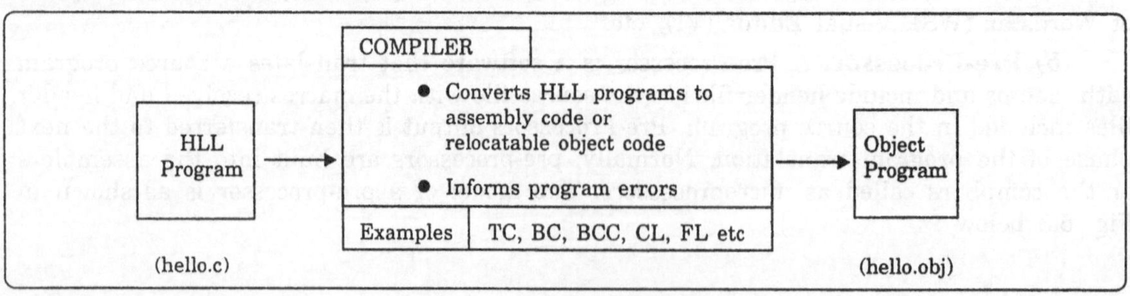

Fig. 5.5 Compiler Model.

Example of compilers are WATEFOR for FORTRAN, Borland or Microsoft compiler for C.Turbo Pascal, etc.

(*e*) **Linker.** A linker is a software that combines relocatable object program modules and library functions into an executable file and combines library functions referenced in the program and other program modules and produces a single executable file. It resolves the external symbols *(variables)* referenced in a module which is defined in other modules next it reports linking errors if external symbols cannot be resolved. The model of a linker is as shown in Fig. 5.6.

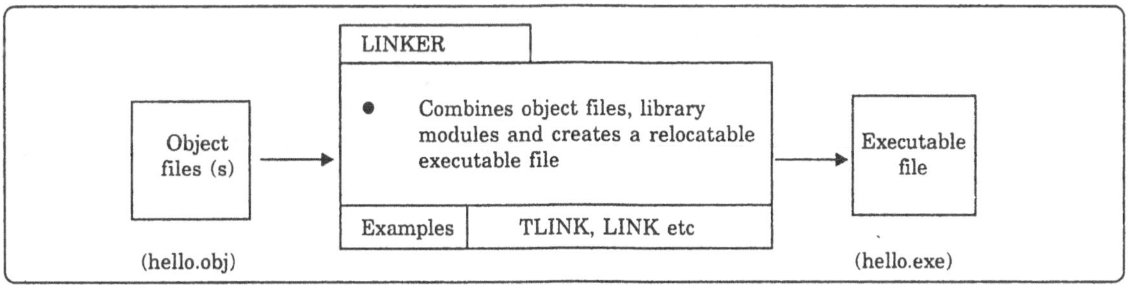

Fig. 5.6 Linker Model.

Examples of linkers are Borland's Turbo linker (TLINK), Microsoft's linker (LINK), etc.

(*f*) **Loader.** A Loader is a software that loads the executable file produced by a linker into the main memory for execution. After loading the program in main memory, the loader transfers control to the program for execution. The model of a loader is shown in Fig. 5.7. The operating system shell functions as a loader.

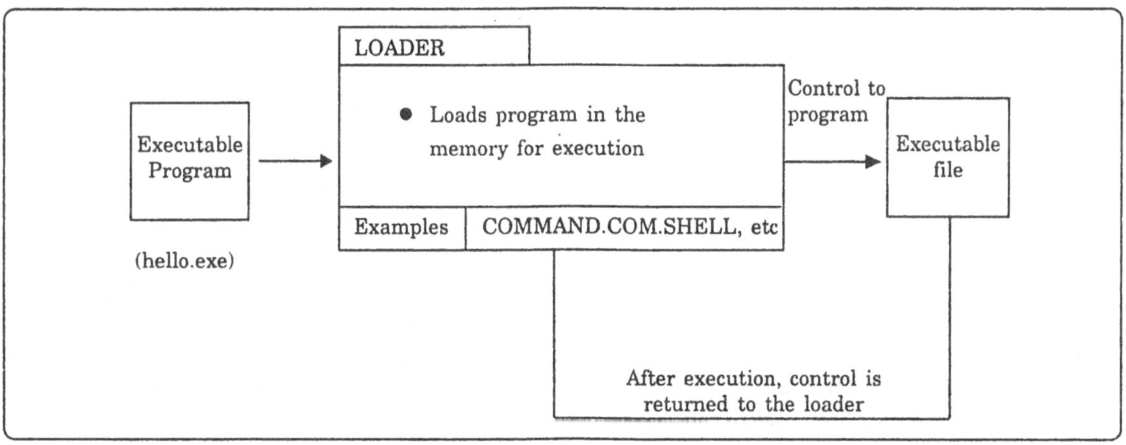

Fig. 5.7 Model of Loader.

The example of loader is Ms.DOS shell COMMAND.COM.

(*g*) **Debugger.** The Debugger is also a software that allows the execution of a program in single step mode under control of the user. The process of locating and correcting an error using a debugger is known as debugging, it executes the object program generated by an assembler or compiler with debug options. The debuggers that facilitates program debugging at source level are known as source level debuggers. The debuggers provide single stepping, break point, and program tracing facility. The model of such a debugger is as shown in Fig. 5.8.

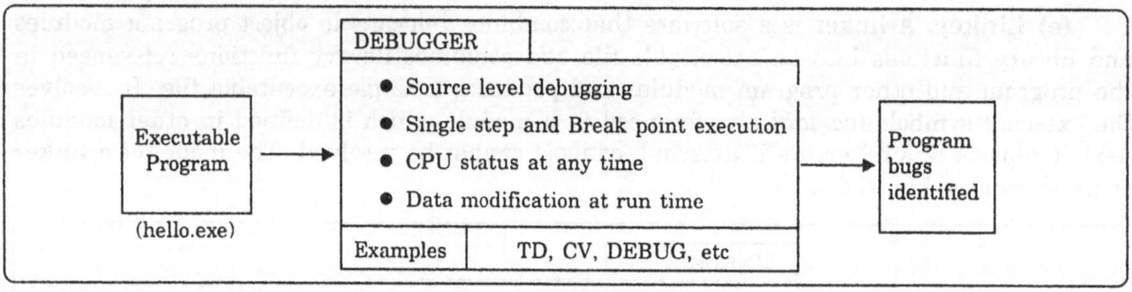

Fig. 5.8 Debugger Model.

Examples of debugger are Borland's Turbo debugger (TD), Microsoft's debugger known as Code View (CV) etc.

***(h)* EXE to COM Converter.** EXE to COM converter is again a software that converts executable file (EXE) to a command file (COM). The model of a COM converter is shown in Fig. 5.9 below :

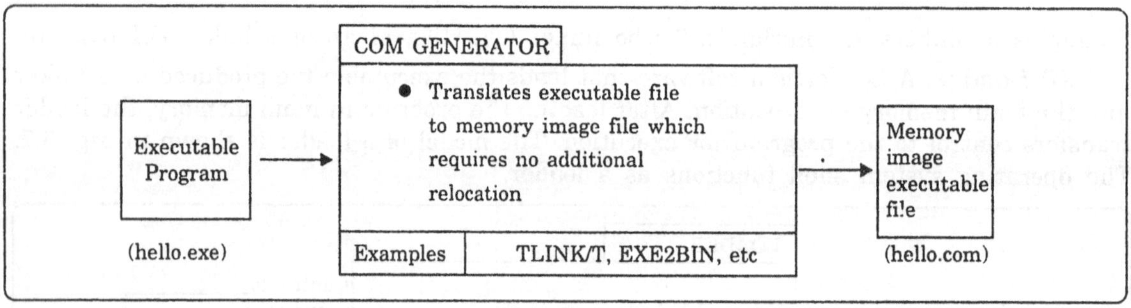

Fig. 5.9 COM Generator Model.

Example of EXE to COM converters are Microsoft's EXE2BIN utility, Borland's Turbo linker with tiny memory model option (TLINK/T), etc.

***(i)* Library Builder.** Library Builder is a software that combines all the program modules into a single file known as library. The library can be shared by multiple users for program development. It includes very frequently used functions. The compiler developers supply the built-in functions in the form of a library. The model of a library builder is as shown below in Fig. 5.10.

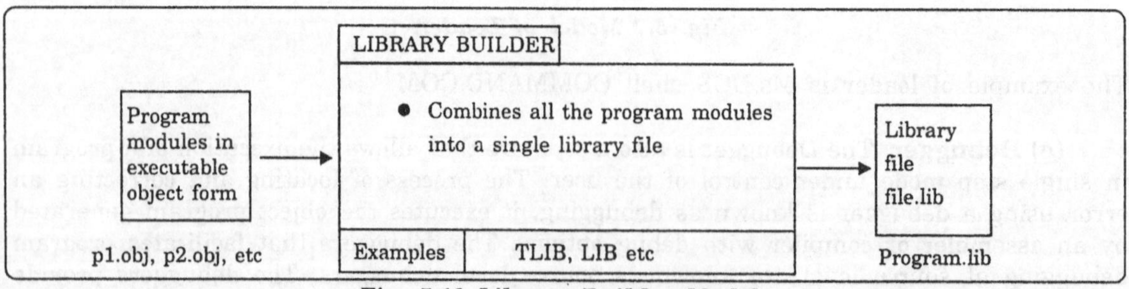

Fig. 5.10 Library Builder Model.

Examples of library builder are Microsoft's LIB, Borland's TLIB utilities, etc.

5.5 PROCESS FOR PROGRAM DEVELOPMENT

The Program development is the process of analysis, design, development, implementation, translation, testing debugging, and maintenance of the program. Which is an interactive and iterative process. There are six steps involved in program development as mentioned below :

(*a*) Source file Creation
(*b*) Object Code Generation
(*c*) Executable file Creation
(*d*) Program Running
(*e*) Program Testing
(*f*) Program Debugging

The Borland C has program development environment known as Integrated Development Environment *(IDE)*, which is a menu driven environment from which the development tools can be invoked. Fig. 5.11 depicts the program development process in the form of flow chart.

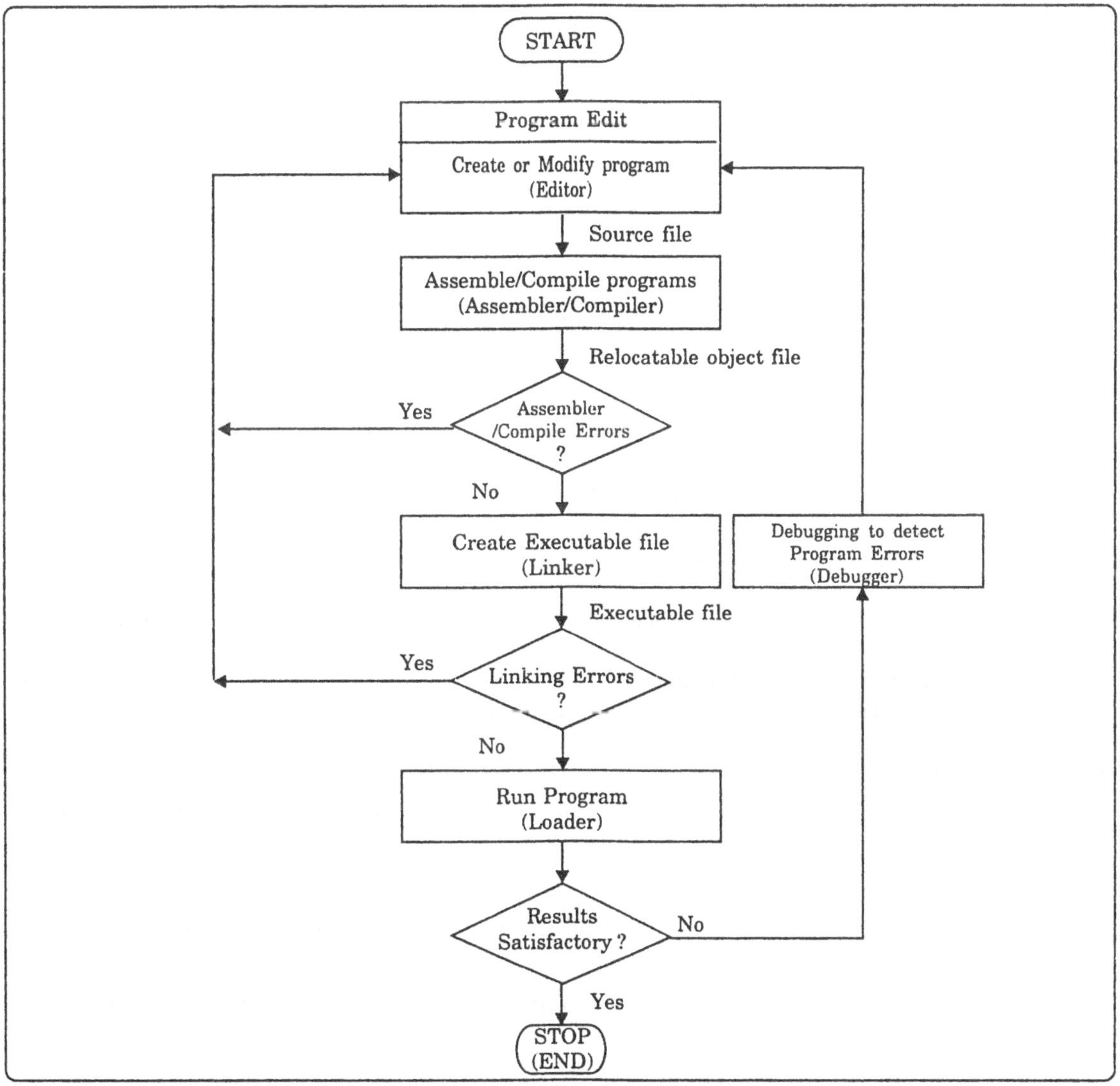

Fig. 5.11 Flow chart for Development process of a program.

Note. Borland program development tools are made used in explaining the program development process. The assembly program name HELLOA.ASM and C program name HELLOC.C are used in the explanation as source files.

Now let us briefly analyse the 6 steps, which are involved in program development.

(*a*) Source file Creation

A file containing program statements in assembly or high level language is known as 'source file'. Which is created and edited using text editors. The assembly language source file has an extension ASM *(Eg. HELLO.ASM)* and C language source file has an extension C (Eg. HELLO.C). *(The C source file can be created in an integrated program development editor).*

(*b*) Object Code Generation

Language translators are made use to translate source program into a relocatable object file. An assembler used in the translation of assembly language source code and the compiler is used in translation of high level language source code such as C. The commands for object code generation are as mentioned below.

Assembly source file : TASM HELLOA.ASM

Borland C source file : BCC HELLOC. C or compile option selection in IDE

An assembler generates a file HELLOA.OBJ and compiler generates HELLOC.OBJ if no errors are encountered during translation.

(*c*) Executable file Creation

A Linker is used to create executable files. The same linker is also used to translate assembly and C language object files into executable files.

Assembly Object file : TLINK HELLOA.OBJ
OR : TLINK HELLOA.OBJ
OR : TLINK HELLOA

Borland C Object file : TLINK HELLOC
OR : TLINK HELLOC.OBJ
OR : TLINK HELLOC

Note. If the object file extension is not specified then OBJ is considered as default extension.

Above commands generate an executable file HELLOA.EXE for assembly and generates an executable file HELLOC.EXE if no errors are encountered during linking.

(*d*) Program Running

The executable files can run by entering the name of the executable file mentioned below :

C> HELLOA

{The message displayed by the program on the screen is shown here}

C> HELLOC

{The message displayed by the program on the screen is shown here}

(e) Program Testing

The results generated by the program have to be tested for their validity and the result must confirm to the program specification. If there are any errors in the result, then the program must be debugged.

(f) Program Debugging

The errors in the program may be located using debuggers. The executable file of the program to be debugged must be created with program debug options. The Turbo assembler uses/ZI option and linker uses N option to create an executable file which can be debugged at the source level.

Assembling : TASM/ZI HELLOA.ASM

Linking : TLINK/ V HELLOA.OBJ

5.6 DEVELOPMENT OF PROGRAM

While developing a program first it is required to know, what is the task of the program ? It is just similar to thinking of a problem. Best method is to write what program has to do and the step/order in which it has to do. Now let us develop few steps in generally *(called as algorithm).* As seen in Fig. 5.11 a flow chart will provide a graphic shape to represent different types of program operation. The steps those required to convert an algorithm to assembly language are :

- Set up and declare data structure for the algorithm required to work with it.
- Write down the instructions required for initialization at starting of the code section.
- Mention the instructions required to implement the major actions taken in the algorithm, and decide how the data must be positioned for these instructions.
- Insert the instructions required to fetch the data in correct position.

5.6.1 Instruction Format for Assembler

In Fig. 5.4 model of assembler is shown the general format of assembler instruction will be :

Label : Opcode and Operand, Mnemonic Operand, Operand ; Comments

Spaces between label, opcode, operands, mnemonics and comments are arbitrary, except that at least one space one space must be inserted if no space would lead to an ambiguity *(e.g., between the mnemonic and first operand).* There may be no spaces within a mnemonic or identifier and spaces within string constants or comments will be included as space characters. Each statement in the program fields.

Now let us define the terms will have its own :

- Label,
- Opcode and Operand,
- Mnemonic,
- Operand, and
- Comments

■ **Label.** It is an identifier which is assigned the address of the first byte of the instruction in which it appears. Presence of a label in an instruction is optional, but, if present, the label provides a symbolic name that can be used in branch instructions to branch to the instruction. When there is no label, then the colon must not be entered. All labels begin with a letter or one of following special character : @, $, _ or ?. A label may be any length from 1 to 35 characters, and appears in a program to identify the name of a memory location for storing data and for other purposes.

■ **Opcode and Operands.** Opcode field is designed to hold the instruction opcode. To the right of the opcode field is the operand field, which contains information used by opcode.

■ **Mnemonic.** All instructions has to contain a mnemonic, which specifies the operation to be executed.

■ **Operand.** Presence of operands depends on the instruction. But some instructions have no operands, some have one operand, and some have two. When there are two operands, they are separated by a comma.

■ **Comments.** The comment file is for commenting the program and may contain any combination of characters, which is optional and if it is deleted the semicolon may also be deleted. *(A comment may appear on a line by itself provided that the first character on the line is a semicolon).*

5.6.2 Program Format and Assembler Directives

Following example shows a typical assembler program which is constructed for 8086/8088 microprocessor.

Example :

```
Line 1  . MODEL SMALL        ; Select small model
Line 3  . Data               ; Indicates data segment
        .......
                }  .......   data declaration
        .......
link k  . Code               ; Indicates start of code segment

                }            Program body

Line n  End                  ; End of file
```

In the above example. MODEL directive selects a standard memory model for the assembly language program. A memory model can be thought of a standard blue print or configuration, which determines the way segments are linked together, where each memory model has a different set of restrictions as to the maximum space available for code and data. Important thing to know about models is that they affect the way that subroutines and data be reached by programs. In Table 1 summary of program memory models are listed.

Program running under DOS is divided into 3 primary segments, each of which holds different types of values.

Code Segment *(pointed to by CS)* contains program code,

the data segment *(pointed to by DS)* contains the program variables, and
the stack segment *(pointed to by SS)* contains the program stack.

- "DATA" directive *(line 2)* indicates the start of the data segment, contains program variables.
- "CODE" directive *(line k)* indicates the start of the code segment and end directive *(line n)* indicates end of the program file.

5.7 THE ASSEMBLER DIRECTIVES

The Pseudo instructions are directives entered into the source code along with the assembly language program. The Directives do not get translated into object code but are used as special instructions to the assembler to perform some special functions. Assemblers can recognize directives that assign memory space, assign address to tables, format the pages of the source code and so on.

Pseudo instruction *(directives)* are normally placed in the opcode field. If any label or data are required by the directives, they are placed in the label or operand field as necessary. Now let us briefly observe the usages of different directives.

■ The SEGMENT and ENDS Directive

These are used to identify a group of data items or a group of instructions that we want to put together in a particular segment. A group of data statements or a group of instruction statements combined between *"SEGMENT"* and *"ENDS"* directives is known as a logical segment. When we set up a logical segment, we give a name of our choice.

Example 1 :

DATA-WHERE SEGMENT

....

DATA WHERE ENDS

Such statement will set up a logical segment named DATA WHERE. Here the name has been chosen as DATA WHERE helps to remember that this logical segment contains data statements. The statements *"CODE-WHERE SEGMENT"* and *"CODE-WHERE ENDS"* set up a logical segment name CODE-WHERE which contains instruction statements.

Normally logical segment is not given a physical starting address when it is declared. After the program is assembled and perhaps linked with other assembled program modules, it is assigned the physical address then it is loaded in memory to run.

■ ASSUME Directive

Assume directive helps to tell the assembler the name of the logical segment it should use for a specified segment.

ASSUME CS : CODE-WHERE tells the assembler that the instructions for a program are in a logical segment named CODE-WHERE. ASSUME DS : DATA-WHERE tells the assembler that for any program instruction which refers to the data segment it should use the logical segment called DATA-WHERE.

Example 2 : MOV AX, [BS]

An assembler will know that the memory location referenced to by [BX] is in the logical segment DATA-WHERE. The assembler has to be told what to assume for any segment used in the program, *(assembler can be told the name of the logical segment used for the stack and extra segments also).*

■ END

END directive is put after the last statement of a program to tell the assembler that it is the end of the program module, then assembler will ignore any statement after the END directive; hence the END directive should be used at the very end of the program module. *(A carriage return is required after the END directive).*

■ EQU

An EQU *(EQUATE)* directive will help to give a name to some value or symbol. Every time the assembler finds the given name in the program it will replace the name with the value or symbol that is equated with the name.

Example 3 :

FACTOR EQU 03H-is written at the start of a program and later in the program, the instruction is ADD AL, FACTOR. When the assembler codes this instruction statement, it will code it as 'ADD AL, 03H'.

Advantage of using EQU in this way is that if FACTOR is used several times in a program, and the value has to be changed, all that has to be done is to change the EQU statement and reassemble the program. Then the assemblers will automatically put the new value each time it finds the name FACTOR. When a constant is used in the program instead of the EQU approach, then the value of the constant has to be found and changed at all the occurrences.

■ Data Definition Directives

The Data definition directives are used to allocate storage, based on the following predefined types are as shown in Table 3 below.

Table 3. Pre-defined Definition Directives

S. No.	Directive	Description	Byte	Attributes
1	DB	Define byte	1	Byte
2	DW	Define word	2	Word
3	DD	Define double word	4	Double word
4	DF, DP	Define Far Pointer	6	Far pointer
5	DQ	Define Quard word	8	Quard word
6	DT	Define Ten bytes	10	Ten byte

■ Define Byte [DB]

DB directive is used to define a byte-type variable, or to set aside one or more storage locations of type in memory. The syntax diagram shows that "name" is optional, and only one initial value will be needed. *(If more values are supplied, they must be separated by commas).*

[name] DB initial Value [, initial value] ....

The initial value can be one or more 8-bit numeric values, a string constant, a constant expression, or a question mark (?). If the value is signed, it has a range of –128 to +127; if unsigned, the range is 0 to 255.

Table 4

1	Char	DB	'A'	; ASCII character
2	Signed1	DB	–128	; Smallest Signed Value
3	Signed2	DB	+127	; Largest Signed Value
4	Unsigned1	DB	0	; Smallest Unsigned Value
5	Unsigned2	DB	255	; Largest Unsigned Value

■ Multiple Values

A list of 8-bit numbers can be grouped under a single label, they are specified by listing their values separated by commas.

Example 4 : *List DB 10, 20, 30, 40*

Let list be stored at offset 0000 : That mean 10 is stored at offset 0000, 20 at offset 0001, 30 at offset 0002, and 40 at offset 0003.

The Characters and integers are one and the same, as far as the assembler is concerned. Each constant can use a different radix. If a list of items is defined, numeric, character and string constants may be freely mixed. *(If a hexadecimal number begins with a letter (A-F), a leading zero is added to prevent the assembler from interpreting it as a label).*

Table 5

1	Char1	DB	'A'	; A ASCII character
2	Char	DB	41h	; Hexadecimal
3	Dec	DB	65	; Decimal
4	Bin	DB	0100001B	; Binary
5	Oct	DB	107q	; Octal
6	List1	DB	10, 20, 41h, 0101010B, 'C', 0FFH	

Variable content may be left undefined by using a question mark (?). Or a numeric expression can initialize a variable with a value that is calculated at assemble time.

Example 5 :

```
Count       DB   ?
Ages        DB   ?,?,?,?
Row size    DB   10*10
```

The string can be assigned to a variable in which case the variable stands for the address of just the first byte of the string. Following example 6 shows one string terminated by a null byte *(containing 0)*, and another string with is length encoded in the first byte.

Example 6 :

```
C_String        DB       "Good Afternoon",0
Pascal_string   DB   14, "Good Afternoon"
```

DB directive is ideal for storing one or more lines of text, where single variable can be continued on multiple lines without the necessity of supplying a label for each. Following string is terminated by an end_of_line sequence and a null byte :

```
A_long_string   DB "This is a long string that"
                DB "clearly is going to take"
                DB "several lines." Odh, Oah, O
```

Assembler can automatically calculate the length of a string, where the "$" character is a special character which indicates the address of the current location. Following Example 7 shows how the '$' character can be used to find the length of a string.

Example 7 :

```
a_string          DB "This is a string"
a_string_length   DB S - a-string
a_string_length is initialized to 16
```

■ Define Word [DW]

The values specified using DW can be 16-bit numeric numbers ranging from 0 to 65,535 *(FFFFH)*. If signed, values are declared the acceptable range is –32,768 (8000H) to +32,767 *(7FFFH)*. A character constant can be stored in the lower half of a word, as observed in example 8.

Example 8 :

NUM	DW 1, 2, 3,	; Define 3 words
	DW 0, 65535	; Lower and Higher Unsigned Values
	DW -32768, + 32767	; Lower and Higher Signed Values
	DW 256*2	; Constant Expression
	DW 1000h, 4096, 'AB',0	; Mixed Notation
	DW ?	; Single Un-initialized Word

■ Pointers

The offset of a variable or subroutine may be stored in another variable, called a pointer, as shown in example 9.

Example 9 :

```
List    DW        256, 257, 258     ; Define 3 words
P       DW list                     ; P points to list
P1      DW P                        ; A pointer to a pointer
```

Note. Assembler sets P to the offset of list P1 contains the address of P.

■ Define Double Word [DD]

DD directive creates storage for one or more 32-bit double words. The syntax of which, will be :

```
[name]    DD initial value [, initial value] ......
```

The initial value can be a binary number up to 0FFFFFFFFH, a segment-offset address a 4-byte encoded real number or a decimal real number.

```
AP      DD      12562926H
RAT     DD      0ABCD111H
```

■ Define Quad Word [DQ]

[name] DQ initial value [, initial value]

This directive will help the assembler to declare a variable 4 words in length or to reserve 4 words of storage in memory.

```
Big-num     DQ      123456789ABCDEF0H
```

■ Define Ten-Byte [DT]

[name] DT initial value [, initial value]

It helps to tell the assembler to define a variable which is 10 bytes in length, or to reserve 10 bytes of storage memory.

```
Pac_bcd     DT      1234567890
```

■ DUP Operator

DUP operator only appears after a storage allocation directive *(DB, DW, DD, DQ, DT)*. With DUP, you may repeat one or more values when allocating storage. *(It is mainly useful when allocating space for a table or array).*

NUM	DB	20 DUP [0]	; 20 bytes, all equal to zero
	DB	20 DUP [?]	; 20 bytes, uninitialized
	DB	4 DUP ['ABC']	; 12 bytes "ABCABCABCABC"

DUP operator can also be nested.

```
ATABLE      DB      4DUP (3 DUP [0], 2 DUP ['X'])
```

The assembler assign values in memory as shown below :

```
00, 00, 00, X, X;
00, 00, 00, X, X;
00, 00, 00, X, X;
00, 00, 00, X, X;
```

■ Type Checking

If a variable is created using DB, DW, DD or any of the other data definition directives, assembler gives it a default attribute *(byte, word, double word)* based on its size. Such type is automatically checked when the variable is referred and an error results if the types do not match.

Example 10 : *MOV AL, COUNT ; Error : operand sizes must match*

:

:

COUNT DW20H

■ Even Directive [Align on Even Address]

Since assembler assembles a section of data declarations or instruction statements, it used a location counter to keep track of how many bytes it is from the start of a segment at all time. EVEN directive tells the assembler to increment the location counter to the next address if it is not already at even address. 8086 μP can read a word from memory in one bus cycle if the word is at an even address position. When the word starts at an odd address, 8086 has to use two read cycles to obtain bytes of the word. Hence a series of words can be much more quickly read if they are at an even address.

If EVEN is used in a data segment, location counter will be incremented to the next even address *(if not at an even location)* and if EVEN is used in a code segment, the location counter will be incremented to the next even address *(if note already at an even location).* A NOP instruction is inserted in the location incremented completion.

Example 11 :

```
DATA-WHERE SEGMENT
AVG DB 9 DUP (?)      : Declare array of 9 bytes
                      : Location Counter (LC) will point to 009: after assembler reads
EVEN                  : increment LC to next even Address array of 100 words
INV DW 100 DUP (0) :    starting on even address
DATA-WHERE ENDS
```

■ Group

GROUP directive helps to tell the assembler, to group logical segments named after the directive into one logical group segment, which the segments to be accessed from the same group segment base. Next the assembler sends a message to the linker telling it to link the segments so that the segments are physically in the same 64K bytes segment.

SMAL_SYS GROUP CODE_WHERE< DATA_WHERE STACK_WHERE

An appropriate ASSUME statement to follow this can be as below :

ASSUME CS : SMAL_SYS, DS : SMAL_SYS, SS : SMAL_SYS

■ LABEL

LABEL directive will help to give a symbolic name to the current value of the LC *(Location Counter).* LABEL directive has to be followed by a term which is specifies by the type associated with that symbolic name. When the label is going to be used as the destination for a jump or a CALL statement then the LABEL must be specified as *'type near'* or as *'type far'*.

Example 12 :

```
ENTRY-POINT LABEL FAR    ; Can be able to jump here from another Segment
```

Note. When the label is going to be used as reference of a data item, then the label must be specified as type byte, word, or doubled word.

Example 13 :

```
STACK_WHERE SEGMENT
    DW 100 DUP (0)             ; Sets aside 100 words for the stack
    STACK_TOP LABEL WORD       ; Give a symbolic name to next location
                               ; after the last word in the stack
STACK_WHERE ENDS
```

■ ORG

The ORG *(Originate)* directive allows to set the location counter to any desired value at any point in the program. Since the assembler assembles a section of data declarations or instruction statements, it uses the LC to keep track of the address of current location from the start of the segment at any time. Then the LC is automatically set to 0000H when the assembler starts reading a segment.

Example 14 :

ORG 2000H ; makes assembler to set the LC to 2000H

A $ is often used as symbolical representation for current value of the LC as such $ actually represents the next available byte location where the assembler can put a data or code bye. (The $ is often used in ORG statements to tell the assembler to change the LC value relative to its current value).

Example 15 :

ORG $ + 100

Tells the assembler and makes to increment the value of the LC by 100 from its current value. (A statement such as this might be used in a data segment to leave 100 bytes of space for future use).

■ OFFSET

The OFFSET is an operator which instructs assembler to determine the offset or displacement of a named data item *(variable)* from the start of the segment which contains it. *(This operator is usually used to load the offset of a variable into a register so that the variable can be accessed with one of the indexed addressing modes).*

Example 16 :

MOV BX, OFFSET PRICES

Assembler will determine the offset of the variable PRICES from the start of the segment in which PRICES is defined and code this displacement as part of the instruction. When the instruction executes, this computed displacement will be loaded into BX. (Any instruction can follow, which is base addressing mode).

■ PTR

PTR *(Pointer)* operator helps to assign a specific type to a variable or to a label. It ins necessary to do this in any instruction where the type of the operand is not clear.

Example 17 :

(*a*) INC [BX]

Assembler will not be knowing whether to increment the byte pointed to by BX or increment the word pointed to by BX. Therefore such ambiguity can be avoided by making use of PTR directive.

```
INC BYTE PTR [BX]     ; This statement tells the assembler to increment the byte
                      ; pointed to by [BX]
INC WORD PTR [BX]     ; Increment the word pointed to by [BX]
```

(*b*) The PTR operator can be used to override the declared type of a variable. An array of words is declared with the statement WORDS

WORDS DW 2345H, 0B987H, 7C41H

Usually an access will be in terms of words. However, if it is required to access a byte in an array, it could be done by writing an instruction with the PTR operation

MOV AL, BYTE PTR WORDS

(c) While indirect jump instruction is made PTR may be used. JMP [BX] instruction does not tell the assembler whether to code the instruction for a near jump or for a far jump. For a near jump the instruction must be written as below :

JMP WORD PTR [BX]

When far jump is required the above instruction must be

JMP DWORD PTR [BX]

Different type of PTR which are used as listed in Table 6 below :

Table 6. Types of PTR

S. No.	Operator	Operand Byte
1	Byte ptr	Byte (8 bits)
2	Word ptr	Word (16 bits)
3	Dword ptr	Double word (32 bits)
4	Qword ptr	Quad word (64 bits)
5	Tbyte ptr	Ten byte (80 bits)

■ PROC

PROC *(Procedure)* directive helps to identify the start of a procedure. The procedure directive follows a name given for the procedure. After the PROC directive the term NEAR or the term FAR is used to specify the type of the procedure.

Example 18 :

WRITE-DEC PROC FAR

The procedure name is WRITE-DEC and informs the assembler that the procedure is far *(in segment with a different name from where the procedure is called)*. The PROC directive is used with the ENDP directive to bracket a procedure.

■ ENDP

The ENDP (END PROCEDURE) directive helps along with the name of the procedure to indicate the end of a procedure to the assembler. This directive, along with the procedure directive PROC, is made used to "bracket" a procedure.

Example 19 : Square-root PROC NEAR

Body of the procedure

Square-root ENDP

■ EXTRN

EXTRN directive is used to inform the assembler that the names or labels following the directive are in some other assembly module.

Example 20 :

EXTRN DIVISOR : WORD

When implicity not evident one must specify whether the label is near or far.

Example 21 :

EXTRN DIVIOR : FAR

This statement informs the assembler that DIVISOR is a label of type far in another assembly module. *(In the first example when DIVISOR is declared as type WORD, type NEAR is implicit).*

Now to call a procedure which is in a program module assembled at a different time from that which contains the CALL instruction, the assembler has to be inform that the procedure is external. Then the assembler will put information in the object code file so that the linker can connect the two modules together appropriately.

Note. The names or labels referred to as external in one module must be declared public with the PUBLIC directive in the module where they are defined. The EXTRN statements will SEGMENT-ENDS directive which identify the segment in which the external name or label will be found.

Example 22 :

PROCEDURE-WHERE SEGMENT

EXTRN SQURT : FAR

EXTRM SUM : NEAR

PROCEDURE-WHERE ENDS

■ PUBLIC

LARGE programs are normally written as several separate modules where each modules is individually assembled, tested and debugged. If all modules are working correctly, their object code files are linked together to form a complete program. For the modules to link together correctly, any variable name or label referred to by any other module when the variable is not defined, must be declared PUBLIC in the module where it is defined. PUBLIC directive is used to inform the assembler that a specified name or label will be accessed from other modules.

Example 23 :

PUBLIC MULTIPLICAND, MULTIPLIER

PUBLIC SQRT, SUM

Such statement makes two variables MULTIPLICAND and MULTIPLIER available to other assembly modules. When an instruction in a module refers to a variable or label in another assembly module, the assembler must be told that it is external with the EXTRN directive.

■ ALIGN directive

General format is : ALIGN number,

(where number is 2, 4, 8 or 16)

'ALIGN 16' directive forces assembler to align the next segment at an address that is divisible by 16. The assembler fills the unused bytes with 0 for data, and NOP instructions for code.

Note. DOS will load always the CODE segment first, and only then loads the DATA segment immaterial of the order in which they are defined in our program.

ALIGN 16 is used at the end of the CODE segment so as to force the beginning of DATA segment at an address divisible by 16. Thus the offset address of N1 will now be 0 in the DATA segment, which is convenient for viewing results using DEBUG utility.

- ALIGN 2 directive may be used to start the DATA segment on a word boundary when the processor is 8086, which speeds up accessing of words.
- ALIGN 4 directive can be used to start the DATA segment on a double word (32 bit) boundary when the processor is 80386, or a later processor, which again speeds up assessing of double words. *(For 8088 processor, ALIGN is not of much advantage as it can only access a byte at a time).*

5.7.1 Summary of the Assembler Directives

Now to summarize the assembler directives it is formated in the form of table as below in Table 7.

Table 7. Assembler Directives

S. No.	Directive	Action of the Directive
1	ALIGN	aligns next variable or instruction to byte which is multiple of operand
2	ASSUME	selects segment register(s) to be the default for all symbol in segment(s)
3	COMMENT	indicates a comment
4	DB	allocates and optionally initializes bytes of storage
5	DW	allocates and optionally initializes words of storage
6	DD	allocates and optionally initializes doublewords of storage
7	DQ	allocates and optionally initializes quadwords of storage
8	DT	allocates and optionally initializes 10-byte-long storage units
9	END	terminates assembly; optionally indicates program entry point
10	ENDM	terminates a macro definition
11	ENDP	marks end of procedure definition
12	ENDS	marks end of segment or structure
13	EQU	assigns expression to name
14	EVEN	aligns next variable or instruction to even byte
15	EXITM	terminates macro expansion
16	EXTRN	indicates externally defined symbols
17	LABEL	creates a new label with specified type and current location counter
18	LOCAL	declares local variables in macro definition
19	MACRO	starts macro definition
20	MODEL	specifies mode for assembling the program
21	ORG	sets location counter to argument
22	PAGE	sets length and width of program listing; generates page break
23	PROC	starts procedure definition
24	PTR	assigns a specific type to a variable or to a level
25	PUBLIC	identifies symbols to be visible outside module
26	TITLE	defines the program listing title

5.8 VARIABLES SUFFIX AND OPERATORS

Now let us briefly define

(*i*) Variables

(*ii*) Suffix, and

(*iii*) Operator

used is 8086 microprocessor.

(*i*) Variable : A variable is an identifier which is associated with the first byte of data item. In assembly language statement : COUNT DB 20H, COUNT is a variable.

Example 1: *Array DB 10, 20, 30, 40, 50*

Here, array is the variable which is associated with the first byte of the data item, i.e., 10.

(*ii*) Suffix : In assembly language programming base of the number of indicated by a suffix as mentioned below :

- B-Binary
- D-Decimal
- O-octal
- H-Hexadecimal

The default will be decimal. The first digit in a hexadecimal number must be 0 through 9 ; that means, if the most significant digit is a letter (A-F), then it must be prefixed with a 0.

Example 2 : *1010 B = 10102*

967 D = 2967 = 296710

3F2AH = 2F2A16

0B129H = B12916

(*iii*) Operators : The Arithmetic operators are : "+", "–", "*", and "/".

The Logical operators : "AND", "OR", "NOT", and "XOR".

(The Logical operators are specially used for binary operands).

5.9 THE CONVENTIONS USED IN WRITING COMMENTS IN THE PROGRAM

An information present on a line after a ';' is normally a comment, and of course assembler does not generate any code for the comment portion. Comments are liberally written to make the program easily understandable to any reader. Normal method used while writing comments makes it easy to trace program execution.

In the following steps, the conventions are briefly made clear.

(*a*) When several instructions are executed in a loop, normally we indicate the value of the contents of the affected register or variable, for each pass through the loop, consider the following line in the program, as example ;

Example 1 : MOV AL, N1[BX] ; AL = F0 |DE |BC |9A |78 |56 |34 |12|

That means, AL gets the value F0 when the instruction is executed for the first time in the loop. In subsequent executions, value of AL will be DE, CX, 9A, 78, 56, 34, 12.

(*b*) When the values of a variable or a register is not convenient to represent on a single line we use several lines and we indicate the information within square brackets.

```
Example 2 : ADDAGN  AL, N1[BX]  ;[CX = 0   | 6   |5   |4   |3   |2   |1   |0   |
                                ;J         | J   | J  | J  | J  | J  | J  |0J  | ]
```

That means, first time in the loop, CX value changes to 7 and hence is a jump case indicated by 'J' in the comment. In subsequent executions, value of CX will be 6, 5, 4, 3, 2, 1, and they all jump cases. At the end CX becomes 0 and is a no jump case indicated by 'NJ' in the comment.

Few times, we write the comments within square brackets for several instructions considered as a group. *i.e.*,

Example 3 :

```
MOV     AX, @SATA       ; [Move to DS the segment
MOV     DS, AX          ; address of DATA area]
```

5.9.1 Working of Program

Information present on a line after a ';' is a comment, and the assembler does not generate any code for the comment portion. The Comments are liberally written to make the program easily understandable, abouts execution. Now the following table provides the provisional or normally and comments for few instruction codes for 6 PAGS after ';' as below :

```
.CODE
; PAGA 1
ADDSUB :
MOV     AX, @DATA      ; [Move to DS the segment
MOV     DS, AX         ; address of DATA area]
; PROGRAM SEGMENT TO PERFORM ADDITION OF 64 BIT NUMBERS
; PAGA 2
MOV     CX,8           ; CX = 8
MOV     BX,7           ; BX = 7
CLC                    ; CY = 0
;................................................
; PAGA 3
ADDAGIN :
MOV     AL,N1 [BX]     ; AL = F0   |DE   |BC   |9A   |78   |56   |34   |12    |
ADC     AL,N2 [BX]     ; [AL = 8A  |CF   |9B   |F1   |AC   |CE   |46   |CE    |
                       ; CY = 1    |1    |1    |0    |0    |0    |0    |0     | ]
MOV     S[BX], AL      ; [S[7]     |S[6] |S[5] |S[4] |S[3] |S[2] |S[1] |S[0]  |
                       ;    = 8A   |=CF  |=9B  |=F1  |=AC  |=CE  |=46  |=CE   | ]
DEC     BX             ; BX = 6    |5    |4    |3    |2    |1    |0    |-1=FF |
```

```
LOOP   ADDAGN        ; [CX = 7  |6    |5    |4    |3    |2    |1    |0      |
                     ;       J  |J    |J    |J    |J    |J    |J    |NJ     ||
; PROGRAM SEGMENT TO PERFORM SUBTRACTION OF 64 BIT NUMBERS
; ........................................................
```

; PAGA 4

```
MOV    CX,8          ; CX = 8
MOV    BX,7          ; BX = 7
CLC                  ; CY = 0
; ........................................................
```

; PAGA 5

```
SUBAGN :
MOV    AL,N1 [BX]    ; AL = F0  |DE   |BC   |9A   |78   |56   |34   |12     |
SBB    AL,N2 [BX]    ; [AL = 56 |EE   |DD   |43   |44   |DE   |21   |56     |
                     ; CY = 0   |1    |1    |0    |0    |1    |0    |1      ||
MOV    D[BX],AL      ; [D[7]    |D[6] |D[5] |D[4] |D[3] |D[2] |D[1] |D[0]   |
                     ; =55      |=EE  |=EE  |=43  |=44  |=DE  |=21  |=56    ||
DEC    BX            ; BX = 6   |5    |4    |3    |2    |1    |0    |-1=FF  |
LOOP   SUBAGN        ; [CX = 7  |6    |5    |4    |3    |2    |1    |0      |
                     ; J        |J    |J    |J    |J    |J    |J    |NJ     ||
```

; PAGA 6

```
INT 3                ; [Use Break point interrupt to
                     ; check results using DEBUG]
ALIGN 16             ; [Align the next segment to start on
                     ; a paragraph (16 byte) boundary . ]
END    ADDSUB        ; [Physical end of program
                     ; start execution from ADDSUB]
```

The comments for these 6 PAGAS as briefly put in the following Table 8 :

Table 8

Paga No.	Type of Writing Comments
1	'MOV AX,@DATA', moves to AX the segment address of DATA ares. *('MOV AX, SEG N1', which moves to AX the segment address of N1, is an alternative way to achieve this goal. Even 'MOV AX, SEG E' also achieves the same, as the segment address of N! and D are equal. 'MOV AX, SEG_DATA' is yet another way of achieving the same as the default data segment name is _DATA, and also even 'MOV AX,_DATA achieves the same)* MOV DS, AX moves segment address of DATA segment in AX to DS. In 8086 it is not to load a segment register with an immediate value In this paga, BX, CX, or DX could also be used instead of AX
2	Move 8 to CX. CX is used as a loop counter here, which is decremented every tine through the loop. Move 7 to BX. BX contains the offset address of the byte in N1 and N2 to be added. So to start with, BX points to the LS byte in N1 and N2. Clear the Carry flag to 0. It is done because in paga 3, we always add a byte of each of the two numbers along with carry.

3	Move to AL the byte of N1 pointed by BX Add AL and the byte of N2 pointed by BX, along with carry The sum in AL is then moved to the byte of S pointed by BX. BX is then decremented thus now BX points to the next higher significant byte in N1, N2 and S. Because of 'LOOP ADDAGN CX is decremented. If CX is not zero, jump to location ADDAGN, to repeat addition of next higher significant byte of N1 and N2. If CX = 0, go to paga 4, to perform subtraction of multi byte numbers at N1 and N2.
4	Move 8 to CX. CX is used as a loop counter Move 7 to BX. BX contains offset address of the byte in N1 and N2 to be subtracted. That means to start with BX points to the LS byte in N1 and N2. Clear the Carry flag to 0. This is done because in paga 5, it is always subtracted a byte of each of the two numbers along with carry.
5	Move to AL the byte of N1 pointed by BX Subtract from AL the byte of N2 pointed by BX, along with borrow and the difference in AL is moved to the byte of D pointed by BX. BX is then decremented. That means now BX points to the next higher significant byte in N1, N2 and D. Because of 'LOOP SUBAGN', CX is decremented. If CX is not zero, jump to location SYBAGN, to repeat subtraction of next higher significant byte of N1 and N2. IF CX = 0, go to paga 6, to terminate the program.
6	INT 3 instruction implements a Break point at this point of the program during the execution of the program. DOS DEBUG utility can then be used to check the results. Use of DEBUG to check the program is discussed in a later section. ALIGN 16 is used to force the beginning of DATA segment at an address divisible by 16. That means, the offset address of N1 will now be 0 in the DATA segment. Which is convenient for viewing results using DEBUG, END ADDSUB indicates that the execution is to begin at the instruction with the label ADDSUB.

5.9.2 Keying in the Assembly Language Program

A program to be entered by the user is indicated below, where the comments and nonessential assembler directives are not present. The user can insert comments and assembler directives to make the program more easily readable as well understandable.

```
            . MODEL     SMALL
            . STACK 64
            . DATA
N1          DB          12H, 34H, 56H, 78H, 9AH, 0BCH, ODEH, OF0H
N2          DB          0BCH, 12H, 78H, 34H, 56H, ODEH, OF0H, 9AH
S           DB          8 DUP (?)
D           DB          8 DUP (?)
            . CODE
ADDSUB :    MOV         AX, @DATA
            MOV         DS, AX
            MOV         CX,8
            MOV         BX,7
            CLC
```

```
ADDAGN :   MOV     AL,N1[BX]
           ADC     AL, N2[BX]
           MOV     S [BX], AL
           DEC     BX
           LOOP    ADDAGN
           MOV     CX,8
           MOV     BX,7
           CLC
SUBAGN :   MOV     AL,N1 [BX]
           SBB     AL, N2 [BX]
           MOV     D [BX], AL
           DEC     BX
           LOOP    SUBAGN
           INT     3
           ALIGN   16
           END     ADDSUB
```

The above program can be entered using any editor, normally a screen editor. While entering the program, the exact columns at which label, operation, operand(s), and comment start is not very important. But, the program is easy to read when various field are properly aligned.

Note. File so created is given with a suitable filename with a ASM extension. For instant, the above program may be entered as a file using Norton editor with ADDSUB. ASM as the file name.

5.9.3 Assembling the Program

ADDSUB.ASM is created using an editor is just a text file which cannot be directly executed. We have to first assemble, and then link-it, before executing. The assembly step involves translating the assembly language program *(called source code)* into machine code *(called object code),* and generating a. OBJ file, before execution.

Note. The examples are mentioned for Microsoft assembler MASM 5.1 version.

The type 'MASM ADDSUB ; <cr>' to perform translation of the source code in ADDSUB.ASM to the object code. The result of which will be placed in ADDSUB.OBJ file. '<cr>' stands for 'carriage return' which is done by typing the key on which 'Enter' is marked.

The same effect is obtained by typing 'MASM ADDSUB.ASM' '<cr>'.

In case of errors were encountered during the assembly process, it is suitable to generate a LIST file, which contains the source code, object code, along with details of errors encountered, if any, present.

To generate list file during assembly, we can type the following codes :

'MASM/L ADDSUB ; <cr>'

This results in generation of ADDSUB.OBJ as well as ADDSUB.LST files. Next using a text editor we can see the contents of ADDSUB.LST and locate the errors in the source file ADDSUB.ASM and rectify them.

Information regarding various options is possible during the assembly process can be obtained by typing 'MASM/H <cr>'.

Note. In the assembly process, or later on, during the linking operation, we can come across warning messages and/or severe error messages. When we come across warning messages, we can ignore them if they are not serious in nature. However severe errors have to be always corrected.

5.9.4 Linking of the Program

Next step is to link the object module ADDSUB.OBJ, which contains the machine code. Now the linker converts the .OBJ file to .EXE file. The .EXE file is the one which can be executed. Linker is also used for combining more than one separately assembled modules into one executable program.

Type 'LINK ADSDSUB ; <cr>' to perform conversion of the object code in ADDSUB.OBJ to the executable code. The result will be placed in ADDSUB.EXE file.

The same effect can also be achieved by typing "LINK ADDSUB.OBJ; '<cr>'.

Note. In addition to .EXE file, a .COM file is also an executable file on the IBM-PC.

5.9.5 Testing of the Program

To test the program normally we use DOS DEBUG utility which enables the programmer to trace the execution of a program. The programmer may step through the program one instruction at a time, or step through several instructions at a time, or even run the entire program in one operation, also.

Type 'DEBUG ADDSUB.EXE <cr>'. The DEBUG prompt '-' appears on the CRT. The information regarding the various DEBUG commands can be obtained by typing '? <cr>' at the '-' prompt of DEBUG.

5.9.6 Running the Entire Program in a Single Operation

At the '-' prompt, type 'G <cr>' then the entire program Upton, but not including the INT 3 break point instruction is executed.

To test the result type 'D DS"0L 20 <cr>'. This commands stands for 'Display memory contents starting from offset value 0 in the DS segment. The length or number of bytes to be displayed is 20H'. Whereas DEBUG always uses only hexadecimal values.

If we wish to display 20H *i.e.,* 32 bytes of information then the first 8 bytes gives value of NUMBER 1, then the next 8 bytes giving value of NUMBER 2, the next 8 bytes giving value of their SUM, and finally last 8 bytes gives value of their difference.

Same can also be achieved by typing any of the following commands :

D DS : 0, 1F <cr>
D DS : 0, IF <cr>
D DS : 0000, 001f <cr>
D xxxx : 0, 1f <cr>, where xxxx is contents of DS as displayed by DEBUG on the CRT.

Here , 0, 1f stands for memory locations from offset value 0 to 1f, a total of 20H or 32 bytes.

We exit DEBUG and return to DOS by typing 'Q <cr>' at the '-' prompt of DEBUG.

Note. In all these DEBUG commands we can use lower case characters also.

5.9.7 Running the Program in Single Step Operation

Normally we resort to single step operation when we find that the program has not yielded desired results after executing it in a single burst. In other words we use it for *'debugging'* the program, and hence the name DEBUG for this utility.

With the '-' prompt, type 'T <cr>'. Only first instruction of the program is executed, the value of various registers as well as the next instruction and its address will be displayed.

When the contents of any of the registers is not as per our calculations, we rectify our program logic, which is done by quitting DEBUG by typing 'Q <cr>' at the '-' prompt and then modifying program logic using a text editor.

For the register contents are as per our calculations, we go ahead and execute the next instruction by typing 'T <cr>' at the '-' prompt.

This process is repeated till the 'bugs' are identified and rectified in the program. We exit DEBUG and return to DOS by typing 'Q <cr>' at the '-' prompt of DEBUG.

5.9.8 Stepping through the Program Several Instructions at a Time

This method is adopted when the program is quite large, and hence single stepping through the program will be tedious.

When we want to step through 5 instructions at a time, we type 'T5 <cr>' at the '-' prompt. Then the value of the various registers at the end of these 5 instructions will be displayed. The instruction to be executed next, as well as its address is also displayed.

When the contents of any of the registers is not as per the calculations, then we rectify our program logic. Which is done by quitting DEBUG by typing 'Q <cr>' at the '-' prompt and then modifying the program logic using a text editor.

Next if the register contents are as per our calculations, we can go ahead and execute several more instructions by typing 'Tn <cr>' at the '-' prompt, where n is any integer value of our choice.

This process is repeated till the 'bugs' are identified and rectified in the program. We exit DEBUG and return to DOS by typing 'Q <cr>' at the '-' prompt of DEBUG.

REVIEW QUESTIONS

1. What is an Assembly Language ? What way it differs from high level and low level language ?
2. Explain briefly program segments.
3. Describe how Assembly language program is developed.
4. Write an ALP in 8086 to add two unpacked BCD strings.

5. Written an ALP in 8086 to multiply two signed numbers.
6. What is Program Development ? List the important program development tools.
7. Briefly explain the advantage of 'EQU' and PTR directives.
8. Delete ALGIN 16 directive in the example program. Run the program and check the results. Explain the inconvenience faced in checking the results.
9. Describe the method of debugging an assembly language program.

OBJECTIVE TYPE QUESTIONS

1. *We have to examine .... file, to known the memory address for each instruction in our program.*
2. *By examining .... file, any error in program syntax can be found, after assembly.*
3. *To name a memory location we have to use a ....*
4. *8086 ALP in source code form has the extention ....*
5. *8086 Microprocessor has .... address lines and .... data lines.*

6 DEVELOPMENT OF MODULAR PROGRAM

6.0 OBJECTIVE

At the end of this chapter you will be :

- Familiar with assemblying process
- Able to understand what are structures
- Knowing to use procedure and Macros
- Able to decide to use Recursive and Iterative action
- Clear about conditional statements in Macros

6.1 INTRODUCTION

While developing or writing lengthy or large program, it is good and convenient to write smaller or sub program called as *'program modules'*. Such program modules or just modules will perform the task which is defined. That means, *'modular programming'* is nothing but the method of developing program. Each of such module may be separately be written, assembled tested and at the end debugged for mistakes. The working modules may be linked together.

While designing modules of a program care must be taken that with how and under what circumstances the modules are entered and exited, the control coupling and how the information is communicated between the modules and the data coupling. Coupling between two modules depends on several factors. Such as including whether they are assembled together or separately and the organization of the data. However general, the task selection and module design must be such that control coupling is kept simple and data coupling is minimized, in the development of ALP is 8086. Now let us analise briefly about;

(1) Assembling Process,

(2) Linking Process and

(3) Debugging Process.

6.1.1 Assembling Process

An assembler translates a source file that was created using the editor into machine language such as binary or object code. The assembler reads the source file of the program from the disk where we have saved it after editing. *(Assembler normally reads our source file more than once).*

While translating, an assembler generates two files on the floppy or hard during these two passes. Where first file is called the *'object file'*, which contains the binary codes for the instructions and information about the address of the instructions. Next the second file generated by the assembler is *'assembler list life'*, which contains the assembly language statements, the binary code for each instruction, and the offset for each instruction.

In the first pass, assembler performs the operations as mentioned below :

(*a*) Reading the source program instructions.

(*b*) Creating a symbol table in which all symbols used in the program, together with their attributes, are stored.

(*c*) Replacing all mnemonic codes by their binary codes.

(*d*) Detecting any syntax errors in the source program

(*e*) Assigning relative address to instructions and data.

With a second pass through the source program, assembler extracts the symbol from the operand field and searches for it symbol table. When the symbol does not appear in the table, the corresponding statement is obviously erroneous, and when the symbol does appear in the table, the symbol is replaced by its address or value.

6.1.2 Linking Process

Linker is a program which helps to join together several object files into one large object file. While writing large programs, it is normally much more efficient to divide the large program into smaller modules. Each module can be individually written, tested and debugged. When all the module work, they can be linked together to form a large functioning program.

A linker produces a link file which contains the binary codes for all the combined modules. The linker also produces a link map which contains the address information about the link files. The linker, however, does not assign absolute addresses to the program, it only assigns relative addresses starting from zero. *(Such form of the program is said to be relocatable, because it can be put anywhere in memory to be run).*

6.1.3 Debugging Process

Debugger is a program that allows to load object code program into system memory, execute the program, and debug it. The benefits of debugger are :

(*a*) The debugger allows to look at the contents of registers and memory locations after our program runs.

(*b*) It allows to change the contents of register and memory locations and return the program.

(*c*) Some debugger allows to stop execution after each instruction so that we can check or alter memory and register contents.

(*d*) A debugger can also allows to set a breakout at any point in the program, if you run a program, the system will execute instructions up to this breakpoint and stop. Then we can examine register and memory contents to see if the results are correct at the point. When the results are correct, we can move the breakpoint to a later point in our program, and when results are not correct, we can check the program up to that point to find out and the mistakes.

(Shortly debugger tools can help us to isolate problems in our program).

The assembly language are to be designed to help develop modular program. The three important components which are used for development of modular program are :

1. Structures
2. Procedures and
3. Macros

6.2 STRUCTURES

The elements allocated by a single storage definition statement has to be of the same type *(bytes, words, or double words).* It is recommended, especially in business data processing applications for a variable to have several fields, with each field having its own type. The data structure for a personnel record might be of the form :

Table 1

Field	Length of the Field
– Initials	– 2 byte character string
– Last name	– 5 byte character string
– ID number	– 4 digit packed BCD number
– Age	– 1 byte binary number

Now to permit a program to define and refer to such data structures as a whole, many high-level languages like C, C++, Pascal etc, include special declarative statements. Therefore, 8086 assembler includes the STRUC directive to serve these needs and allows instruction operands to be modified so that they can point to a specific field within a structure, whose definition provides the pattern of structure and can have the simplified form as below :

Structure name STRUC

:

Sequence of DB, DW, AND DD directives

:

Structure name ENDS

When DB, DW OR DD statements includes a variable identifier, it can be denote by the beginning of a field and is referred to as a field identifier. The structure of above personnel record can be preserved as following :

```
PERSONNEL_DATA STRUC
    INITIALS          DB      'XX'
    LAST_NAME         DB      5 DUP (?)
    ID                DB      0, 0
    AGE               DB      ?
PERSONNEL_DATA        ENDS
```

Definition. Structure definition does not reserve storage or directly pre-assign values; but it merely defines a pattern. That means, to reserve the necessary it must be accompanied by a statement for invoking the structure, which is of the form :

Variable Structure name < pre-assignment specifications >

Here the name is the structure name in the STRUCT directive and the variable is an identifier associated with the beginning of the structure, and variable can also be used with the field names to refer to the various fields within the structure. Eventhough, structure definition may indicate pre-assigned values, they are only default values and, except when multiple pre-assignments are made in a single statement, they can be overridden by the pre-assignment specifications.

The pre-assignment specification consists of a sequence of constant expressions, one expression per field, separated by commas. When a sequence begins with one or more commas or if there are consecutive commas, it depicts the absence of one or more expressions is indicated. Therefore for the corresponding fields pre-assigned values in the structure definition are to be used. There may be more fields than the implied by the expressions and commas, such case only structure definition pre-assignments for the indicated fields are overriden. When there are no expressions or commas (*i.e. only <> appears),* then all of the pre-assignment values in the structure definition are made used of.

Apart from the structure PERSONNEL_DATA one can use the statements as below :

- employee_1 PERSONNEL_DATA <'NK',,,35>
- employee_2 PERSONNEL_DATA <'KA', 'RAMU', 22,26>

6.3 PROCEDURES

During the usages of group of instructions several times throughout a program there are two methods we can avoid having to write the group of instructions each time we want to use them. First method is to write the group of instructions as a separate procedure, then we can just CALL the procedure whenever we need to execute that group of instructions. To call the procedure we have to store the return address onto the stack, which takes sometime. If the group of instructions is big enough then this overhead time is negligible with respect to execution time. Therefore if the group of instructions is too short, the overhead time and execution time are comparable. In such cases, it is not desirable to write procedures. For which, we can use 'macros' which is also a group of instructions. Each time we *"CALL"* a macro in our program, the assembler will insert the defined group of instructions in the place of *"CALL"*. One important point to remember here is that the assembler generates machine codes for the group of instructions each time macro is called. So there is not overhead time involved in calling and returning from a procedure. But disadvantage of macro is that it generates inline code each time when the macro is called which takes more memory.

Now it is clear that the procedure is a group of instructions stored as a separate program in the memory and it is called from the main program whenever required. The type of procedure will depends on where the procedure is stored in the memory needed if it is in the same code segment where the main program is stored then it is called near procedure else it is referred to far procedure. For near procedure CALL instruction pushes only the IP register contents on the stack, since CS register contents remains unchanged for main program and procedure, and for far procedure CALL instruction pushes both IP and CS on stack.

Procedure to process some data or address variable from the main program, it is necessary to pass these address variables or data, normally referred as passing parameters to the procedure. There are four ways to pass parameters to and from the procedure, they are

(*a*) Pass parameter using registers

(*b*) Pass parameter using general memory

(*c*) Pass parameter using pointers

(*d*) Pass parameter using stack

(*a*) Passing Parameter Using Registers

Here the data, to be passed is stored in the registers and these registers are accessed in the procedure to process the data.

Example 1: (Using Registers)

```
; Main program
CODE SEGMENT
MOV   AL, DATA;   Data to be passed is loaded in the AL register
  .
  .
  .
CALL  PRO1
  .
  .
  .
;   Procedure
PRO1  PROC  NEAR
MOV   INPUT, AL ; Procedure access the data from AL register
  .
  .
  .
RET
PRO1  ENDP
CODE  ENDS
```

(*b*) Passing Parameter Using General Memory

For the cases where we need to pass few parameters to and from a procedure, registers are a convenient way to do it. But, in cases where we need to pass a large number of parameters to procedures we use memory, which can be a dedicated section of general memory or a part of stack.

Example 2 : (Using general memory)

```
; Main program
DATA SEGMENT
BCD_INPUT  DB 42     ; Storage for BCD value
HEX_VALUE  DB ?      ; Storage for HEX value
DATA ENDS
CODE SEGMENT
        .
CALL PRO1

; Procedure
PRO1 PROC NEAR
MOV AL, BCD_INPUT
        .
MOV HEX_VALUE, AL
        .
RET
PRO1   ENDP
CODE  ENDS
```

(c) Passing Parameter Using Pointers

Example 3 : (Using pointers)

```
; Main program
DATA SEGMENT
BCD_INPUT   DB 42           ; Storage for BCD value
HEX_VALUE   DB  ?           ; Storage for HEX value
DATA ENDS
CODE SEGMENT
        .
MOV  SI, OFFSET BCD_INPUT
MOV  DI, OFFSET HEX_VALUE
CALL PRO1
        .
; Procedure
```

```
PRO1  PROC  NEAR
      .
MOV   AL,  [SI]          ; Get BCD number pointed by SI in accumulator
      .
MOV   [DI],  AL          ; Store result in location pointed by DI
RET
PRO1  ENDP
CODE  ENDS
```

(*d*) Passing Parameters Using Stack

Passing the parameter using stack we push them on the stack before the call for the procedure in the main program. Here the instructions in the procedure read these parameters from the stack. If stack is used to pass parameters we have to keep track of what is pushed on the stack and where the stack points all the time in our program.

Example 4 : (Using stack)

```
; Main program
DATA SEGMENT
BCD_INPUT  DW  4209          ; Storage for BCD value
HEX_VALUE  DW  ?             ; Storage for HEX value
DATA ENDS
CODE SEGMENT
      .
MOV   AS,  OFFSET BCD_INPUT
PUSH  AX
CALL  PRO1
      .
; Procedure
PRO1   PROC   NEAR
      .
MOV   AX, [BP + 2]           ; Base pointer is decremented by 2 when the
                             ; procedure is called. So BCD_INPUT
                             ; is available at location [BP + 2].
```

.

```
MOC    [BP + 2], AX  ; Store the result on the stack
RET
PRO1   ENDP
CODE   ENDS
```

6.3.1 Recursive Procedure

It is a procedure which calls by itself. A Recursive procedure is used to work with complex data structures known as trees. When the procedures is called with N *(recursion depth)* = 3. Then the n is decremented by one after each procedure CALL and the procedure is called unit n = 0. Fig. 6.1 depicts the flow diagram and pseudo-code for recursive procrdure as below :

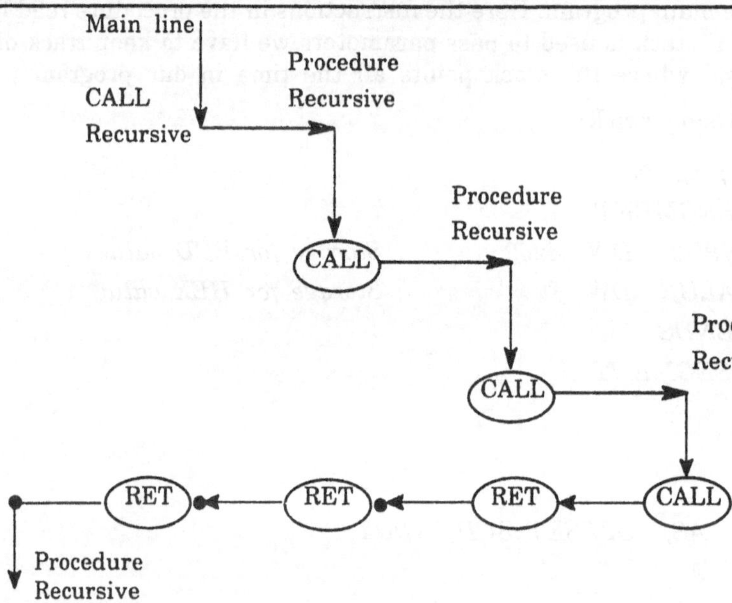


```
IF   N = 0
     DECR  N
     CALL Recursive
ELSE
     RETURN
```

Fig. 6.1 Flow diagram and pseudo-code for recursive procedure.

6.3.2 Reentrant Procedure

Sometime the situation may demand that procedure1 is called from main program, procedure2 is called from procedure1 and procedure1 is again called from procedure2. In such situation program execution flow reenters in the procedure1. Such type of procedure are called **reentrant procedures.** The reentrant procedure is shown in Figure 6.2 as below with program execution.

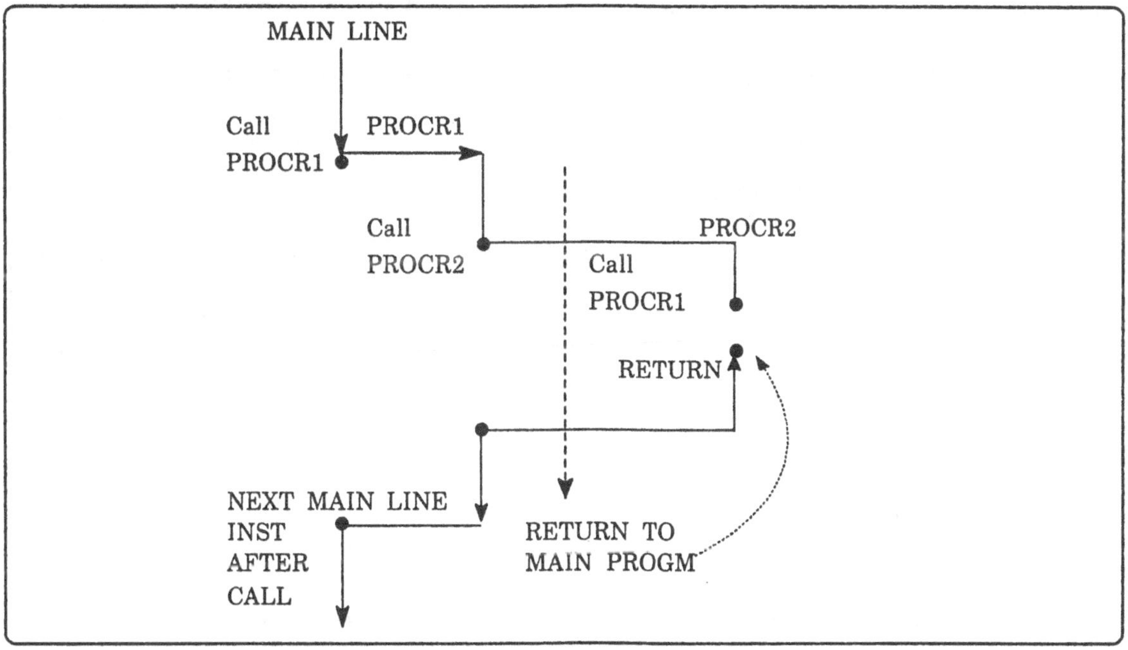

Fig. 6.2 Flow of Program execution for Re-entrant Procedure

6.4 MACROS

Macro is a group of instructions which perform one task, like a procedure performing one task. But the difference is that a procedure is accessed via a CALL instruction, while a macro is inserted in the program at the point of usage as a new sequence of instructions. To create a macro it is very similar to creating a new opcode that can be used in the program. Macro sequences execute faster than procedures because there is no CALL and RET instructions to execute, *(macro instructions are placed in program by the assembler at the point that they are invoked).*

Example 1: *Macro definition for initialization of segment registers*

```
INIT MACRO         ; Define macro
MOV AX, @data      ;  }
MOV DA             ;  }  Body of macro definition
MOV ES, AS         ;  }
ENDM               ; End macro
```

Note. Macro sequences execute faster than procedures because there are no CALL and RET instruction of execute. The assembler places the macro instructions in the program each time when it is invoked, such is known as **Macro expansion.**

6.4.1 Comparison between Procedure and Macro

By the above discussion we know that there are some statements which can give comparison between procedure and Macro which is listed in following Table 2.

Table 2. Comparison between Procedure and Macro

S. No.	Procedure	Macro
1.	Accessed by CALL and RET instruction during program execution.	Accessed during assembly with name give to macro when defined.
2.	Machine code for instructions is put only once in the memory.	Machine code is generated for instructions each time when macro is called.
3.	With procedures less memory is required.	With macros more memory is required.
4.	Parameters can be passed in registers memory locations, or stack.	Parameters passed as part of statement which calls macro.

6.4.2 Concentration in MACRO

8086 assembler can also permit dummy parameters to be concentrated to form a single character string. Ampersand (&) character informs the assembler to join *(Concatenate)* test or symbols as explained in the following program on concatenation.

Example 2: Write a Macro to provide MOVS OR MOVW instruction

```
MOVE MACRO TAG              ; Macro definition
        REP MOVS & TAG      ; Macro instruction with concatenation
ENDM                        ; End of macro
. MODEL SMALL               ; Program directive
. CODE
        MOV B               ; Calling the macro in the program with the
        MOVE W              ; Parameter B and W for TAG
        MOV AH, 4CH
        INT 21H
        END
```

Above program calls macro instruction MOVE B and MOVE W and the assembler will concatenate the parameter with the REP MOVS instruction, to produce as follow :

```
REP MOVSB
and
REP MOVSW
```

The concentration is also known as Passing Parameters in Macro.

Example 3 :

```
PROMPT MACRO MASSAGE        ; Define macro with MESSAGE as a parameter
       MOV    AH, 09H
       LEA    MESSAGE
       INT    21H
       ENDM                 ; End macro
       DATA
       MES1   DB   10,  13,  'Student Name : $'
       MES2   DB   10,  13,  'Student Address : $'
. CODE
START : MOV AX,   @ data          ; [Initialize]
        MOV DS,   AX              ; [data segment]
        PROMPT    MES1            ; Display MES1
        PROMPT    MES2            ; Display MES2
        MOV AH,   4CH             ; Return to DOS
        INT 21H
        END START
```

Again in this example it shows that parameters can be passed in macro with the help of dummy argument, which tells assembler to match its name with any occurrence of the same name in the macro body. For instant the dummy argument MESSAGE also occurs in the LEA instruction. Macro instruction "PROMPT MES1" passes the MESI as a parameter and macro accepts that as an argument.

6.4.3 Local Variables in Macro

A local variable defined in Macro is available in the Macro but it is not available outside the Macro. To define a local variable, LOCAL directive is used Example 4 shows that how local variable is used as a jump address, and if this jump address is not defined as a local, assembler gives an error message on the second and subsequent attempts to use the Macro as can be observed below :

Example 4 :

```
DISPLAY   MACRO A                 ; Displays ASCII character in uppercase
          LOCAL J_LABEL           ; Defines J_LABEL as local
          PUSH DX
          CMP AL, '2'
          JBE J_LABEL             ; Check if uppercase
          SUB AL, 20H             ; Convert to uppercase
J_LABEL : MOV DL, AL
          MOV AH, 02H
          INT 21H
          POP DX
          ENDM
```

Above Macro accepts ASCII code for character. (A-Z or a-z). For lowercase character, Macro converts it to uppercase character and displays the uppercase character on video screen.

Note. Local variable or variables must be defined using LOCAL directive immediately after MACRO directive.

Few macros require to define data items and instruction labels within the macro definitions. If we use the macro more than once in the same program, and the assembler defines the data item or label for each occurrence, the duplicate names would cause the assembler to generate an error message. To ensure that each generated name is unique, code the LOCAL directive immediately after the MACRO statement, even before comments, the general format will be as below :

LOCAL dummy 1, dummy 2, ..., One or more dummy arguments

Example 5 :

A Macro program shows segment declaration within a macro definition

```
SAY   MACRO MESSAGE
      MOV AH, 09
      MOV DX, OFFSET MESS_TEXT
      INT 21H
DATA  SEGMENTS
      MESS_TEXT DB MESSAGE, ODH, OAH, '$'
      DATA ENDS
SAY   ENDM
CODE  SEGMENT
      MAIN PROC
      MOV AX, DATA
      MOV DS, AX
      SAY 'HELLO!'
      SAY 'GOOD-BUY'
      MOV AH, 4C00H
      INT 21H
      MAIN ENDP
      CODE ENDS
      END
```

Same macro *"SAY"* is invoked twice in the same program in Example 5. Each invocation of the macro "SAY" allocates, initialized, and displays contents of a variable named MESS_TEXT. First time "SAY" in invoked, MESS_TEXT is initialized with the value "HELLO!", and the second time "GOOD_BYE".

6.4.4 Nested Macros

In a macro it is possible to call to appear within a macro definition which is referred as macro nesting and has the limitation that all macros included in the definition of a given macro must be defined before the given macro is called.

Example 6 :

To compute (diff1 - diff2)2 ; First macro finds the difference

```
DIV MACRO VAR1, VAR2
        MOV AX, VAR1
        SUB  AX, VAR2
        ENDM
; Second macro which calls the first and computes the difference square.
        DIFSQR MACRO OPR1, OPR2, RESULT
        PUSH DX
        PUSH AX
        DIV OPR1, OPR2
        IMUL AX
        MOV RESULT, AX
        POP AX
        POP DX
        ENDM
; Main program starts which calls the nested macro.
            :
DIFSQR VAR1, VAR2, RES
            :
```

To include macro definitions within macro definitions also possible. Defining macros for performing a variety of binary operations and storing the result in an arbitrary location is as depicted in above example.

Example 7 : *Demonstrate a nested macro that can be used for different add and subtract operations.*

```
DEFMAC  MACRO MACNAM, OPERATOR
        MACNAM MACRO X, Y, Z
        PUSH AX
        MOV AX, X
        OPERATOR AX, Y
        MOV Z, AX
        POP AX
        ENDM
        ENDM
            :
; Define the macro addition
; Assembled portion of source with WIDTH = FALSE and LENGT = TRUE
IF WIDT                    ; 72 columns
ELSE WIDE DB 80            ; 80 columns
```

```
ENDIF
IF LENGT;              If continuous LONG
    DB-1
ELSE ENDIF
```

6.4.5 Conditional Statements in Macros

An Assembler supports the use of conditional statements in the assembly language program which also allows to use them in macro sequences. Such conditional statements control the flow of the program execution. Let us see the conditional statements in the assembly language program and conditional statements used in macros.

The Macro sequences have their own set of conditional instructions that differ somewhat from the one used with the assembler. For instant, macros can use REPEAT and WHILE, but they do so without the period in front of the keywords REPEAT and WHILE statement. The REPEAT has a corresponding UNTIL, and the WHILE statement which has no corresponding ENDW when used in a macro. The Table 3 given the list of statement for operators.

Table 3

S. No.	Operator	Function
1	EQ	Equal
2	NE	Not equal
3	LE	Less than or Equal
4	LT	Less than
5	GT	Greater than
6	GE	Greater than or equal
7	NOT	Logical inversion
8	AND	Logical AND
9	OR	Logical OR
10	XOR	Logical XOR

6.4.6 IF, ELSE and ENDIF Statement

Conditional statements are implemented in the assembly language program using. .IF, .ELSE, ENDIF structure found in higher level language and MASM version 6-X supports the same. But the earlier versions of the assembler does not support. If statement the general format for the .IF conditional statement will be :

```
. IF (condition)          ⎫
        ...               ⎪
. ELSE (optional)         ⎬ Conditional Block
        ...               ⎪
. ENDIF (end of IF)       ⎭
```

As observed above every .IF directive must have a matching ENDIF to terminate a tested condition. ELSE is optional, which provides an alternate action.

The assembly also allows to use relational operates with .IF statement as shown in Table 4.

Table 4. Relational Operators

S. No.	Operator	Function
1	==	Equal to
2	!=	Not equal to
3	>	Greater than
4	>=	Greater than or equal to
5	<	Less than
6	<=	Less than or equal to
7	&	Bit test
8	!	Logical inversion
9	&&	Logical AND
10	||	Logical OR

Now example 8 shows source program which contains. IF -.ELSE-.ENDIF statement. This example program accepts the character input from keyboard. When the character is a uppercase alphabet, the program converts it to lowercase by adding 20H to the ASCII code of the character *(For instant : ASCII code for 'A' = 41H, 41H + 20H = 61H is the ASCII code for 'a'.).* But IF statement is used to check whether the read character is a uppercase alphabet or not.

Example 8 :

```
        [TITLE 8086 ALP to convert uppercase to lowercase].
, MODE   SMALL
. CODE
START :  MOV AH, 07H              ; [Read character
         INT 21H                  ; from keyboard in AL]
         .IF AL >='A' && AL <='Z' ; [If character is
             ADD AL, 20H          ; lowercase add 20h]
         . ENFIF
         MOV AH, 02               ; [Display character on
         MOV DL, AL               ; video screen whose ASCII
         INT 21H                  ; code is stored in dl]
         MOV AH, 4CH              ; [Exit
         INT 21H                  ; To DOS]
         END START
         END
```

6.4.7 Global Identifier

Where the identifiers defined within macros are local to the macros within which they are defined, identifiers that are defined in the main body of a source file outside of macros are global with respect to all macros. The Global identifiers may be accessed by code within any macros. Global label or proc name can be jumped to or called by code within a macro. A global variable name can be referenced from inside a macro.

When a potential for conflict is observed from the fact the same identifier is used both globally and locally, code within a macro that defines that identifier locally will refer to the local definition of the identifier, and code elsewhere will refer to its global definition only.

6.5 LIBRARIES

The Library files are the collection of procedures which can be used in other programs. Such procedures are assembled and compiled into a library file by the LIB program. Library file is invoked when a program is linked with the linker program, then, only the procedures required by the program are copied from the library file into the program. The use of the library files increase software reuseability and reduces software development time. A library file is created with the help of LIB command at the DOS prompt and the library file is a collection of assembled. OBJ files that each perform one procedure or task.

Note : The name of the procedure has to be declared in a library file.

6.6 PROGRAM STRUCTURE

DOS supports two types of load modules : EXE file and COM file, where the load module is can be loaded into the primary memory for execution.

- **EXE File :** EXE file is a standard load module which can have more number of program code and data segments limited only by the memory model selected for organization of the program. The EXE file contains a special header that directs the loader to perform adjustments to segment references in a load module. The far jump/procedure call and data reference using far pointer is possible only in an EXE file. When an EXE file is loaded, the address-relocation of segment is performed. After relocation, control is transferred via a far jump to the CS : IP that was read from the EXE header. The programs PRGI.ASM and PRG2.ASM listed above are example programs of EXE file.
- **COM File :** COM file has an extension COM. For instant : MORE.COM. The COM files are designed to be self-contained programs the DOS can load and start as fast as possible. A COM file is a binary image of the program code and data, whose maximum size must be less than or equal to 64K bytes and hence it does not require relocation of the segment-address while loading the program.

6.6.1 COM File Program Designing

Here the source program has to be designed according to a certain criteria. The Program assembling and linking is performed in the usual manner. The DOS command

EXE2BIN is used to convert the EXE load module to COM load module. In Turbo Linker, the COM file is created with the linker option/T.

■ COM file Design Procedure

COM file is a self contained program organized as a single segment. The entire program *(code, data and stack)* must fit into a single segment of size 64Kb and the memory model used by the COM file is known as tiny memory model. The components of COM program file are listed below :

- Program Segment Prefix *(PSP, the load module header)*
- The Data Segment
- The Code Segment
- The Stack Segment
- The Extra Segment

Note. The program must be organized such that it uses only one segment.

Some Guidelines for designing COM files are :

(*a*) Structure the entire program as one large code segment

(*b*) The assembler directives are used to inform the assembler that all the segment registers will point to same segment. The DOS will set up all the four segment registers to point to the beginning of the code segment after loading the program.

(*c*) The size of PSP created by DOS is 100H bytes and loaded directly in first 100H locations (0 to 99) from the beginning of the segment. Therefore program must be organized in such a manner that the first 100H bytes are not used in the program. It is performed by using the directive ORG as mentioned below :

```
ORG        100H
```

(*d*) The program must start immediately after the Program Segment Prefix (PSP). Therefore a label is placed after the directive ORG 100H and directive END must refer to this label.

(*e*) There are no separate data and extra segments, that means data must be somewhere in the code segment. Use JMP instruction to jump over the data.

(*f*) All code and data references are near.

(*g*) DOS will automatically set up a stack. It sets stack pointer to the end of the 64K area. In 8086 processor, the stack grows down towards the program. It is the responsibility of the program to see that stack does not grow into the program code.

The COM program can be terminated by invoking the program terminate function of the DOS.

The format of COM file will be :

```
0       ->      Program Starting Point          ; PSP starts here
                ..
                . Program Segment Prefix (psp)
99H     ->      ..                              ; PSP ends here
100H    ->      JMP     EXESTART                ; program code starts here
```

```
                ; Program Dat'a

EXESTART:       ..
                ..
                ; Program code can be up to 64K location

64K      ->     END of COM program segment
```

■ **Program Listing :** The steps involved in building a COM file are depicted in the following program [HELLO4.ASM]

```
        ; HELLO4.AM    -  To display one message one screen using COM file.
        ;
        ; Build        -  TASM HELLO4.ASM
                          TLINK /T HELLO4
        ; USAGE        -  HELLO4
_CODE segment                                ; _CODE - name of user defined segment
        assume cs : _CODE, ds : _CODE        ; Initialize CS and DS to respective segments
        org  100H                            ; skip Program Segment Prefix area
start   : jmp  skip                          ; skip to make a room for program data
        msg  db  'Welcome to Assembly Programming',  '$'
skip    mov  ax,  cs
        mov  ds,  ax
        ;  Display message
    mov  ah,  09h                            ; display string function
    mov  dx,  offset msg                     ; initialize DX to offset of msg
    int     21h                              ; call DOS service
    ; Program  terminate
    mov  ah,  4ch                            ; program terminate function
    mov  a1,  00h                            ; return code for error level setting
    int   21h                                ; call DOS service
_CODE ends                                   ; End of segment
    end  start                               ; End of program
```

Then it can be RUN

6.6.2 Creating COM Files

Borland/Microsoft software development tools can be made used in building a COM file. The process of building a COM file involve are listed in the forms of steps as below :

(*a*) Using Turbo Tools

(*i*) Create program object file :

TASM <source filename >

Example 1 : *TASM HELLO4.ASM*

It creates an object file HELLO4.OBJ. The extension to the input here is optional. The assembler assumes the extension as .ASM by default.

(*ii*) Create program executable file :

TLINK/T < object filename >

Example 2 : *TLINK / T HELLO4*

It takes HELLO4.OBJ as an input file and produces HELLO4.COM as an output file.

(*b*) Using Microsoft Tools

(*i*) Create program object file :

MASM <source filename>

Example 3 : *MASM HELLO4.ASM*

It creates an object file HELLO4.OBJ

(*ii*) Create program executable file :

LINK < object filename >

Example 4 : *LINK HELLO4*

It takes HELLO4.OBJ as an input file and produces HELLO4.EXE as an output file.

(*iii*) Convert EXE file to COM file

EXE2BIn <EXE filename> <COM filename>

Example 5 : *EXE2BIN HELLO4. EXE HELLO4. COM*

It takes HELLO4.EXE as an input file and produces HELLO4.COM as an output file.

6.7 THE RECURSION

The recursion is the action of a subroutine by calling it self. Following example brings out the concept of recursion and iteration.

Example 1 : (*i*)
```
RECURSIVE PROC
    :
    :
    CALL RECURSIVE
    RET
RECURSIVE ENDP
```
(*ii*)
```
INTERATIVE PROC
    :
    :
    JMP ITERATIVE
    RET
ITERATIVE ENDP
```

In each of these sequences an infinite loop developed.

(*i*) In recursive loop, the instruction at the bottom of the loop directs program flow back to the top of the loop with a CALL instruction.

(*ii*) In iterative loop, the instruction at the bottom of the loop directs program flow back to the top of the loop with a jump instruction.

In both the case, RET instruction at the end of the procedure will never executed. Only difference between two sequences is that the CALL instruction in sequence (*i*) PUSHES the address of the instruction which follows it onto the stack with each pass through its loop, which the jump instruction in sequence (*ii*) does not.

Because neither of these sequences is very useful as an infinite loop, now let us presuppose the existence of a test routine called TESTER that adjust carry flag according to some set or protocols. By introducing a CALL to TESTER and a branch, we can convert each sequence from an infinite loop into a controlled loop.

That means both loops are usable, but they still differ by one aspect *i.e.*, the recursive procedure uses a CALL, as a result of which the address of the statement EXIT is saved on the top of the stack, whereas, the iterative procedure does not save anything on the stack, as can be observed in example 2; with (*iii*) and (*iv*) as below :

Example 2 :

(*iii*)
```
RECURSIVE PROC
CALL TESTER
JC DONE
CALL RECURSIVE
DONE : RET
RECURSIVE ENDP
```

(*iv*)
```
ITERATIVE PROC
CALL TESTER
JC DONE
JMP ITERATIVE
DONE : RET
ITERATIVE ENOP
```

The practical difference between (*iv*) and sequence is that sequence (*iii*) is that sequence strings (*iii*) out a list of return addresses in the stack as it loops and the rolls them back as it exists. That means sequence (*iv*) is more efficient than sequence (*iii*) which is generally valid observation. Now Concerning the comparison of an iterative loop and the recursive loop, both of which are designed to do the same thing. Iterative loop takes less time to execute and at least during it execution, occupies stack space.

However recursive loops are quite popular in programming because several useful algorithms are much more easily expressed and coded in recursive form than that in iterative forms.

Note. The Recursion is really useful when the iterative solution of a problem is much more complex than its recursive counterpart.

REVIEW QUESTIONS

1. What is meant by Modular Programming.
2. Briefly explain Assembling Process.
3. Explain the steps in development and execution of a program.
4. Explain the function of Linker.
5. Why NEAR or FAR attribute appear in PROC statement.
6. What is a Macro ? What is must be used ? What are its advantage.
7. With the help of example explain the structure of Macro.
8. Bring out the Comparison between Procedure and Macro.
9. Briefly explain Program structure.
10. Briefly compair Recursive and Iterative action.

7 ASSEMBLY LANGUAGE PROGRAMMING (ALP)

7.0 OBJECTIVE

After completion of this chapter you will be able to know the steps involved in writing ALP. Will be able to write ALP in proper formate with algorithm for the programs.

7.1 INTRODUCTION TO ALP

In earlier chapters we have seen the instruction set of 8086, learnt techniques for ALP and we have observed the development of Modular Program. Now after knowing instruction set, generating machine code for an instruction using instruction format, it is clear that generating machine code for ALP, manually is difficult. That means we are suppose to use assembly language tools like, assembler, like, loader and debugger to develop and execute ALPs, as seen earlier.

Therefore now let us analise few aspects to develop ALP, such as :

(*i*) Writing ALP (*assembly language Program*)

(*ii*) Assembling ALP,

(*iii*) Linking the program, and

(*iv*) Testing the program

Fig. 7.1 depicts the steps involved in developing and executing an ALP, as developing and executing, an ALP, as shown on the next page.

From fig 7.1, it can be observed that in the development process the first step is to write an assembly language program, which can be written with an ordinary text editor such as word star, edit and so on. ALP text is an input to the assembler, which translates assembly language statements to their binary equivalents, usually known as object code. The time required to translate assembly code to object code is called **assemble time**. During assembling process the assembler checks for syntax errors and displays them before giving object code moudle.

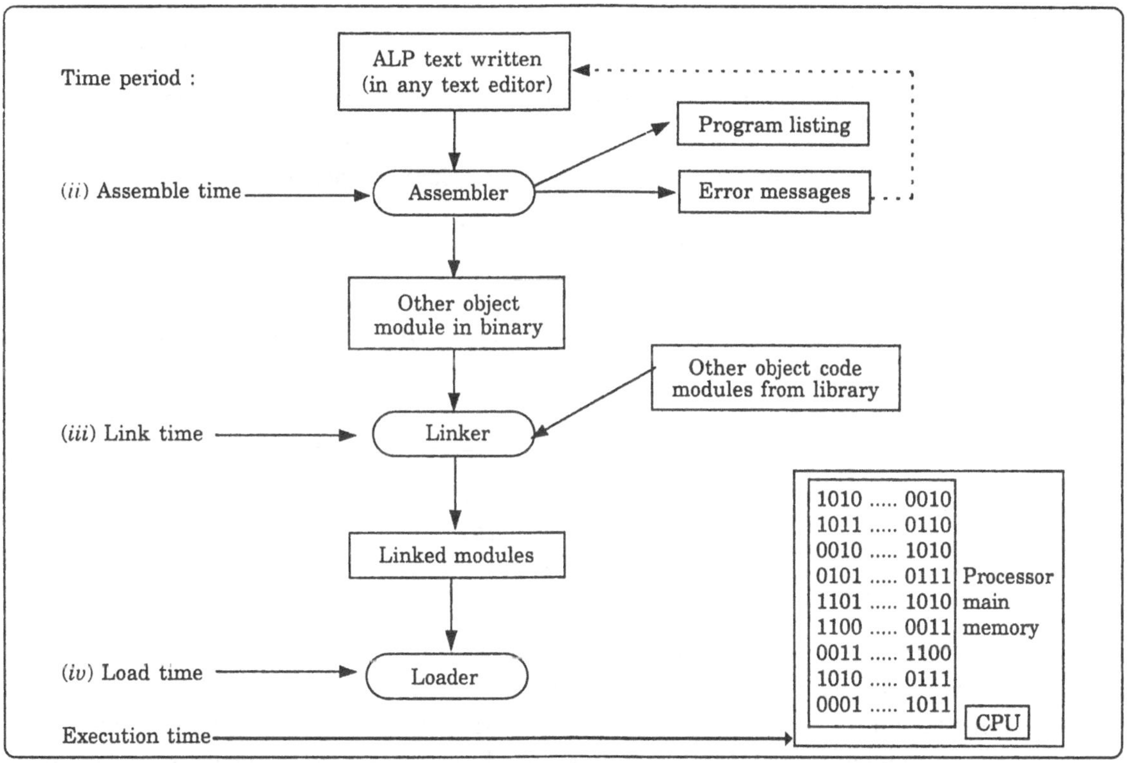

Fig. 7.1 Program development and execution steps

Next object code module contains information about where the program or module to be loaded in memory, and it is to be linked with other separately assembled modules then it contains additional linkage information. At link time, separately assembled modules are combined into one single load module, by the linker, which also adds any required initialization or finalization code to allow the operating system to start the program running and to return control to the operating system after the program has completed. Normally all, linkers allow assembly language modules to be linked with object code modules compiled from from high-level languages as well. That means, the programmer can insert a time-critical assembly language routines, library modules into a program.

Finally at load time, the program loader copies the program into the computer's main memory, and at execution time, program execution will begin.

7.2 Structure of ALP Instruction

Normally each assembler will have its own syntactical structure, such as requiring upper case, or requiring lower case, or requiring colons after label definitions, but let us analyse most common features that an assembler shares in the processor. An assembly text is normally divided into fields, which is separated by spaces and tabs. Now let us observe the format for ALP line as below :

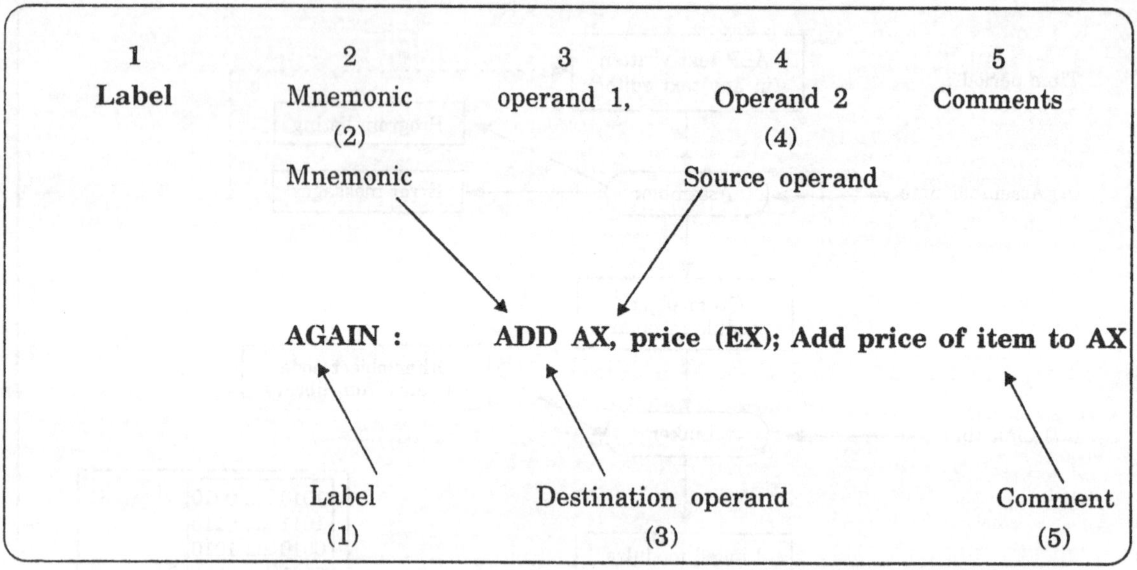

Fig. 7.2 Assembly language Instruction of 8086

In the format shown above, first field in optional, *i.e.*, label field, which is required to specify symbolic labels, which is an identifier that is assigned to the address of the first byte of the instruction in which it appears. That means, the presence of a label is optional, but if present, the label provides a symbolic name that can be used in branch instructions to branch to the instruction.

Next second filed is mnemonic, which is required and compulsory, that is all instructions will contain a mnemonic. The third and following fields are operands, the presence of which depends on the instruction. Some instruction have no operands, but as shown in Fig. 7.2 some have one, and some have two. If there are two operands, they are separated by a comma, as shown in Fig. 7.2

Finally the last field is a comment field, which begins which with a delimiter such as the semicolon and continues to the end of the line. Such comments will tell us what the program is trying to accomplish and how the program is executing.

7.3 A SAMPLE PROGRAM

Now before going to write ALP in 8086, let us take a simple sample program for addition and analyse it, as below :

Example 1 : Write an ALP to 8086 to add two 16-bits number stored in CX and DX respectively and store the result in location addressed by DI.

Label/Address	Mnemonics and Operand (1 and 2)	Comments
xxx 1	.MODEL SMALL	
xxx 2	.DATA	
xxx 3	D1 DW 100 H	; First data is initialized
xxx 4	D2 DW 200 H	; second data is initialized
xxx 5	D3 DE ?	; To store the result
xxx 6	CODE	
xxx 7	MOV AX, @ DATA	; The process of initializing
xxx 8	MOV DS, AX	; DS register
xxx 9	MOV CX, D1	; First data into CX register
xxx 10	MOV DX, D2	; 2nd data into DX register
xxx 11	ADD CX, DX	; Perform addition
xxx 12	MOV DI, offset D3	; Initialize the destination address in D
xxx 13	MOV [DI] CX	; Save the result
xxx 14	MOV AH, 4CH	;
xxx 15	INT 21 H	; Exit to DOS
xxx 16	END	

In example 1, XXX 1, XXX 2 are assembler directives.

and XXX 3, XXX 4, XXX 5 are data definition with labels D, D_2 and D_3.

Here xxx5 had D_3 lable with one word, which has not initialized and reserved for result. xxx6 is assembler directive.

The address line xxx7 and xxx8 helps to set DS register to address data segment, where xxx7 uses immediate addressing mode, which cannot be loaded directly into the Data segment. Therefore data is first moved into AX register and then the data is transferred into DS register.

Next line xxx9 and xxx10 are used to load contents of memory locations labelled D1 and D2 into CX and DX registers respectively. The line xxx11 adds contents of CX and DX register and result will be stored back into CX register. (*The lines xxx10 and xxx11 can be replaced by single instruction as ADD CX, D2*).

Next line xxx12 loads offset address off the result into D1 register xxx13 saves result stored in CX into memory location by the help of register indirect addressing mode. The lines xxx14 and xxx15 exists to DOS, by using interrupt service routine, and finally line xxx16 end the program as assembler directive.

The procedure for excuting the program (*of example 1*), will be as below :

(*i*) Save the above assembly program in a file name as FIRST.ASM (*ii*) Assemble the program using MASM assembler, which generates FIRST.OBJ file as;

MASM FIRST;

The output is : Microsoft Macro Assembler Version 4.0

0Waring errors 0 Severe errors

(*For any errors, they are displayed with the line number and the type of error*).

(*iii*) Link the .OBJ file and create an.EXE file

C>LINK FIRST :

The output :

IBM Personal Computer Linker

Version 1.10

Warning : No stack segment

Note. Even though the linker warns us that there is no stack segment, we need not worry as are not in need of it, and the FIRST.EXE file generated.

7.3.1 Debugging

It is often or always required to debugg any ALP for mistakes and to view the result. Now let us observe the debugging process for the above sample program *i.e.,* for example 1.

DEBUG FIRST.EXE, will take into DEBUG which can be listed using unassemble, 'U' command. There are few differences by program listed by debug with respect to list-file of FIRST.ASM as below in Table 1.

Table 1

First.ASM	-U (Command)		
	2757:0000	B8582F	(MOV AX,3F00)
	2757:0003	8ED8	(MOV DS,AX)
Symbolic label names are replaced by absolute elements	2557:0005	8B0E0800	(MOV CX, [0000])
Assembler directives are eliminated	2757:0009	8B160A00	(MOV DX, [0002])
Comments are not displayed	2757:000D	03CA	(ADD CX, DX)
Address, Opcodes and Mnemonics are displayed	2757:000F	BF0C00	(MOV DI, 0004)
	2757:0012	890D	(MOV [DI], CX)
	2757:0014	B44C	(MOV AH, 4C)
	2757:0016	CD21	(INT 21)
	(Here data values and address values are listed in HEX)		

Now by debugging we can see that :

(*i*) The Data can be viewed by using dump command, the beginning address of the DS must be mentioned as a parameter of the dump command to view the appropriate data.

```
-D 3F00:0000
3F00:0000 00 01 00 02 00 00 00 00 - 53 74 61 63 6B 20 20 20 00
3F00:0010 50 51 30 05 50 20 20 00 - 53 72 62 23 6B 20 20 20 20
```

(*In the case of word data (16-bits) the lower order byte appears first*).

(*ii*) Execute the program by giving the G command.

-G The above command results in the following message

program terminated usually

(*iii*) View the result using dump command. D 3F00:000

```
3F00:0000 00 01 00 02 00 03 00 00 - 53 74 61 63 6B 20 20 20 20
3F00:0010 50 51 30 05 50 20 20 00 - 53 72 62 23 6B 20 20 20 20
```

(*The result is stored in location DS:0004 and 0005*).

7.4 SAMPLE ALPS

Here let us observe few sample ALPs in 8086/8088.

ALP 1 : Write a program to add two 8 bytes of data available in memory locations arry 1 and arry 2 and store the result in arry3. Assume the least significant byte is available in the first location arr1 and most significant byte in location arr1 + 7. Use modular programming.

(*i*) ; Main .ASM;

; The name of file in which following instructions are typed.

; MODEL SMALL

```
Label      Mnemonics     Operand                Comments
.DATA
    PUBLIC ARRY1, ARRY2, ARRY3          ; Declaration of data and making it
    ARRY1 DB 01,02,03,04,05,06,07,08    ; public
    ARRY2 DB 08,02,05,04,06,07,01,08
    ARRY3 DB 8 DUP ( ? )                ; Declared space for the result
.CODE
    EXTRN PROADD:FAR                    ; Procedure is available in another file
        MOV AX, @DATA
        MOV DS, AX
        MOV SI, OFFSET ARRY1            ; Initializing the registers with
        MOV DI, OFFSET ARRY2            ; the starting address of the
        MOV BX, OFFSET ARRT3            ; data and the result
```

```
        CALL PROADD                    ; Calling a procedure which is far
        MOV AH, 4CH                    ; Normal termination of the
        INR 21H                        ; program to DOS
        END
```

(*ii*) ;SEC.ASM :

; The name of the second file in which the following codes are types.

. MODEL SMALL

```
Label   Mnemonics    Operand   Comments
. DATA
     EXTRN ARRY1, ARRY2, ARRY3 ; Declaring that these variables are
. CODE                         ; Available in another file
     PUBLIC PROADD             ; Making this procedure available to other files
     PROADD PROC
     MOV CX, 08                ; Initialize count to 8 CLC cy flag is make 0
NEXT :   MOV AL, [SI]          ; Fetches the first operand
         ADC AL, [DI]          ; Adds with the next operand
         MOV [BX], AL          ; Stores the result in memory
         INC SI                ; Point to next data
         INC DI
         INC BX
         LOOP NEXT             ; Repeat till all data is added
         RET
         PROADD ENDP
         END
```

In above two modules they are assembled separately with MASM.

- MASM main;
- MASM sec;

Both the files must be linked to obtain the executable file, which below :

LINK MAIN + SEC ;

Note. By Using DEBUG the result is seen on the screen.

ALP 2 : Write an ALP in 8086 to find the average of two numbers

```
   NAME     Average
   TITLE    8086/8088 ALP to find average of two numbers.
.mode  small
. stack 100
. data
   Label          Mnemonics          Comments
      No 1        DB 63H             ; First number storage
      No 2        DB 2EH             ; Second number storage
      Avg         DB ?               ; Average of two numbers
```

```
 code
 START :  MOV AX, @data            ; Initialises
          MOV DX,AX                ; data segment]
          MOV AL, No1              ; Get first number in AL
          ADD AL, No2              ; Add second to it
          ADC AH, 00H              ; Put carry in AH
          SAR AX, 1                ; Divide sum by 2
          MOV AVG, AL              ; Copy result to memory
          END START
```

ALP 3 : Write an ALP in 8086/8088 to find maximum number in the array of 10 numbers.

```
 NAME      Maximum number
 TITLE 8086 ALP to find maximum in the array.
. model small
. stack 100
. data
 ARRAY DB 63H, 32H, 45H, 75H, 12H, 42H, 09H, 14H, 56H, 38H; Array of; ten numbers
 Labels              Mnemonics           Comments
    MAX              DB 0                ; Maximum number
 .code
 START :  MOV AX, @data            ; [Initialises
          MOV DS, AX               ; data segment]
          XOR DI, DI               ; Initialise pointer
          MOV CL, 10               ; Initialise counter
          LEA BX, ARRAY            ; Initialise base pointer for array
          MOV AL, MAX              ; Get maximum number
 BACK :   CMP AL, [BX + DI]        ; Compare number with maximum
          JNC SKIP
          MOV DL, [BX + DI]        ; [If number > MAX
          MOV AL, DL               ; MAX = number]
 SKIP :   INC DI                   ; Increment pointer
          DEC Cl                   ; Decrement counter
          JNZ BACK                 ; IF count = 0 stop; otherwise go BACK
          MOV MAX, AL              ; Store maximum number
          END START
```

Note. Above two ALPs will not accept any input from keyboard—and will not display result on screen. For display we need to use DOS.

ALP 4 : Write a recursive program in 8086/8088 ALP to find the sum of the first "*n*" integers.

Recursive algorithm (*for summing the first 'n' integers in PASCAL A*)

```
BEGIN
        IF N = 0 THEN SUM : = 0
        ELSE
                SUM = SUM + SUM (N-1);
        END;
                ALP :
; Program in assembly :
.MODEL SMALL
. DATA
        N EQU X
. CODE
   MAIN PROC
        MOV BX, N
        CALL SUM_OF_N
        MOV AX, 4C00H
        INT 21H
MAIN ENDP
SUM_OF_N_PROCEDURE
        CPM BX, 00
        JZ BX_ZERO
        PUSH BX
        DEC BX
        CALL SUM_OF_N
        POP BX
BX_ZERO : ADD AX, BX
        RET
SUM_OF_N END P
        END
```

ALP 5 : Write an ALP in 8086 to move a block of N bytes of data from source to destination and display the result.

Soln : To write this progam and make it clear, we divide the solution into three parts as

(*i*) ALP-explanation

(*ii*) ALP-Algorithm and

(*iii*) The ALP

(i) ALP-Explanation

Let, Block 1 data is available from memory location block 1, and destination memory location be block 2. The number of bytes in block 1 be 10 *i.e.*, 0AH. To start, initialize counter CX register.

Block 1 address be in SI register and Block 2 address in DI register. Clear the direction flag. By using string instruction move the data from the memory pointed by SI register to memory pointed to DI register, (*it is assumed that the data is to be moved within the same segment*). Therefore DS and ES are initialized to the same segment value. Finally execution of the program, and the contents of destination memory location can be viewed by using debug DUMP command.

(ii) ALP-Algorithm

Step No	Action/Process
1.	Initialize data in the Source memory
2.	Initialize SI, and DI and CX registers with the source address, destination address and the count
3.	Initialize the DF (*Direction Flag*) to zero
4.	Use move string instruction to transfer the data
5.	Check whether all data is transferred
6.	If not transferred go to step 4
7.	else terminate the program normally using interrupt 21 function 4CH

(iii) The ALP

```
        .MODEL  SMALL
        .DATA
        BLOCK1 DB 01, 02, 03, 04, 05, 06, 07, 08, 09.0AH
        BLOCK2  DB 10 DUP (?)
        COUNT DW 0AH
        .CODE
            MOV AX, @DATA           ; Initialize the DS and EX with the segment address
            MOV DS, AX
            MOV ES, AX
            MOV SI.OFFSET BLOCK1    ; Initialize the SI and DI with the address
            MOV DI, OFFSET BLOCK2   ; of the source and destination
            MOV CX, COUNT           ; Initialize CX with the count
            CLD                     ; Initialize the DF = 0
AGAIN :     REP MOVSB               ; String move the data from source to destination
                                    ; till the CX register is zero
            MOV AH, 4CH             ; Terminate the program
            INT 21H
            END
```

(*Result : 01,02, 03, p4, 05, 06, 07, 08, 0AH is moved to BLOCK 2 as required*).

ALP 6 : Write an ALP in 8086/8088 to add 5 bytes of data in an array by making use of Procedure.

Soln : Here let us write the program directly with normal format.

```
Label     Mnemonics and Operand              Comments
.MODEL SMALL
.DATA
          ARR DB 5,6,7,0A, 0FFH        ; Declaration of variable and data
          COUNT DW @-ARR               ; Count gives number of bytes in the ARR
          SUM DW ?
.CODE
          MOV AX, @DATA
          MOV DS, AX
          XOR AX, AX                   ; Clearing AX register to 0
          MOV SI, OFFSET ARR
          MOV CX, COUNT                ; Initializing the CX register with count
NEXT :    PUSH CX
          CALL NEAR PTR PROADD         ; Procedure which is with in the same
          INC SI                       ; Segement is called
          POP CX                       ; Procedure is repeated till the
          LOOP NEXT                    ; Count in CX is zero
          MOV SUM, AX                  ; Result is stored in memory
          MOV AH, 4CH                  ; Normal termination to DOS
          INT 21H
PROADD    PROCNEAR                     ; Near procedure add the contents
          ADD AL, [S1]                 ; In Al with contents pointed by the
          ADC AH, 00                   ; Index register, and the carry is stored in AH
          RET
          PROADD ENDP
          END
```

Note. In this program, the procedure *"PROADD"* can refer to the variable AR, COUNT, and SUM directly, because the main calling program and the procedure are in the same file.

ALP 7 : Write an ALP in 8086 to add two 16-digit packed BCD numbers.

Soln : The Packed BCD numbers have two digits in each byte, therefore 16-digit packed BCD we need 8 locations, and count is initialized into CX register. The first set of BCD numbers are brought into AL register and added to the second set of BCD numbers. Observe that we get only two digit numbers at a time since we decimal adjust the result after addition only in AL register. The order in which the addition takes place is by bringing only lower order BCD first followed by higher order bytes. After decimal adjust is completed the result is stored back into memory. When all8 bytes of BCD are added the result is stored, and the program is terminated, as usual.

Algorithm for ALP 7 :

Step No	Process
1.	Initialize data memory with data
2.	Initialize data segment register with the data segement address
3.	Initialize CX register with the count of 0008. (*i.e.*, number of bytes that has to be added)
4.	Initialize SI.DI and BP with the address pointing to first operand, second operand and the result respectively
5.	Clear the carry flag
6.	Move first operand byte into AL register
7.	Add with carry the contents of AL to the second operand byte
8.	Decimal adjust to obtain BCD number
9.	Increment to SI, DI and BP to point to the next byte of the first operand second operand and the result respectively
10.	Decrement CX register and go to step 6 is CX register is not zero.
11.	If the result is zero terminate the program.

The ALP for ALP 7

```
      .MODEL SMALL
      .DATA
            SOURCE1   BD   01H, 08H, 99H, 42H, 11H, 46H, 32H, 55H
            SOURCE2   DB   55H, 44H, 66H, 41H, 83H, 21H, 41H, 84H
            RESULT    DB   8DUP (?)
      .CODE
Label   Mnemonic and Operand       Comments
        MOV AX, @DATA              ; Initialize the DS register
        MOV DS, AX                 ;
        MOV CX, 08                 ; Initialize the Counter with 08
        MOV SI, OFFSET SOURCE1     ; Initialize SI, DI and BP with addresses
        MOV DI, OFFSET SOURCE2     ;
        MOV BX, OFFSET RESULT      ;
        CLC                        ; Clear carry flag
BEG :   MOV AL, [S1]               ; Add two numbers after fetching the data first into
        ADC AL, [DI]               ; AL and then perform decimal adjust
        DAA                        ;
        MOV [BX], AL,              ; Save the result into memory
        INC SI                     ; Point to the next data
        INC DI                     ;
        INC BX                     ; Point to the next location to save
        LOOP BEG                   ; Have all the bytes computed ?
                                   ; If 'NO' go to BEG
```

```
        MOV AH, 4CH                  ; If 'YES' terminate the program and return to
        INT 21H                      ; DOS
        END
```

[The Result : 56, 52, 65, 84, 94, 67, 73, 39, 56 in same sequence].

ALP 8 : Write an ALP sequence to convert a 16-bit packed BCD number beginning at PACKED to unpacked BCD number and store result beginning at UNPACKED location.

Soln : As mentioned in ALP 7, the Packed numbers have two digits in 8 bit byte, that means there are four digits in 16-bit packed BCD number. The Unpacked numbers have one digit per byte, where the higher nibble is always zero. Therefore16-bit packed number requires four location to store it as unpacked number. Each time the least significant number is separated and is ANDed with 0FH, and such data is stored in memory, and other 120bits are rotated right. The process continued four times to get the unpacked number at the end.

```
Label   Mnemonic and Operand         Comments
        .MODEL SMALL
        .DATA
          PACKED DW 1234H            ; Declare packed number
          UNPACKED DB 4 DUP (?)      ; Allocate the space for unpacked Data
        .CODE                        ;
          MOV AX, @DATA              ; Initialize ES with data area
          MOV DS, AX                 ;
          MOV DI, OFFSET UNPACKED    ; Number  is loaded
          MOV AX, PACKED             ; Packed number is loaded into AX
NEXT :    MOV CX, 04                 ; The number of digits is initialized
          MOV BX, AX                 ;
          PUSH CX                    ; Save CX on stack
          ADD AX, 0000               ; Add 00 to Ax register
          AAA                        ; Adjust to obtain unpacked number in AL
          MOV [DI], AL               ; Save the data
          INC DI                     ; Point to the next address to save unpacked
DATA
          MOV CL,04                  ;
          SHR BX, CL                 ; Obtain the previous digit to first digit position
          MOV AX, BS                 ;
          POP CX                     ; Reload the count into CL
          LOOP NEXT                  ; Loop till all 4 digits are converted
          MOV AH, 4CH                ; Terminate the program
```

```
        INT 21H
        END
```

[*The Result : 04, 03, 02, 01 is the output for the ALP8*]

7.5 COMPUTATION OF LCM

Here let us developed a ALP using 8086 to compute LCM of two 16 bit unsigned integers, and store result in word location LCM.

7.5.1 Approach Methodology

Let the two 16 bit numbers be 25 and 15. First let us divide the first number (25) by the second number (15). When there is no remainder, then the first number is the LCM. But there is a remainder, in this case.

Therefore we try if 2 × 25 = 50 is the LCM. Again, 50 does not divide exactly, and so is not the LCM.

So we try if 3 × 25 = 75 is the LCM. In this case 75 divides exactly by 15. Thus 75 is the LCM of 25 and 15.

7.5.2 Explanation of New Directives Encountered

DW Directive : DW stands for 'Define Word', which defines the data item size as a 16 bit value.

'*VALUES DW 0025,0015*' means that at symbolic memory location the word 0025 is stored. It is stored using 16 bits *i.e.*, two bytes.

In location VALUES is stored 19H, (*i.e., 0025 = 0019H*) which the LS byte and in location VALUES +1 is stored 00H, which is the MS byte. (*In 8086 programs, always LS byte is stored at a lower address and the MS byte at a higher address in memory*).

In location Values + 2 is stored 0FH and in location Values + 3 is stored 00H, as 0015 is 000FH.

Some of the several possible equivalent representations for 0025 in the above example are as below :

25 19H 0019H 31 31Q 31q 00031Q 11001B

7.5.3 Few More Conventions Used in Writing Comments

Here we consider three cases as below :

(*a*) Consider following lines of the program :

```
JE      EXIT    ;NJ             | NJ     |
POP     DX      ;[
POP     AX      ;DX_AX = 00,25  | 00,50  | -      | ]
```

Here, DX_AX stands for the contents of DX and AX.

DX_AX = 00, 25 means DX = 00 and AX = 25

DX_AX = - means DX and AX are not affected, because POP DX and POP AX are not executed in view of control getting transferred to location EXIT.

(*b*) Consider following lines of the program

```
AGAIN :   PUSH AX   ;[
POP       AX        ;s_top = 00,25      |00, 50   |00,75     |]
```

Here, s_top = 00, 25 means the topmost element on the stack is 00, and below that we have 25. 's_top' stands for stack top.

(*c*) Consider following lines of the program

```
          CMP DX,0   ;10 Vs 0          |5 Vs 0   |0 Vs 0   |
          JE EXIT    ;NJ               |NJ       |J        |
```

Here, '*10Vs 0*' stands for '*compare 10 versus 0*'. The '*NJ*' below '10 vs *0*' *stands for* '*No Jump*'.

7.5.4 Working of the Program

The program find LCM of two 16 bit unsigned numbers at word location VALUES and VALUES + 2. Initially, the 32 bit result will be accumulated in DX (*MS word*) and AX (*LS word*). The result will be stored in word locations LCM (*LS Word*) and ICM + 2 (*MS Word*) at the end.

7.5.5 ALP to compute LCM

The program is split into 5 PAGS's. First we will write ALP and next the comment for each PAGA can be written.

```
          .NAME       LCM
          PAGE        60, 80
TITLE     PROGRAM TO FIND LCM OF 216 BIT UNSIGNED NUMBERS
          .MODEL SMALL
          .STACK      64
          .DATA
VALUES    DW          0025, 0015
LCM       DW          2 DUP (?)
```

PAGA No./Label	Mnemonics		Operand			
	.CODE					
PAGA 1						
L_C_M :	MOV	AX, @DATA	;			
	MOV	DS, AX	;			
;PAGA 2						
	MOV	DX, 0	;[			
	MOV	AX, VALUES	; DX_AX - 25]			
	MOV	BX, VALUES + 2	; BX = 15			
;PAGA 3						
AGAIN	PUSH	AX	;[			
	PUSH	DX	; s_top = 00,25	00,50	00,75	]
	DIV	BX	; AX = 1	3	5	
			; DX = 10	5	0	]
	CMP	DX,0	; 10 Vs 0	5 Vs 0	0 Vs 0	
	JE	EXIT	; NJ	NJ	J	
	POP	DX	; DX_AX = 00,25	00,50	-	]
;PAGA 4	POP	AX	; [			
	ADD	AX, VALUES	; ax = 50	75	-	
	JNC	DX	; J	J	-	
	INC	BX	; -	-	-	
NOINCDX :	JMP	AGAIN	; J	J	-	
;PAGA 5						
EXIT :	POP	LCM + 2	; -	-	LCM + 2 = 0	
	POP	LCM	;-	-	LCM = 75	]
	INT	3	;			
	ALIGN	16				
	END	L_C_M	;			

PAGA No	Comments
1.	Move to DS the segment address of DATA area
2.	First number is moved to AX, and the second number is moved to BX, as working with register is faster compared to working with memory. To DX we move 0.
3.	We store on stack DX_AX for later use. We divide DX_AX by BX using DIV BX. Then we compare remainder in DX with 0. If DX equals 0, we go to para 5. Otherwise, we restore to DX_AX the value stored on the stack top.
4.	We add to AX the first number. When this results in a carry we increment DX. (*In other words, we add the first number to DX_AX, to have in DX_AX the next multiple of the first number*). Then, we go to para 3.
5.	The accumulated result in DX_AX which is on the stack top is now popped to word locations LCM + 2 and LCM. The Program is then terminated using break point interrput type number 3.

7.5.6 Changing the Data Values for the Program

Assume that after the program has run correctly for the data 25 and 15, we want to test the program with the values as 40H and 60H. Which can be done in two ways, below.

- **Changing data values in the source program :** Assume that, the name of the source file was LCM.ASM. Using a text editor, we can change the line 'VALUES DW 0025, 0015' in LCM.ASM to 'VALUES DW 40H, 60H'.

  Next, we can assemble this program, link it, and finally run this program to get the LCM of 40H and 60H.

Note. Such exercise of assembling and linking can be avoided if we change the data in memory using DEBUG.

- **Changing data values in memory using DEBUG :** Assume that, we have just tested the program with data values 25 and 15. Now we want to change the data in memory to 40H, 00H, 60H, 00H from the original 19H, 000H, 0fH, 00H.

First we type 'L< cr >' while we are still in the DEBUG mode, which results in Loading the LCM.EXE program once again for execution.

Next we run the program in single step till MOV DS,AX is executed. Thus now DS is loaded with the segment address of DATA area.

Now we type 'e ds:0<cr>'. Here, 'e' stands for '*Enter a new value*'. DEBUG will display the contents of ds:0 which is 19H and we type '40<sb>' to change it to 40H. Here, <sb> stands for pressing the SPACE BAR key.

So, now automatically the next byte, which is 00, is displayed, since we are not wanting any change in this we simply type <sb>. That means the value is not alterd and next it displays 0F. Now we type '60<sb>' to change it to 60H. The next byte 00 is now displayed. Since we are not interested in changing this, and we are not interested in making changes for the subsequent bytes we type <cr> to terminate the Entry process. All we have to do now is to type 'G<cr>' to run the program and to test the results.

7.6 SAMPLE ALP ON ASCII AND STRINGS. HERE LET US DEVELOP PROGRAMS ON ASCII CODES

ALP-ASCII (1) Algorithm Write an ALP in 8086/8088 to multiply an ASCII string of 16 bytes by a single ASCII digit and the result be a string of unpacked BCD digit.

Algorithm for ASCII (1) :

Step No.	Explanation
1.	Initialize data in data memory area
2.	Initialize DS and ES with data area addresses
3.	Initialize SI & DI with the offset addresses of multiplicand and the result
4.	Load multiplier into DL register
5.	Initialize CX register as counter
6.	Initialize the destination location with 00
7.	Convert multiplier from ASCII to a singled digit number decimal number
8.	Fetch string bytes from memory and convert it to decimal number

9.	Multiply the two numbers and ASCII adjust the result
10.	Add the lower bytes of the result to the memory pointed to DI and ASCII adjust
11.	Store the result in the destination and increment the destination pointer
12.	Store the higher bytes of the result in the memory
13.	Have all the 16 bytes been multiplied
14.	If No, go to step 8
15.	Terminate the program normally.

```
.MODEL  SMALL
.DATA
   STRI      DB 32, 34, 31, 32, 31, 30, 32, 33, 37, 31, 32, 34, 37, 30, 32 }
   MULTIP  DB 36                                                         } Initialization
   RESULT  DB 16 DUP (?)                                                 }
.CODE                                                                    }
Label   Mnemonics Operand                 Comments
        MOV AX, @DATA
        MOV DS.AX
        MOV ES, AX
        MOV SI, OFFSET STRI        ; Initialize with string Address
        MOV DI, OFFSET RESULT      ; Initialize with result Address
        MOV DL, MULTIP             ; Multiplier is loaded into Register
        MOV CX, 10H                ; Counter is initialized to 16 Dec.
        AND DL, OFH                ; Seperate the lower-nibble of DL
LOOP1:  MOV AL, [SI}               ; Access the string data and adjust the
        INC SI                     ; pointer to next string data
        AND AL, OFH                ; Clear the higher nibble of ASCII String
        MUL DL                     ; Multiply with the Multiplier and save the result
                                   ; after ASCII- adjust
        AAM
        ADD AL, [DI]
        AAA
        MOV [DI], AL               ; Store the result back in memory
        INC DI                     ; Adjust the pointer to the next memory address
        MOV [DI], AH               ; and save the higher byte of the result in memory
        LOOP LOOP1                 ; Have all 16 ASCII string multiplied ?
        MOV AH, 4CH                ; If no, go to loop
        INT 21H                    ; Else terminate the program normally
        END
```

The Result for ASCII (1) : 02, 05, 00, 01, 03, 07, 00, 02, 09, 03, 00, 03, 05, 04, 04, 02, 01

Note : Here each digit of ASCII multiplicand is brought into AL register and the single digit multiplier is stored in a register and after each multiplication, the result is ASCII adjusted after multiplication and result is stored back into the memory. When there is carry, each operation, the carry is taken care during the next multiplication.

ALP-ASCII (2) : Write an ALP in 8086/8088 to add two 2-digit numbers and display the result and develop a procedure to convert result into ASCII code and display.

Algorithm for ASCII (2) :

Step No.	Explanation of Task
1.	Initialize DS with the data memory area
2.	Initialize Data for addition
3.	Initialize SI with the address of data
4.	Fetch the first data in AL
5.	Increment SI to point to next data and add
6.	Call the procedure ASCII - Code for display
7.	Terminate the program
8.	Move the result into DH-register
9.	Mask the higher nibble in AL and make lower nibble zero
10.	Move the higher nibble into lower nibble position
11.	Check whether AL is less than OAH. If yes, go to step 13
12.	No.ADD 07 to AL
13.	Add 30H to AL
14.	Save DX register on top of stack
15.	Display the contents of AL by moving it to DL and calling interrupt 21H function 02
16.	Pop the contents from the top of the stack into DX
17.	Mask the lower nibble and clear the higher nibble of DH - register
18.	Check whether DH is less than OAH. If yes, go to step 20
19.	Else add 07 to DH
20.	Add 30H to DH
21.	Display the contents of DH by moving it into DL and by calling INT 21 H function 02
22.	Return to main program

```
        .MODEL TINY
        .DATA
            VALUE DB 32H, 64H
        .CODE
Label   Mnemonics and Operand        Comments
        MOV AX, @DATA               ; Initialize data segment
        MOV DS, AX                  ;
        MOV SI, OFFSET VALUE        ; Initialize SI with the address of the data
        MOV AL, [SI]                ; Move the first data in AL
```

```
        ADD AL, [SI + 1]          ; Add with the second data
        CALL WRITE_CHAR           ; Display data on screen
        MOV AH, 4CH               ; Terminate the program normally
        INT 21H
        WRITE_CHAR PROC
        MOV DH, AL                ; Save the result in DH register
        AND AL, OFOH              ; Mark lower nibble and convert the higher
                                  ; nibble to ASCII value
        MOV AL, 04H
        ROR AL, CL
        CMP AL, 04H
        JB NEXT
        ADD AL, 07H
NEXT : ADD AL, 30H
        PUSH DX
        MOV DL, AL                ; Display the digit on the screen using DOS
                                  ; interrupt function 02
        MOV AH, 02H
        INT 21H
        POP DX                    ;
        AND DH, OFH               ; Now convert lower nibble to ASCII character
        JB NEXTI
        ADD AL, 07h
NEXTI : ADD DH, 30H
        MOV DL, DH                ; Display it on the screen
        MOV AH, 02H
        INT 21H
        RET
        WRITE_CHAR ENDP           ; End the procedure and return to main program
        END
```

The Result for ASCII (2) : 96 ; is displayed on screen

ALP-String (1) : Write an ALP in 8086 to display the string Information Science on the screen.

Explanation to write ALP for ALP string (1) : By using DOS interrup, INT21; and function 09, it is possible to display a string of data on the screen. The function 09 needs the starting address of the function 09 needs the starting address of the string to be initialized in DI register, and using $ string in the data area has to be terminated.

```
.MODEL SMALL
.DATA
```

```
        MSG DB "INFORMATION SCIENCE $"
    CODE
        MOV AX, @DATA           ; Initialize the Data Segment
        MOV DS, AX              ;
        MOV DX, OFFSET MSG      ; Initialize DX with the starting of the message
        MOV AH, 09              ; Call interrupt 21 function 09
        INT 21H                 ; To display the string
        MOV AH, 4CH             ; Terminate the program
        INT 21H
        END
```

The Result for ALP-string (1) : INFORMATION SCIENCE is displayed.

7.7 SAMPLE ALP ON SORTING

Let us see few program on sorting the given set of numbers.

ALP-Sort (1) : Write an ALP in 8086 to sort a given set of 8-bit unsigned integers into accending order by bubble sort.

Algorithm for bubble sort :

Step No.	Explanation
1.	Initialize the data in memory
2.	Initialize size of data
3.	Initialize the DS register
4.	Initialize BX with [size-1]
5.	Copy contents of BX to CX register
6.	Load first element into AL and by using a pointing to next element
7.	Compair this element with next element
8.	If AL is greater exchange two elements in the memory
9.	Increment points and decrement CX
10.	If all elements compared in the array (Is CX = 0)
11.	If No, go to step 6
12.	If yes decrement BX, if BX content is not zero go to step 5.
13.	If BX is zero terminate program

Label Mnemonics and Operand Comments

```
        .MODEL SMALL
        .DATA
            A DB 23H, 45H, 55H, 22H, 14H, 64H
            SIZE1 DW $-A            ; Calculate size of numbers
        .CODE
```

```
            MOV AX, @DATA          ; Initialize data segment
            MOV BX, SIZE1          ; Number of data byte is initialized in BX
            DEC BX                 ; Load BX with (N-I)
OUT LP :    MOV CX, BX             ; Save count in CX
            MOV SI, O              ; Initialize pointer
INLP        MOV AL, A[SI]          ; Load data into AL pointed
                                     by SI
            INC SI                 ; Increment the point
            CMP AL, A[SI]          ; Is contents of AL less than
                                     that of SI pointed data ?
            JB NEXT                ; If yes, go to NEXT
            XCHG AL, A[SI]         ; If No, exchange two data in the memory
            MOV A[SI], AL          ; Repeat till end of memory
NEXT :      LOOP INLP              ; Hass all comparison is over
                                     in this pass ?
                                   ; If No, go to INLP to continue
            DEC BX                 ; Has all passes completed ?
            JNZ OUTLP              ; If No, go to OUTLP
            MOV AH, 4CH            ; Terminate program
            INT 21H
            END
```

The result for ALP-sort (1) : 14, 22, 23, 45, 55, 64

ALP-Sort (2) : Write an ALP in 8086 to sort a given set of 16-bit unsigned integers into ascending order using insertion sort algorithm.

Algorithm for—ALP sort (2) :

Step No.	Explanation
1.	Initialize the data in data memory location
2.	Initialize to calculate number of integers in location (size 1)
3.	Initialize the DS register with the data memory
4.	Initialize the CX register with count 2 (to sort first-two integers)
5.	Check whether all numbers up to the count indicated by CX is stored
6.	If No, sort them by comparing and inserting it in right position
7.	If Yes, include next number and increment the count in CX
8.	Has all numbers include and stored
9.	If No go to steps
10.	If Yes, terminate program

```
Lable       Mnemonics and Operand           Comments
            .MODEL SMALL
            .DATA
               LIST DW 0078H, 0058H, 0092H, 1124H
                                            ; Initialize data
               SIZEL DW (S-LIST)/2          ; number of integers is calculated
            .CODE
               MOV AX @DATA                 ; Initialize data segement
               MOV DS, AX
               MOV CX, 02                   ; Initialize register to sort first two integers
UP_LP :        MOV DX, CS                   ;
               DEC DX
               MOV SI, SI                   ; Index register point to the data
               MOV AX, LIST (SI)            ; The pointed data is brought to AX
IN_LP :        CMP LIST (SI-2), AX          ; Is previous data greater than AX ?
               JBE EXIT-IN                  ; If No, exit inner loop
               MOV DI, LIST (SI-2)          ; If Yes, move previous data
                                              to the present position
               MOV LIST (SI-1), DI          ;
               SUB SI,01                    ; Point SI to previous data
               DEC DX,                      ; Check all data has been compared
               JN2 IN_LP                    ; If No go to IN_LP
EXIT_IN :      MOV LIST (SI), AX            ; If Yes, save data into its position
               INC CX,                      ; Increment counter
               CMP CX, SIZE 1               ; Has all data been checked ?
               JBE UP_LP                    ; If No go to UP_LP
               MOV AH, 4CH                  ; If Yes terminate program
               INT 21
               END
```

The result for ALP-SORT (2) : 58,00, 78,00, 92,00, 24,11.

Note. The data is 16 bit, therefore lower byte is first stored in memory and higher byte.

REVIEW QUESTIONS

1. Write an ALP in 8086/8088 to find the maximum number in a given string of 16 bytes length.
2. Write an ALP in 8086 to divide a string of unpacked ASCII digits by a single ASCII digit.
3. Write an ALP in 8086 to display the string *"Welcome"* on the screen.
4. Write an ALP in 8086 to find cube of an 8-bit hexadecimal number.
5. Write an ALP to compare two strings and display whether they are equal or not. And it should display the length of the two strings at the end.

8 8086 INTERRUPTS AND P/C (8259A)

8.0 OBJECTIVE

In this chapter the reader will be made clear about Interrupt and its purpose. Different interrupts in 8086 will be discussed. 82C55 keyboard interrupt is also explained. The 8259A programmable, Interrupt controller is explained in details. At the end the real-time clock in discussed.

8.1 INTRODUCTION TO INTERRUPTS

Many times it will be required by computer system to execute one of the collection of special routines whenever certain conditions exists within the program. Most normal method of servicing such device is polled approach, where processor test each device in sequence and in effect ask each one if it needs communication with processor. In a large portion of main program it is easy with looping through continuous polling cycle, which many have serious and determinal effect system to limit the task. It may be assumed up μP to reduce cost effectiveness by such devices in use.

Therefore the preferable and suitable method will be one that allows μP to execute its main program and only stop to service peripheral devices where it is asked to do so by the device itself such method provide an external asynchronous input that would inform the μP to complete whatever instruction that is currently being executed and fetch new routine that will service the requesting device. After such servicing is completed, μP would assume at the same portion where it was left off and the method is called as ***'Interrupt Method'*** and processes is called ***"Interrupt Service Routine"*** - (ISR)

Such Interrupts increase processor system efficiency by letting an I/O device request CPU time only when that device needs immediate action. In Main-line programs can then perform routine tasks without continually checking I/O device status. In routine duties, interrupts generally accomplish programmed I/O upon device demand; respond to time-critical events; establish a time base with a time or clock; or count external events.

An Interrupt is a subroutine call initiated by external hardware. When an I/O device requires service, it sets interrupt request through a flip-flop as shown in Fig. 8.1 below.

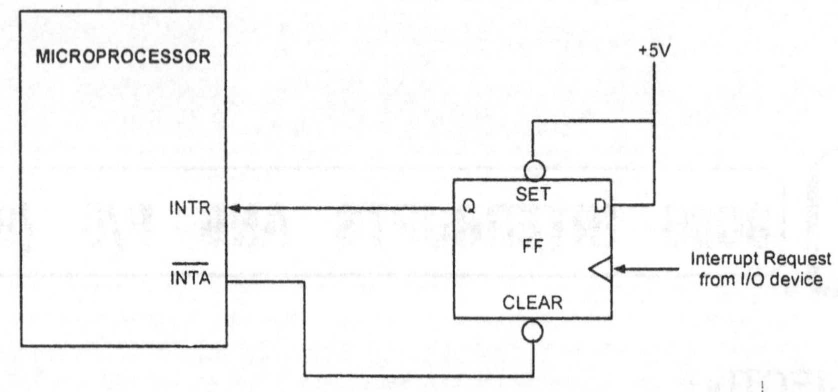

Fig. 8.1 Interrupt though FF

Such interrupt request is connected to the interrupt pin of the microprocessor as shown in Fig. 8.1 which stores the I/O device's interrupt request until it is acknowledged by the microprocessor.

As interrupt request is asynchronous, it may occur at any point in a program's execution. That means when an interrupt occurs, the execution of the current instruction is completed, the interrupt is acknowledged by the microprocessor, and control is transferred to a subroutine that services the interrupt. If the microprocessor responds to the interrupt, the interrupt request flip-flop is cleared by a signal directly from the microprocessor or by a device select pulse generated by the service routine.

8.2 PURPOSE OF INTERRUPTS

Interrupts are particularly useful when interfacing I/O devices that provide or require data at relatively low data transfer rates. We know that in a keyboard example using strobed input operation of the 82C55, the software polled 82C55 and its IBF bit to decide if data were available from the keyboard. If a person using the keyboard typed one character per second, the software for the 82C55 waited an entire second between each keystroke for the person to type another key. Such process is such a tremendous waste of time that designers have developed another process called *interrupt processing* to handle such situation, in the program.

Unlike the polling technique, interrupt processing allows the microprocessor to execute other software while the keyboard operator is thinking about what key to type next. Immediately after a key is pressed, the keyboard encoder de-bounces the switch and puts out one pulse that interrupts the microprocessor. In this way, the microprocessor executes other software until the key is actually pressed when it reads a key and returns to the program that was interrupted. Finally the microprocessor can print reports or complete any other task while the operator is typing a document and thinking about what to type next.

The Fig. 8.2 shows a time line that indicates a typist typing data on a keyboard, a printer removing data from the memory, and a program executing.

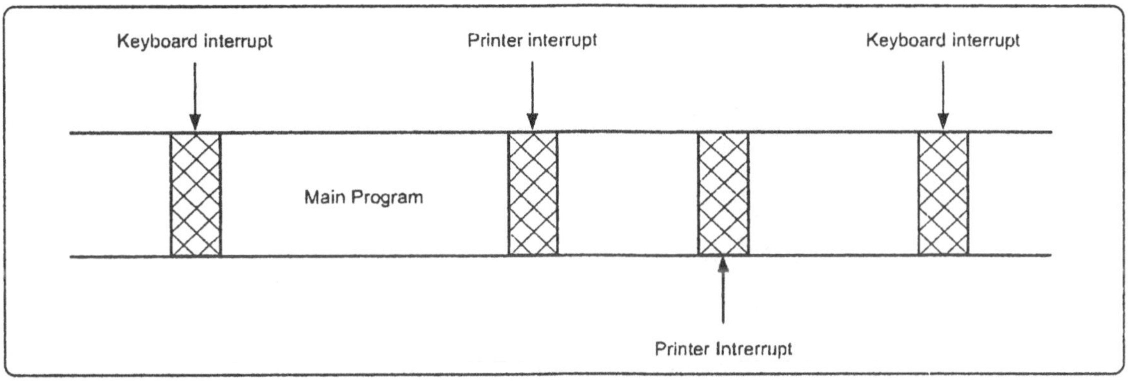

Fig. 8.2 Time line indicating Interrupt usage in a typical system

The program is the main program that is interrupted for each keystroke and each character that is to print on the printer. Observe that the keyboard interrupt service procedure, called by the keyboard interrupt, and the printer interrupt service procedure each take little time to execute.

8.3 THE INTERRUPT OPERATION

In the program of an 8086 processor, the clock will check at the end of each instruction cycle for the request of interrupts. The 8086 will respond to interrupt with following steps, when it is requested by an interrupt.

Steps	**Operation**
(*a*)	Decrements stack pointer by 2 and pushes the flag register on the stack.
(*b*)	Disables the INTR interrupt by clearing interrupt flag in the flag register.
(*c*)	It rests the trap flag in the flag register
(*d*)	Decrements stack pointer by 2 and pushes the current IP contents on the stack.
(*e*)	Decrements the SP again by 2 and pushes the current CS contents on the stack.
(*f*)	It performs an indirect far jump to the start of the interrupt service routine.
(*g*)	If there is an NMI then the microprocessor once again proceeds to step 'a'
(*h*)	If the TF = 1, (*for single step*), the program jumps to step 'a'.
(*i*)	If there is no nested interrupt, execution is completed.
(*j*)	Pops the IP first and then the CS upon encountering an IRET instruction
(*k*)	It pops the flag register.
(*l*)	Program execution is continued with the interrupted instruction in calling program.

The interrupt operation is still made clear by the following flow Chart of Fig. 8.3. The same can be analyzed in different ways.

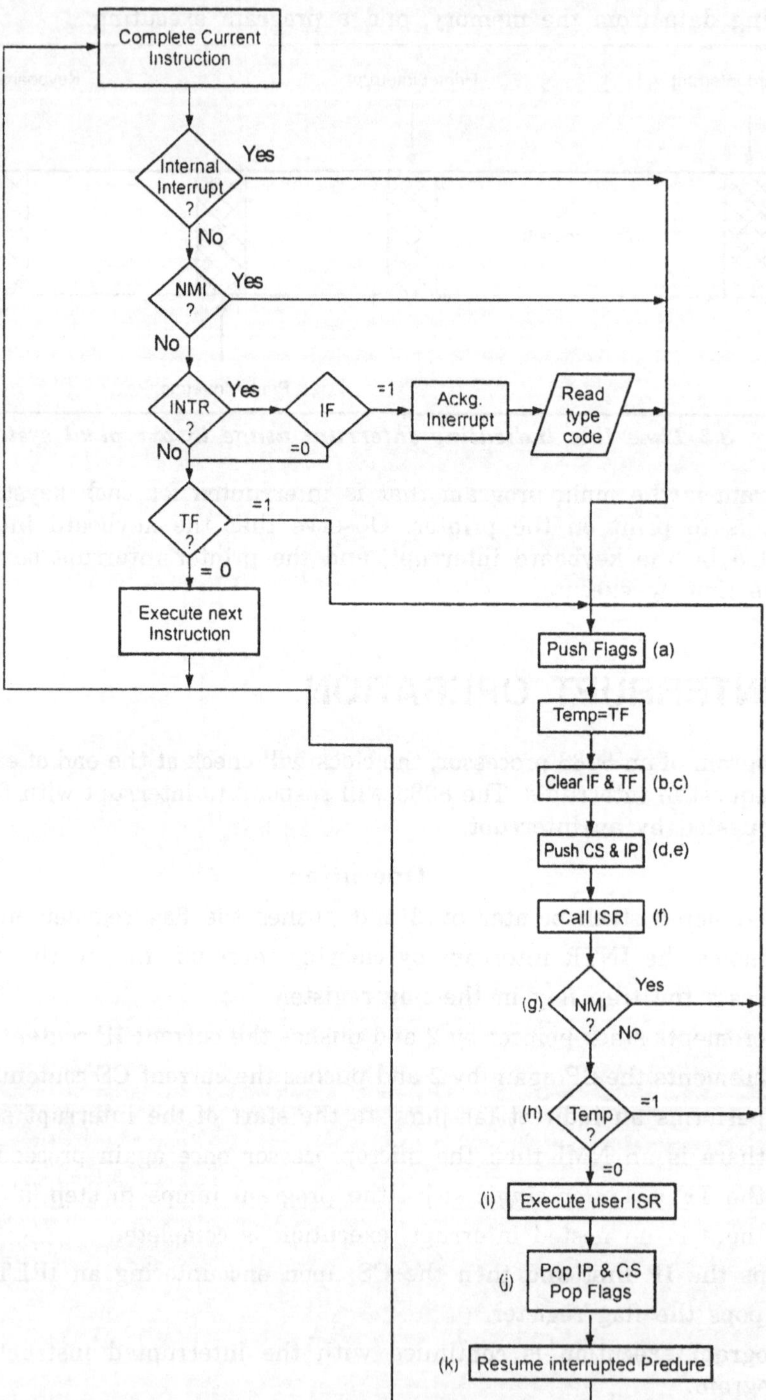

Fig. 8.3 The interrupt operation flow chart

8.4 THE INTERRUPTS

The interrupts of entire Intel family of microprocessors include two hardware pins that request interrupts (*INTR and NMI*) and one hardware pin $\overline{INTA}$ that acknowledges the interrupt requested through INTR. In addition to the pins, the microprocessor also has software interrupts INT, INTO, INT3, and BOUND. Two flag bits, IF (*interrupt flag*) and TF (*trap flag*), are also used with the interrupt structure and a special return instruction IRET.

8.4.1 Interrupt Vectors

Interrupt vectors and vector table are crucial to an understanding of hardware and software interrupts. The **interrupt vector table** is located in the first 1,024 bytes of memory at addresses 0000000H-0003FFH. It contains 256 different 4-byte interrupt vectors. An interrupt vector contains the address (*segment and offset*) of the interrupt service procedure.

The Fig. 8.4 depicts the interrupt vector table for μP.

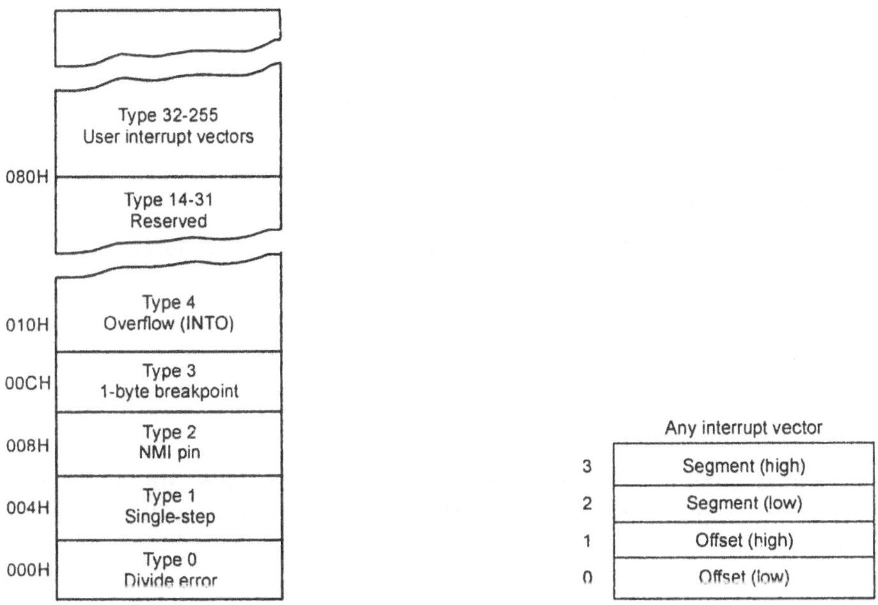

(*a*) For microprocessor **(*b*) Contents of an interrupt vector**

Fig. 8.4 Interrupt vector table

The interrupt vectors shown in Fig. 8.4 are identical in all Intel microprocessor family members from 8086 to Pentium. Each vector is four bytes long and contains the starting address of the interrupt service procedure. The first two bytes of the vector contain the offset address, and the last two bytes contain the segment address.

The following Table 1 describes the function of each dedicated interrupt in the microprocessor :

Table 1. Function of Interrupt

Type No.	Function
0	Divide Error-Occurs whenever the result of a division overflows or whenever an attempt is made to divide by zero
1	Single-Step or Trap-Occurs after the execution of each instruction if the trap TF flag bit is set. Upon accepting this interrupt, the TF-bit is cleared so that the interrupt service procedure executes at full speed. More detail is provided about this interrupt later in this section of the chapter.
2	Non-maskable Hardware Interrupt–A result of placing a logic 1 on the NMI input pin to the microprocessor. This input is non-maskable, which means that it cannot be disabled.
3	One-Byte Interrupt–A special 1-byte instruction (INT 3) that uses this vector to access its interrupt service procedure. The INT 3 instruction is often used to store a breakpoint in a program for debugging.
4	Overflow–A special vector used with the INTO instruction. The INTO instruction interrupts the program if an overflow condition exists, as reflected by the overflow flag (OF).

8.4.2 The 8086 Interrupts

In 8086 the interrupts have the possibility of appearing from any one of the four sources mentioned below :

1. An external signal applied on NMI pin or the interrupt INTR input pin (*user defined hardware interrupt*)
2. Execution of interrupt instruction INTR (*user defined software interrupt*)
3. As a result of execution of some software interrupt instructions
4. Program execution will automatically be interrupted if an attempt to divide by zero is executed (*predefined interrupt*)

8.4.3 External Interrupts

As mentioned earlier, 8086/8088 microprocessor have two lines for which external devices may use to signal interrupts *i.e.,* INTR and NMI, where NMI being with higher priority over that of INTR.

- **INTR Interrupt**

Once INTR is active, the processor takes different actions depending on the state of the interrupt enable flag (IF). When 'IF' flag is cleared (*INTR is masked or disabled*), the CPU ignores the interrupt request and processes the next instruction. (*INTR is level triggered*). INTR signal is not latched by the μP, so it must be held active until response is received or the request is withdrawn. The microprocessor acknowledges the interrupt request only after completions of the currently executing instruction, by executing two consecutive interrupt acknowledge bus cycles (INTR). When a bus hold signal arrives (*via the HOLD or request/grant lines*) during the INTR cycles, which is not honoured until these cycles have been completed. Apart when the microprocessor is configured in maximum mode it activates the LOCK signal during these cycles to indicate to the other processors that they should not attempt to take control of the bus.

First INTA cycle is to indicate that the request has been honoured. Next during the 2nd INTA cycle, the device (*requesting the interrupt*) has to place on the data bus a byte containing the interrupt type (0-255). The μP reads this type code and uses it to call the corresponding interrupt procedure. The μP saves the Flag register and the address of the next instruction by pushing the code segment and Instruction pointer contents on to the stack and clears IF and TF.

Once reading the interrupt type number, the CPU μP multiplies it by 4. The number obtained (*in Hexadecimal*) is the address pointing to the vector interrupt table. The CPU accesses the IP and CS addresses and loads them into the respective registers. (*Processors then proceeds with the Interrupt service routine*).

▪ NMI INTERRUPT

The NMI is edge triggered and is generally used to signal the μP of a "*catastrophic*" event, such as an imminent loss of power, memory error detection or bus parity error. Interrupt requests arriving on NMI cannot be disabled; they are latched by the microprocessor and have higher priority than interrupt requests on INTR. The Non-maskable interrupts are predefined ax TYPE-2; the processor does not need to be supplied with a type code to call the NMI procedure, and it does not run INTA bus cycles in response to a request on NMI.

CPU (*microprocessor*) saves the flag-register and the address of the next instruction on the top of the stack (CS & IP). There are then replaced by the 2nd and 1st words of the interrupt vector table element, thus transferring control to the ISR.

Software Interrupts

Type 0 - 255 :

8086 INT instruction can be used to cause the 8086 to do one of the 256 possible interrupt types. The interrupt type is specified by the number as a part of the instruction. You can use an INT2 instruction to send execution to an NMI interrupt service routine. This allows you to test the NMI routine without needing to apply an external signal to the NMI input of the 8086.

With software interrupts it is possible to call the desired routines from many different programs in a system, (e.g. BIOS in IBM PC). The IBM PC has in its ROM collection of routines, each performing some specific function such as reading character from keyboard, writing character to CRT. This collection of routines referred to as Basic Input Output System or BIOS.

Here the BIOS routines are called with INT instructions, the summary of interrupt response and about its service can be observed as under :

(*a*) The 8086 pushes the flag register on the stack.

(*b*) It disables the single step and INTR input by clearing the trap flag and interrupt flag in the flag register.

(*c*) Next it saves the current CS and IP register contents by pushing them on the stack.

(*d*) It performed an indirect far jump to the start of the routine by loading the new values of CS and IP register from the memory whose address calculated by multiplying 4 to the interrupt type, (e.g. If interrupt type is 4 then memory address is $4 \times 4 = 10_{10}$ = 10H. So 8086 will read new value of IP from 00010H and CS from 00012H).

(*e*) Once these values are loaded in the CS and IP, 8086 will fetch the instruction from the new address which is the starting address of interrupt service routine.

(*f*) An IRET instruction at the end of the interrupt service routine gets the previous values of CS and IP by popping the CS and IP from the stack.

(*g*) Finally at the end, the flag register contents are copied back into flag register by popping the flag register form stack.

INTR Interrupt 0 - 255

8086 INTR input can be used to interrupt a program execution, which is provided with a maskable handshake interrupt. This interrupt is implemented by using two pins - INTR and $\overline{\text{INTA}}$. This interrupt can be enabled or disabled by STI (IF = I) or CLI (IF = 0), respectively. When the 8086 is reset, the interrupt flag is automatically cleared (IF = 0). That means after reset INTR is disabled (User has to execute STI instruction to enable INTR interrupt).

The respond of 8086 to INTR interrupt is discussed as under :

(*a*) 8086 first does two interrupt acknowledge machine cycles as depicted in the Fig. 8.5 to get the interrupt type from the external device. In first interrupt acknowledge machine cycle the 8086 floats the data bus lines AD_0-AD_{12} and sends out an $\overline{\text{INTA}}$ pulse on its $\overline{\text{INTA}}$ output pin, which indicates an interrupt acknowledge cycle in progress and the system is ready to accept the interrupt type from the external device. During the second interrupt acknowledge machine cycle the 8086 sends out another pulse on its $\overline{\text{INTA}}$ output pin. In response to this second $\overline{\text{INTA}}$ pulse the external device puts the interrupt type on lower 8 bits of the data bus.

(*b*) Once 8086 receives the interrupt type, it pushes the flag register on the stack, clears TF and IF, and pushes the CS and IP values of the next instruction on the stack.

(*c*) 8086 then gets the new value of IP from the memory address equal to 4 times the interrupt type (number), and CS value from memory address equal to 4 times the interrupt number plus 2.

The interrupt acknowledge cycle is explained below in Fig. 8.5

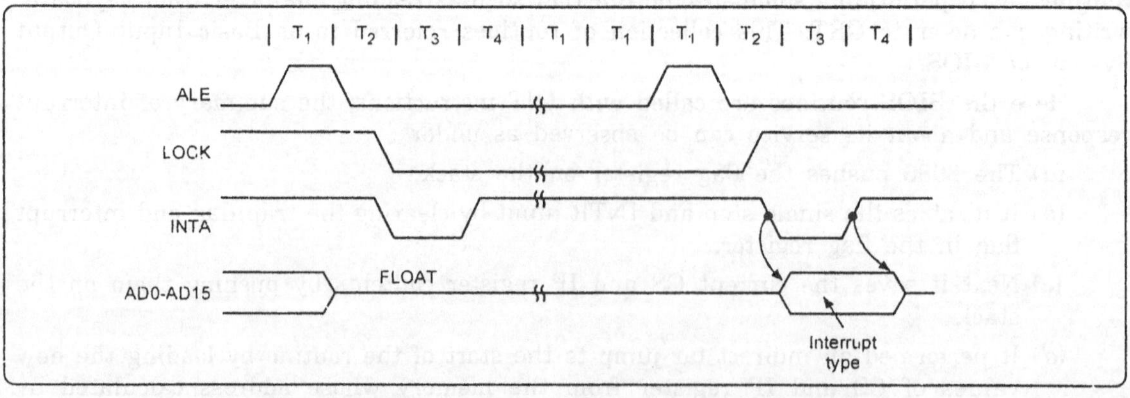

Fig. 8.5 The interrupt acknowledge M/C cycle

In minimum mode, the $M/\overline{IO}$ is low, indicating I/O operation during the $\overline{INTA}$ bus cycles. The 8086 activates $\overline{LOCK}$ signal by making it low from T_2 of the first cycle until T_2 of the second bus cycle to avoid the BIU from accepting a hold request between the two $\overline{INTA}$ cyles.

In maximum mode, the staus lines S_0 to S_2 enable the $\overline{INTA}$ output for each cycle via the 8288. 8086 activates $\overline{LOCK}$ from T_2 of the first cycle until T_2 of the second to prevent the 8086 from accepting a hold request on either $\overline{RQ}/\overline{GT}$ input and to prevent bus arbitration logic from releasing the bus between $\overline{INTA}$s in multiprocessor system.

8.4.4 Interrupt Instructions (INTO, INT and IRET)

Among these software interrupt instruction available to µP, INT and INT 3 are similar, INTO is conditional and IRET is a special interrupt return instruction.

- INTO instruction checks overflow flag (OF). If OF = 1, the INTRO instruction calls the procedure whose address is stored in interrupt vector type number 4. For Of = 0, then the INTRO instruction performs no operation and the next sequential instruction in the program executes.
- INT n instruction calls the interrupt service procedure that begins at the address represented in vector number n. For instant an INT 80Hor INT 128 call the interrupt service procedure whose address is stored in vector type number 80H (000200H-00203H). To determine the vector address, just multiply the vector type number (n) by 4, which gives the beginning address of the 4-byte long interrupt vector. For instant, an INT 5 = 4 × 5 or 20 (14H). The vector for INT 5 begins at address 000014H and continues to 000017H. Each INT instruction is stored in two bytes of memory with the first byte containing the opcode and the second the interrupt type number. The only exception to this is the INT 3 instruction, a 1-byte instruction. The INT 3 instruction is often used as a breakpoint interrupt because it is easy to insert a 1-byte instruction into a program. Breakpoints are often used to debug faulty software.
- IRET instruction is a special return instruction used to return for both software and hardware interrupts. The IRET instruction is much like a normal far RET, since it retrieves the return address from the stack. Which is unlike the normal return because it also retrievers a copy of the flag register from the stack. An IRET instruction removes six bytes from the stack two for the IP two for CS and the for the flags.

8.5 OPERATION OF REAL MODE INTERRUPT

If microprocessor completes executing the current instruction, it determines whether an interrupt is active by checking (1) instruction executions, (2) single-step, (3) NMI, (4) coprocessor segment overrun, (5) INTR and (6) INT instruction in the order presented. If one or more of these interrupt conditions are presents, the following sequence of events occurs :

- The contents of the flag register are pushed onto the stack.

- Both the interrupt (IF) and trap (TF) flags are cleared. This disables the INTR pin and also the trap or single-step feature.
- The contents of code segment register (CS) are pushed onto the stack.
- The contents of the instruction pointer (IP) are pushed into the stack.
- The interrupt vector contents are fetched and placed into both IP and CS so that the next instruction executes at the interrupt service procedure addressed by the vector.

If an interrupt is accepted, the microprocesor stacks the contents of the flag register, CS and IP; clears both IF and TF; and jumps to the procedure addressed by the interrupt vector. After the flags are pushed onto the stack, IF and TF are cleared. These flags are returned to the state prior to the interrupt when IRET instruction is encountered at the end of the interrupt service procedure. Hence, if interrupts were enabled prior to the interrupt service procedure, they are automatically re-enabled by the IRET instruction at end of the procedure.

The return address (in CS and IP) is pushed onto the stack during the interrupt. Once a while the return address points to the next instruction in the program, and sometimes it point to the instruction or point in the program where the interupt occurred.

8.6 OPERATION OF A PROTECTED MODE INTERRUPT

In protected mode, interrupts have similar assignments as in real mode, but the interrupt vector table is different. In place of interrupt vectors, protected mode uses a set of 256 interrupt descriptors stored in an interrupt descriptor table (IDT), which is 256 × 8 (2K) bytes long with each descriptor containing eight bytes. The interrupt descriptor table is located at any memory location in the system by the interrupt descriptor table address register (IDTR).

The IDT contains the address of the interrupt service procedure in the form of a segment selector and a 32-bit offset address in each its entary. Apart it contains the P-bit (present) and DPL-bits to describe the privilege level of the interrupt. The Fig. 8.6 shows the contents of the interupt descriptor below :

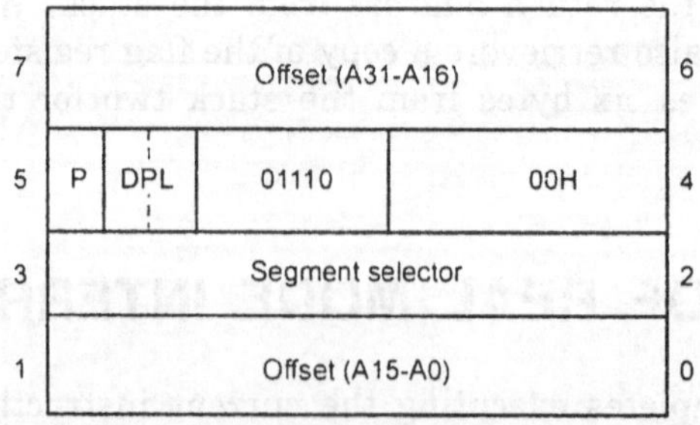

Fig. 8.6 Protected m ode interrupt descriptor

The Real mode interrupt vectors can be converted into protected mode interrupts by copying the interrupt procedure address from the interrupt vector table and converting them to 32-bit offset addresses that are stored in the interrupt descriptors. A single selector and segment descriptor can be placed in the global descriptor table that identifies the first IM byte of memory as the interrupt segment.

Note. Apart IDT and interrupt descriptors, the protected mode interrupt functions similarly to real mode interrupt. We can return from both interrupts using the IRET or IRETD instruction. Only difference in protected mode is that the microprocessor accesses the IDT instead of interrupt vector table.

8.7 THE HARDWARE INTERRUPT

Almost all microprocessors has two hardware interrupt inputs : i.e., non-maskable interrupt (NMI) and interrupt request (INTR). When NMI input is activated, a type 2 interrupt occurs because NMI is internally decoded. The INTR input must be externally decoded to select a vector. Any interrupt vector can be chosen for the INTR pin, but we usually use an interrupt type number between 20H and FFH. $\overline{\text{INTA}}$ singal is also an interrupt pin on the microprocessor, but it is an output that is used in response to the INTR input to apply a vector type number to the data bus connections D7-D0. The Fig. 8.7 shows the three user interrupt connections on the Intel microprocessor.

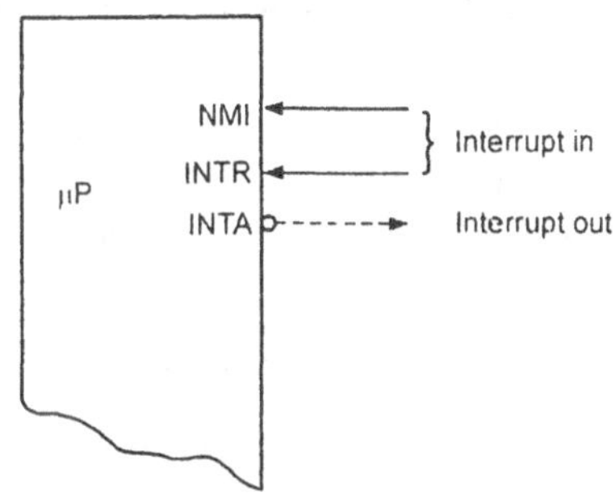

Fig. 8.7 Interrupt pins on all versions of the intel microprocessor

The **non-maskable interrupt** (NMI) is an edge-triggered input that request an interrupt on the positive edge (0-to-1 transition). After a positive edge, NMI pin must remain at logic 1 unit it is recognized by the microprocessor. Remember that before the positive edge is recognized, the NMI pin must be a logic 0 for at least two clocking periods.

NMI input is frequently used for parity errors and other major system faults such as power failures. The power failures can be easily detected by monitoring the AC power line and causing an NMI interrupt whenever AC power drops out. In response to such type of interrupt, the microprocessor stores all the internal register in a battery-backed-up memory or an EEPROM. The Fig. 8.8 depicts a power failure detection circuit that provides a logic 1 to the NMI input whenever AC is interrupted.

In the circuit of fig 8.8 an optical isolator provides isolation from the AC power line, output of which is shaped by a Schmitt-trigger invester that provides a 60 Hz pulse to the trigger input of the 74LS122 retriggerable monostable multivibrator. The values of R and C are selected so that the 74LS122 has an active pulse width of 33 ms or 2 AC input periods. Since the 74LS122 is retriggerable, as long as AC power is applied, the Q output remains triggered at a logic 1 and $\overline{Q}$ remains a logic 0.

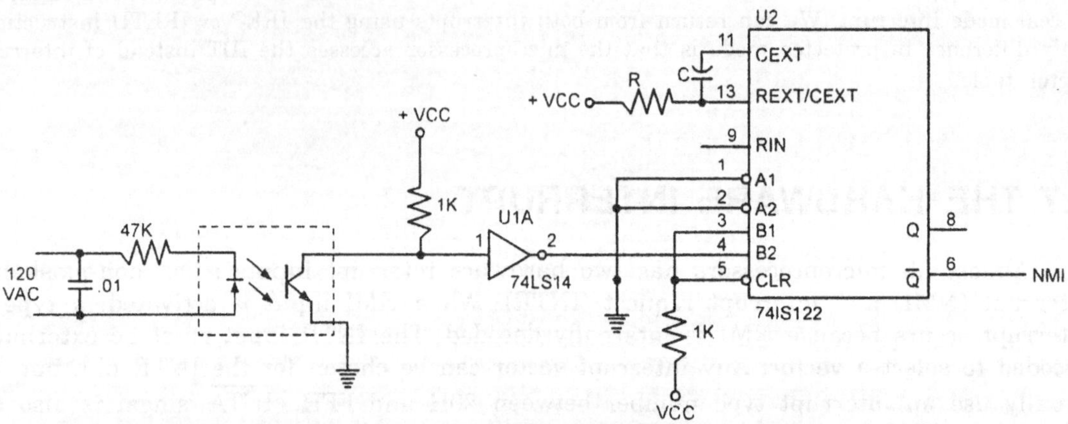

Fig. 8.8 Power failure defection circuit

If AC power fails, the 74LS122 no longer receives trigger pulses from the 74ALS14, which means that Q returns to a logic 0 and $\overline{Q}$ returns to a logic 1, interrupting the microprocessor through the NMI pin. Interrupt service procedure (*not shown here*), stores the contents of all internal registers and other data into a battery-backed-up memory. Such system assumes that the system power supply has a large enough filter capacitor to provide energy for at least 75 ms after the AC power ceases.

Now Fig. 8.9 depicts a circuit that supplies power to a memory after the DC power fails, where the diodes are used to switch supply voltages from the DC power supply to the battery.

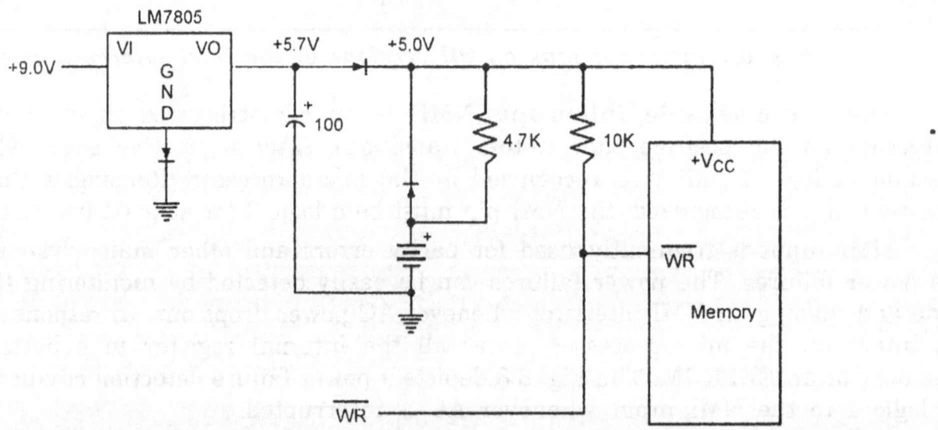

Fig. 8.9 Battery backed-up memory system using a NiCad, lithium, (or get cell)

In Fig. 8.8 diodes used are standard silicon diodes because the power supply to this memory circuit is elevated above + 5.0V to + 5.7V. Also notice that the resistor is used to trickle-charge the battery, which is either NiCAD, Lithium, (*or a gel cell*).

Note. For DC power failure the battery provides a reduced voltage to the V_{CC} connection on the memory device. Most memory devices will retain data with V_{CC} voltages as low as 1.5V so the battery voltage does not need to be + 5.0V. The $\overline{WR}$ pin is pulled to V_{CC} during a power outage so no data will be written to the memory.

8.7.1 The INTR and $\overline{INTA}$ Interrupts

The interrupt request input (INTR) is level-sensitive, which means that it must be held at a logic 1 level until it is recognized. INTR pin is set by an external event and cleared inside the interrupt service procedure such input is automatically disabled once it is accepted by the microprocessor and re-enabled by the IRET instruction at the end of the interrupt service processor.

Microprocessor responds to the INTR input by pulsing the $\overline{INTA}$ output in anticipation of receiving an interrupt vector type number on data bus connection D_7-D_0. The Fig. 8.10 shows the timing diagram for the INTR and $\overline{INTA}$ pins of the microprocessor. There are two $\overline{INTA}$ pulses generated by the system that are used to insert the vector type number on the data bus, as shown.

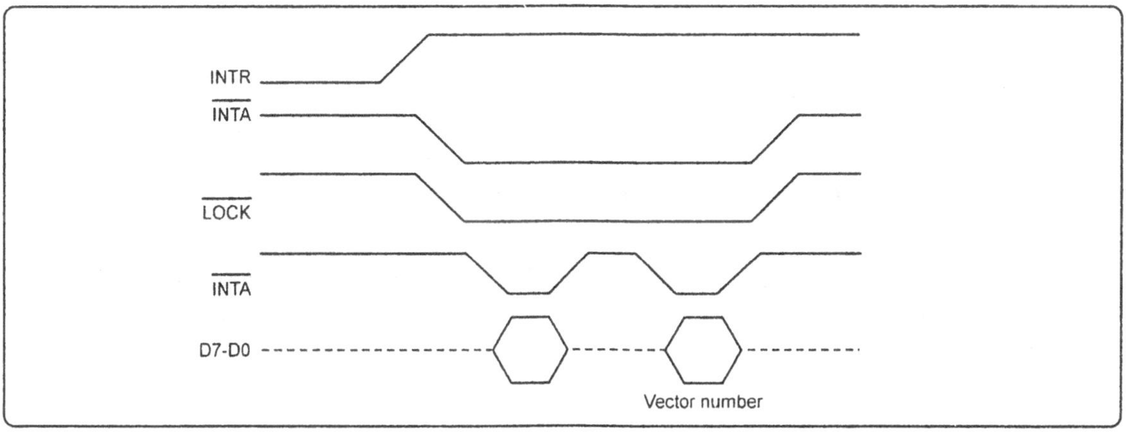

***Fig. 8.10 Timing of the INTR input and INTA OUTPUT** (A portion of the data bus is ignored and usually contains the vector number)*

Now Fig. 8.11 illustrates a simple circuit that applies interrupt vector type number FFH to data bus in response to an INTR. Observe that the $\overline{INTA}$ pin is not connected in this circuit.

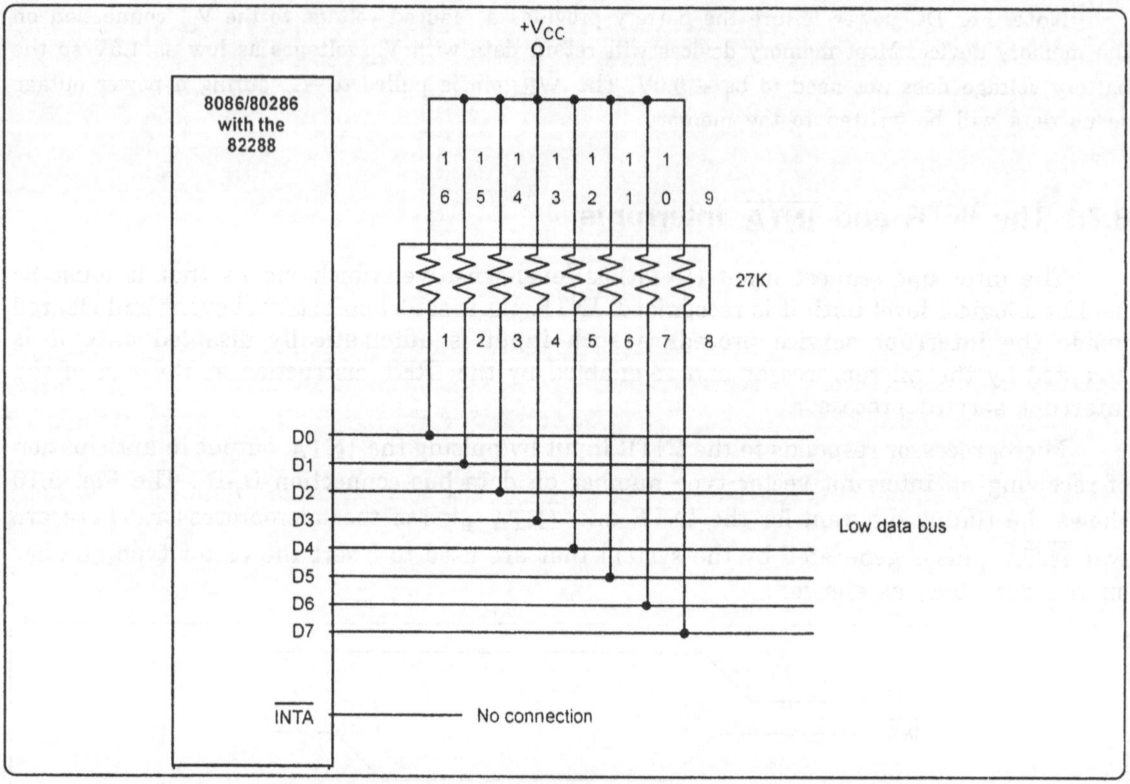

Fig. 8.11 Simple method for generating interrupt vector type number FFH in response to INTR

In Fig 8.11, since resistors are used to pull the data bus connections (D_0 - D_7) high, the microprocessor automatically sees vector type number FFH in response to the INTR input, which is possibly the least expensive way to implement the INTR pin on the microprocessor.

A Three-state Buffer for INTA

The Fig 8.12 shows interrupt vector type number 80H is applied to the data bus (D_0 - D_7) in response to an INTR. In response to the INTR, the microprocessor outputs the $\overline{\text{INTA}}$ that is used to enable a 74ALS244 three-state octal buffer.

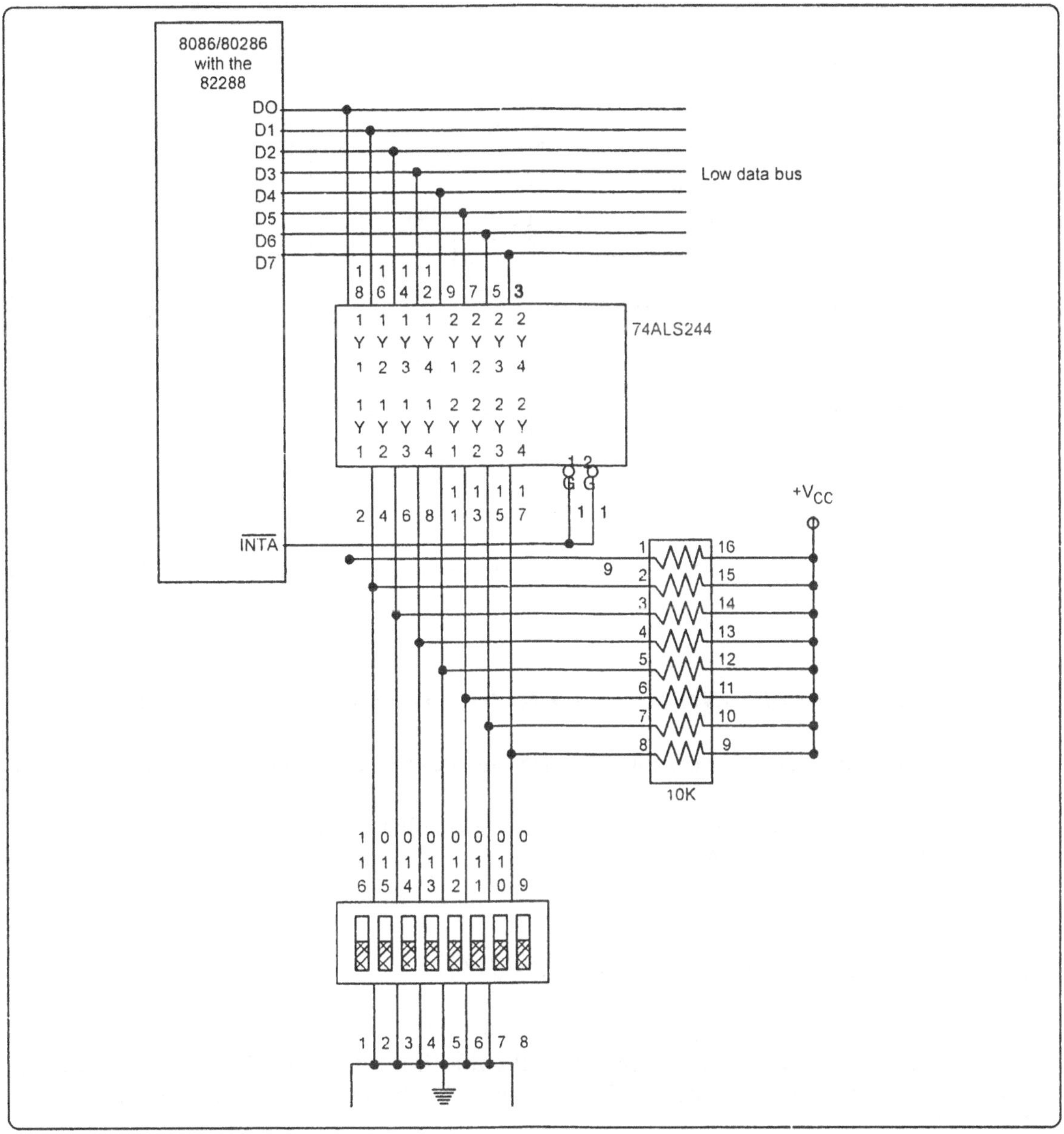

Fig. 8.12 A circuit that applies any interrupt vector type number in response to INTA.
(Here the circuit is applying type number 80H)

Here the octal buffer applies interrupt vector type number to the data bus in response to the $\overline{\text{INTA}}$ pulse. The vector type number is easily changed with the DIP switches that are shown here

Making INTR Input Edge-triggered

Frequently we require an edge-triggered input instead of a level-sensitive input. The INTR input can be convered to an edge-triggered input by using a D-type flip-flop, as shown in Fig 8.13 below :

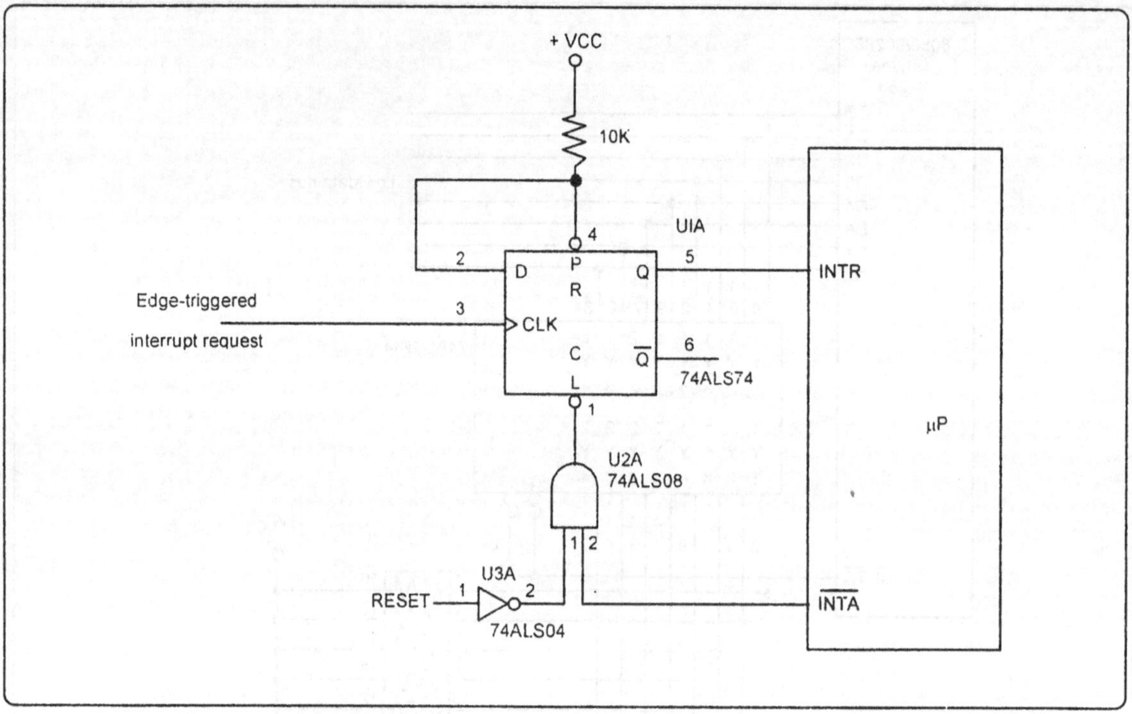

Fig. 8.13 Converting INTR into an edge-triggered interrupt request input

Note. In such arrangement clock input becomes an edge-triggered interrupt request input, and the clear input is used to clear the request when the $\overline{\text{INTA}}$ singal is output by the microprocessor. Also observe that the RESET signal initially clears the flip-flop so that no interrupt is requested when the system is first powered.

8.8 82C55 KEYBOARD INTERRUPT

Fig. 8.14 illustrates the interconnection of the 82C55 with the microprocessor and the keyboard which also shows how a 74ALS244 octal buffer is used to provide the microprocessor with interrupt vector type number 40H in response to the keyboard interrupt during the $\overline{\text{INTA}}$ pusle.

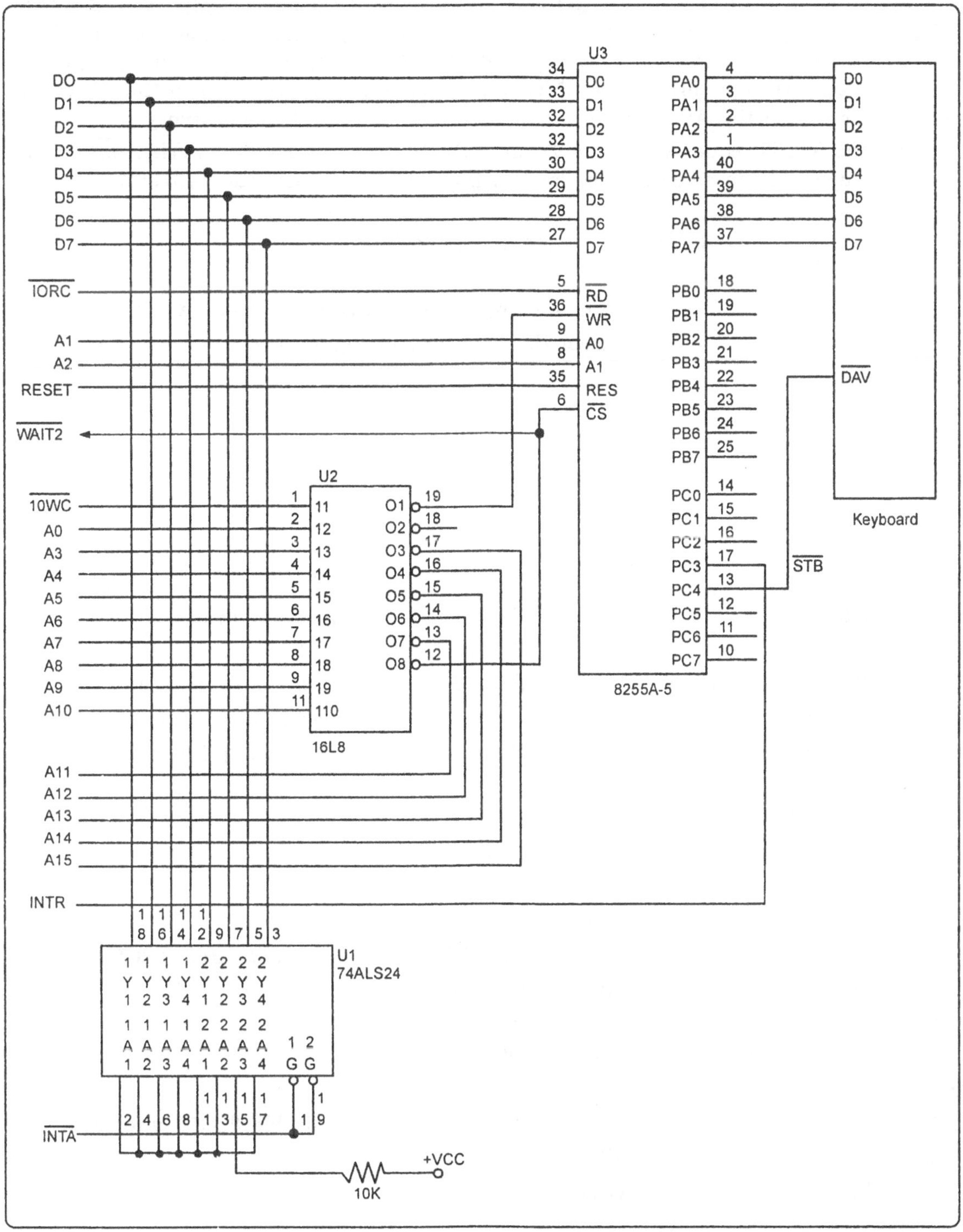

Fig. 8.14 An 8255A-5 interfaced to a keyboard from the microprocessor system using interrupt vector 40H

82C55 is decoded at 80386SX I/O port address 0500H, 0502H, 0504H, and 0506H by a PAL 16L8 (*the program is not illustrated*). The 82C55 is operated in mode 1 (*strobed input mode*) that means whenever a key is typed, the INTR output (PC_3) becomes a logic 1, requesting an interrupt through the INTR pin on the microprocessor. Then the INTR pin remains high until the ASCII data are read from port. A that means every time a key is typed, the 82C55 requests a type 40H interrupt through the INTR pin. The $\overline{DAV}$ signal from the keyboard causes data to be latched into port A and also causes INTR to become a logic 1.

Example 1 : The interrupt service procedure for keyboard

Procedure :

```
                                  ; An interrupt service procedure that reads a key
                                  ; from the keyboard in Figure 11-12
                                  ;
= 0500                            PORTA  EQU    500H
= 0506                            CNTR   EQU    506H
0000  0100 [                      FIFO   DB     256   DUP (?)  ; queue
              00
                ]
0100  0000                        INP    DW     ?       ; Input pointer
0102  0000                        OUTP   DW     ?       ; output pointer
0104                              KEY    PROCH  FAR   USES  AX  BX  DI  DX
0180  2E : 8B 1E 0100 R           MOV    BX, CS : INP   ; load input pointer
010D  2E : 8B 3E 0102 R           MOV    DI, CS : OUTP  ; load output pointer
0112  FE C3                       INC    BL             ; test for queue = full

0114  3B    DF                    CMP    BX, DI
0116  74    11                    JE     FULL           ; if queue is full
0118  FE    CB                    DEC    BL
011A  BA    0500                  MOV    DX, PORTA
011D  EC                          IN     AL, DX         ; get data from 82C55
011E  2E :  88 07                 MOV    CS : [BX], AL  ; save data in queue
0121  2E :  FE 06 0100 R          INC    BYTE PTR  INP
0126  EB    07 90                 JMP    DONE
0129                   FULL :
0129  B0    08                    MOV    AL, 8          ; disable 82C55 interrupt
012B  BA    05 06                 MOV DX, CNTR
012E  EE                          OUT DX, AL
012F                   DONE :
012F                              IRET
0134                   KEY        ENDP
```

It is to be noted that all registers affected by an interrupt be saved before they are used. In the software required to initialize 82C55 (*not shown here*). the FIFO is initialized so that both pointers are equal, the INTR request pin is enabled through the INTE bit inside the 82C55, and the mode of operation is programmed.

The procedure is a bit short since the 80386SX already known that keyboard data are available when the procedure is called. Data are input from the keyboard and then stored in the FIFO (*first-in, first-out*) buffer. Most of the keyboard interfaces contain a FIFO that is at least 16 bytes in depth. The FIFO in this example is 256 bytes, which is more than adequate for a keyboard interface. Observe how the INC BYTE PTR INP is used to add one to the input pointer and also make sure that it always addresses data in the queue.

Such procedure first checks to see if the FIFO is full which is indicated when the input point (INP) is one byte below the output pointer (OUTP). When the FIFO is full, the interrupt is disabled with a bit set/reset command to the 82C55, and a return from the interrupt occurs. And if the FIFO is not full, the data are input from port A, and the input pointer is incremented before a return occurs.

Example 2 : The procedure that removes data from FIFO

Procedure :

```
                              ; A procedure that reads data from the queue of
                              ; Example 11-5 and returns with it in AII
                              ;
0134                          READ  PROC   FAR   USES   BX   DI   DX
0137                          EMPTY :
0137  2E : 8B 1E 0100 R             MOV BX, CS : INP       ; load input pointer
013D  2E : 8B 3E 0102 R             MOV DI, CS : OUTP      ; load output pointer
0142  3B DF                         CMP BX, DI
0144  74 F2                         JE     EMPTY           ; if queue is empty
0146  2E : 8A 25                    MOV AH, CS : (DI)      ; get data
0149  B0 09                         MOV AL,9               ; enable 82C55 interrupt
014B  BA 0506                       MOV DX, CNTR
014E  EE                            OUT DX, AL
014F  2E : FE 06 0102 R             INC BYTE PTR CS:OUTP
RET
0157                          READ END ?
```

This procedure first determines whether the FIFO is empty by comparing the two pointers when the pointers are equal, the FIFO is empty, and the software waits at the EMPTY loop where it continuously tests the pointers. The EMPTY loop is interrupted by the keyboard interrupt, which stores data into the FIFO so it is no longer empty. Finally this procedure returns with the character in register AH.

8.9 THE 8259A PROGRAMMABLE INTERRUPT CONTROLLER

8259A programmable interrupt controller (PIC) adds eight vectored priority encoded interrupts to the microprocessor. Such controller can be expanded without additional hardware to accept up to 64 interrupt request input. This expansion requires a master 8259A and eight 8259A slaves.

8.9.1 General Description of the 8259A

The Fig. 8.15 depicts the pin-out of the 8259A, which is easy to connect to the microprocessor because all of its pins are direct connections except the $\overline{CS}$ pin, which must be decoded, and the $\overline{WR}$ pin, which must have an I/O bank write pulse.

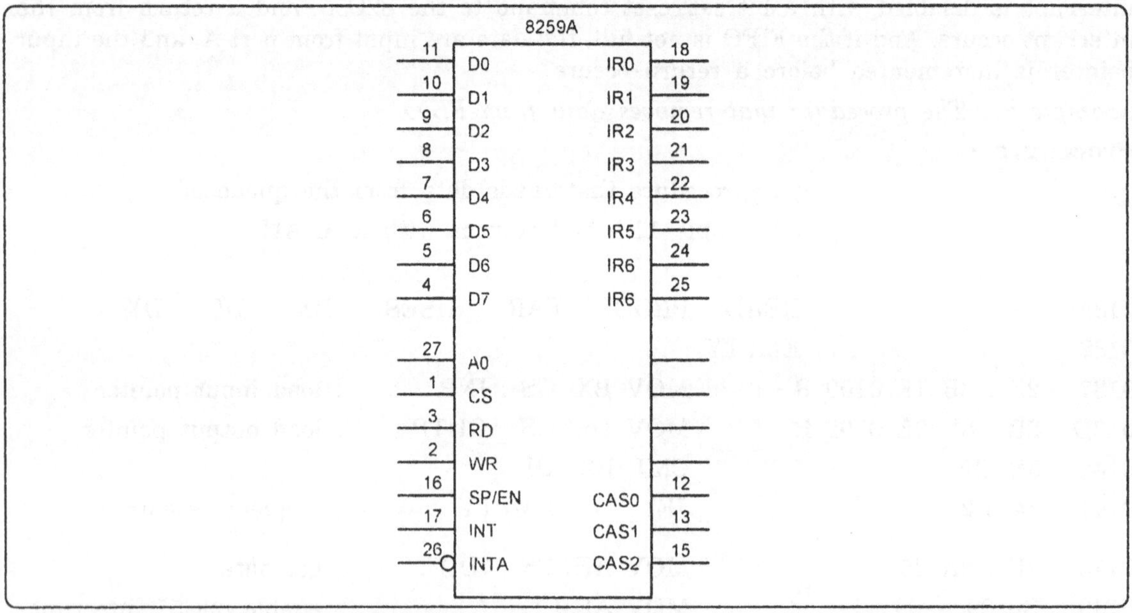

Fig. 8.15 Pin-out of the 8259A programmable interrupt controller (PIC)

The description of each pin on 8259A programmable interrupt controller is listed in the following Table 2.

Table 2. 8259A Pin Description

Pin-out	Description
D_7 - D_0	The **bi-directional data connections** are usually connected to either upper or lower data bus on the 80386SX microprocessor or the data on the 8088. If an 80486 or Pentium/Pentium Pro is used, then they connect to any 8-bit bank.
IR_7 - IR_0	**Interrupt request** inputs are used to request an interrupt and to connect to a slave in a system with multiple 8259As
$\overline{WR}$	The **write** input connects to either the lower or upper write strobe signal in a 16-bit system or any other bus write strobe in any size system

$\overline{RD}$	The **read** input connects to the $\overline{IORC}$ signal.
INT	The **interrupt** output connects to the INTR pin on the microprocessor from the master, and is connected to a master IR pin on a slave
INTA	**Interrupt acknowledge** is an input that connects to the $\overline{INTA}$ signal on the system. In a system with a master and slaves, only the master $\overline{INTA}$ signal is connected.
A_0	The A_0 address input selects different command words within the 8259A
$\overline{CS}$	The **Chip select** enables the 8259A for programming and control.
SP/$\overline{EN}$	**Slave program/enable buffer** is dual-function pin. If 8259A is in buffered mode, this pin is an output that controls the data bus transceivers in a large microprocessor-based system. When the 8259A is not in buffered mode, this pin programs the device as a master (1) or a slave (0).
CAS_2 - CAS_0	The cascade lines are used as outputs from the master to the slaves for cascading multiple 8259As in a system.

An interrupt can be used for a variety of applications, which requires a separate interrupt input. If we are working with an 8086, we get only two interrupt inputs INTR and NMI. For applications where we have multiple interrupt sources, we use external device called a priority controller (PIC). The Fig 8.16 depicts the connection between 8086 and 8259.

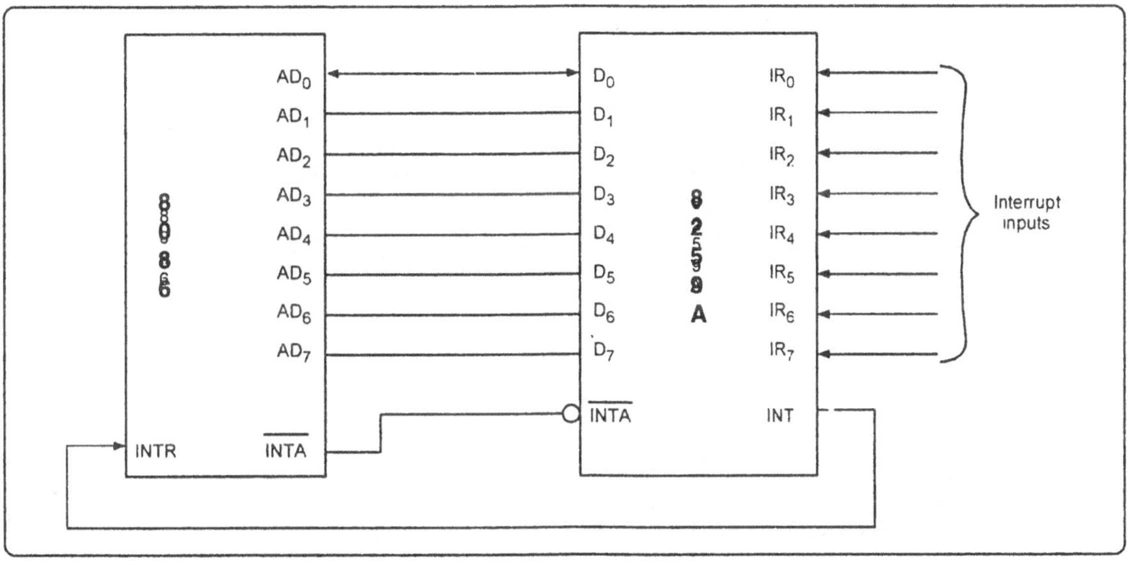

Fig 8.16 The connections between 8086 and 8259A

8259A is very commonly used priority interrupt controller, which is specifically designed for use with interrupt signals INTR and INTA of Intel series. It is packaged in a 28 pin DIP. It makes use of NMOS technololgy and requires a single + 5V supply to enerzise.

8.9.2 8259A PIC's Features

The 8259A PIC has many features which are useful to interrupt microprocessor. The main features are :

- It can manage eight priority interrupts. This is equivalent to provide eight interrupt pins on the processor in place of INTR pin.
- It is possible to locate vector table for these additional interrupts anywhere in the memory map. Of course all eight interrupts are spaced at the interval of either four or eight locations.
- By cascading 8259s it is possible to get 64 priority interrupts.
- Interrupt mask register makes it possible to mask individual interrupt request.
- 8259A can be programmed to accept either the level triggered or the edge triggered interrupt request.
- By making use of 8259A user can get the information of pending interrupts, in-service interrupts and masked interrupts.
- 8259A is designed to minimize the software and real time overhead in handling multilevel priority interrupts.

8.9.3 The Block Diagram of 8259, PIC

The Fig. 8.17 illustrate internal block diagram of the 8259A, which includes eight blocks : data bus buffer, read/write logic, control logic, three registers (*IRR, ISR and IMR*), priority resolver, and cascade buffer, as shown.

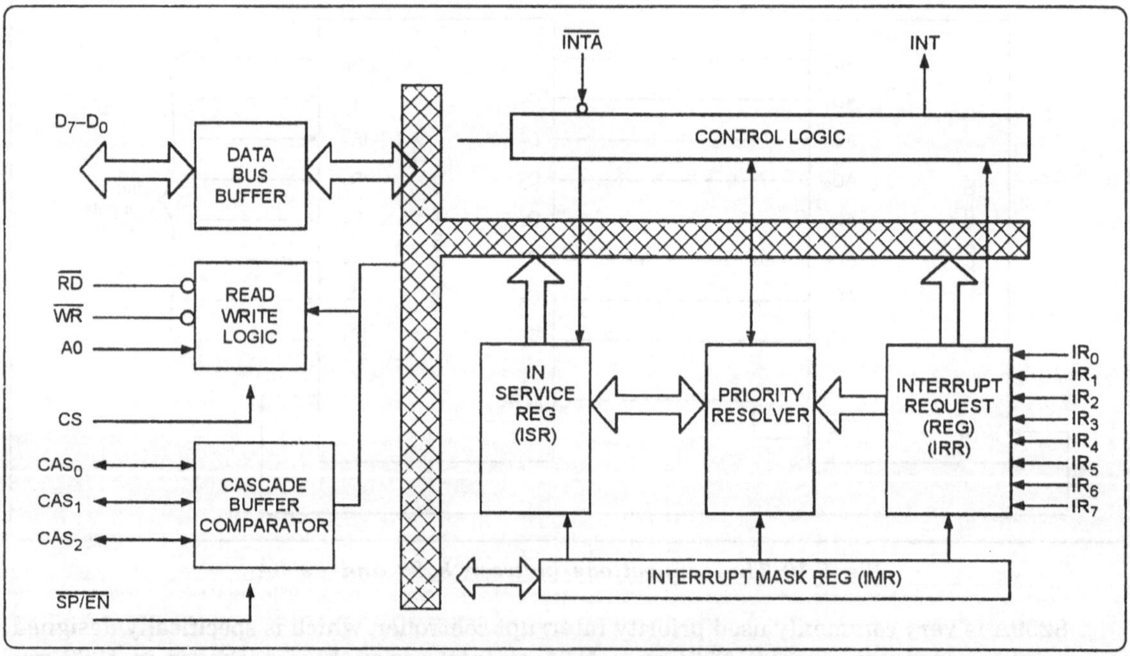

Fig. 8.17 Block diagram of 8259A PIC

Now let us briefly explain eight block of 8259A PIC :

1. Data Bus Buffer

Data bus allows the 8086 to send control words to the 8259A and read a status word from the 8259A and read a status word from the 8259A. The 8-bit data bus also allows the 8259A to send interrupt types to the 8086.

2. Read/Write Logic

The $\overline{RD}$ and $\overline{WR}$ inputs control the μ data flow on the data bus when the device is selected by asserting its chip select $(\overline{CS})$ input low.

3. Control Logic

This block has an input and an output line. When 8259A is properly enabled the interrupt request will cause the 8259A to assert its INT output pin high. If this pin is connected to the INTR pin of an 8086 and if the 8086 interrupt flag is set, then this high signal will cause 8086 to respond INTR as explain earlier.

4. Interrupt Request Register (IRR)

IRR helps to store all the interrupt levels which are requesting service. The eight interrupt inputs set corresponding bits of the interrupt Request Register.

5. Interrupt Service Register (ISR)

Interrupt Service Register (ISR) stores all the levels that are currently being serviced.

6. Interrupt Mask Register (IMR)

The Interrupt Mask Register (IMR) stores masking bits of the interrupt lines to be masked. This register can be programmed by an OCW. An interrupt which is masked by software will not be recognized and serviced even if it set the corresponding bits in the IRR.

7. Priority Resolver

Priority resolver determines the priorities of the bits set in the IRR. The bit corresponding to the highest priority interrupt input is set in the ISR during the $\overline{INTA}$ input.

8. Cascade Buffer Comparator

Finally this section generates control signals necessary for cascade operations. Which also generates Buffer-Enable signals. The 8259 can be cascaded with other 8259s in order to expand the interrupt handling capacity to sixty-four levels. In such, a case, the former is called a **master**, and the latter are called slaves. The 8259 can be set up as a master or a slave by the $\overline{SP}/\overline{EN}$ pin. The function of the pins of this block are :

(*a*) CAS 0 - 2

(*b*) $\overline{SP}/\overline{EN}$

(*a*) CAS 0 - 2

For master 8259, the CAS_0 - CAS_2 pins are outputs, and for slave 8259s, these are inputs. If 8259 is a master (that is, when it accepts interrupt requests from other 8259s), the CALL opcode is generated by the Master in response to the first $\overline{INTA}$. The vectoring address must be released by the slave 8259. The master sends an identification code of three-bits (*to select one out of the eight possible slave 8259s*) on the CAS_0 - CAS_2 lines and the slave 8259s accept these three signals as inputs (*on their* CAS_0 - CAS_2 *pins*) and compare the code sent by the master with the codes assigned to them during initialization. The slave thus selected (*which had originally placed an interrupt request to the master 8259*) then puts out the address of interrupt service routine during the second and third $\overline{INTA}$ pulses from the CPU.

(b) $\overline{SP}/\overline{EN}$ (Salve Program/Enable Buffer)

The $\overline{SP}/\overline{EN}$ signal is tied high for the master, but it is grounded for the slave.

In the case of large systems where buffers are used to drive data bus, the data sent by the 8259 in response to $\overline{INT}$ cannot be accessed by the CPU (*due to the data bus buffer being disabled*).

When 8259 is used in buffered mode (*Buffered or non-buffered modes of operation can be specified at the time of initialising the 8259*), the $\overline{SP}/\overline{EN}$ pin is used as an output which can be used to enable the system data bus buffer whenever the 8259's data bus outputs are enabled (*when it is ready to send data*).

That means, in non-buffered mode, the $\overline{SP}/\overline{EN}$ pin of an 8259 is used to specify whether the 8259 is to operate as a master or as a slave, and in the buffered mode, the $\overline{SP}/\overline{EN}$ pin is used as output to enable data bus buffer of the system.

8.9.4 The Interrupt Sequence

In 8086 systems the events of interrupt occurs in the sequence mentioned below :

- One or more of INTERRUPT REQUEST lines (IRO-IR7) are raised high, setting the corresponding IRR bit (s).
- The priority resolver checks three registers i.e., IRR for interrupt requests, the IMR for masking bits, and the ISR for the interrupt request being served. It resolves the priority and sets the INT high when appropriate.
- CPU acknowledge the INT and responds with an $\overline{INTA}$ pulse.
- After receiving an $\overline{INTA}$ form CPU, the highest priority ISR bit is set and the corresponding IRR bit is reset. The 8259A does not drive data bus during this cycle.
- A selection of priority modes is available to programmer so that the manner in which the requests are processed by the 8259A can be configured to match his system requirements. The priority modes can be changed or reconfigured dynamically at any time during the main program. So the complete interrupt service structure can be defined as required, based on the total system environment.
- 8086 will initiate a second INTA pulse. During this pulse, the 8259A releases a 8-bit pointer (*interrupt type*) onto the Data Bus where it is read by the CPU.
- It completes interrupt cycle. In AEOI mode the ISR bit is reset at the end of the second INTA pulse. Otherwise, the ISR bit remains set until an appropriate EOI command is issued at the end of the interrupt subroutine.

8.9.5 Operation Modes of 8259A

The 8259A PIC can operate on various modes as mentioned below :

(*i*) Fully Nested Mode

(*ii*) Special Fully Nested Mode

(*iii*) Rotating Priority Mode

(*iv*) Special Masked Mode, and

(*v*) Polled Mode

(*i*) Fully Nested Mode

Once initialization is done the 8259A operates in fully nested mode and called as default mode. The 8259 continues to operate in the Fully Nested Mode until the mode is changed through Operation Command Words. In such mode, IR0 has highest priority and IR7 has lowest priority. When the interrupt is acknowledged, it sets the corresponding bit in ISR. This bit will prevent all interrupts of the same or lower level, however, it will accept higher priority interrupt requests. The vector address corresponding to this interrupt is then sent. The bit in the ISR will remain set until an EOI command is issued by the microprocessor at the end of interrupt service routine. But if AEOI (*Automatic End of Interrupt*) bit is set, the bit in the ISR resets at trailing edge of the last $\overline{\text{INTA}}$.

■ End of Interrupt (EOI)

Is bit can be reset by an End Of Interrupt command issued by the CPU, normally just before existing from the interrupt routine.

In Fully Nested Mode, highest level in the ISR would necessarily correspond to the last interrupt acknowledged and serviced. Hence, a nonspecific EOI command may be issued by the CPU.

But when FNM is not used, 8259 may not be able to determine the last interrupt acknowledged. In such a case, a specific FOI command will have to be issued by the CPU.

Note. In cascade mode, the EOI command must be issued twice, once for the master and once for the slave.

■ Automatic End of Interrupt (AEOI)

When AEOI mode is set, 8259 will perform a nonspecific EOI on its own on the trailing edge of the third $\overline{\text{INTA}}$ pulse. (*AEOI mode can only be used for a master 8259 and not for a slave*).

(*ii*) Special Fully Nested Mode (SFNM)

In FNM on the acknowledgement of an interrupt, further interrupts from the same level are disabled. Hence, in large systems which use cascaded 8259s and where the interrupt levels within each slave have to be considered. An interrupt input to a slave, in turn causes the slave to place an interrupt request to the master on one of the master's inputs. Next interrupts to the slave will cause the slave to place requests to the master on the same input to the master, but these will not be recognized because further interrupts on the same input level are disabled by the master.

Special Fully-Nested Mode is used to avoid such problem. SFNM is set up ICW4 during initialization, which it is similar to the FNM except for the following two differences. i.e.,

(*a*) When an interrupt request from a slave is being serviced, the slave is allowed to place further requests (*these requests are of a higher priority than the request currently being serviced*). Such interrupts are recognized by master and it initiates interrupt requests to the CPU.

(*b*) Before existing from interrupt service routine, a non-specific EOI has to be sent to the slave and its ISR must be read to determine if it was the only interrupt to the slave. When the ISR is empty, a non-specific EOI command can be sent to the master, and when it is not empty, it implies that the same IR level input to the master is to be serviced again due to more than one interrupt being presented to the slave, and an EOI must be sent to the master.

(*iii*) Rotating Priority Mode

Rotating Priority mode can be set in

(*a*) Automatic Rotation, and (*b*) Specific Rotation

(*a*) Automatic Rotation. In this mode, a device, after being serviced, receives the lowest priority. Assuming that IR_3 has just been serviced, it will receive the seventh priority as shown in Table 3.

Table 3

IR_0	IR_1	IR_2	IR_3	IR_4	IR_5	IR_6	IR_7
4	5	6	7	0	1	2	3

(*b*) Specific Rotation. In Automatic Rotation mode, the interrupt request last serviced is assigned the lowest priority, but in Specific mode, the lowest priority can be assigned to any interrupt input (IR_0 to IR_7) thus fixes all other priorities.

For instant when the lowest priority is assigned to IR_2, other proprieties are as shown in Table 4.

Table 4

IR_0	IR_1	IR_2	IR_3	IR_4	IR_5	IR_6	IR_7
5	6	7	0	1	2	3	4

(*vi*) Special Mask Mode

If any interrupt is in service then corresponding bit is set in ISR and the lower priority interrupts are inhibited. Few applications may require an interrupt service routine to dynamically alter the system priority structure during its execution under software control, for instant the routine may wish to inhibit lower priority requests for a portion of its execution but enable some of them for another portion. In these cases we have to go to special mask mode.

Note. In special mask mode it inhibits further interrupts at that level and enables interrupts from all other levels (*lower as well as higher*) that are not masked.

Thus any interrupt may be selectively enabled by loading the mask register.

(*v*) Polled Mode

In this mode INT output is not used but the microprocessor checks the status of interrupt requests by issuing poll command. The microprocessor reads contents of 8259A after issuing poll command. During this read operation the 8259A provides polled word and sets ISR bit of highest priority active interrupt request FORMAT as shown on the next page.

1	X	X	X	X	W_2	W_1	W_0

1 = 1 → One or more interrupt requests activated

1 = 0 → No interrupt request activated

W_2 W_1 W_0 → Binary code of highest priority active interrupt request

8.10 INTERFACING SINGLE 8259A WITH 8086

The Fig. 8.18 illustrates a single 8259A connected to the 8086 microprocessors where SP/$\overline{\text{EN}}$ pin is pulled high so as to indicate that it is a raster.

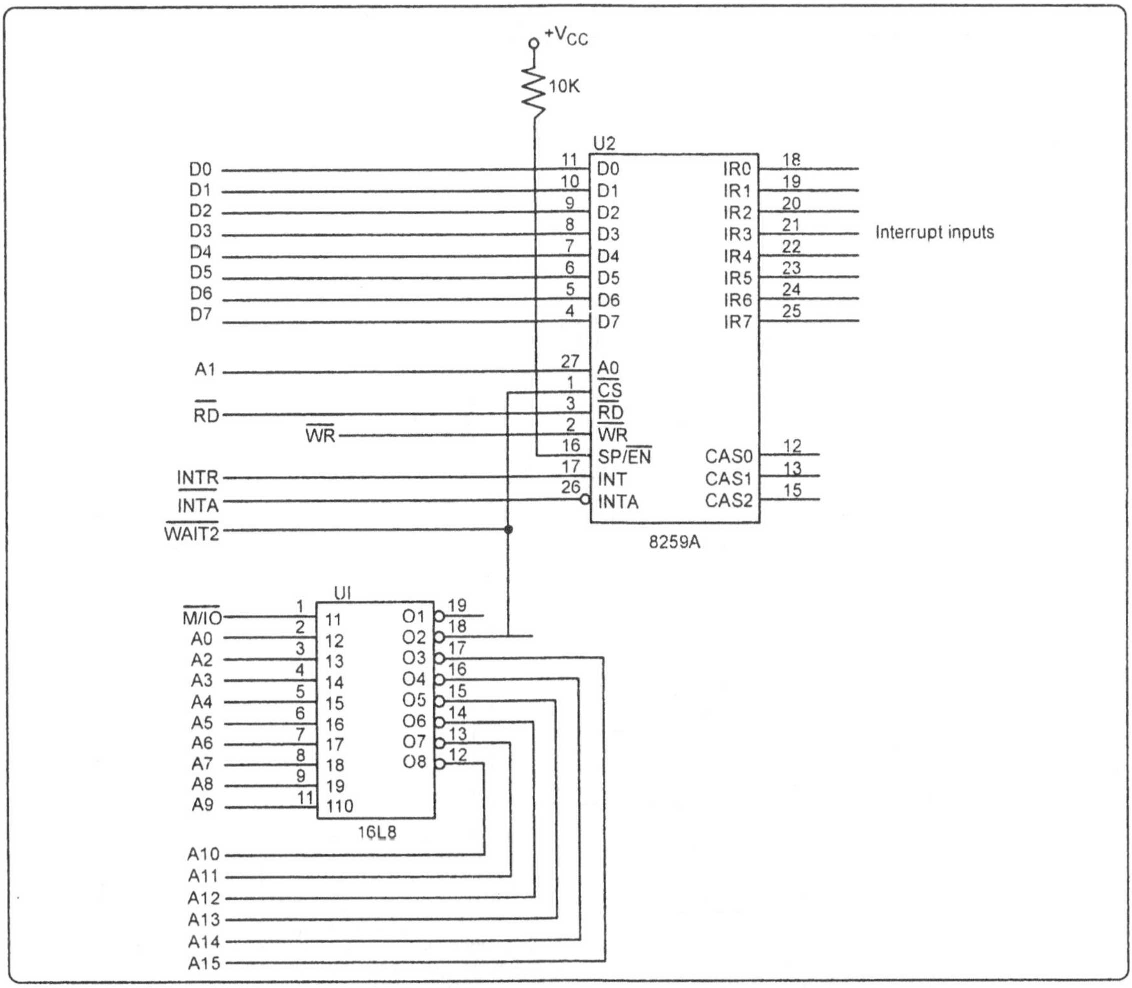

Fig. 8.18 Single interfaced to the 8086 microprocessor

It can be observed from Fig 8.18 that 8259A is decoded at I/O ports 0400H and 0402H by the PAL16L8 (*no program shown*). The 8259A requires four wait states for it to function properly with a 16MHz 80386SX and more for some other versions of the Intel Microprocessor family.

8.10.1 Cascading Multiple 8259A's to Microprocessor

The Fig. 8.19 illustrate two 8259As connected to the 80386SX microprocessor in a way which is often observed in the AT-style computer, with 8259As for interrupts.

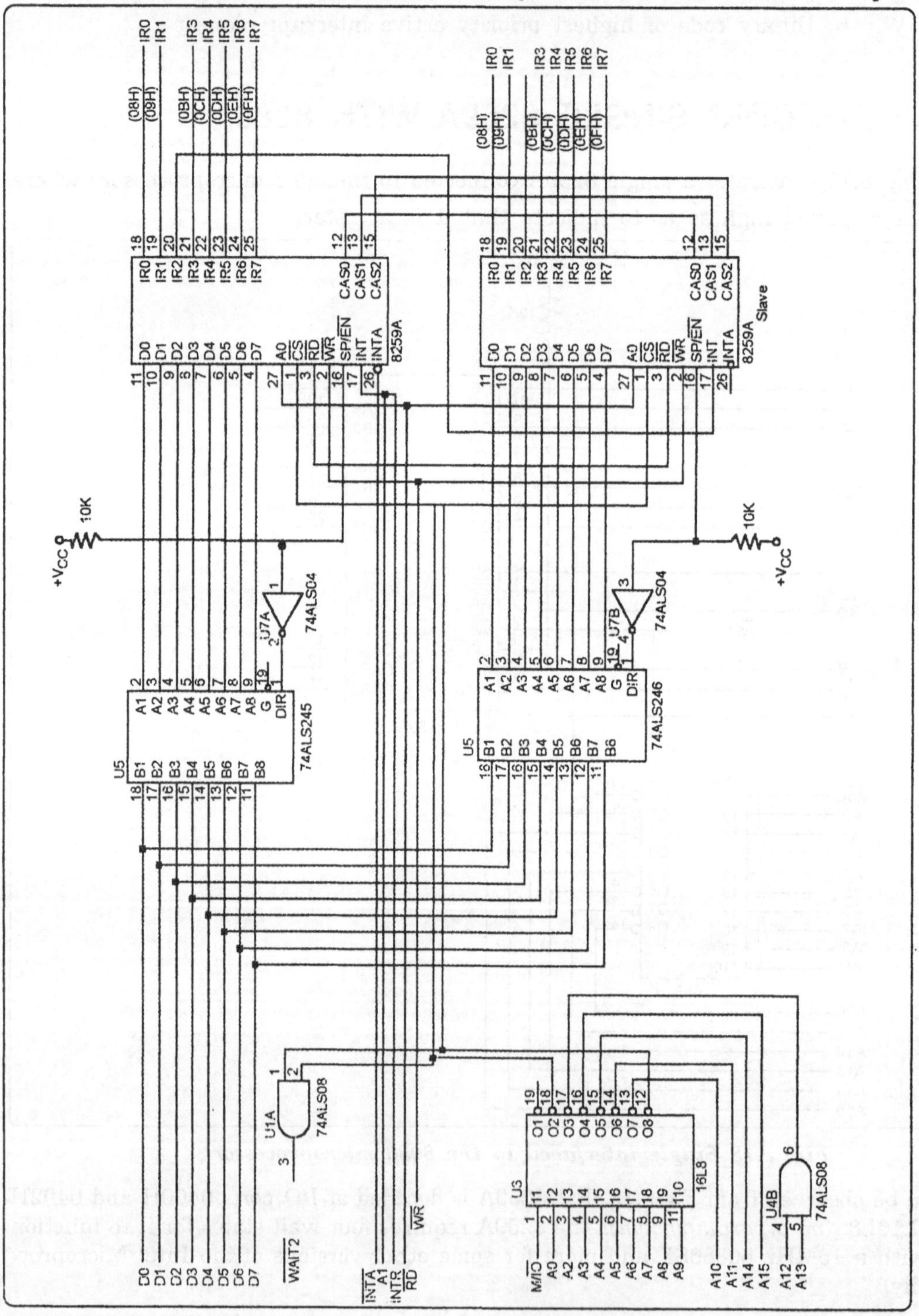

***Fig. 8.19** Two 8259As interfaced to the 8259A at I/O ports 0300H and 0302H for the master and 0304H and 0306H for the slave*

In the case of XT-or PC-style computer they uses one 8259A at interrupt vectors 08H-0FH. The At-style computer uses interrupt vector 0AH as a cascade input from a second 8259A located at vectors 70H-77H.

Note. Such circuit uses 08H-0FH and I/O ports 0300H and 0302H for UI, the master, and vectors 70H-77H and I/O ports 0304H and 0306H for U2, the slave. Observe data bus buffers one added to illustrate the use of the SP/EN pin on the 8259A. Such buffers are used only in vary large systems that have many devices connected to their bus connections. In practice, we seldom find these buffers.

8.11 PROGRAMMING 8259A

8259A can be programmed by initialization and operation command words that is **Initialization command words** (ICWs) are programmed before the 8259A is able to function in the system and dictate the basic operation of the 8259A. **Operation command words** (OCWs) are programmed during the normal course of operation and allow the 8259A to function properly.

8.11.1 Initialization Command Words

There are four initialization command words (ICWs) for the 8259A that are selected when the A_0 pin is a logic 1. Which is shown in the form of flow chart in Fig. 8.20 below :

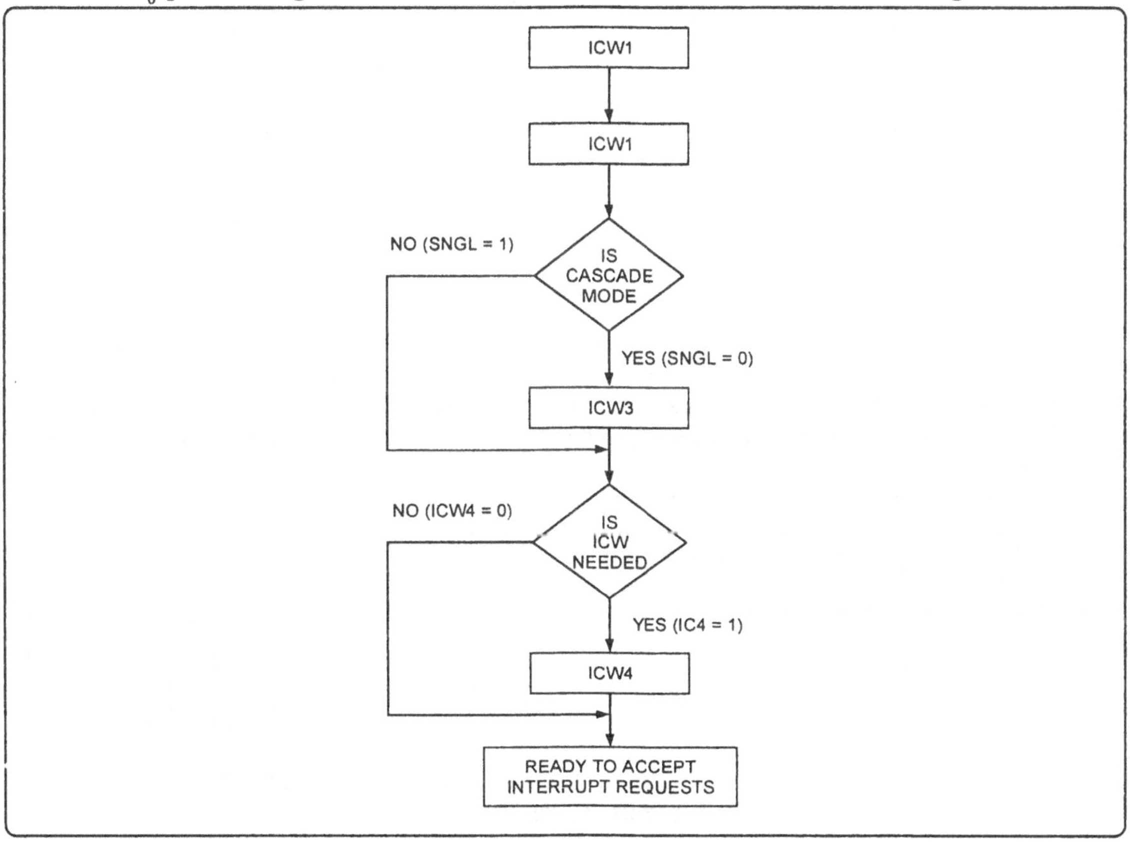

Fig. 8.20 Initilization flow chart for 8259

As shown in flow chart when 8259A is first powered up, it must be sent ICWI, ICW2 and ICW4. If the 8259A is programmed in cascade mode by ICWI then we also must program ICW3. So if a single 8259A is used in a system. ICWI, ICW2 and ICW4 must be programmed. When cascade mode is used in a system, then all four ICWs must be programmed.

- **ICW1** Programs the basic operation of the 8259A. To program this ICW for 8086 we place a logic 1 in bit IC4. Bits AD1, A7, A6 and A5 are don't cares for microprocessor operation and only apply to the 8259A when used with an 8-bit 8085 microprocessor. Fig. 8.21 illustrates the format for ICWI.

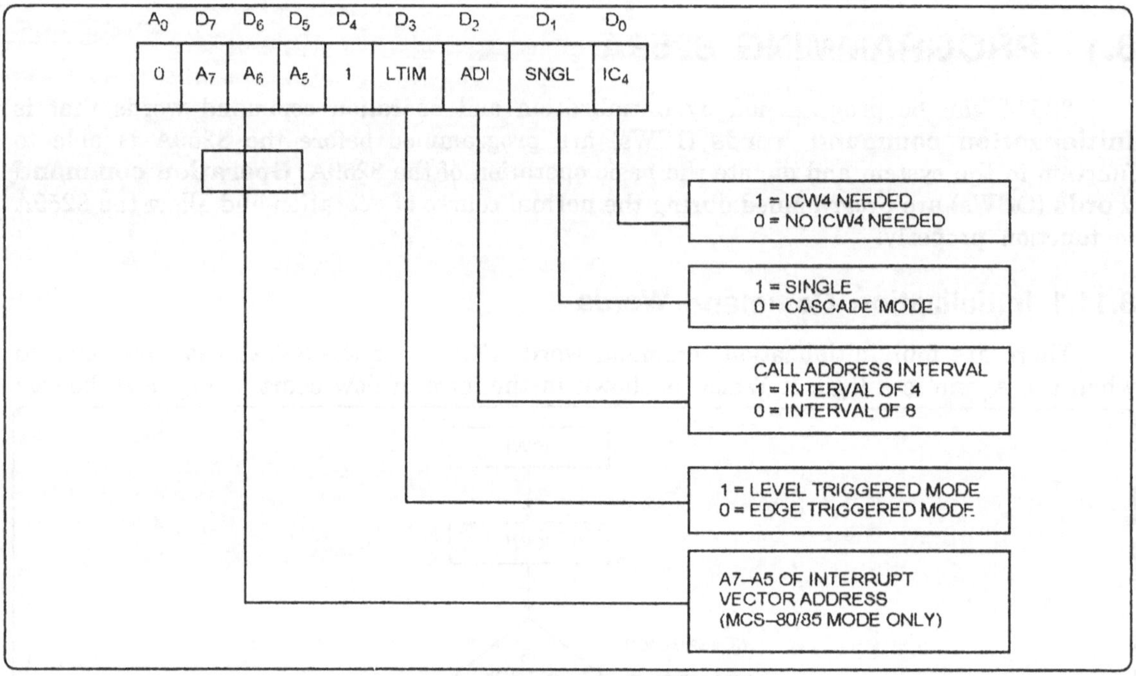

Fig. 8.21 ICWI Format

Such ICW selects single or cascade operation by programming the SNGL bit. When cascade operation is selected, we must also program ICW3. The LTIM bit determines whether the interrupt request inputs are positive edge-triggered or level-triggered.

- **ICW2** Selects the vector number used with the interrupt request inputs. Fig. 8.22 illustrates the format for ICW2.

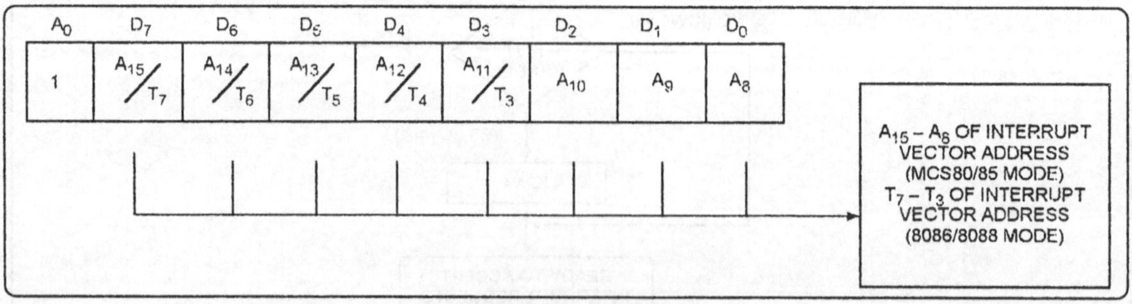

Fig. 8.22 ICW2 Format

Now for instant when we decide to program the 8259A to that it functions at vector locations 08H-0FH we place a 08H into his command word. Similarly when we decide to program the 8259A for vectors 70H-77H. We place a 70H in this ICW.

■ **ICW3** Is used if ICWI indicates that the system is operated in cascade mode. This ICW indicates where the slave is connected to the master. The format of ICW3 is as shown in Fig. 8.23.

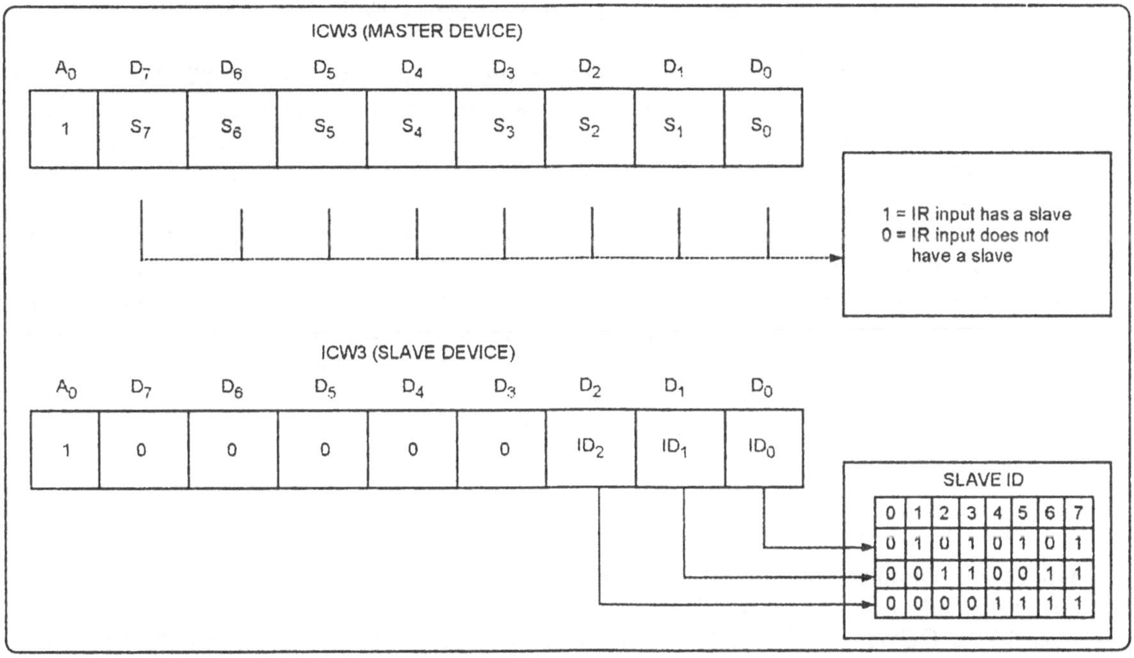

Fig. 8.23 ICW3 format

As observed from Fig. 8.23, as an instant we connected a slave to IR2. To program ICW3 for this connection in both master and slave, we place a 04H in ICW3. Suppose we have two slaves connected to a master using IR0 and IRI. The master is programmed with an ICW3 of 03H; one slave is programmed with an ICW3 of 01H and other with an ICW3 of 02H.

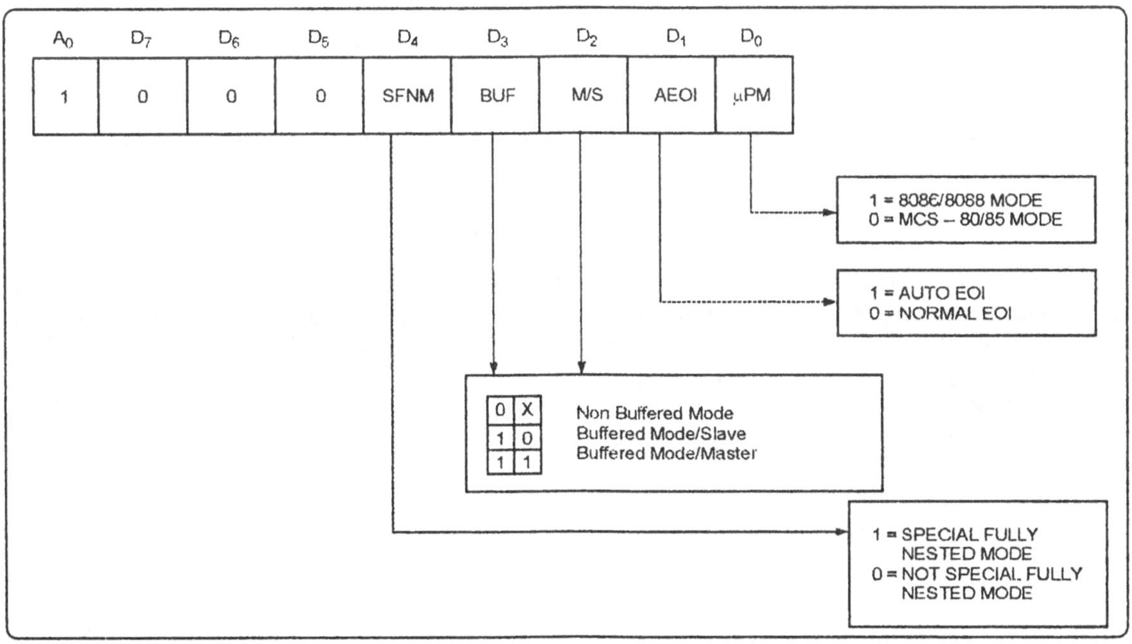

Fig. 8.24 ICW4 format

- **ICW4** Is programmed for use with the 8086—such ICW is not programmed in a system that functions with the 8086 microprocessors. The Fig 8.24 illustrates the format for ICW4. As observed in Fig. 8.24 the rightmost bit must be a logic 1 to select operation with the 8086 microprocessors and the remaining bits are programmed as mentioned below :
- SFNM-Selects the special fully nested mode of operation for the 8259A if a logic 1 is placed in this bit which allows highest priority interrupt request from a slave to be recognized by the master while it it processing another interrupt from a slave usually, only one interrupt request is processed at a time and others are ignored until the process is complete.
- BUF and M/S – Buffer and master slave are used together to select buffered operation or nonbuffered operation for the 8259A as a master or a slave
- AEOI – Selects automatic or normal end of interrupt (*discussed more fully under operation command words*). EOI commands of OCW2 are used only if the AEOI mode is not selected by ICW4. When AEOI is selected, the interrupt automatically resets the interrupt request bit and does not modify priority. This is the preferred mode of operation for the 8259A and reduces the length of the interrupt service procedure.

8.11.2 Operation Command Words

Operation command words (OCWs) are used to direct the operation of the 8259A once it is programmed with the ICW. The OCWs are slected when the A_0 pin is at a logic 0 level, except for OCWI, which is selected when A_0 is a logic 1. There are three OCWs, which may be loaded anytime after 8259's initialization so as to dynamically alter the priority modes.

- **OCW1** is used to set and read the interrupt mask register. The Fig. 8.25 illustrates the format of operation command word 1.

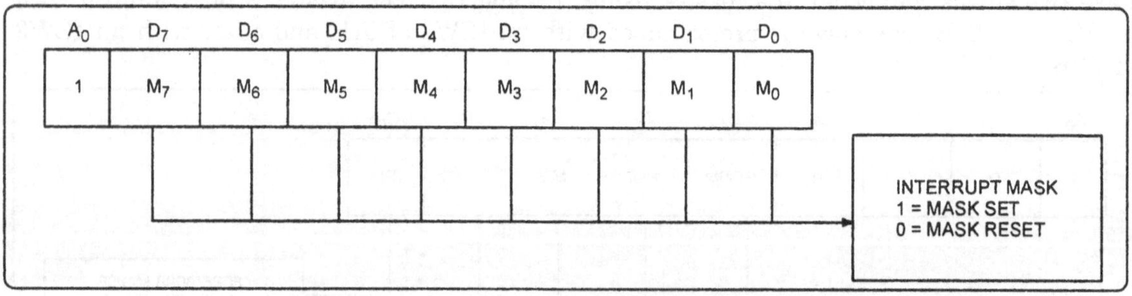

Fig. 8.25 OCW1 format

In this case, when a mask bit is set, it will turn off (mask) the corresponding interrupt input. The mask register is read when OCW1 is read. Because the state of the mask bits is unknown when the 8259A is first initialized. OCW1 must be programmed after programming the ICW upon initialization.

- **OCW2** is programmed only when the AEOI mode is not selected for the 8259A, Fig 8.26 illustrates the format for OCW2.

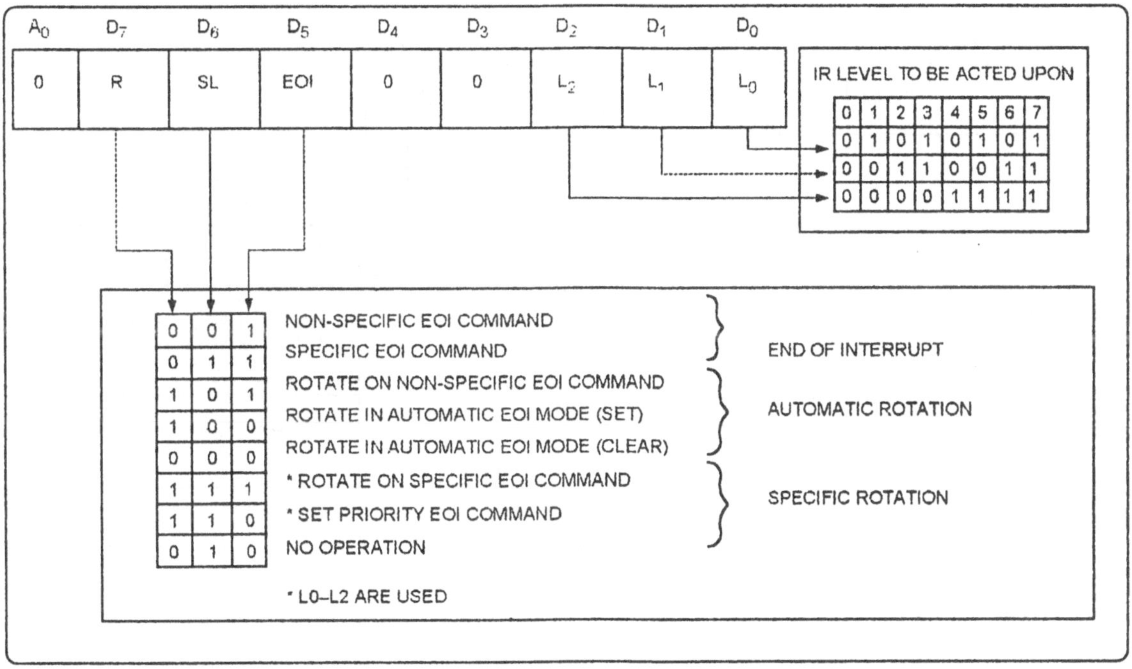

Fig. 8.26 OCW2 format

This OCW selects how the 8259A responds to an interrupt, such modes are depicted as below :

- Nonspecific End-of-Interrupt – A command sent by the interrupt service procedure to signal the end of the interrupt. The 8259A automatically determines which interrupt level was active and resets correct bit of the interrupt status register. Resetting the status bit allows the interrupt to take action again or a lower priority interrupt to take effect.
- Specific End-of-interrupt – A command which allows a specific interrupt request to be reset. The exact position is determined with bits L2 - L0 of OCW2.
  Rotate-on-Nonspecific EOI –A command which functions exactly like the nonspecific end-of-interrupt command except that it rotates interrupt priorities after resetting the interrupt status register bit. The level reset by this command becomes the lowest priority interrupt. For instant, if IR4 was just serviced by this command, it becomes the lowest priority interrupt input and IR5 becomes the highest priority.
- Rotate-on-Automatic EOI – A command that selects automatic EOI with rotating priority. Such command must be sent to the 8259A only once if this mode is desired. If this mode must be turned off, use the clear command.
- Rotate-on-Specific EOI – Functions as the specific EOI, except that it selects rotating priority.
- Set Priority – Allows the programmer to set the lowest priority interrupt input using the L2-L0 bits.
- **OCW3** Selects the register to be read, the operation of the special mask register, and the poll command, Fig 8.27 illustrates format for OCW3.

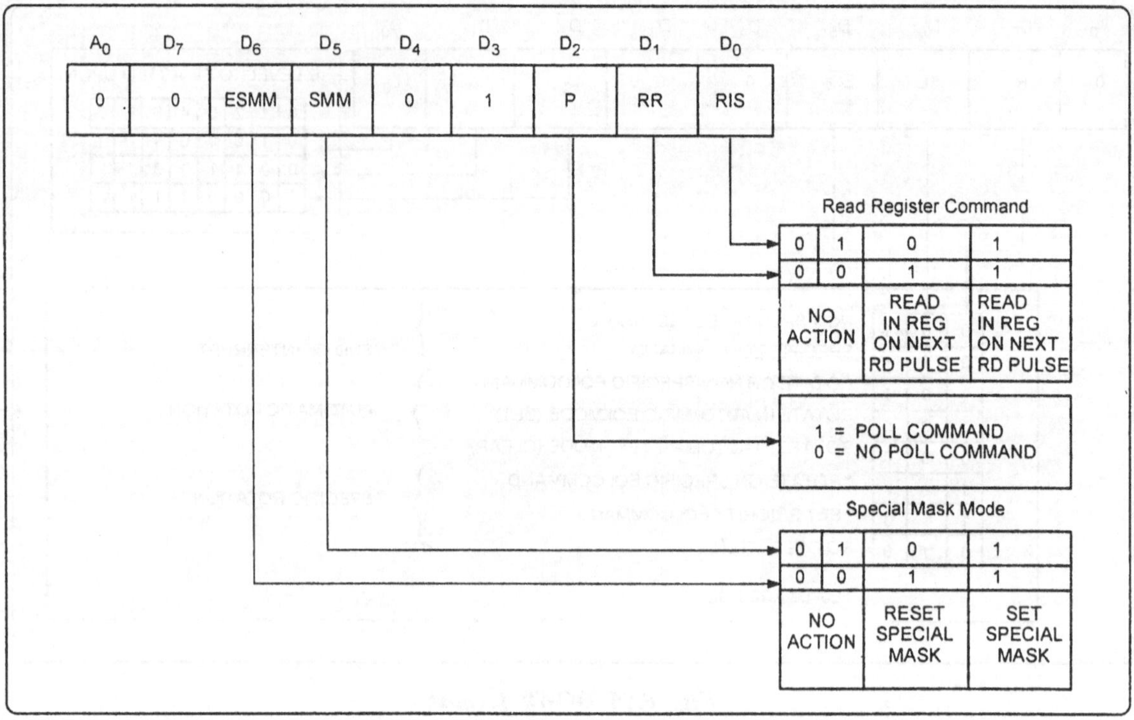

Fig. 8.27 OCW3 format

In this case when polling is selected, the P-bit must be set and then output to the 8259A. The next read operation will read the poll word. The rightmost three bits of the poll word indicate the active interrupt request with the highest priority. The leftmost bit indicates whether there is an interrupt, and must be checked to determine whether the rightmost three bits contain valid information.

8.11.3 The Status Register of 8259A

Three status registers are readable in the 8259A : interrupt request register (IRR), in service register (ISR), and interrupt mask register (IMR). The Fig. 8.28 illustrates for there status registers which have same bit configuration.

IRR is an 8-bit register that indicates which interrupt request inputs are active. The ISR is an 8-bit register that contains the level of the interrupt request inputs are active. The ISR is an 9-bit register that contains the level of the interrupt being serviced. The IMR is an 8-bit register that holds the interrupt mask bits and indicates which interrupts are masked off.

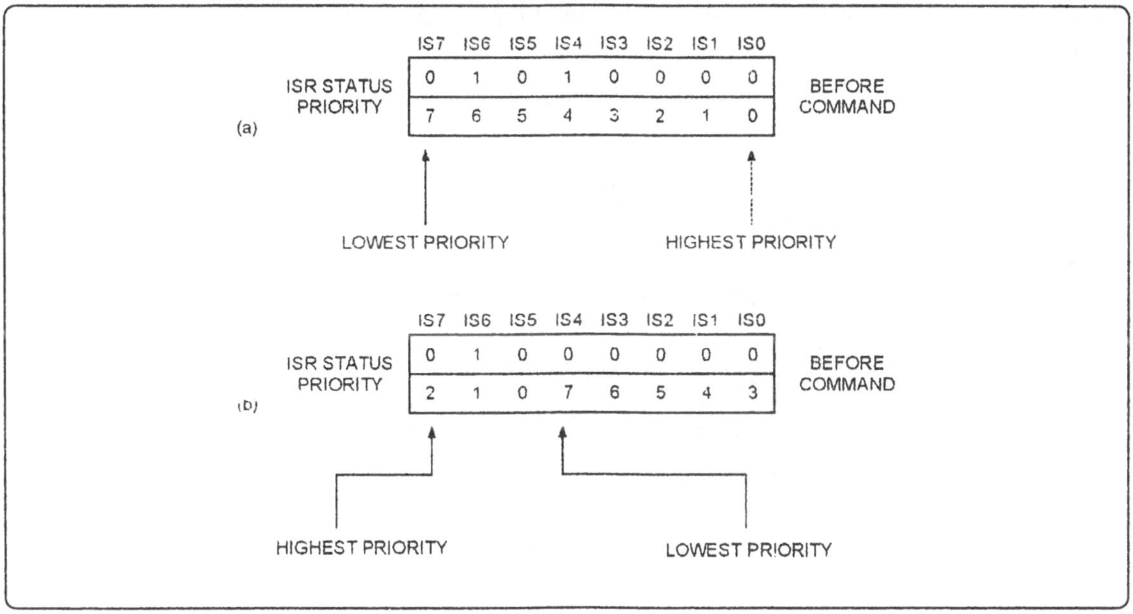

Fig. 8.28 The 8259A in-service register (ISR). (a) Before IR4 is accepted, and (b) after IR4 is accepted.

Both IRR and ISR are read by progamming OCW3, and IMR is read through OCW1. To read the IMR, A0 = 1; to read either IRR or ISR, A0 = 0. The Bit positions D0 and D1 of OCW3 selected which register (IRR or ISR) is read when A0 = 0.

Example 3 : Write initialization instructions for 8259A interrupt controller to meet the following specifications.

(i) Interrupt type 32 (ii) Edge triggered, single and ICW4 needed (iii) Mask interrupts IRI and IR3.

Solution : First we have to write four ICWs and OCW1.

- **ICW1**

A_7	A_6	A_5	1	LTIM	ADI	SNGL	IC_4	
0	0	0	1	0	0	1	1	= 13H

It must be remembered that when we use with 8086 microprocessor, the bits D_4, D_5, D_6 and D_7 will be don't cares i.e., simply they can be made as 0's.

- **ICW2**

In 8086 system ICW2 is used to tell 8259A the type number to send in response to an interrput signal on the IR0 input.

B_7	B_6	B_5	B_4	B_3	B_2	B_1	B_0	
0	0	1	0	0	0	0	0	= 20H = 32 Decimal

The ICW2 for sending interrupt type 32 to the 8086 in response to an IR0 interrupt will be 20H.

(For an IRI input the 8259A will send 00100001 binary (33 decimal) and so on for the other IR inputs).

Note. As we are not using a slave in this example, ICW3 is not sent.

■ **ICW4**

Here only reason why we need to send an ICW4 is to let the 8259A know that it is operating in an 8086 system. Which is done by making bit D_0 of the ICW4 one.

■ **OCW1**

The OCW1 must be sent to an 8259A to unmask any IR inputs. Here we want to mask IR1 and IR3, so we put 1's in these two bits and 0's in the rest of the bits.

M_7	M_6	M_5	M_4	M_3	M_2	M_1	M_0	
0	0	0	0	1	0	1	0	= 0AH

Program :

Opcode	Explanation
MOV AL, 13H	; edge triggered, single, ICW4 needed
OUT 40H, AL	; Send ICW1
MOV AL, 20 H	; type 32 is first 8359A type
OUT 41H, AL	; send ICW2
MOV AL, 0 1H	; ICW4, 8086 mode
OUT 41H, AL	; send ICW4
MOV AL, 0AH	; OCW1 to mask IR1 and IR3
OUT 41H, AL	; send OCW1

Example 4 : Write the initialization instructions for master and slave configuration to meet the following specifications :

(a) INTR of slave is routed through IR2 of the master 8259A to the 8086.
(b) Master and slave are both level triggered.
(c) First interrupt types for master and slave are 32 and 64 respectively.
(d) Modes : automatic rotation and auto end of interrupt.
(e) Addresses of the master are 40H and 41H and the slave are 80H and 81H.
(f) Buffers are not used.

Solution : *Initialization command words for Master ICW1 (Master)*

ICW1 (master)

A-	A_6	A_5	1	LTIM	ADI	SNGL	IC4	
0	0	0	1	1	0	0	1	= 10H

ICW2 (master)

B-	B_6	B_5	B_4	B_3	B_2	B1	B_0	
0	0	1	0	0	0	0	0	= 20H

ICW3 (master)

S_7	S_6	S_5	S_4	S_3	S_2	S_1	S_0	
0	0	0	0	0	0	1	0	0 = 04H

ICW4 (master)

0	0	0	SFNM	BUF	M/S	AEOI	μPM	
0	0	0	0	0	0	1	1	= 03H

Program :

Opcode	Explanation
MOV AL, 19H	; level triggered, cascaded, ICW4 needed
OUT 40H, AL	; send ICW1 (master)
MOV AL, 20H	; type 32 is first 8259A type
OUT 41H, AL	; send ICW2 (master)
MOV AL, 04H	; slave at IR2
OUT 42H, AL	; send ICW3 (master)
MOV AL, 03H	; ICW4, 8086 mode, and set AEOI
OUT 41H, AL	; send ICW4 (master)
MOV AL, 19H	; level triggered, cascaded, ICW4 needed
OUT 80H, AL	; send ICW1 (slave)
MOV AL, 40H	; type 64 is first 8259A type
OUT 81H, AL	; send ICW3 (slave)
MOV AL, 02H	; ID for slave connected to IR2
OUT 81H, AL	; send ICW2 (slave)
MOV AL, 01H	; ICW4, 8086 mode
OUT 81H, AL	; send ICW4
MOV AL, 80H	; OCW2 (rotate in auto EOI mode set command)
OUT 80H, AL	; send OCW2 (slave)

8.11.4 The Special Mask Mode (SMM)

This mode will enable interrupt from all level except in service level, which is done by masking the level which is in service and next issuing special mask mode command. After special mask mode is set, it will be in effect till it is reset, again.

Now the Fig 8.29 illustrates how to enable lower priority interrupts by making use of SMM.

Here let us consider that IR0 has highest priority when main program is interrupted by IR4. In the IR4 service routine an enable interrupt instruction is executed, then it only allows higher priority interrupt requests to interrupt IR4 in the normal full nested mode.

Later in IR routine, bit 4 of the IMR is masked and the special mask mode is entered. Priority operation is no longer in the fully nested mode. All interrupt levels are enabled except for IR4. To leave the special mask mode, the sequence is executed in reverse.

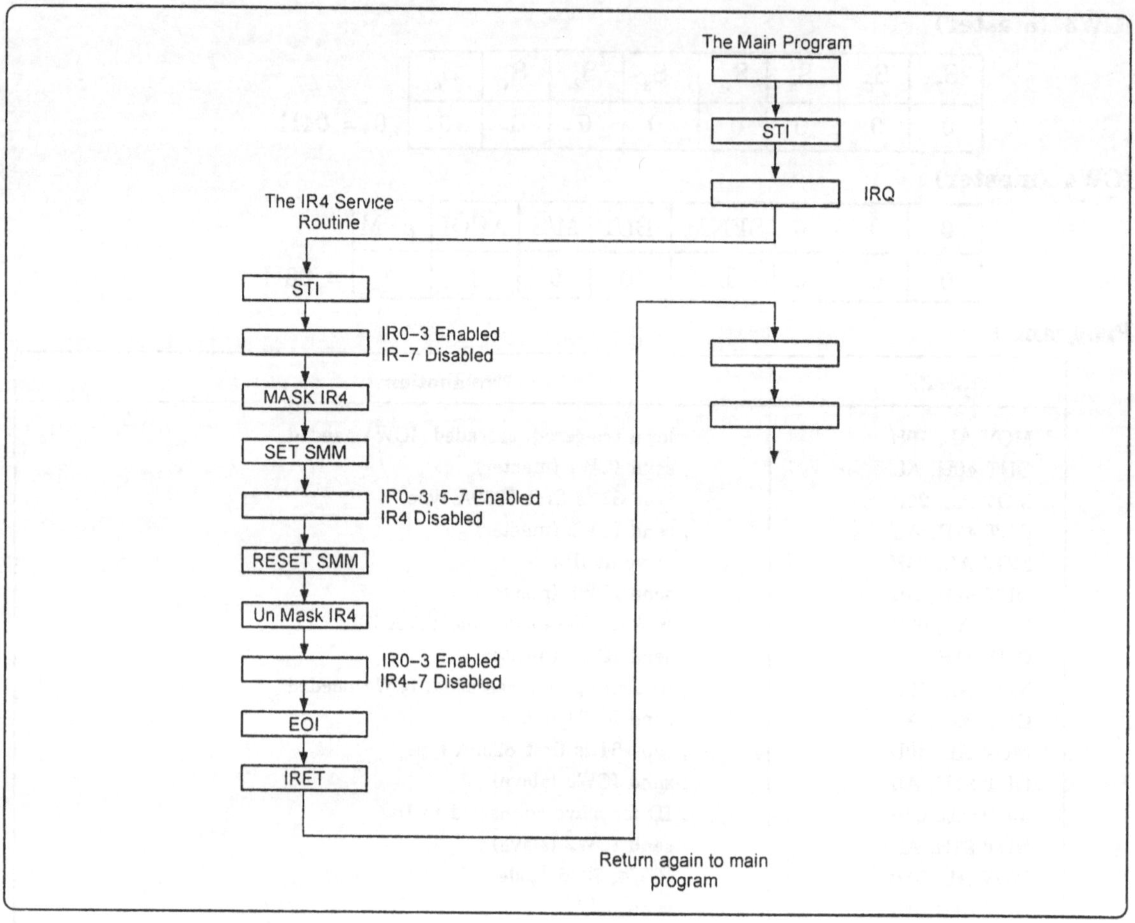

Fig. 8.29 Priority interrupt enabling by using SMM

(NESOI command cannot be used in the special mask mode, because a non-specific end of interrupt won't clear an ISR bit of an interrupt that is masked when in the special mask mode, i.e., actually the bit will appear invisible).

8.11.5 The Real-Time Clock

The timer chip 8254, is used in the design so as to keep time in hours, minutes and seconds using three memory locations to hold the BCD time of day. The Fig. 8.30 shows a simple circuit that uses the PCLK clock of 8284A clock generator to generate a periodic interrupt signal assuming the PCLK clock as 2.5 MHz.

Here the counter-0 of 8254 is initialized with C350H to generate a clock frequency of 50Hz (mode-3), and output is fed as an input to CLK-1 of counter-1. The counter-1 is programmed to generate a clock output of 1Hz by initalizing it to 50D, and its output is fed to 8259 IR0 to which interrupts 8086 microprocessor.

Next Output1 of counter-1 is given as an input to CLK-2 of counter-2 and initialized the counter-2 with 60D to generate 1/60 of a second. Whose output is fed into IR1 of 8259. The interrupt IR1 is generated for every one minute. All 3 counters are in mode-3 to generate a clock generator at equal interval. The 3 programs to generate real-time clock are as mentioned below :

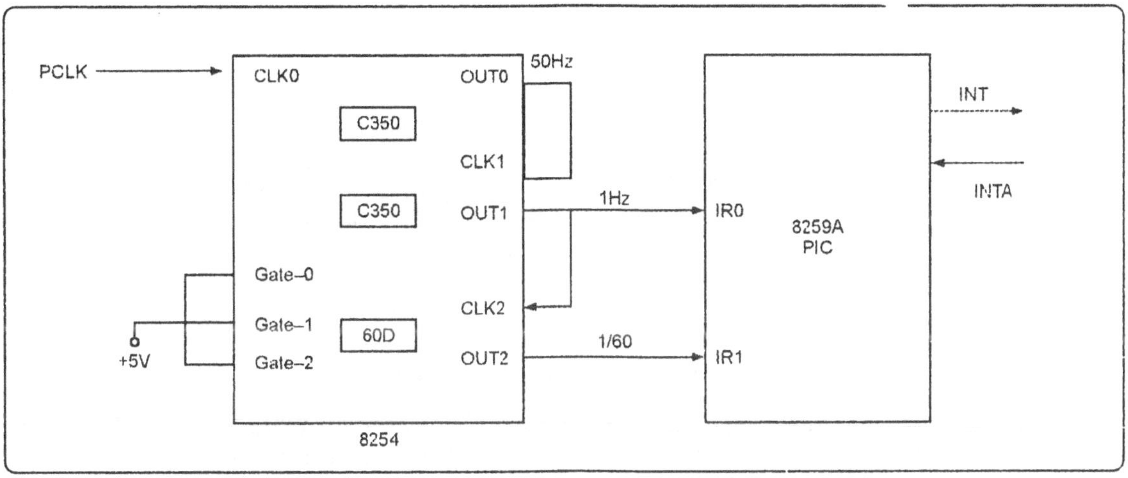

Fig. 8.30 Clock generator arrangement using 8254

(*a*) Main program to initialize 8259 and 8254
(*b*) ISR for IRO to update and keep Seconds
(*c*) ISR for IR1 to update minutes and hours

Now let us see the application of Real-time clock for *i.e.*,

(*a*) The Real-Time clock from AC power
(*b*) The Power Failure Protection

(*a*) The Real-Time clock from AC power :

Simple circuit is shown in as a Fig 8.31 real time clock using 50Hz AC power line to generate a periodic interrupt request signal for the NMI input pin below :

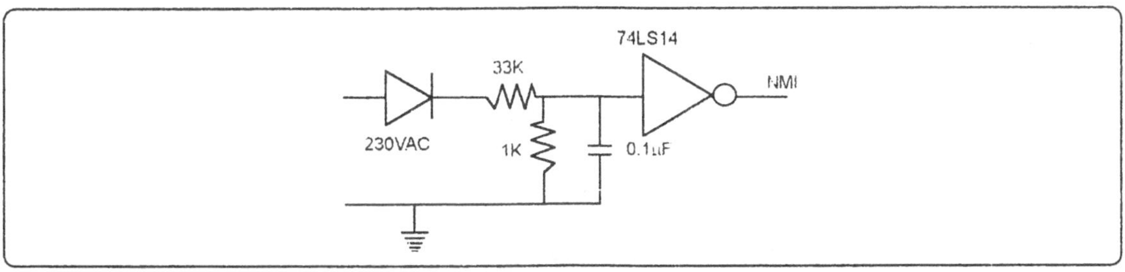

Fig.8.31 Circuit for real-time clock from AC power

Eventhough we use signal from the AC power line, which varies slightly in frequency from time to time it is accurate over a period of time. Here circuit uses a signal from 230 VCC power line which is conditioned by a Schmitt-trigger inverter before it is applied to the NMI interrupt input. The interrupt is generated 50 times every second. The software has to be developed so as update the count located in four memory locations. Where each location has 50-counter, seconds, minutes and hours.

(*b*) The Power Failure Protection :

While Monitoring AC power line and causing a NMI interrupt whenever AC power drops out easily detect power failures. Hence the microprocessor stores all the internal registers in a battery-backed up memory or EEPROM. The Fig. 8.32 shows power failure detection circuit which provides logic to NMI input whenever AC power is interrupted.

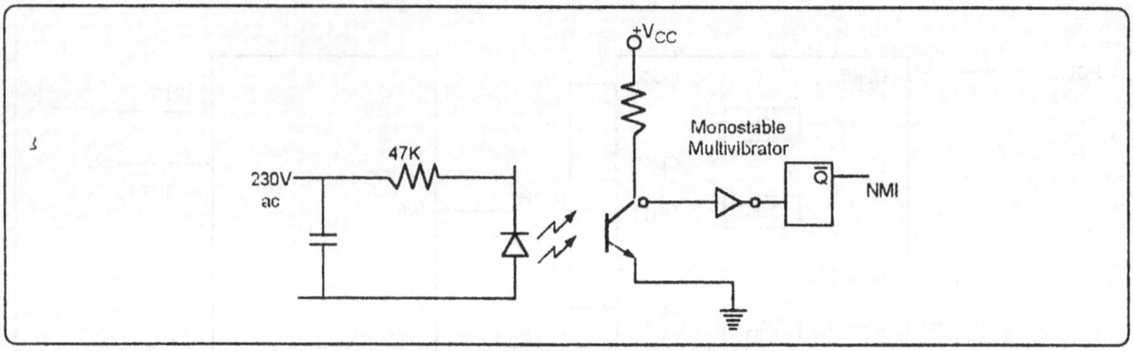

Fig. 8.32 Power failure detection circuit

If can be observed from Fig. 8.32 that an optical isolator provides isolation from the AC power line, and the output of the isolator is shaped by a Schmitt-trigger inverter that provides 50Hz pulse input of the monostable multivibrator. When AC power fails, the multivibrator no longer receives trigger pulse from the inverter, therefore Q returns to a logic zero and $\overline{Q}$ returns to logic one, interrupting microprocessor through the NMI pin. The interrupt service procedure, written for NMI must store the contents of all internal registers and other data into a batter-backed up memory. The memory assumes that the system power supply has a large enough filter capacitor to provide energy for at least 75 msec after the AC power is failed.

REVIEW QUESTIONS

1. What do you understand by the term Interrupt ? What is the purpose of it ?
2. What are the interrupts in 8086/8088 ?
3. Why do interrupts free up time for the microprocessor ?
4. What is an interrupt vector ?
5. Explain how a type 0 interrupt occur in 8086 ?
6. What is 8259A ? Explain ?
7. What is the purpose of IR0-AR7 pins on 8259A ?
8. What is (*a*) ICW and (*b*) OCW ? Explain ?
9. What is the purpose of IRR in 8259A ?
10. List the features of 8259A ?
11. With the help of a neat diagram explain the interfacing of 8259A with microprocessor ?
12. With the help of a block diagram explain 8259A ?

9 BASIC I/O INTERFACING

9.0 OBJECTIVE

After completion of this chapter, reader will be able to

- Know the operation basic input and output interfaces.
- Know the difference between, Isolated and Memory-Mapped I/O.
- Know the decoding of 8-bit and 16-bit I/O device
- Know about Handshaking.
- Know about Address decoding.

9.1 INTRODUCTION

A microprocessor is best system to solve problems. To do so the microprocessor must communicate between humans, or machines (*i.e., external circuitary).* The user must be able to given information to μP based system with the help of keyboard and must be able to see the result on display system. Therefore transfer of data between keyboard and μP and between μP and display devices. Such process is know as Input/Output (I/O) data transfer which can be performed by I/O ports available.

9.1.1 Input/Output Port

The input port is required to read data from input device like keyboard. The buffer is a simplest form of input port. That means input device is connected to the μP through buffer, which is a tri-state one and output is available when enable signal is active. Similarly the output port is required to send data to output device, such as display from μP. The simplest form is a latch, which connect to μP with output device. Fig. 9.1 (*a*) and 9.1 (*b*) depicts the Input and Output port in simplest form.

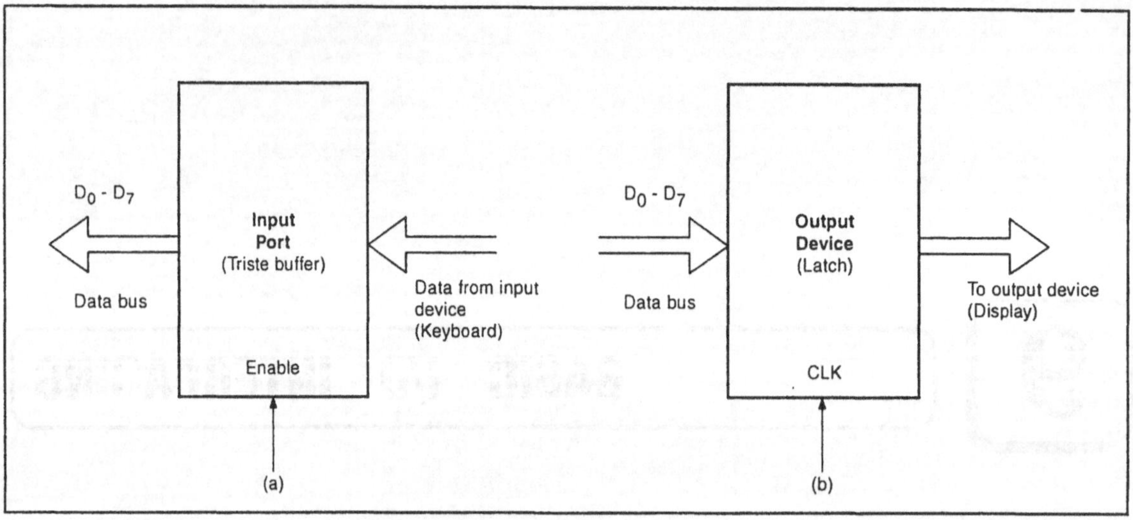

Fig. 9.1 Input and Output Port.

In Fig. 9.1 (*a*). if microprocessor requires to read data from the input device *(keyboard)*, the control signals from the microprocessor activates the buffer by asserting enable input of the buffer. When the buffer is enabled, data from the input device is available on the data bus μP reads this data by initiating read command. Similarly in Fig. 9.1 (*b*) if microprocessor wants to send data to the output device it puts the data on the data bus and activates the clock signal of the latch, the data from data bus at the output of latch, then it is available for output device to read.

9.2 I/O INSTRUCTIONS

To understand I/O interface, the operation of I/O instructions : IN, INS, OUT and OUTS are discussed. The instruction set contains one type of instruction that transfers information to an I/O device (OUT) and another to read information from an I/O device (IN). The Instructions *(INS and OUTS found on all versions except the 8086/8088)* are also provided to transfer strings of data between the memory and an I/O device. The Table 1 lists all versions of instruction available in the microprocessor's instruction set.

IN and OUT instructions both will transfer data between an I/O device and the microprocessor's accumulator (AL, AX, or EAX). The I/O address is stored in register DX as a 16-bit I/O address or in byte (p8) immediately following the opcode as an 8-bit I/O address. The Intel calls the 8-bit from (p8) a fixed address because it is stored with the instruction, usually in a ROM. The 16-bit I/O address in DX is called a variable address because it is stored in a DX then used to address the I/O device. Other instructions that use DX to address I/O are INS and OUTS instructions.

Whenever the data are transferred by making use of IN or OUT instruction, the I/O address, often called a *port number,* appears on the address bus. External I/O interface decodes it in the same manner that it decodes a memory address. The 8-bit fixed port number (p8) appears on address bus connections $A_7 - A_0$ with bits $A_{15} - A_8$ equal to 000000002. The address connections above A_{15} are undefined for an I/O instruction. The 16-bit variable port number (DX) appears on address connection $A_{15} - A_0$. That is the first

Table 1 : Input/Output Instructions

S. No.	Instruction	Data Width	Function
1.	IN AL, p8	8	A byte is input from port p8 into AL
2.	IN AX, p8	16	A word is input from port p8 into AX
3.	IN EAX, 08	32	A doublewords is input from port p8 into EAX
4.	IN AL, DX	8	A byte is input from the port addressed by DX into AL
5.	IN AX, DX	16	A word is input from the port addressed by DX into AL
6.	IN EAX, DX	32	A word is input from the port addressed by DX into AL
7.	OUT p8, AL	8	A byte is output from AL to port p8
8.	OUT p8, AX	16	A word is output from AX to port p8
9.	OUT p8, EAX	32	A doubleword is output from EAX to port p8
10.	OUT DX, AL	8	A byte is output from AL to the port addressed by DX
11.	OUT DX, AX	16	A word is output from AX to the port addressed by DX
12.	OUT DX, EAX	32	A doubleword is output from EAX to the port addressed by DX

256 I/O port address (00H – FFH) are accessed by both the fixed and variable I/O instructions, but any I/O address from 0100H – FFFFH is accessed only by the variable I/O address. In most of the dedicated task systems, only rightmost 8-bits of the address are decoded, therefore reducing the amount of circuitry required for decoding. In the case of a PC computer, all 16 address bus bits are decoded with locations 00XXh – 03XXH, which are the I/O addresses used for I/O inside the PC.

9.3 ISOLATED AND MEMORY-MAPPED I/O

There are fully different methods of interfacing I/O to μP, which are :

(*a*) Isolated I/O, and

(*b*) Memory – Mapped I/O

In the case of isolated I/O, the IN, INS, OUT and OUTS instructions transfer data between the microprocessor accumulator or memory and the I/O device. Whereas in memory-mapped I/O, any instruction that references memory can accomplish the transfer. However isolated and memory-mapped I/O are in use.

(*a*) Isolated I/O

Most common I/O transfer technique used in the Intel microprocessor-based system is isolated I/O. The term isolated describes how the I/O locations are isolated from the memory system in a separate I/O address space, as shown in Fig. 9.2.

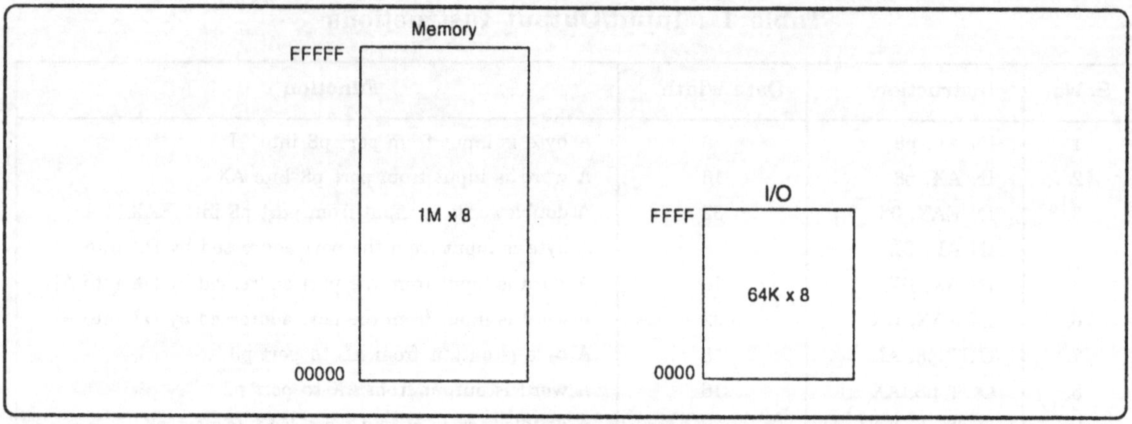

Fig. 9.2 The memory and I/O maps for the 8086/8088 microprocessors for Isolated I/O.

As observed from Fig. 9.2 the addresses for isolated I/O devices, called ports, are separate from the memory, that means the user can expand the memory to its full size without using any of this space for I/O devices. One disadvantage of isolated I/O is that the data transferred between I/O and the microprocessor must be accessed by the IN, INS, OUT and OUTS instructions. Separate control signals for the I/O space are developed (using $M/\overline{IO}$ and $W/\overline{R}$) that indicate an I/O read $\left(\overline{IORC}\right)$ or an I/O write $\left(\overline{IOWC}\right)$ operation. Such signals indicate that an I/O port address appears on the address bus that is used to select the I/O devices.

Note. **As a rule, an 8-bit port address is used to access devices located on the system board, such as the timer and keyboard interface, while a 16-bit port is used to access serial and parallel ports as well as video and disk drive systems.**

(*b*) Memory-Mapped I/O

Unlike isolated I/O, memory-mapped I/O does not use the IN, INS, OUT or OUTS instructions, but it uses any instruction that transfers data between the microprocessor and memory as shown in Fig. 9.3 below :

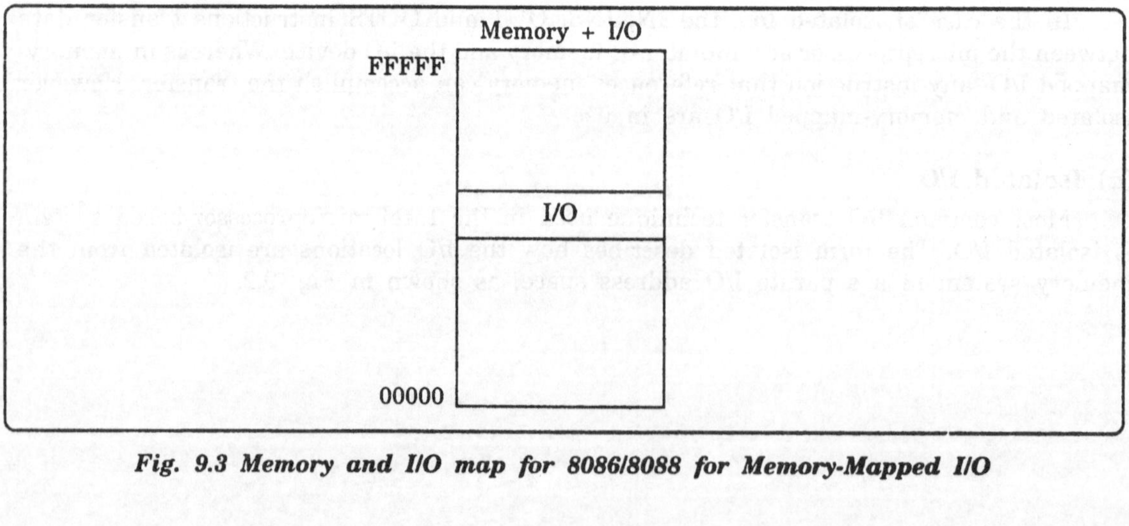

Fig. 9.3 Memory and I/O map for 8086/8088 for Memory-Mapped I/O

Memory-mapped I/O device is treated as a memory location in the memory map. The main advantage of memory-mapped I/O is that any memory transfer instruction can be used to access the I/O device and the main disadvantage is that a portion of the memory system is used as the I/O map. This reduces the amount of memory available to applications. One more advantage is that the $\overline{IORC}$ and $\overline{IOWC}$ signals have no function in a memory-mapped I/O system and may reduce of circuitry required for decoding.

9.4 BASIC INPUT AND OUTPUT INTERFACES

A basic input device is a set of three-state buffers, and basic output device is a set of data latches. The term IN refers to moving data from the I/O device into the microprocessor, and the term OUT refers to moving data out of the microprocessor to the I/O device.

9.4.1 The Basic Input Interface

Three-state buffers are used to construct the 8-bit input port as shown in Fig. 9.4. Observe that the external TTL data *(simple toggle switches in this example)* are connected to the inputs of the buffers.

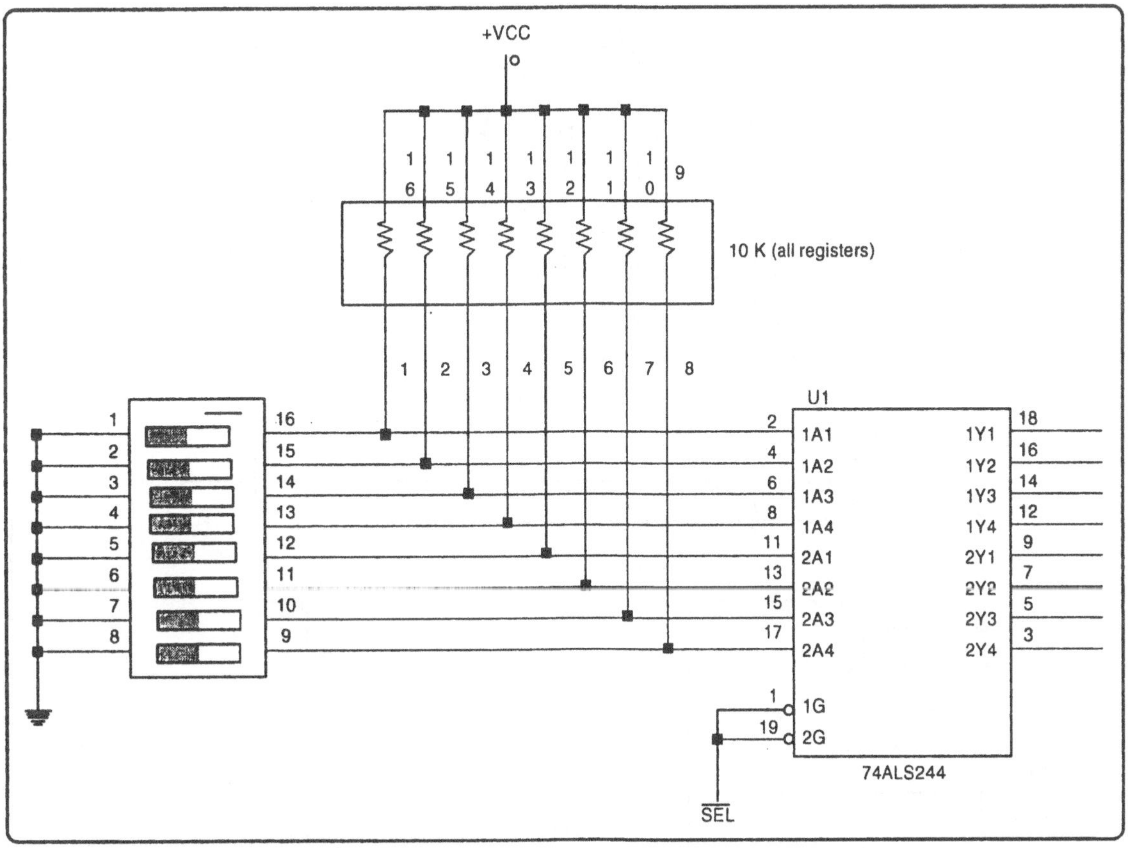

Fig. 9.4 Basic input interface illustrating the connection of eight switches

Note. The 74ALS244 is a three-state that controls the application of the switch data to the data bus.

Outputs of the buffers connect to the data bus, and exact data bus connections depend on the version of the microprocessor. For example, the 8088 has data bus connections $D_7 - D_0$, while the 80486 has $D_{31} - D_0$ and the Pentium and Pentium Pro have $D_{63} - D_0$. The circuit of Figure 9.4 allows microprocessor to read the contents of the eight switches that connect to any 8-bit section of the data bus when the select signal $\overline{SEL}$ becomes a logic 0.

If the microprocessor executes an IN instruction, the I/O port address is decoded to generate the logic 0 on $\overline{SEL}$. A logic 0 places on the output control inputs (1 $\overline{G}$ and 2 $\overline{G}$) of the 74ALS244 buffer causes the data input connections (A) to be connected to the data output (Y) connections. When a logic 1 is placed on the output control inputs of the 74ALS244 buffer, the device enters the three-state high-impedance mode that effectively disconnects the switches from the data bus.

Now let us observe the 8-bit and 16-bit input port interfacing with simple circuit, since 8086 accepts 8-bit or 16-bit data from input port. The Fig. 9.5 illustrate 8-bit input port interfacing which is detailed circuit of Fig. 9.5.

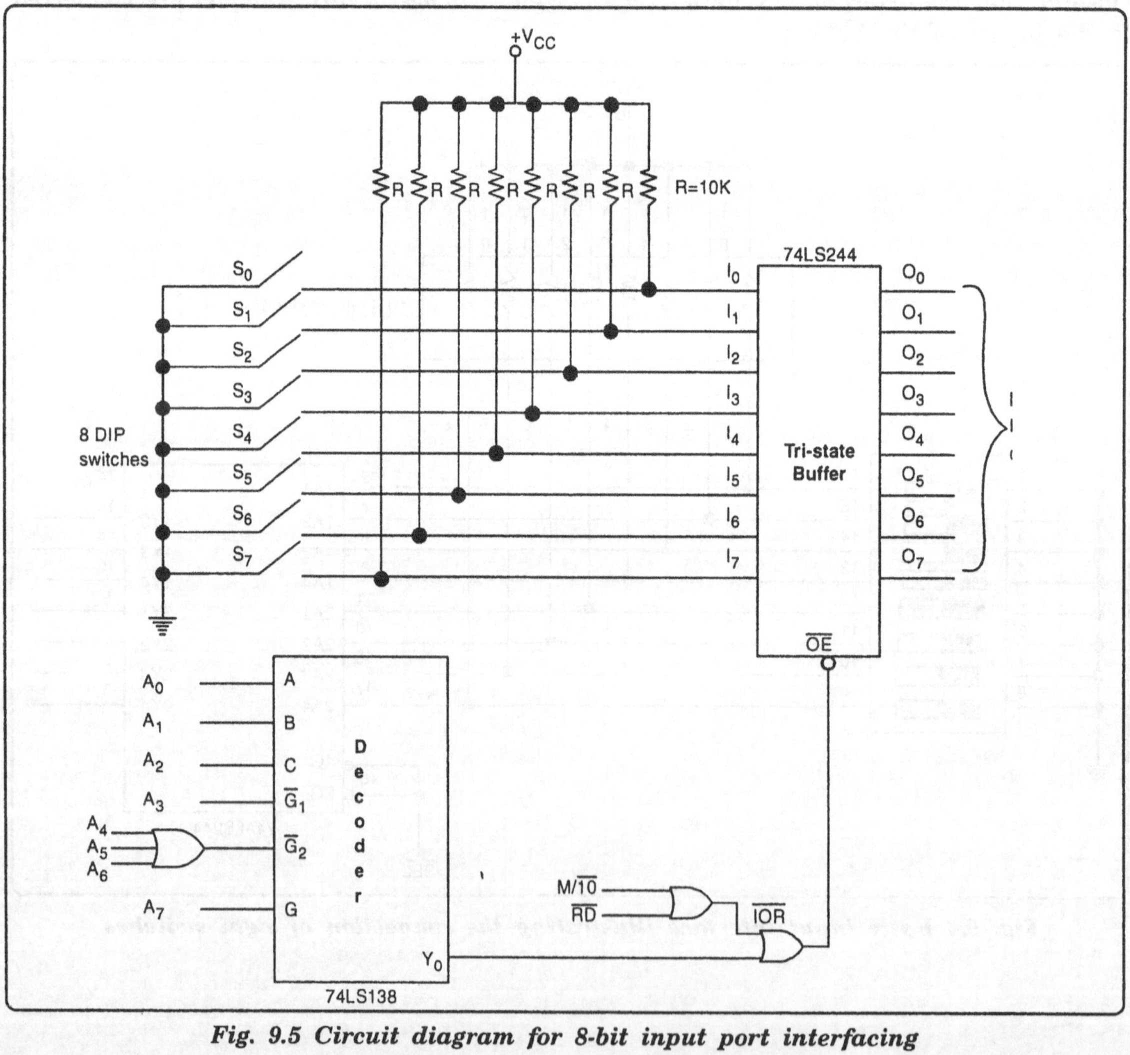

Fig. 9.5 Circuit diagram for 8-bit input port interfacing

The Fig. 9.5 shows circuit diagram to interface input port (buffer) which is used to read the status of 8 switches. The address for input device is 80H as device select signal goes low when address is 80H. If the switch is in the released position, the status of line is high otherwise status is low, that means microprocessor can check a particular key is pressed or not.

Example : Program to check switch 2 (pressed or not)

```
IN   AL, 80H        ; Read status of all switches
AND  AL, 02H        ; Mask bit positions for other switches
JZ   LOOP1          ; If Z = 0, i.e. switch 2 is pressed then
                    ; program control is transferred to label Loop1
```

Now Fig. 9.6 illustrate interfacing of 16-bit input port.

Fig. 9.6 shows the circuit diagram to interface 16-bit port (buffer) which is used to read the status of 16 switches *i.e.,* two 8 DIP switch. The address for this input port is 80H as device select signal goes low when address is 80H.

Example 2 : Program to check switch 15 (pressed or not)

```
IN   AL, 80H        ; Read status of all switches
AND  AX, 8000H      ; Mask bit positions for other switches
JZ   LOOP2          ; If Z = 0, i.e. switch 15 is pressed then
                    ; program control is transferred to label Loop2
```

Such basic circuit is not optional and must appear any time that input data are interfaced to the µP. As we have seen to interface 16-bit data the circuit of 9.4 or 9.3 was doubled, with two. 74ALS244, similarly for 32-bits of data, the circuit is to be expanded by a factor 4.

9.4.2 The Basic Output Interface

The basic output interface receives data from the microprocessor and must usually hold it for some external device, whose latches, like the buffers found in the input device, are often built into the I/O device. The Fig. 9.7 depicts how eight simple light-emitting diodes (LEDs) connect to the microprocessor through a set of eight data latches.

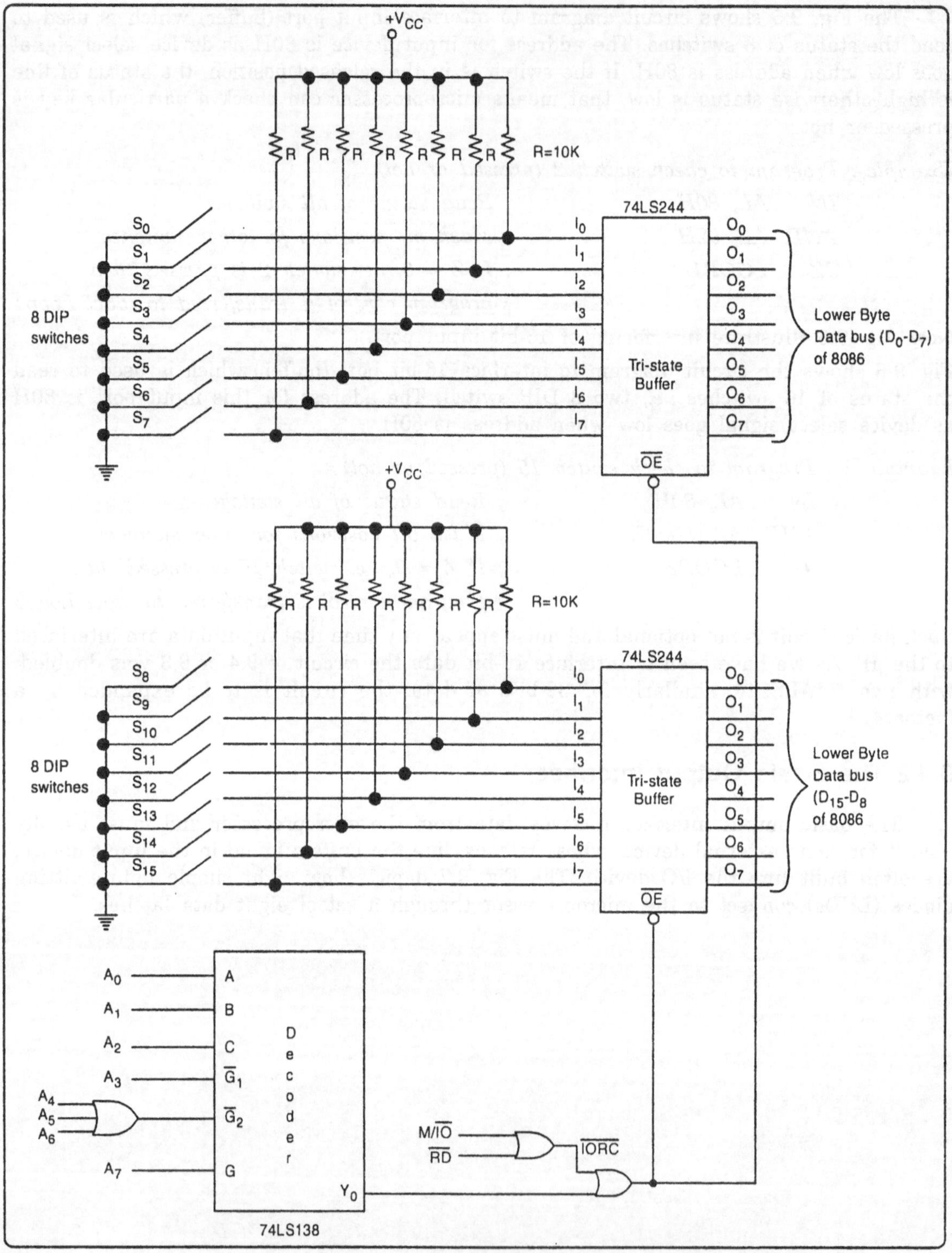

Fig. 9.6 Circuit diagram for interfacing 16-bit input port.

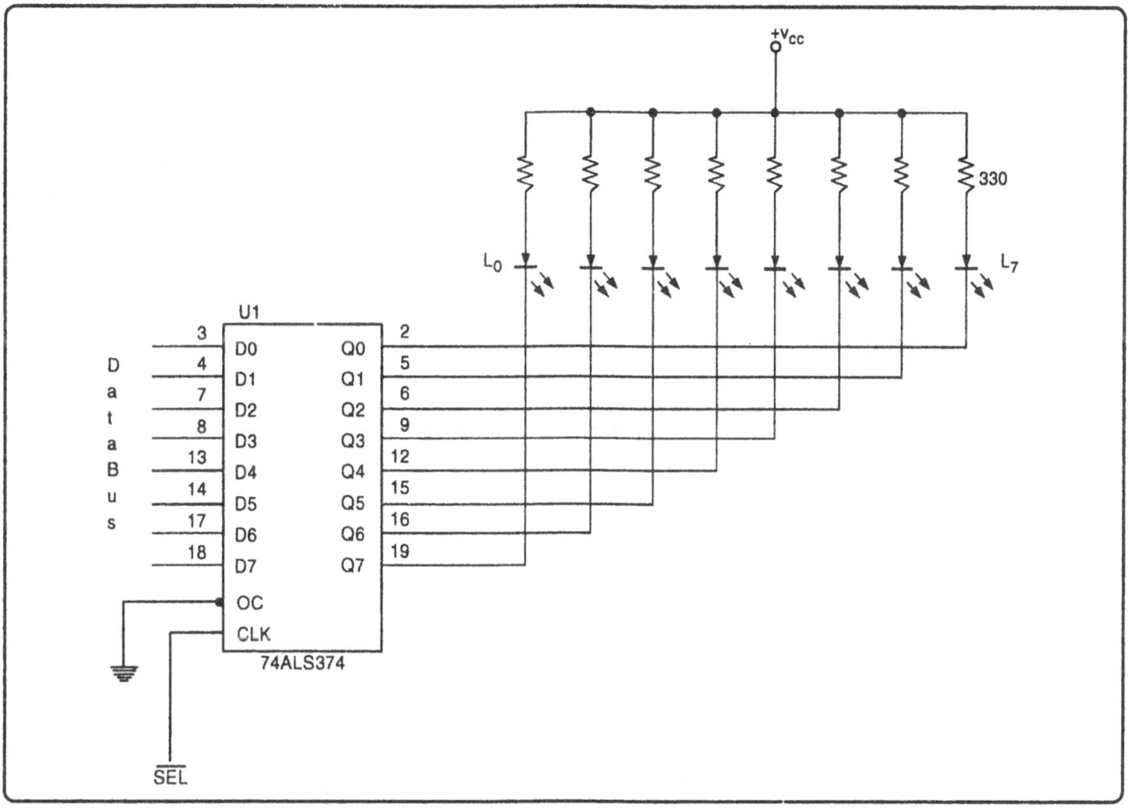

Fig. 9.7 Basic output interface connected to a set of LED displays.

In Fig. 9.7 the latch stores the number output by the microprocessor from the data bus so that the LEDs can be lit with any 8-bit binary number. Here the latches are required to hold the data because when the microprocessor executes an OUT instruction, the data are only present on the data bus for less than 1.0 μs. Without a latch, the viewer would never see the LEDs illuminate at all.

Now let us analyze with some more detailed circuit that how 8086 can send 8-bit or 16-bit data to output port as per the instruction used for data transfer. Fig. 9.8 depicts circuit for interfacing 8-bit output port latch to glow 8 LEDs. *(Which is extension of fig. 9.7 above).*

The LED can glow when output pin status is low. The IC 74LS138 and OR gate are used to generate device select signal. The latch enable signal is active high. So NOR gate is used to generate latch enable signal, which goes high when Y_2 and $\overline{IOWC}$ both are low.

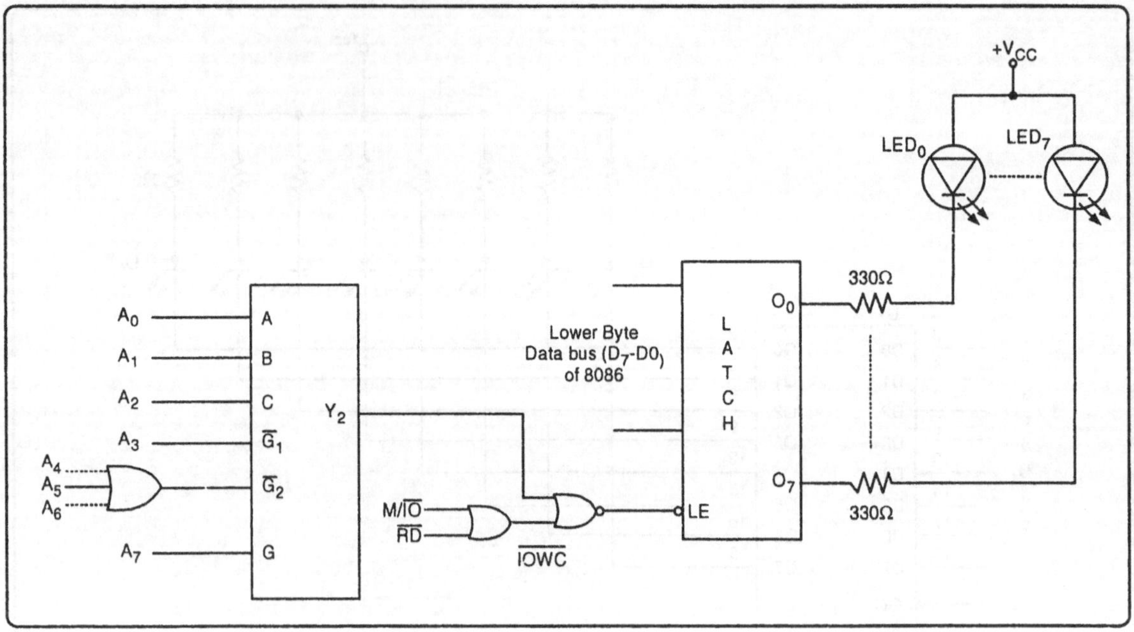

Fig. 9.8 Circuit arrangement for interfacing 8-bit output port.

Decoder select					Decoder input		
A_7	A_6	A_5	A_4	A_3	A_2	A_1	A_0
1	0	0	0	0	0	1	0

= 82 H

Example : Program to glow LED's LED_0, LED_2 and LED_0

```
MOV  AL, 0DAH     ; Loads the data in the accumulator
OUT  82H, AL      ; sends the data on the latch
```

L_7	L_6	L_5	L_4	L_3	L_2	L_1	L_0
1	1	0	1	1	0	1	0

= DAH

(The code (data) DAH is sent on the latch to glow sLEDs L_0, L_2 and L_5)

Fig. 9.9 depicts the circuit diagram to interface 16-bit output port (latch) which is used to send the signal for glowing 16 LEDs. LED will glow when output pin status is low.

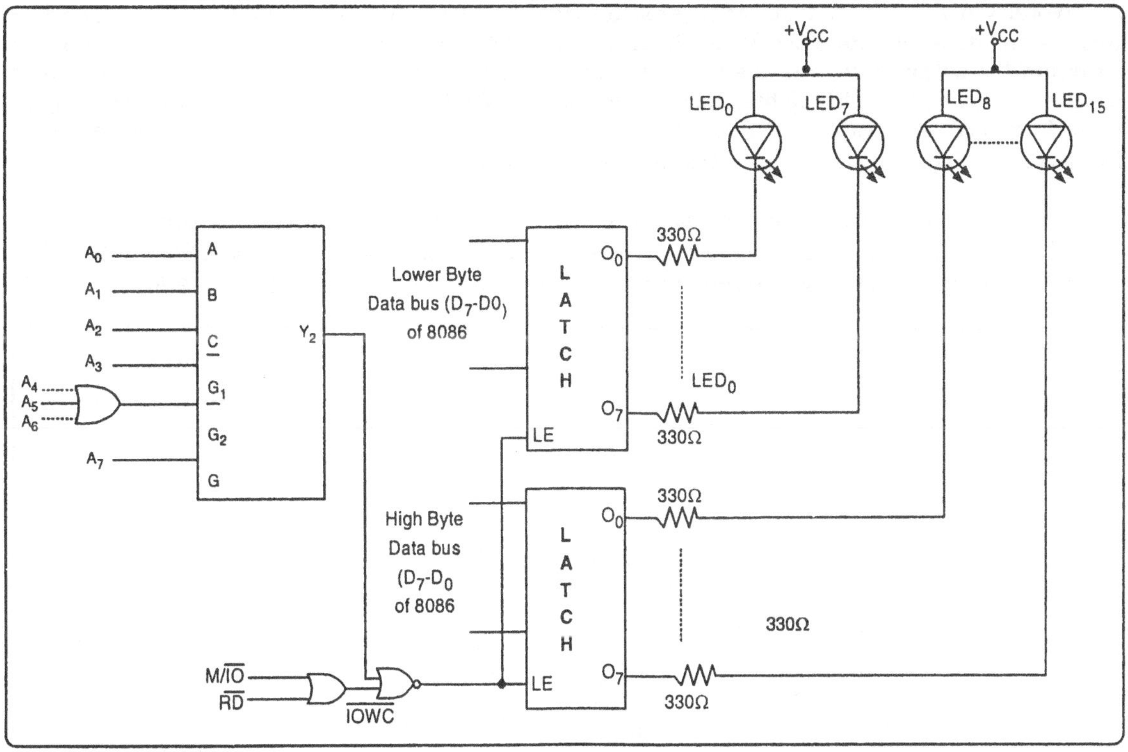

Fig. 9.9 Circuit diagram interfacing 16-bit output port

In this circuit arrangement IC 74LS138 and OR gate are used to device select signal. The latch enable signal is active high, hence NOR gate is used to generate latch enable signal, which goes high when Y_2 and $\overline{IOWC}$ both are low.

Decoder select (A_7 – A_3)   Decoder input (A_2 – A_0)

A_7	A_6	A_5	A_4	A_3	A_2	A_1	A_0	
1	0	0	0	0	0	1	0	= 82 H

Example : Program to glow L_{14}, L_{11}, L_8, L_5, L_2 and L_0

```
MOV AX, 0B6DAH        ; Loads the data in the accumulator
OUT 82H, AX           ; sends the data on the latch
```

L_{15}	L_{14}	L_{13}	L_{12}	L_{11}	L_{10}	L_9	L_8	L_7	L_6	L_5	L_4	L_3	L_2	L_1	L_0	
1	0	1	1	0	1	1	0	1	1	0	1	1	0	1	0	= B6DAH

(code (data) B6DAH is sent on the latch to glow LEDs L_{14}, L_{11}, L_8, L_5, L_2 and L_0)

Note. When the OUT instruction executes, the data from AL, AX or EAX are transferred to latch via the data bus. Here the D inputs of a 74ALS374 octal latch are connected to the data bus to capture the output data, and the Q outputs of the latch are attached to the LEDs. If a Q output becomes a logic 0, the LED lights. Each time that the OUT instruction executes, the SEL signal to the latch activates, capturing the data output to the latch from any 8-bit section of data bus. *(The data are held until the next OUT instruction executes).*

Example : For combined circuit for I/O interfacing. Here address of input port will be 80H and address of output port is 82H. Fig. 9.10 illustrates the circuit of I/O interfacing using I/O mapped I/O. *(i.e., combination explanation of Fig. 9.4 and Fig. 9.8).*

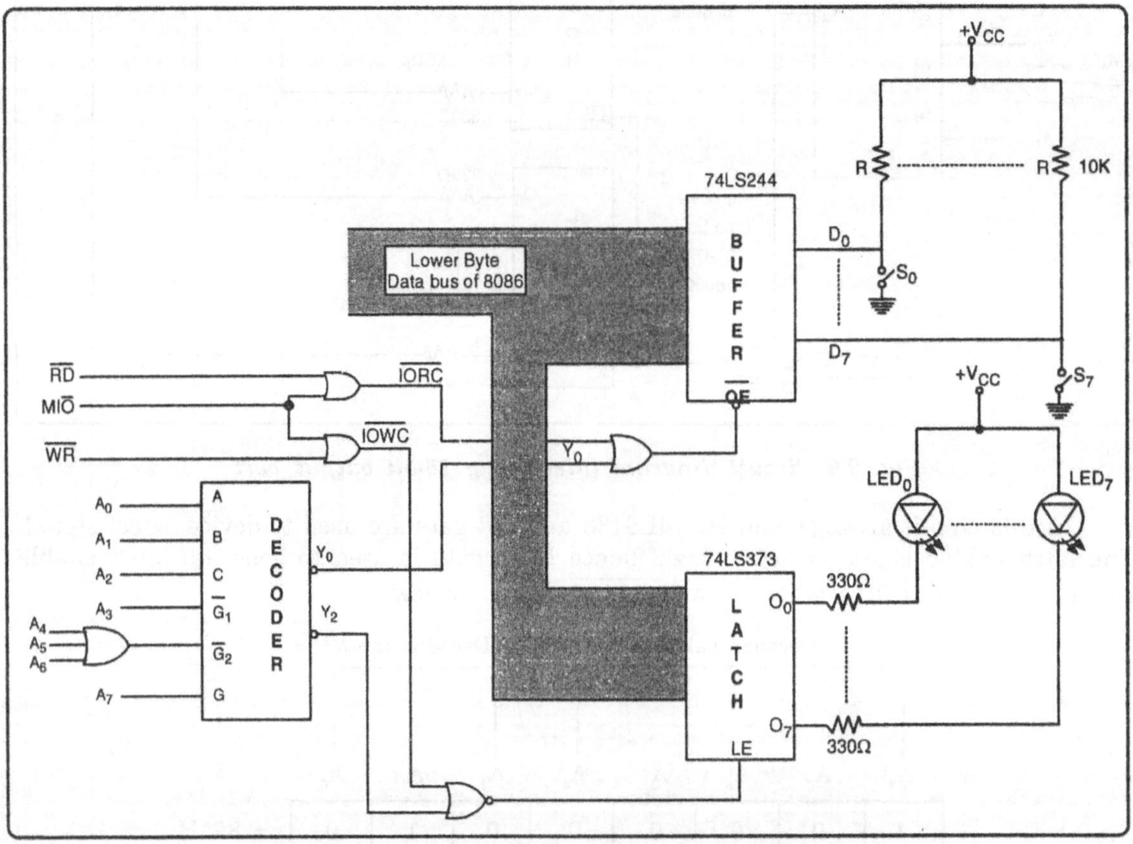

Fig. 9.10 Circuit for I/O interfacing I/O mapped I/O

Program to display status of switches on the LEDs

Opcode	**Comments**
IN 80 H	; Read status of all switches
OUT 82H	; Send status on output port

9.5 THE HANDSHAKING

Almost all I/O devices accept or release information at a much slower rate than the microprocessor. One more method of I/O control, called handshaking or polling, synchronizes the I/O device with the microprocessor. An example device that requires handshaking is a parallel printer that prints 100 characters per second (CPS). It is obvious that the microprocessor can definitely send more than 100 CPS to the printer so a way to slow the microprocessor down to match speeds with the printer must be developed. Now the Fig. 9.11 shows the typical input and output connections found on a normal printer.

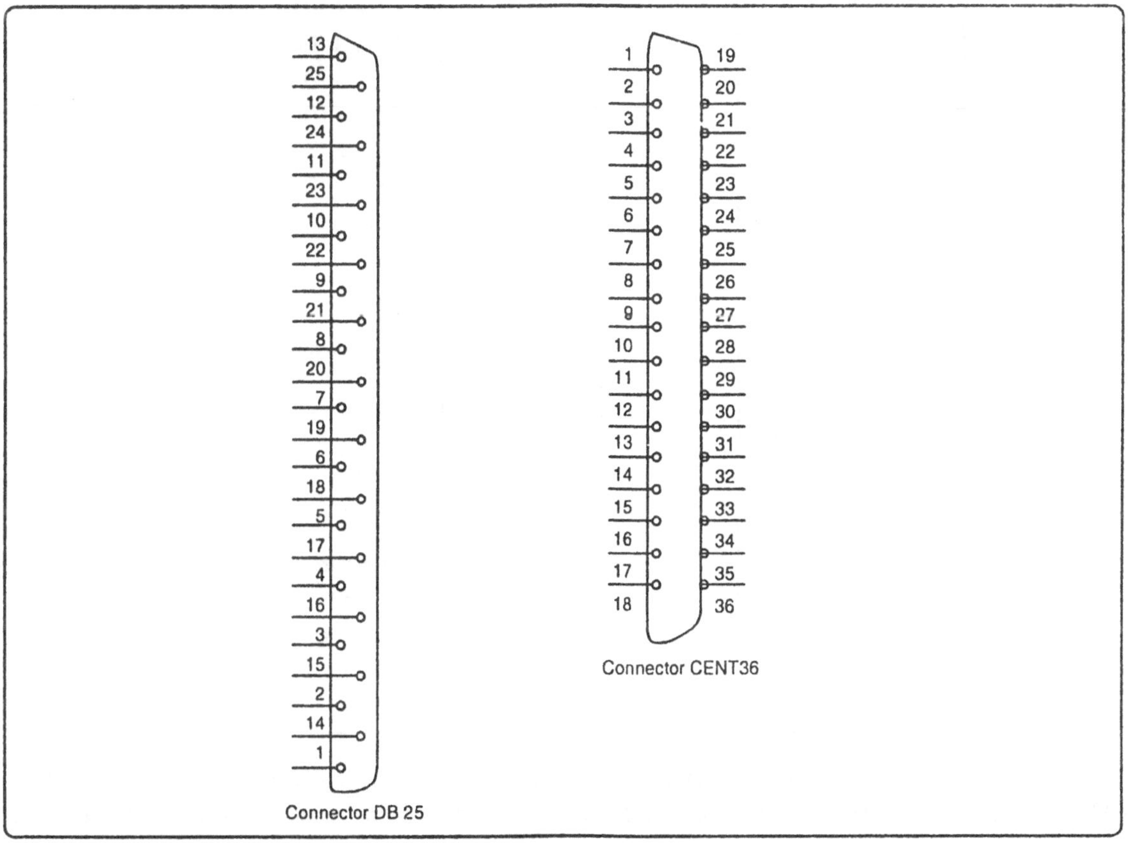

Fig. 9.11 DB25 connector found on computers and the Centronics 36-pin connector found on printers for the Centronics parallel printer interface.

Table 2 gives pins and functions for fig. 9.11.

Table 2

DB 25 Pin number	CENT 36 Pin number	Function
1	1	$\overline{\text{Data Strobe}}$
2	2	Data 0 (D0)
3	3	Data 1 (D1)
4	4	Data 2 (D2)
5	5	Data 3 (D3)
6	6	Data 4 (D4)
7	7	Data 5 (D5)
8	8	Data 6 (D6)
9	9	Data 7 D7)
10	10	$\overline{\text{Ack}}$
11	11	Busy
12	12	Paper empty
13	13	Select
14	14	Afd
15	32	$\overline{\text{Error}}$
16	–	$\overline{\text{RESET}}$
17	31	Select in
18 – 25	19 – 30	Ground
–	17	Frame ground
–	16	Ground
–	33	Ground

In this case the data are transferred through a series of data connections ($D_7 - D_0$), BUSY indicates that printer is busy, and $\overline{\text{STB}}$ is a clock pulse used to send data into the printer for printing.

For ASCII data to be printed by the printer are placed on $D_7 - D_0$ and a pulse is applied to the STB connection. The strobe signals send the data into the printer so it can be printed. Immediately after the printer receives the data, it places a logic 1 on the BUSY pin, indicating that it is busy printing data. The microprocessor polls or tests the BUSY pin to decide if the printer is busy. When the printer is busy, the microprocessor waits; if it is not busy, the microprocessor sends another ASCII character to the printer. Such process of interrogating the printer is called handshaking or polling. Example 10-1 illustrate a simple procedure that tests the printer BUSY flag and sends data to the printer if it is not busy. The PRINT procedure prints the ASCII-coded contents of BL only if the BUSY flag is a logic 0, indicating that the printer is not busy.

Example : Procedure to print ASCII contents of BL

```
                        ; A procedure that prints the ASCII contents of BL
                        ;
0000                    PRINT PROC NEAR

0000  E4  4B                    IN      AL, BUSY          ; get BUSY flag
0002  A8  04                    TEST    AL, BUSY_BIT      ; test BUSY bit
0004  75  FA                    JNE     PRINT             ; if printer busy
0006  8A  C3                    MOV     AL, BL            ; get data from BL
0008  E6  4A                    OUT     PRINTER, AL       ; send data to printer
000A  CB                        RET                       ; return from procedure
000B                    PRINT   ENDP
```

9.6 ADDRESS DECODING

The I/O port address decoding is just similar to memory address decoding, mainly for memory-mapped I/O devices. Here we memory-mapped I/O decoding because it is treated exactly the same as memory, except that the $\overline{IORC}$ and $\overline{IOWC}$ are not used, since there is no IN or OUR instruction. The decision to use memory-mapped I/O is often determined by the size of the memory system and the placement of the I/O devices in the system.

However main difference between memory decoding and isolated I/O decoding is the number of address pins connected to the decoder. We decode $A_{32} - A_0$, $A_{23} - A_0$, or $A_{19} - A_0$ for memory and $A_{15} - A_0$ for isolated I/O. But when the I/O devices use only fixed I/O addressing, we decode only $A_7 - A_0$. One more difference is that we use the $\overline{IORC}$ and $\overline{IOWC}$ to activate I/O devices for a read or write operation. *(In earlier versions of the microprocessor,* $IO/\overline{M}$ *= 1 and* $\overline{RD}$ or $\overline{WR}$ *are used to activate I/O devices. In the newest versions of the microprocessor, the* $M/\overline{IO}$ *= 0 and* $W/\overline{R}$ *are used to activate I/O devices.*

9.6.1 Decoding 8-Bit I/O Addresses

We know that the fixed I/O instruction uses an 8-bit I/O port address that appears on $A_{15} - A_0$ as 0000H – 00FFH. Since a system will never contain more than 256 I/O devices, we often decode only address connections $A_7 - A_0$ for an 8-bit I/O port address. That means we ignore address connections $A_{15} - A_8$. But you cannot ignore these address connections in a personal computer—all 16-bits must be used there. Note that the DX register can also address I/O ports 00H – FFH and also note that if the address. Fig. 9.12 shows a 74ALS138 decoder that decodes 8-bit I/O ports F0H-F7H.

Note. Assume that the system uses only I/O ports 00H – FFH for this decoder

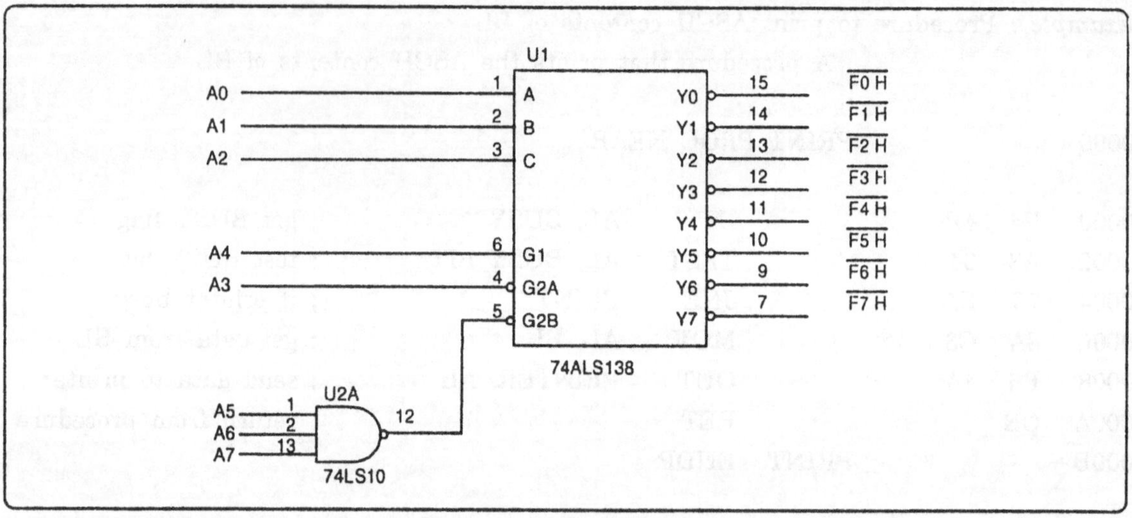

Fig. 9.12 Port decoder that decodes 8-bit I/O ports. (This decoder generates active low outputs for ports F0H - F7H)

This decoder is identical to a memory address decoder except we only connect address bits $A_7 - A_0$ to the inputs of the decoder.

Now fig. 9.13 illustrates PAL version of such decoder.

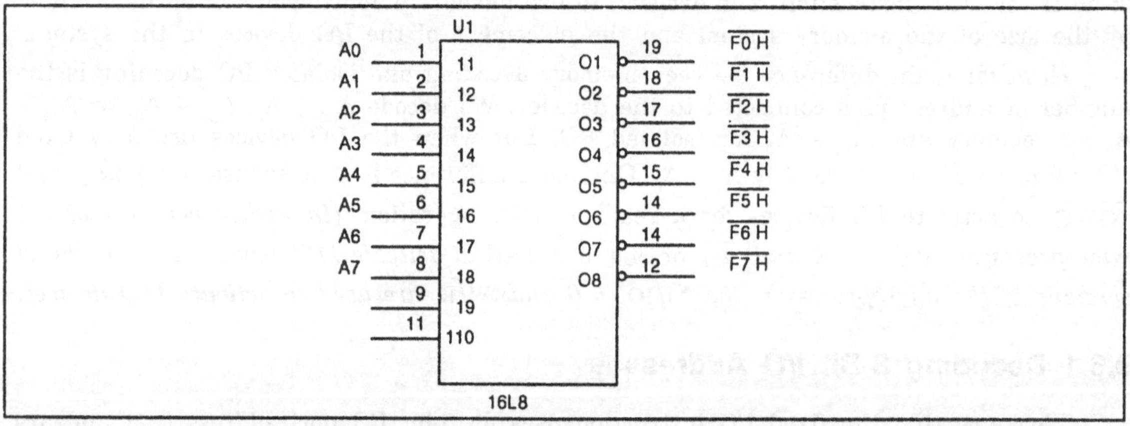

Fig. 9.13 PAL16L8 decoder that generates I/O port signals for port F0H-F7H

From Fig. 9.13 it can be observed that, this is a better decoder circuit since number of ICs are reduced on a single device.

Example : Program for PAL

CHIP			*DECODER*				*PAL16L8*			
; pins	1	2	3	4	5	6	7	8	9	10
	A0	A1	A2	A3	A4	A5	A6	A7	NC	GND
; pins	11	12	13	14	15	16	17	18	19	20
	NC	F7	F6	F5	F4	F3	F2	F1	F0	VCC

```
EQUATIONS
/F0 = A7 * A6 * A5 * A4 * A3 * /A2 * /A1 * /A0
/F1 = A7 * A6 * A5 * A4 * A3 * /A2 * /A1 *  A0
/F2 = A7 * A6 * A5 * A4 * A3 * /A2 *  A1 * /A0
/F3 = A7 * A6 * A5 * A4 * A3 * /A2 *  A1 *  A0
/F4 = A7 * A6 * A5 * A4 * A3 *  A2 * /A1 * /A0
/F5 = A7 * A6 * A5 * A4 * A3 *  A2 * /A1 *  A0
/F6 = A7 * A6 * A5 * A4 * A3 *  A2 *  A1 * /A0
/F7 = A7 * A6 * A5 * A4 * A3 *  A2 *  A1 *  A0
```

9.6.2 Decoding 16-Bit I/O Addresses

It is possible to decode 16-bit I/O addresses, mainly in a personal computer system. The main difference between decoding an 8-bit I/O address and a 16-bit I/O address is that eight additional address lines (A_{15} – A_8) must be decoded. The Fig. 9.14 depicts a circuit which contains a PAL16L8 and an 8-input NAND gate used to decode I/O ports EFF8H-EFFFH, which are common I/O port assignments in a PC used for serial communications port.

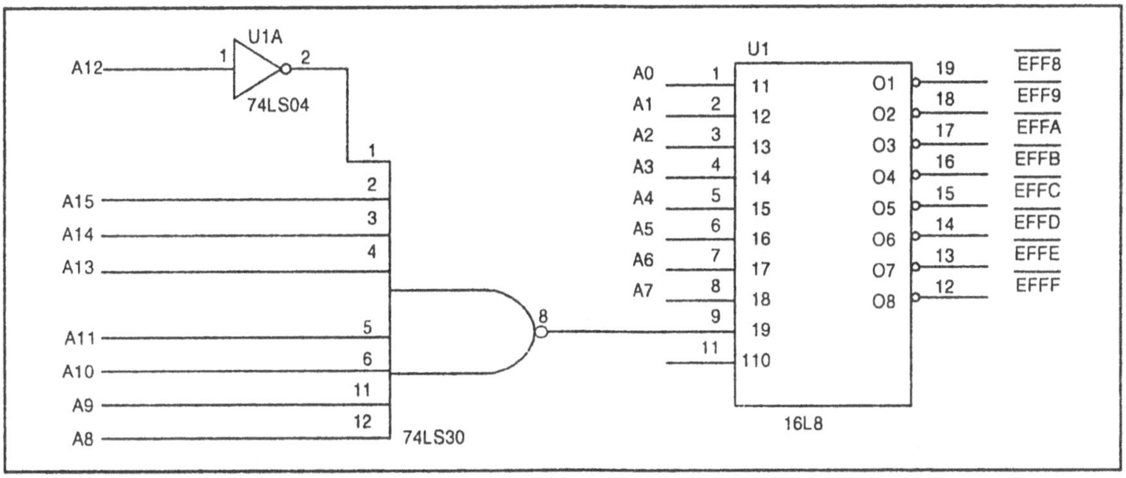

Fig. 9.14 PAL16L8 decoder that decodes 16-bit address EFF8H-EFFFH.

Here the NAND gate decodes the first 8-bits of the I/O port address (A_{15} – A_8) so that it generates a signal to enable the PAL16L8 for any I/O address between EF00H and EFFFH. Next PAL16L8 further decodes the I/O address to produce eight active low output strobes $\overline{\text{EFF8H}}-\overline{\text{EFFFH}}$.

Example : Program for PAL16L8

```
CHIP        DECODER9        PAL16L8
; pins  1     2     3     4     5     6     7     8     9     10
        A0    A1    A2    A3    A4    A5    A6    A7    NAND  GND
; pins 11    12    13    14    15    16    17    18    19    20
       NC  EFFFH EFFEH EFFDH EFFCH EFFBH EFFAH EFF9H EFF8H   VCC
```

EQUATIONS

```
/EFF8H = A7 * A6 * A5 * A4 * A3 * /A2 * /A1 * /A0 */NAND
/EFF9H = A7 * A6 * A5 * A4 * A3 * /A2 * /A1 *  A0 */NAND
/EFFAH = A7 * A6 * A5 * A4 * A3 * /A2 * /A1 * /A0 */NAND
/EFFBH = A7 * A6 * A5 * A4 * A3 * /A2 * /A1 *  A0 */NAND
/EFFCH = A7 * A6 * A5 * A4 * A3 *  A2 * /A1 * /A0 */NAND
/EFFDH = A7 * A6 * A5 * A4 * A3 *  A2 * /A1 *  A0 */NAND
/EFFEH = A7 * A6 * A5 * A4 * A3 *  A2 *  A1 * /A0 */NAND
/EFFFH = A7 * A6 * A5 * A4 * A3 *  A2 *  A1 *  A0 */NAND
```

Now let us observe the I/O interfacing with port address and Memory-Mapped I/O. That is :

(*a*) I/O Interfacing with 16-bit Port Address and

(*b*) I/O Interfacing with memory-mapped I/O

After discussion now we are clear that for (*a*), 16-bit port addressing we need to decode lower sixteen address lines to generate device select signal, and for (*b*) the 8086 makes use

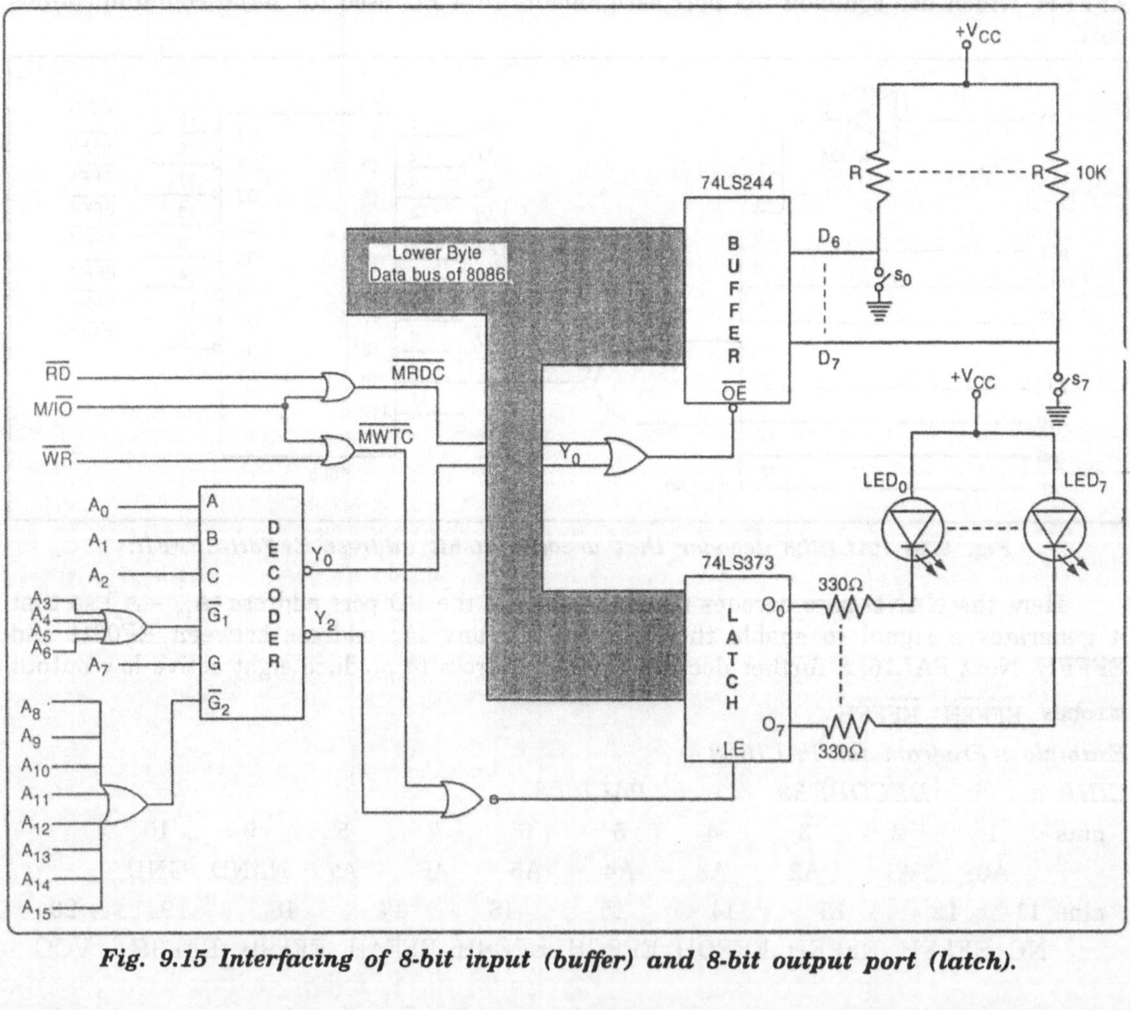

Fig. 9.15 Interfacing of 8-bit input (buffer) and 8-bit output port (latch).

of 20 address lines to identify an I/O device where I/O device will be connected as a memory register.

(a) I/O Interfacing with 16-bit Port Address

The Fig. 9.15 illustrates interfacing of 8-bit input port (*i.e.* buffer) and 8-bit output port (*i.e.* latch) with 16-bit address.

(Here we can recall the Fig. 9.4, Fig. 9.8 and Fig. 9.10) while writing the circuit of Fig. 9.15. Now here address of input port (buffer) is 0080H and address of output port (latch) is 0082H, as considered earlier.

Example : Program to display status switches on LEDs.

```
MOV  DX, 0080H        ; Load 16-bit address of the input port in DX
IN   AL, DX           ; Read byte from input port in AL
MOV  DX, 0082H        ; Load 16-bit address of the output port in DX
OUT  DX, AL           ; Send byte to output port from AL
```

Note. For different address Input and Output 0080H and 0082H will vary.

(b) I/O Interfacing with Memory Mapped I/O

The fig. 9.16 illustrates the circuit arrangement for I/O interfacing with memory mapped I/O.

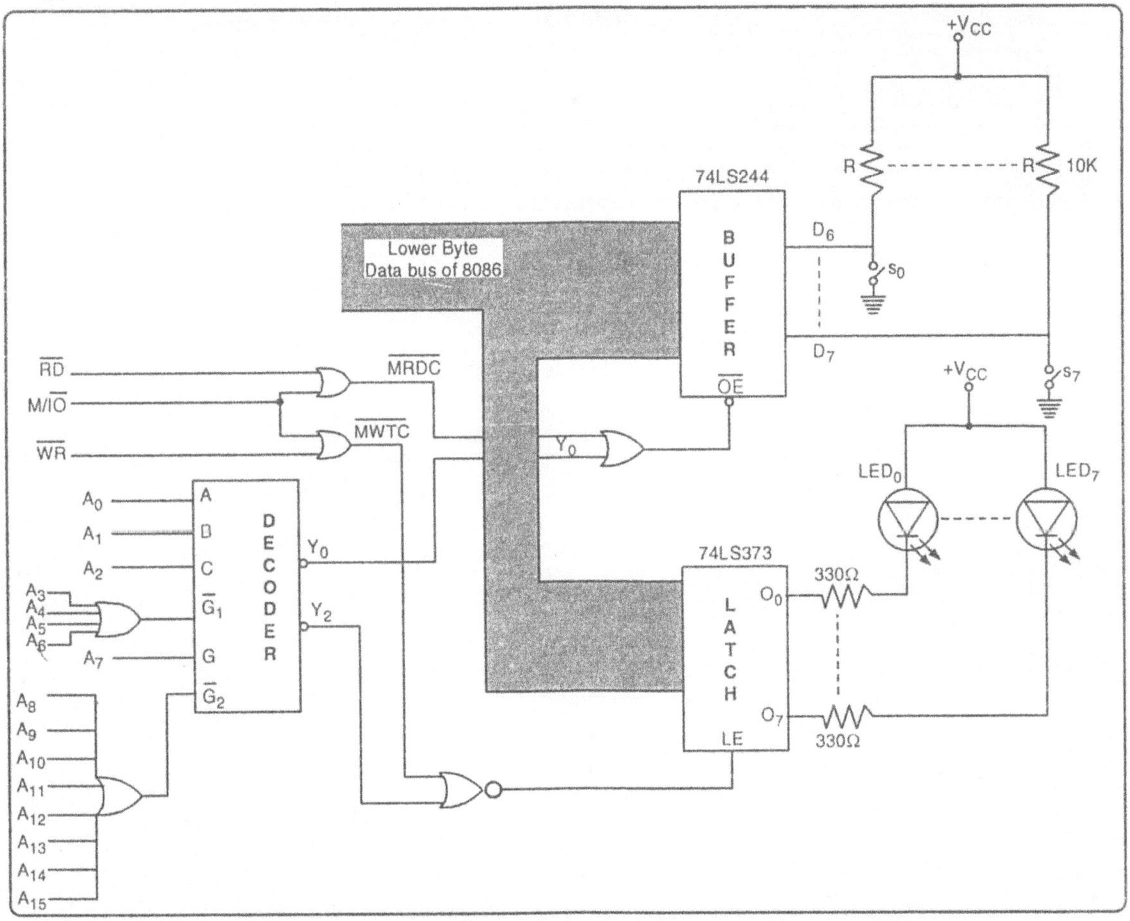

Fig. 9.16 Circuit for I/O Interfacing with memory mapped I/O.

As observed in Fig. 9.16 8086 uses same control signals and instructions to access I/O as those of memory and $\overline{RD}$ and $\overline{WR}$ signals are activated when $M/\overline{IO}$ signal is high, indicating memory bus cycle.

Example : Program to display statics switch on LEDs.

Opcodes	**Comments**
MOV AL, [0080H]	*; Read byte from input port in AL*
MOV [0082H], AL	*; Send byte to output port from AL*

(Here also 0080H and 0082H0 can be changed for different input and output)

REVIEW QUESTIONS

1. Briefly explain the way of data flow for an IN and an OUT instruction ?
2. Explain the term Input port and Output port ?
3. What is the basic input interface and output interface ?
4. Explain I/O data transfer using I/O ports with a next diagram ?
5. Explain the term Handshaking as it applies to computer I/O system ?
6. Draw the circuitary requirement to generate upper and lower I/O write strober ?
7. Draw and explain the circuit for
   (*a*) Interfacing 8-bit output port,
   (*b*) Interfacing 16-bit output port.
8. Draw and explain the circuit for I/O interfacing I/O mapped I/O ?
9. Explain I/O interfacing with memory mapped I/O with the help of a circuit ?

10 8086 SAMPLE PROGRAMS

There is no limit of developing programs in 8086. Depending upon the problem it is possible to write ALP in 8086 to obtain the result. While writing ALP it is recommended to follow the steps and procedure explained in the earlier chapters. A proper format in writing program with address Label, opcode comments etc, makes the user clear about the logic.

10.1 WRITE AN ALP IN 8086 TO ADD TWO NUMBER

TITLE 8086 assembly language program to add two numbers.

```
Label               Mnemonics                   Comments
. MODEL SMALL
. stack 100
. DATA
    No1             DB  63H                     ; First number storage
    No2             DB  2EH                     ; Second number storage
    Result          DW  ?                       ; Double byte reserved for result
. code
START : MOV  AX, @data                          ; [ Initializes
        MOV  DS, AX                             ; data segment ]
        MOV  AL, No1                            ; Get first number in AL
        ADD  AL, No2                            ; Add second to it
        ADC  AH, 00H                            ; Put carry in AH
        MOV  Result, AX                         ; Copy result to memory
        END  START
```

10.2 WRITE AN ALP IN 8086 TO FIND SUM OF NUMBERS STORED IN AN ARRAY

TITLE 8086 ALP to find sum of numbers in the array.

```
Label     Mnemonics                          Comments
.MODEL SMALL
.DATA
          ARRAY  DB 12H, 24H, 26H, 63H, 25H, 86H, 2FH, 33H, 10H, 35H
          SUM    DW0
.CODE
START :   MOV    AX, @data                   ; [ Initialize
          MOV    DS, AX                      ;   data segment ]
          MOV    CL, 10                      ; Initialize counter
          XOR    DI, DI                      ; Initialize pointer
          LEA    BX, ARRAY                   ; Initialise array base pointer
BACK :    MOV    AL, [BX + DI]               ; Get the number
          MOV    AH, 00H                     ; Make higher byte 00h
          ADD    SUM,AX                      ; SUM = SUM + number
          INC    DI                          ; Increment pointer
          DEC    CL                          ; Decrement counter
          JNZ    BACK                        ; if not 0 go to back
          END    START
```

10.3 WRITE AN ALP IN 8086 TO MULTIPLY TWO 32-BITS NUMBER AND TO STORE RESULT OF 64-BIT IN MEMORY

Algorithm for Program 10.3

1. Initialize data memory with data.
2. Initialize data segment register with the data segment address
3. Initialize SI, DI and BP registers with the address of lower word multiplier (X), multiplicand (Y) and the multiplied result (Z) respectively
4. Move first multiplier into AX register (contents of X and X + 1)
5. Multiply AX with the multiplicand word (Y and Y + 1) and store the result (Z and Z + 1) in memory and the higher order 16-bit result from DX register into CX register
6. Multiply next multiplier word (X + 2 and X + 3) with the multiplicand word (Y and Y + 1)
7. Save the higher result word result in BX register
8. Add double precision (DX : AX) and BX : CX) registers
9. Store the lower order result in Z + 2 and Carry in CX
10. Multiply (X + 2 and X + 3) with (Y + 2 and Y + 3)
11. Add lower order of this result to higher order of previous result

12. Add higher order of this result with intermediate carry. Store the result in Z + 4 and Z + 6
13. Terminate the program

ALP for Program 10.3 :

```
Label        Mnemonics                Comments
.MODEL SMALL
.DATA
    X       DB 0AFFH, 0AFFH
    Y       DB 0330H, 4002H
    Z       DB 8 DUP ( ? )
.CODE
    MOV     AX, @DATA               ; Initialize the DS and ES register
    MOV     DS, AX                  ;
    MOV     ES, AX                  ;
    MOV     SI, OFFSET X            ; Initialize the SI and DI with the address pointing
    MOV     DI, OFFSET              ;
    MOV     BP, OFFSET Z            ; The respective data memory area
    MOV     AX, [SI]                ; Load the least significant 16-bits of first data, X into AX
    MUL     [DI]                    ; Multiply with the second data, Y (16-bits)
    MOV     DS: [BP], AX            ; The least significant 16-bits into memory
    MOV     CX, DX                  ; Save the higher 16-bits in CX register
    MOV     AX, [SI + 2]            ; Multiply (X + 2) with Y
    MUL     [DI]
    MOV     BX, DX                  ; Save DX into BX
    ADD     CX, AX                  ; Add the partial results and save it in CX
    ADC     BX, 0000                ; The carry is added to BX
    MOV     AX, [SI]                ; Multiply X with Y + 2
    MUL     [DI + 2]                ;
    ADD     CX, AX                  ; Add partial results and save it in CX
    ADC     BX, DX                  ; Add with carry higher order partial result.
    MOV     DS : [BP + 2], CX       ; Save the 2nd word of result in memory (Z + 2)
    MOV     CX, 0000                ; Clear the CX register
    ADC     CX, 0                   ; Save the carry into CX
    MOV     AX, [SI + 2]            ; The last 16-bit multiplication is performed : (X + 2)*
    MUL     [DI + 2]                ; (Y + 2)
    ADD     AX, BX                  ; Add all the partial results
    ADC     DX, CX                  ; Save the resultant 3rd word into memory (Z + 4)
    MOV     DS ; [BP + 4], AX       ;
```

```
MOV  DS : [BP + 6], DX  ; Save the last part of 16-bit result in memory (Z + 6)
MOV  AX, 4C00H          ; Terminate the program
INT  21H
END
```

(The Result for program 10 : 3) D0, 0C, FI, E0, D8, BF, 02

10.4 DEVELOP A PROGRAM TO DISPLAY A MESSAGE ON SCREEN USING MACROS

```
        , HELLO 3. ASM   —Display a message on screen using MACROS.
Label            Mnemonics                                Comments
;  Build      - TASM HELLO3. ASM
;               TLINK HELLO3
;  USAGE      - HELLO3
                Prinstring macro msg                 ; prinstring user defined macro
          mov       ah, 09h                          ; display string function
          mov       dx, offset msg                   ; initialize DX to msg
          int 21h                                    ; call DOS service
          endm
   _DATA  segment                                    ; _DATA - user defined segment
          cr   equ 0dh                               ; ASCII code for carriage return
          If   equ 0ah                               ; ASCII code for line feed
          msg1   db 'HELLO WORLD', cr, If, '$'       ; msg1 - pointer to the string
          msg2   db 'HELLO TEJA', '$'                ; msg2 - pointer to the string
   _DATA ends                                        ; end of _DATA segment
          _CODE segment
; _CODE - name of user defined segment
                assume cs : _CODE, ds : _DATA ; Initialize CS and DS to segments
          start :
                mov   ax, _DATA                 ; Initialize DS to the data segment
                mov   ds, ax
                Prinstring msg1                 ; print message 1
                Prinstring msg2                 ; print message 2

                ; program termination
                mov   ah, 4ch                   ; program terminate function
                mov   al, 00h                   ; return code for error level setting
```

```
        int 21h                         ; call DOS service
_CODE   ends                            ; End of segment
        end start                       ; End of program
```

10.5 WRITE AN ALP TO CALCULATE FACTORIAL OF A NUMBER

Flow chart for the program 10.5

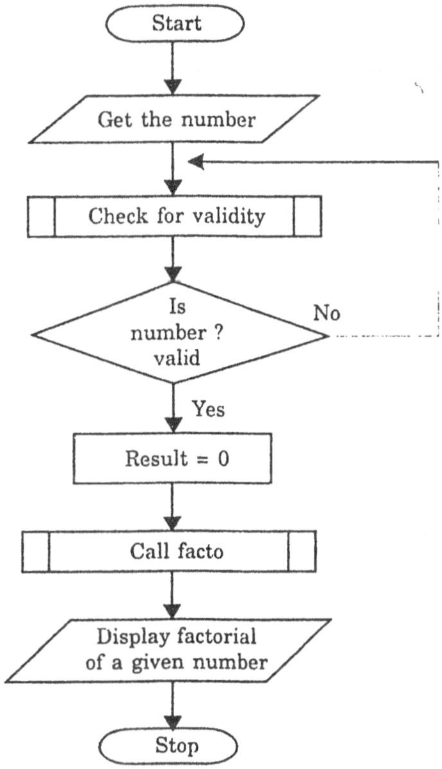

Lable **Mnemonics** **Comments**

```
. MODEL SMALL
. stack 100
. DATA
    MS1         DB 10, 13, 'ENTER THE NO. : $'
    MS2         DB 10, 13, 'THE FACTORIAL IS ; $'
    NUM         DW 0
    ANS         DW 0
.CODE
START :   MOV   DX, @DATA                       ; [Initialize
          MOV   DS, DX                          ; data segment]
```

```
ERROR :     LEA    DX, MSI
            MOV    AH, 09H                  ; Display message MSI
            INT    21H
            MOV    AH, 01H                  ; Input number with echo
            INT    21H
            CMP    AL, 30H                  ; If zero display 1
            JE     DISPLY2
            CMP    AL, 30H                  ; If < 30 then input
            JB     ERROR                    ; Next no
            CMP    AL, 39H                  ; If > 39 then input
            JA     ERROR                    ; Next no
            SUB    AL, 30H                  ; Convert to HEX
            MOV    AH, 00H
            SUBSP, 0004H                    ; Space in stack for
            PUSH   AX                       ; Factorial
            CALL   FACTO
            ADD    SP, 0002                 ; After execution
            POP    AX                       ; Of facto space for
            POP    DX                       ; Result
            MOV    DX, 0010                 ; Convert HEX to BCD
            MOV    CX, 0006                 ; Max input no is 9
BACK :      DIV    DX                       ; To get remainder
            OR     DX, 0030H                ; Convert to ASCII
            PUSH   DX
            XOR    DX, DX                   ; Clear DX
            LOOP   BACK
            LEA    DX, MS2                  ; Output MS2
            MOV    AH, 09
            INT    21H
            MOV    CX, 0006
DISPLAY1 :  POP    DX
            MOV    AH, 02H                  ; Output factorial
            INT    21H
            LOOP   DISPLAY1
            JMP    LAST
DISPLAY2 :  MOV    AH, 09
            LEA    DX, MS2                  ; Display factorial of
            INT    21H                      ; Zero = 1
            MOV    AH, 02H
```

```
            MOV    DL, 31H
            INT    21H
LAST :      MOV    AH, 4CH                        ; Terminate and
            INT    21H                            ; Exit to DOS]
FACTO :     PROC
            PUSHF
            PUSH   AX
            PUSH   DX
            PUSH   BP
            MOV    BP, SP                         ; Point BF at TOS
            MOV    AX, [BP + 10]                  ; Copy no from stack to
            CMP    AX, 000IH                      ; AX, and if no not = 1 then
                   ; GO_ON
            JNE GO_ON                             ; To compute factorial
            MOV    WORD PTR [BP + 12], 000IH
                                                  ; Else load FFACT
            MOV    WORD PTR [BP + 14], 0000H
                                                  ; 0 and 1 in stack
            JMP EXIT
GO_ON :     SUBSP, 0004H                          ; Space for preliminary
            DEC    AX                             ; Factorial
            PUSH   AX
            CALL   FACTO
            MOV    BP, SP
            MOV    AX, [BP + 2]                   ; Last (N - I) ! from
                                                  ; stack to AX
            MUL WORD PTR [BP + 16]                ; Multiply by previous N
            MOV    [BP + 18], AX                  ; Copy new facto to stack
            MOV    [BP + 20], DX
            ADD    SP, 0006H                      ; Point SP at pushed REGR

EXIT :      POP    BP
            POP    DX
            POP    AX
            POPF
            RET
FACTO :     ENDP
            END    START
```

10.6 DEVELOP A PROGRAM IN ALP 8086 TO DIVIDE A STRING OF UNPACKED ASCII DIGITS BY A SINGLE ASCII DIGIT

Note. While dividing unpacked ASCII by another ASCII digit it is required to load unpacked number into AL register and divisor in DL register. Next each digit is converted to unpacked BCD and division is performed and result is stored in memory finally process is to be repeated to the string of data.

```
Label       Mnemonics                        Comments
.MODEL TINY
.DATA
    DIVED   DB    05, 06, 09, 08, 01, 04, 07
    DIVISR  DB    06
    RESULT  DB    7 DUP ( ? )
.CODE
    MOV     AX, @DATA
    MOV     DS, AX
    MOV     ES, AX
    MOV     DL, DIVISR              ; Divisor is loaded into DL
    MOV     SI OFFSET DIVED         ; LSB of the ASCII string address loaded
    MOV     DI, OFFSET RESULT;
    MOV     CX, 0010                ; Initialize the length of the string
    AND     DL, 0FH                 ; Convert ASCII to decimal value
    XOR     AH, AH                  ; Byte is loaded into AL
    AND     AL, 0FH                 ; Convert ASCII to decimal
    AAD                             ; Adjust AL before division
    DIV     DL                      ;
    STOSB                           ; Save string byte into memory
    LOOP    NEXT                    ; Have all bytes been divided ?
    MOV     AH, 4CH                 ; If no, go to next
    INT 21H                         ; If yes, terminate the program
    END
```

(The result for program 10.6 = 00, 09, 04, 09, 06, 09, 01)

10.7 WRITE AN ALP IN 8086 TO REVERSE THE WORDS IN STRING

```
Label           Mnemonics                              Comments
. MODEL SMALL
. stack 100
. DATA
TITLE REVERSE THE WORDS IN STRING
     M1         DB 10, 13 'ENTER THE STRING ; $'
     M2         DB 10, 13 THE REVERSE STRING ; $'
     BUFF       DB 80
                DB 0
                DB 80 DUP ( 0 )
COUNTER1        DW 0
COUNTER2        DW 0
 CODE
START :         MOV     AX, @DATA                      ; [ Initialize
                MOV     DS, AX                         ;   data segement ]
                MOV     AH, 09H                        ; Display message M1
                MOV     DX, OFFSET M1
                INT     21H
                MOV     AH, 0AH
                LEA     DX, BUFF                       ; I/P the string
                INT     21H
                MOV     AH, 09H
                MOV     DX, OFFSET M2                  ; Display message M2
                INT     21H
                LEA     BX, BUFF
                INC     BX
                MOV     CH, 00H                        ; [Take character
                MOV     CL, BUFF + 1                   ;  count in
                MOV     DI, CX                         ; DI ]
BACK :          MOV     DL, [BX + DI]                  ; Point to the and
                                                       ; character and read it
                MOV     AH, 02H
                INT 21H                                ; Display the character
                DEC     DI                             ; Decrement count
                JNZ BACK                               ; Repeat until count is 0
EXIT :          MOV     AH, 4CH                        ; [Terminate
                INT 21H                                ; Exit to DOS]
                END     START
```

10.8 WRITE AN ALP IN 8086 TO DRIVE AN ELEVATOR INTERFACE FOR THE MENTIONED METHOD

Consider the elevator is initially at ground floor and all the service requests are cleared.

The elevator stays in ground floor as long as there is no service request. If a service request is detected, the elevator moves to that floor. While it is travelling to the floor, all other service requests are ignored. After servicing that floor, it scans for further request for sometime. When no request is detected within this time, the elevator travels to the ground floor ignoring any further request. On the other hand if a request is is detected while it is scanning for requests, it moves to that floor, again ignoring any other intermediate requests during the travel. If more than one request is detected at the same time, a floor is selected arbitarily.

```
Label       Mnemonics               Comments
. MODEL SMALL
. CODE
            MOV AL82H               ; Initialize Port A Output
            MOV DX 127H             ; Port B input
            OUT DX, AL
            MOV AX, 00              ; Clear all the request from different floors
            MOV DX, 124H            ;
            OUT DX, AL              ; Retain the higher nibble and set the to
            MOV AL, OFOH            ; ground floor
            OUT DX, AL              ;
LOOP1 :     CALL CHECK              ; Is there any request ?
            JZ LOOP1                ; If No, go to check again
            SHR AL, 01              ; If Yes, is it ground floor ?
            JNC GROUND              ; If Yes, go to service
            SHR AL, 01              ; Is request from Ist floor ?
            JNC FIRST               ; Yes, go to service
            SHR AL, 01              ; Is request from 2nd floor ?
            JNC SEC                 ; Yes, go to service
            JMP THIRD               ; No, go to service 3rd floor
GROUND :    CALL GF_CLEAR           ; Clear the request
            JMP LOOP1               ; Check for request
FIRST :     MOV BL, 03              ; Initialize the number of steps
            MOV AL, OFOH            ; Mask the nigner nibble
            MOV CX, 03              ; Step is initialized
            CALL MOVE_UP            ; Go to first floor
            CALL FF_CLEAR           ; Clear the request
            CALL DELAY              ; Service
```

```
            CALL DELAY
            MOV BH, AL
            CALL CHECK          ; Check if there is any other request
            JZ BACK             ; No go back
            SHR AL, 01          ; Yes, is it ground floor ?
            JNC BACK            ; Yes, go to ground floor
            SHR AL, 01          ;
            SHR AL, 01          ; Is it 22nd floor ?
            JNC SECOND1         ; Yes, go to service it
            JMP THIRD1          ; No, get to service 3rd floor
BACK :      MOV AL, BH          ; Save the registers
            MOV CL, BL
            MOV AL, BL
            OR AL, 0F0H
            CALL GO_DOWN        ; Go to ground floor
            CALL GF_CLEAR       ; Clear the request
TERMINATE : MOV AL, BH          ;
            MOV AH, 4CH         ; Terminate the program
            INT 21H
SECOND1 :   MOV AL, BH          ; Initialize the register the number of steps
            ADD BL, 03          ;
            MOVCX, 03
            CALL MOVE_UP        ; Go to second floor
            CALL SF_CLEAR       ; Clear the request
TERM :      CALL DELAY          ; Service the request
            CALL DELAY
            MOV CL, BL
            CALL GO_DOWN        ; Return to ground floor
            JMP TERMINATE       ; Terminate the program
THIRD1 :    MOV AL, BH
            ADD BL, 06          ; Initialize the registers with the number of steps
            MOV CX, 06          ;
            CALL MOVE_UP        ; Go to third floor
            CALL TF_CLEAR       ; Clear the request
            JMP TERM            ; Terminate the program
            SEC : MOV BL, 06    ; Initialize the registers with the number of steps
            MOV CX, 06          ;
            MOV AL, 0F0H
            CALL MOVE_UP        ; Go to the required floor
            CALL SF_CLEAR       ; Clear the request
```

```
            CALL DELAY          ; Service the request
            CALL DELAY
            MOV BH, AL
            CALL CHECK          ; Is there any other request ?
            JZ BACK             ; No, go to ground floor
            SHR AL, 01          ; Yes, is it ground floor ?
            JNC BACK            ; Yes, go to ground floor
            SHR AL, 01          ; Is it first floor ?
            JNC FIRST2          ; Yes, go to service
            JMP THIRD2          ; Go to service 3rd floor
FIRST2 :    MOV AL, BH          ; Save the register contents
            SUB BL, 03          ; Set the number of steps to ground floor and
                                  number
            MOV CX, 03          ; of steps to next floor
            CALL GO_DOWN        ; Go to next floor and clear request
            CALL FF_CLEAR       ;
            JMP TERM            ; Go to terminate the program
THIRD2 :    MOV AL, BH          ;
            ADD BL, 03          ; Set the number of steps to ground and to next
                                  floor
            MOV CX, 03          ;
            CALL MOVE_UP        ; Go to next floor
            CALL TF_CLEAR       ; Clear the request
            JMP TERM            ; Go to service and terminate
THIRD :     MOV BL, 09          ; Set the number of steps to ground
            MOV CH, 00          ;
            MOV CL, BL          ;
            MOV AL, 0F0H        ; Retain the request
            CALL MOVE_UP        ; Go to the next floor
            CALL TF_CLEAR       ; Clear the request
            CALL DELAY          ; Service the request
            CALL DELAY
            MOV BH, AL
            CALL CHECK          ; Check is there any request
            JZ BACK1            ; No, go to return
            SHR AL, 01          ; Is request from ground
            JNC BACK1           ; Yes, return to ground
            SHR AL, 01          ; Is request from first floor ?
            JNC FIRST3          ; Yes, go to service first floor ?
            JMP SECOND3         ; Else go to second floor
```

```
BACK1 :      JMP BACK             ; Go to terminate
FIRST3 :     MOV AL, BH           ; Service at first floor
             MOV CX, 06           ; Set the number of steps
             SUB BL, 06
             CALL GO_DOWN         ; Go down to first floor
             CALL FF_CLEAR        ; Clear the request
             JMP TERM             ; Go to service
SECOND3 :    MOV CX, 03           ; Service at second floor
             MOV AL, BH           ; Initialize the count
             SUB BL, 03           ;
             CALL GO DOWN         ; Go down the floor
             CALL SF, CLEAR       ; Clear the request
             JMP TERM             ; Go to service request
```

10.9 WRITE AN ALP IN 8086 TO SORT AN ARRAY USING BUBBLE SORT METHOD

```
Label        Mnemonics                          Comments
. MODEL SMALL
. DATA
    M1       DB   10, 13, 'Enter number of elements in the array "$'
    M2       DB   10, 13, 'The sorted array is : "0dh, 0ah, '$'
    M3       DB   10, 13, '$'
    ARRAY    DB   100 DUP ( 0 )
. CODE
START :      MOV  AX, @DATA                     ; [ Initialize
             MOV  DS, AX                        ; data segment ]
             MOV  AH, 09H                       ; Display message M1
             MOV  DX, OFFSET M1
             INT  21H
             MOV  AH, 01H                       ; Input the number of elements
                                                ; in the array
             INT  21H
             SUB  AL, 30H                       ; Get the number from its
                                                ; ASCII code
             MOV  BH, AL                        ; Save the number of elements
             MOV  BL, Al                        ; Save another copy of the
                                                ; number of elements
```

```
          XOR   DI, DI                  ; Clear pointer
          MOV   CL, 4                   ; Initialize rotation count
BACK :    MOV   AH, 09H                 ; Carriage return and next line
          MOV   DX, OFFSET M3
          INT   21H
          MOV   AH, 01
          INT   AH, 01                  ; [Read 2 digits of a
          INT   21H                     ; number
          SUB   AL, 30H                 ; from
          SAL   AL,  CL                 ; keyboard
          MOV   CH,AL                   ; and
          MOV   AH, 01                  ; from the
          INT   21H                     ; packed number]
          SUB   AL, 80H
          OR    CH, AL
          MOV   [ARRAY + DI], CH        ; Save the packed number
          INC   DI                      ; Increment pointer
          DEC   BL                      ; Decrement count
          JNZ   BACK                    ; If count not = 0 go to BACK
          MOV   CH, BH                  ; Get the number of elements
                                        ; as a counter
AGAIN1 :  MOV   DI, 0                   ; Initialize pointer to 0
AGAIN :   MOV   DI, [ARRAY + DI]        ; Get the number
          CMP   CL, [ARRAY + DI + 1]    ; Compare if with the next
                                        ; number
          JBE   SKIP                    ; If it is less or equal
                                        ; go to skip
          MOV   AH, [ARRAY + DI +1]     ; [ Otherwise
          MOV   [ARRAY + DI], AH        ; exchange
          MOV   [ARRAY + DI + 1], DL    ; two numbers]
SKIP :    INC   DI                      ; Increment poiner
          MOV   AH,0                    ; [ Check for
          MOV   AL, BH                  ; last number
          DEC   AL                      ; in the array ]
          CMP   AX, DI
          JNZ   AGAIN                   ; if no go to AGAIN
          DEC   CH                      ; Decrement counter
          JNZ   AGAIN1                  ; If counter not = 0 go to
                                        ; AGAIN1
```

```
            MOV   AH, 09H                   ; Display message M1
            MOV   DX, OFFSET M2
            INT   21H
            MOV   BL, BH                    ; Get the number of
                                            ; elements as a counter
            XOR   DI, DI                    ; Clear pointer
BACK1 :     MOV   AH, 09H                   ; Carriage return and next
                                            ; line
            MOV   DX, OFFSET M3
            INT   21H
            MOV   DL, [ARRAY + DI]          ; [ Display
            AND   DL, 0F0H                  ; the
            SAR   DL, CL                    ; sorted
            ADD   DL, 30H                   ; numbers
            MOV   AH, 02                    ; in
            INT   21H                       ; the
            MOV   DL, [ARRAY + DI]          ; Array ]
            AND   DL, 0FH
            ADD   DL, 30H
            MOV   AH, 02
            INT   21H
            INC   DI
            DEC   BL
            JNZ   BACK1
TER :       MOV   AH, 4CH                   ; [Terminate
            INT   21H                       ; and Exit to DOS]
            END   START
```

10.10 WRITE AN ALP IN 8086 TO DISPLAY A 4-DIGIT BCD NUMBER ON SEVEN SEGMENT DISPLAY USING DISPLAY INTERFACE

```
Lable            Mnemonics                        Comments
.MODEL SMALL
.DATA
  CODE1 DB 099H, 0B0H, 0A4H, 0F9H  ; Code for seven segment display
.CODE
         MOV AX, @ DATA            ; Initialize Data segment
         MOV DS, AX                ;
```

```
        MOV SI, OFFSET CODE1     ; Load the starting address of the code
        MOV AL, 80H              ; Initialize 8255A in Port - A, Port - B
        MOV DX, 127H               and Port - C output mode
        OUT DX, AL
        MOV CX, 04               ; No. of digits to be displayed
LOOP2 : MOV BL, 08               ; A counter to sent code bits for each segment
        MOV AL, [SI]             ; Code is brought from memory
NEXT :  ROL AL, 01               ; Most significant bit is brought to A0 bit position
        MOV DX, 125H             ; The bit data is sent out through Part-B
        OUT DX, AL
        PUSH AX                  ; Save AX register on stack
        MOV AL, 0FFH             ; Give a clock pulse first high and then low
        MOV DX, 0126H            ;
        OUT DX, AL
        MOV AL, 00
        OUT DX, AL
        DEC BL                   ; Has all the 8-bits sent ?
        JZ NEXT1                 ; Yes. Go to fetch next data code
        POP AX                   ; No, Restore AX value
        JMP NEXT                 ; Go to next
NEXT1 : INC SI                   ; Point to the next code
        LOOP LOOP2               ; Has all the 4 digits displayed ?
        MOV AH, 4CH              ; No, Go to loop 2
        INT 21H                  ; Yes. Terminate the program
        END
```

10.11 WRITE AN ALP IN 8086 TO GENERATE TRIANGULAR WAVE FORM WITH THE HELP OF DAC INTERFACE WITH IT

```
Lable   Mnemonics                Comments
.MODEL SMALL
CODE
        MOV AL, 80H              ; Initialize the control word with Port - A as output
        MOV DX, 127H             ;
        OUT DX, AL               ;
        MOV DX, 124H             ; Output AL contents through Port - A
        MOV AL, 00
```

```
LOOP1 : OUT DX, AL
        INC AL
        CMP AL, 0FFH          ; BAL contents FF ?
        JNZ LOOP1             ; No go to output the data to DAC
LOOP2 : OUT DX, AL            ; Yes, output the data to DAC
        DEC AL                ; Decrement ALL till zero. Is it zero ?
        JNZ LOOP2             ; If No, go to loop 2
        JNZ LOOP1             ; If Yes, start from the beginning
        END
```

10.12 WRITE AN ALP IN 8086 TO CHECK WHETHER A STRING IS PALINDROME OR NOT

```
Label          Mnemonics                           Comments
.MODEL SMALL
.DATA
    M1      DB   10,  13,  'Enter the string : $'
    M2      DB   10,  13,  'String is palindrome $'
    M3      DB   10,  13,  'String is not palindrome $'
    BUFF    DB   80
            DB   0
            DB   80 DUF (0)
.CODE
START :     MOV  AX, @DATA                        ; [Initialize
            MOV  DS, AX                           ; data segment]
            MOV  AH, 09H
            MOV  DX, OFFSET M1
            INT  21H                              ; Display message M1
            MOV  AH, 0AH                          ; Input the string
            LEA  DX, BUFF
            INT  21H
            LEA  BX, BUFF + 2                     ; Get starting address of string
            MOV  CH, 00H
            MOV  CL, BUFF + 1
            MOV  DI, CX
            DEC  DI
            SAR  CL, 1
            MOV  SI, 00H
```

```
BACK :    MOV  AL, [BX + DI]        ; Get the right most character
          MOV  AH, [BX + SI]        ; Get the left most character
          CMP  AL, AH               ; Check for palindrome
          JNZ  LAST                 ; If not exist
          DEC  DI                   ; Decrement end pointer
          INC  SI                   ; Increment starting pointer
          DEC  CL                   ; Decrement counter
          JNZ  BACK                 ; If count not = 0, repeat
          MOV  AH, 09H              ; Display message 2
          MOV  DX, OFFSET M2
          INT  21H
          JMP  TER
LAST :    MOV  AH, 09H
          MOV  DX, OFFSET M3        ; Display message 3
          INT  21H
TER :     MOV  AH, 4CH              ; [ Terminate and
          INT  21H                  ; Exit to DOS ]
          END  START
```

11 8087 NUMERIC DATA PROCESSOR

11.0 OBJECTIVE

After completion of this chapter, the reader will be able to analyse :

- Need of numeric data processor.
- Features of 8087.
- Functional pin diagram of 8087.
- 8087 Data types.
- Internal architecture of 8087
- 8086 – 8087 operations.
- Functional registers of 8087.
- 8087 Instruction set.
- Programming in 8087.

11.1 INTRODUCTION

The Science, engineering, computer graphics, computer aided design, and image processing applications involve intensive number crunching operations. Such applications require high computational power and parallel execution of operations. Most of the operations in scientific and engineering applications are performed on the floating point data. In this context, dedicated processors have been developed for these applications known as the coprocessors. The coprocessors operate in parallel with the main processor. It uses the same system bus and instruction byte stream. The most popularly used coprocessors with Intel 8086 series of microprocessors are 8087 series of coprocessors. (The Intel coprocessor series 8087 is similar to the microprocessor 8086 series, *i.e.*, the coprocessor 8087 works with 8086.)

11.2 IMPORTANCE OF NUMERIC DATA PROCESSOR

A numeric data processor is also variously termed as arithmetic coprocessor, a numeric co-processor, math co-processor, numeric processor extension, floating point processor etc.

8086 microprocessor is basically an integer processing unit and works directly on a variety of integer data types. Most of the programs used in engineering, science, business, need to perform mathematical operations like logarithm of a number, square root of a number, sine of an angle etc. But it may be needed to perform computations with very large numbers like 10^{+56}, or very small numbers like 10^{-67}. There are no instructions in 8086 to directly find sine of an angle etc. Also 8086 can only perform computations on 16 bit fixed point numbers, with a range of –32768 to +32767. In other words, 8086 does not provide any intrinsic support for operations on floating point numbers.

To solve this problem : Any calculation can be performed using only the basic operations of add, subtract, multiply, and divide. For instant, for the sine of an angle there is a series expasion using only these basic operations as follows.

$$\text{Sin } x = x - x^2/2\,! + x^4/4\,! - \qquad \ldots (11.1)$$

That means it is possible to write procedures to perform any operation like sine of an angle etc using only the instruction set of 8086. But such approach takes a long time for the required operations, as a large number of instructions that constitute the procedure have to be fetched and executed by the 8086, processor.

To represent numbers with magnitude much larger than 32767, we should interpret the 16 bit pattern in a different way, called the floating point notation. But then, even basic arithmetic operations, like addition, on such floating point numbers are no longer trivial. Therefore, it has to be performed using a procedure which uses a large number of instructions. Floating point emulator is a practical example using this principle. However it will be a slow computation.

Hence, it is possible to perform any calculations using only 8086. But if speed becomes important, it is necessary to use the dedicated **Numeric coprocessor Intel 8087,** to speed up the matters. Typically it provides a 100 fold speed increase for floating point operations, compared to 8086 alone.

11.3 8087 COPROCESSOR'S FEATURES

As mentioned 8087 is good coprocessor with 8086 µP. It is having many features like :

— 8087 can operate on data of the integer, decimal and real types, with lengths ranging from 2 bytes to 10 bytes.

— The instruction sets of 8087 not only include various forms of addition and substraction, but also provides many useful functions like square root, exponential, tangent etc.

— 8087 is a high performance numeric data/coprocessor which can multiply two 64 bit real numbers in 27 µ sec and calculate square root in 36 µ sec only.

— 8087 is multibus compatible.

— 8087 coprocessor follows IEEE floating point standards.

11.4 OVERVIEW OF 8087 NUMERIC DATA PROCESSOR

It may be possible to have a microcomputer with 8086 as the only CPU but it is not possible to have a microcomputer with only 8087 as the CPU. Hence the 8087 is termed a coprocessor rather than a processor.

8087 numeric data coprocessor is designed to specifically work with 8086 and 8088 processors. Addition of 8087 in a 8086/8088 systems gives the programmer the feeling that powerful arithmetic floating point instructions are added to the instruction set. Actually, 8087 adds 68 mnemonics to the instruction set of 8086. It is fabricated using HMOS III technology and packaged in a 40 pin CERDIP.

8087 has its own internal registers and information is always stored inside 8087 using floating point notation. Its instruction set includes simple floating point operations like add, divide etc, as well as more complex operations like logarithm of a number etc.

There are instructions to transfer integer operand, BCD operand or floating point operand from memory to 8087. But non-floating point formats are always automatically converted to floating point format when moved to 8087. And there are instructions to transfer integer result. BCD result, or floating point result from 8087 to memory. But the internal floating point format for result is always automatically converted to the desired and floating point format when moved to memory, if needed.

Note. There are no instructions in the instruction set of 8086 or 8087 to perform data transfer directly between the registers of 8086 and 8087.

Each processor in the 8086 family, has a corresponding numeric data processor, as shown in the Table 1.

Table 1

S. No.	Processor	Compatible numeric data processor
1	8086 and 8088	8087
2	80286	80287 and 80287XL
3	80386SX	80387SX
4	80386DX	80287 and 80387DX
5	80486SX	80487SX
6	80486DX	Numeric coprocessor is built-in.

11.5 PIN DESCRIPTION OF 8087

Fig. 11.1 (*a*) and 11.1 (*b*) shows the functional pin diagram and conventional pin diagram of 8087.

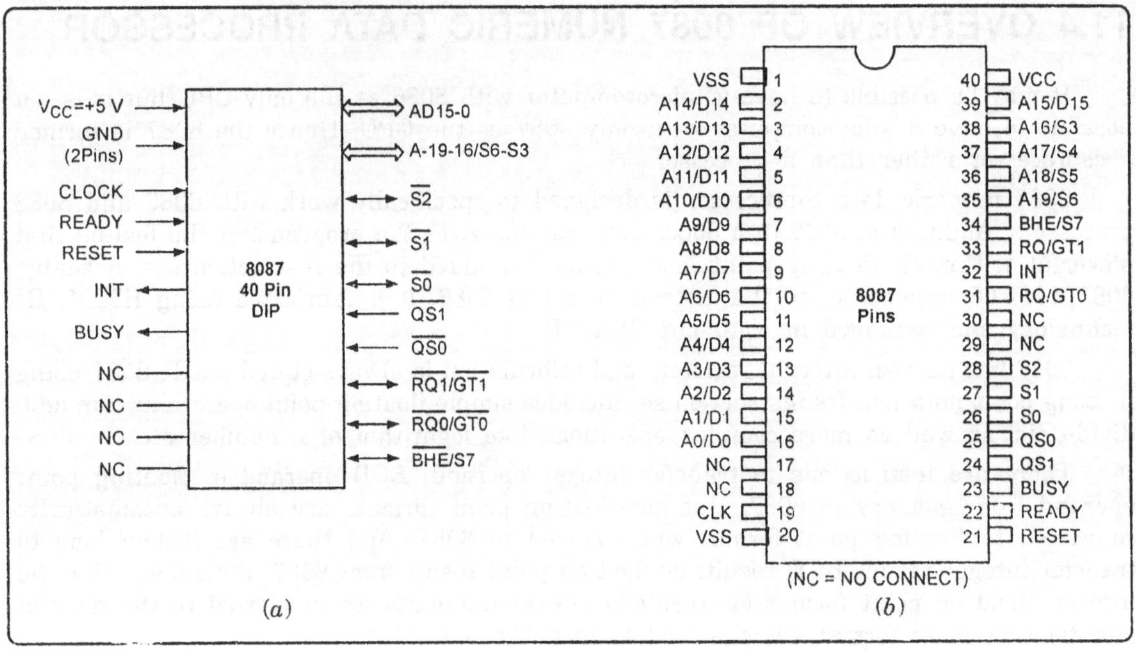

Fig. 11.1 Pin diagrams of 8087.

8087 is packaged in a 40 pin DIP (Dual In-line Package), and requires a single +5V power supply. It has 2 pins which have to be connected to Ground.

As observed few pins of 8087 are similar to 8086, however the importance of pins are :

— **Clock.** A clock is used for timing purposes. Generally, the 8087 receives the same clock input as 8086.

— **Ready.** This is an input signal which has the same purpose as the ready pin of 8086. As such, the ready pin of 8086 and 8087 are tied together, and they are driven by the same circuitry to generate wait states, if required.

— **AD15-0.** These signals have the same purpose as the AD15-0 pins of 8086. During T1 state of a machine cycle, address is present on these pins. In subsequent clock cycles, they are used for data. However the AD15-0 pins of 8086 and 8087 are tied together. If 8086 is in control of the local bus, the LS 16 bits of address information sent out by the 8086 is received on these pins by the 8087. And if 8087 is having control of the local bus, these pins are used for sending out from 8087 the LS 16 bits of address of a memory operand during T1 of a machine cycle.

— **A19-16/S6-3.** These signals have the same purpose as the A19-16 pins of 8086. During T1 state of a machine cycle, address will be present on these pins. In subsequent clock cycles, they are used for status information. As such, the A19-16/S6-3 pins of 8086 and 8087 are tied together. During 8086 is in control of the local bus, the MS 4 bits of address information sent out by the 8086 during T1 is received on these pins by the 8087. So also, the status information sent out by the 8086 during subsequent clock cycles is received by the 8087 on these pins.

Note. When 8087 is having control of the local bus, these pins are used for sending out from 8087 the MS 4 bits of address of a memory operand during T1 of a machine cycle. During subsequent clock cycles, the 8087 sends out a 0 on S5 and 1's on the other status pins.

— **$\overline{\text{BHE}}$/S7.** This signal has the same purpose as the $\overline{\text{BHE}}$/S7 pin of 8086. As such, the $\overline{\text{BHE}}$/S7 pins of 8086 and 8087 are tied together. When 8086 is in control of the local bus, the $\overline{\text{BHE}}$ information sent out by the 8086 during T1 is received on this pin by the 8087. Similarly, the 0 value sent out as status information by the 8086 during subsequent clock cycles is received by the 8087 on this pin.

Note. When 8087 is having control of the local bus, this pin is used for sending out from 8087 the $\overline{\text{BHE}}$ signal during T1 of a machine cycle. During subsequent clock cycles, the 8087 sends out status information on this pin.

— **INT.** This is an active high output pin. The 8087 activates this pin whenever an exception occurs during 8087 instruction execution provided the 8087 interrupt system is enabled and the relevant exceptions are not masked using the 8087 control register. (Control register of 8087 is discussed later.)

(The INT output of 8087 is connected directly to NMI or INTR input of 8086. Subsequently, INT output of 8087 is connected to an interrupt request input of 8259 Interrupt controller, which in turn interrupts the 8086 on its INTR input.)

— **Reset.** When a 1 is sent into the 8087 on the Reset input pin, for at least 4 clock cycles, the 8087 gets reset. That is, it starts from a clearly defined initial state. Generally, this pin is driven by the same circuit that drives the Rest input of 8086 or 8088.

Note. During reset, the 8087 checks pin number 34 of the CPU, which may be an 8086 or an 8088. Pin 34 of these processors are $\overline{\text{BHE}}$/S7 and $\overline{\text{SSO}}$ respectively. For 8086, the S7 status will be 0, and for 8088, the $\overline{\text{SSO}}$ signal will be a 1 in the maximum mode. Thus the 8087 can identify the processor with which it is required to work. Accordingly, the 8087 sets its own queue length to 6 bytes if the CPU is 8086, and to 4 bytes if the CPU is 8088.

— **Busy.** Assume that 8086 is used in maximum mode and is required to wait for some result from the coprocessor 8087 before proceeding with the next instruction. Then we can make the 8086 execute the WAIT instruction. Then the 8086 enters an idle state, where it is not performing any processing. The 8086 will stay in this idle state till $\overline{\text{Test}}$ input of 8086 is made 0 by the coprocessor, indicating that the coprocessor has finished its computation.

Note. While 8087 is busy executing an arithmetic instruction, its Busy output line will be in the 1 state. This pin is connected to $\overline{\text{Test}}$ pin of 8086. That means when the Busy pin is made 0 by the 8087 after the completion of execution of an arithmetic instruction, the 8086 will carry on with the next instruction after the WAIT instruction.

That is the $\overline{\text{Test}}$ input of 8086 and Busy output of 8087 are used for synchronizing the 8086 with 8087 data-processor.

11.6 THE MAXIMUM MODE SIGNALS OF 8086 AND 8087 SIGNALS (WITH SAME NAME)

The 7 maximum mode signals of 8086, and 7 signals of 8087 with the same function are discussed here :

— $\overline{S_2}$, $\overline{S_1}$, $\overline{S_0}$. These signals indicate the local bus cycle status. $\overline{S_2}$, $\overline{S_1}$ and $\overline{S_0}$ pins of 8086 and 8087 are tied together. When 8086 is in control of the local bus, the 8288 bus controller generates bus control signals using these signals sent out by the 8086. The status information sent out by the 8086 is monitored by the 8087. In the maximum mode, $\overline{S_2}$, $\overline{S_1}$ and $\overline{S_0}$ sent out by 8086 provide the complete bus status are shown in Table 2.

Table 2

$\overline{S_2}$	$\overline{S_1}$	$\overline{S_0}$	Bus cycle
0	0	0	Interrupt Acknowledge
0	0	1	Read I/O port
0	1	0	Write I/O port
0	1	1	Halt
1	0	0	Access Code
1	0	1	Read Memory
1	1	0	Write Memory
1	1	1	Passive

If 8087 is having control of the local bus, $\overline{S_2}$, $\overline{S_1}$ and $\overline{S_0}$ pins of 8087 are used for sending out from 8087 the following bus status information. Observe that 8087 can only communicate with memory, but not with I/O ports, observe Table 3.

Table 3

$\overline{S_2}$	$\overline{S_1}$	$\overline{S_0}$	Bus cycle
0	0	0	Unused
0	0	1	Unused
0	1	0	Unused
0	1	1	Unused
1	0	0	Unused
1	0	1	Read Memory
1	1	0	Write Memory
1	1	1	Passive

— QS_1, QS_0. These signals indicate the queue status. QS_1 and QS_0 pins of 8086 and 8087 are tied together. When 8086 is in control of the local bus these signals sent out by the 8086 allow external tracking of the internal 8086 instruction queue by the 8087. For this purpose, the status information sent out by the 8086 is monitored by the 8087. That means these signals are output from the 8086 and are input to the 8087. In maximum mode, QS_1 and QS_0 sent out by 8086 provide the queue status which are shown in Table 4.

Table 4

QS_1	QS_0	Queue Status
0	0	No operation
0	1	First byte of opcode from queue
1	0	Empty the queue
1	1	Next byte of opcode from queue

— $[\overline{RQ_0}/\overline{GT_0}]$ and $[\overline{RQ_1}/\overline{GT_1}]$. These are input as well as output pins for both the 8086 and 8087, which are active low pins. These pins have internal pull-up resistors so that if they are left without any connection, they will be in the 1 state, which is the inactive state. These pins are used for gaming control over the local bus.

11.7 THE CIRCUIT CONNECTION FOR 8087

The circuit connection for 8087 numeric data processor with 8086 is illustrated in Fig. 11.2.

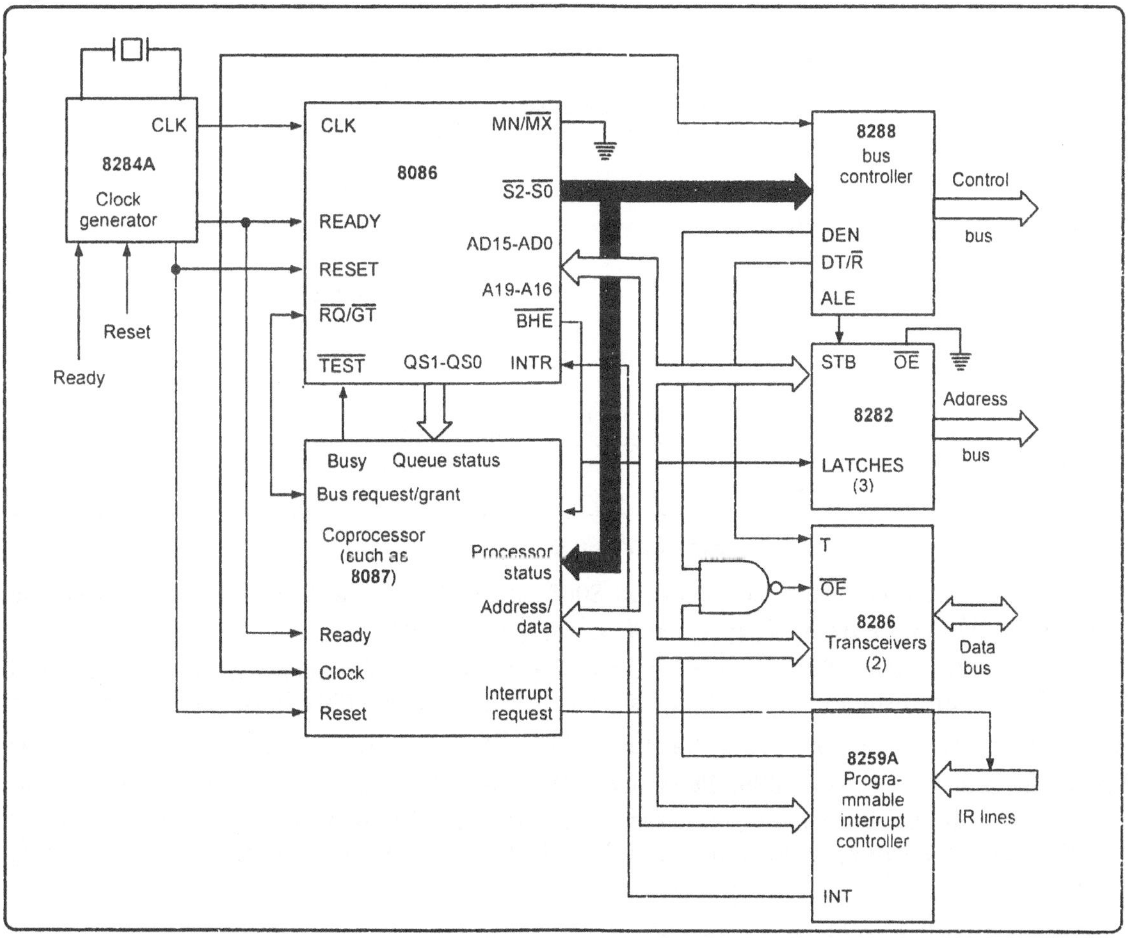

Fig. 11.2. 8087 Configuration.

In Fig. 11.2 observe that $MN/\overline{MX}$ pin of the 8086 is grounded, that means 8086 is operating in its maximum mode. However in maximum mode the 8086 sends encoded control signals on the status lines $\overline{S2}$, $\overline{S1}$ and $\overline{S0}$ and the queue status lines QS1 and QS0 instead of generating the control signals directly. As observed in the Fig. 11.2, in maximum mode system an external controller such as the 8288 decodes these status signals to produce the control signals. (These status signals also go directly from the 8086 to the 8087).

As observed upper address lines A_{16}-A_{19} are also connected directly from 8086 to the 8087, and 8087 receives the same clock and reset signals. The Bus request/grant signal from 8087 is connected to $\overline{RQ_0/GT_0}$ or $\overline{RQ_1/GT_1}$ signal of the 8086 and Busy signal from 8087 is connected to the TEST input of the 8086. If the 8086 must have result of some computation that the 8087 is doing before it can execute its further instructions, we tell the 8086 with a WAIT instruction to keep looking at its TEST pin until it finds the pin low. A low on the TEST pin, (*i.e.,* on the BUSY output) indicates 8087 has completed the computation.

11.8 8086-8087 OPERATION

The Fig. 11.3 illustrates the interaction or operation between 8086 μp and 8087 coprocessor.

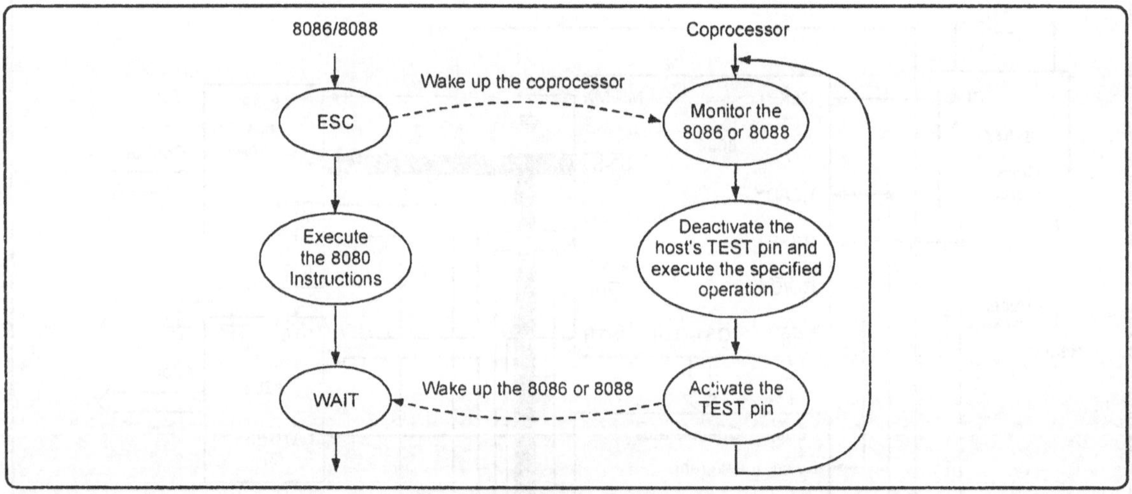

Fig. 11.3. 8086 and 8087.

The operation or interaction between 8086 and 8087 can be discussed while observing the use of the two pins $\overline{RQO/GTO}$ and $\overline{RQ_1/GT_1}$ in them.

Assume that 8086 fetches an instruction from memory and stores it in its queue, and 8087 also simultaneously stores it in its own internal queue. This is possible because the 8087 constantly keeps monitoring the status lines $\overline{S2}$, $\overline{S1}$, $\overline{S0}$, as well as the Queue status lines QS1 and QS0 of the 8086. Then both 8086 and 8087 decode the instruction present in their own queue. 8087 simply ignores the instruction if it is a 8086 instruction. If 8086 decodes the instruction, and finds it is a 8087 instruction, it performs one of the following actions :

(*a*) For a 8087 instruction which does not need any operand from memory, the instruction is simply ignored by the 8086.

(*b*) For a 8087 instruction which needs an operand from memory, the 8086 initiates a memory read bus cycle. The 8086 sends out a 20 bit address based on the operand address specified in the 8087 instruction. This address sent out by the 8086 is received by the 8087 also in an internal register, for later use if needed. The 8086 also activates the $\overline{S2}$, $\overline{S1}$ and $\overline{S0}$ signals appropriately to indicate to the bus controller that a memory read operation is desired. The 8288 bus controller sends out a memory read signal using the $\overline{S2}$, $\overline{S1}$ and $\overline{S0}$ status signals emitted by the 8086.

Next in bus cycle, if memory supplies the operand to the bus, the 8087 and not 8086, receives this information, when this 8087 instruction needs only this one word operand, the 8087 proceeds with the instruction execution. However, if the 8087 needs an operand whose size is larger than 16 bits, like say 32, 64 or 80 bits, the subsequent memory read cycles are performed by the 8087 itself. In this way the 8087 directly accesses memory for the required operand. This is termed as Direct Memory Access or DMA (For this purpose, the 8087 needs to have control over the local bus). This is where $\overline{RQ}/\overline{GT}$ lines come into the picture.

Now $\overline{RQ}/\overline{GT}$ of 8087 is connected to $\overline{RQ1}/\overline{GT1}$ of 8086, then 8087 activates $\overline{RQ}/\overline{GT}$ pin requesting for control over the local bus. Then the 8086 which receives this request on the $\overline{RQ1}/\overline{GT1}$ pin will send out the bus grant signal by activating the $\overline{RQ1}/\overline{GT1}$ pin and then 8086 floats its data and address lines so that the 8087 can take control of the local bus.

8087 receives the bus grant signal from 8086 on $\overline{RQ0}/\overline{GT0}$ pin and becomes the local bus master. Then it will increment the address it has received earlier in an internal register and sends it on the address bus. It then activates its $\overline{S_2}$, $\overline{S_1}$ and $\overline{S_0}$ signals to indicate a memory read operation. The 8288 bus controller then generates the memory read signal so that the 8087 can receive the data from memory. This process is repeated till all the data words required by the instruction are transferred to the 8087.

Then the 8087 activates the $\overline{RQ}/\overline{GT}$ pin to indicate to the 8086 that the DMA data transfer is complete. This results in the 8086 gaining control over the local bus once again in the next clock cycle.

So also, when 8087 requires to store the result of a computation in memory, it gains control over the local bus and uses DMA to store the result in memory.

However in 8086, if bus requests come at the same time from two different local bus masters on $\overline{RQ0}/\overline{GT0}$ and $\overline{RQ1}/\overline{GT1}$ pins, bus grant signal will be issued by the 8086 on $\overline{RQ0}/\overline{GT0}$. This is because, $\overline{RQ0}/\overline{GT0}$ has higher priority than $\overline{RQ1}/\overline{GT1}$.

In the case of 8087, $\overline{RQ1}/\overline{GT1}$ pin is used by another local bus master to request the 8087 to grant the bus control. There are two cases to be considered here, *i.e.,* action taken by the 8087 depends on whether 8087 is presently having control over the local bus or not.

(*i*) When 8087 is presently having control over the bus, it will issue a bus grant on the same $\overline{RQ1}/\overline{GT1}$ pin at the end of the current bus cycle. Then the other local bus master takes control of the local bus and performs the required data transfer with memory. Once the data transfer is complete, this bus master again activates $\overline{RQ1}/\overline{GT1}$ pin to indicate to the 8087 to take control of the local bus again.

(*ii*) When 8087 is presently not having control over the bus, it will issues a bus request signal on its own $\overline{RQ0}/\overline{GT0}$ pin to the 8086, which is done by the 8087 on behalf of the other local bus master. Then the 8086 issues a bus grant signal at the end of the current bus cycle, which reaches the 8087 on $\overline{RQ0}/\overline{GT0}$ pin. The 8087 in turn, activates bus grant signal on $\overline{RQ1}/\overline{GT1}$. Then the other local bus master takes control of the local bus and performs the required data transfer with memory. Once the data transfer is complete, this bus master again activates $\overline{RQ1}/\overline{GT1}$ pin to indicate to the 8087 the completion of data transfer. Then the 8087 in turn, activates $\overline{RQ0}/\overline{GT0}$ to inform the 8086 to take control of the local bus again.

Note. When 8087 is executing an instruction it asserts BUSY pin high. After completing its instruction it drops its BUSY pin low. The BUSY pin from 8087 is connected to the TEST pin of the 8086. Therefore 8086 can check this pin to see if 8087 has executed its instruction. (8086 checks TEST pin with the help of WAIT instruction.)

11.9 8087 DATA TYPES

The 8087 works on 80 bit data internally. That, 80 bit floating, point format is termed as **'Temporary Real'** format. Even though the coprocessor works with seven types of numeric data, they are divided into three main classes as :

(1) Integer Data Types :
- (*a*) Word integer—16 bit size
- (*b*) Short integer—32 bit size
- (*c*) Long integer—64 bit size.

(2) Packed decimal number (BCD)—80 bit size.

(3) Floating point/Real numbers :
- (*a*) Short real—32 bit size
- (*b*) Long real—64 bit size
- (*c*) Temporary real—80 bit size.

The formats designed for different types of number in 8087 coprocessor is as illustrated in Fig. 11.4.

In Fig. 11.4, few abbriviations are to read as :

I. S = sign bit (0 = positve, 1 = negative)

d_n = Decimal digit (two per byte)

X = Bits have no significance : 8087 ignores when loading zeroes when storing

i = Position of implicit binary point

I = Integer bit of significand : stored in temporary real, implicit in short and long real

II. Exponent Bias (normalized values) :

Short Real : 127 (7FH)

Long Real : 1023 (3FFH)

Temporary Real : 16383 (3FFFH)

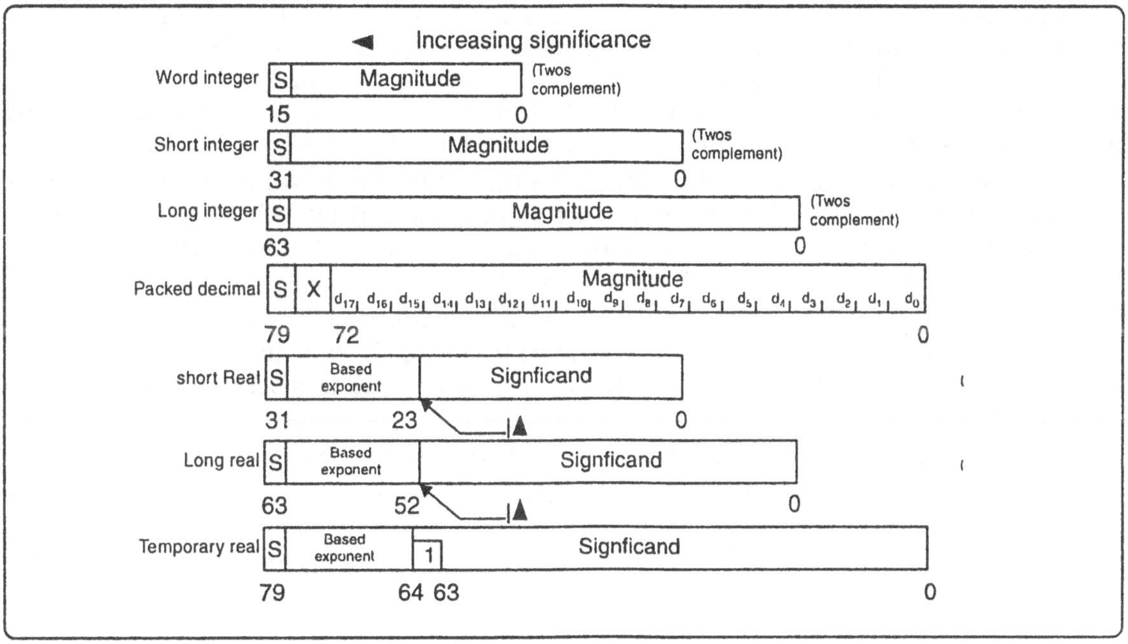

Fig. 11.4. 8087 Coprocessor Data Format.

The most significant bit of a number represents the sign bit. If it is zero, it represents a positive number and if it is one, it represents a negative number. In memory, the least significant byte is stored in the lowest address, in the Fig. 11.4 above. This number read from memory is internally converted to the 80 bit temporary real format before performing any computations. Similarly, the result is converted automatically by the 8087 to one of the formats mentioned above before storing it in memory.

11.9.1 Integer Data Types

The positive numbers are represented with a 0 in the MS bit position, and the remaining bits provide its magnitude. For instant +3 is represented using 4 bits as 0011 = +3.

The largest magnitude positive number using 4 bits will be 0111 = +7 = $+(2^3 - 1)$. That means, if numbers are represented using N bits, the largest magnitude positive number would be $+(2^{N-1} - 1)$.

The negative numbers are represented in 2's complement notation. That is, –3 is represented as 2's complement of the notation for +3, as :

Notation for –3 = 2's complement of notation for +3
= 2's complement of 0011
= 1101.

Observe that negative numbers always start with a 1 in the MS bit position. The other bits provide the 2's complement of the magnitude. For example, in the notation for –3 which is 1 101, the 1 in the MS bit position indicates it is a negative number. The remaining bits, namely 101, is the 2's complement of the magnitude 011.

One more method of finding the value of a negative number in 2's complement can be; similar to the way the value of an unsigned number represented in binary is computed. In the case of unsigned numbers, the value of a series of 0's and 1's is calculated by assigning weights to each bit position, like 1, 2, 4, 8 etc. starting from the LS bit. That means 1101 as an unsigned number will have the value 8 + 4 + 0 + 1 = 13.

Similar technique can be used for obtaining the value of a signed number represented as a series of 0's and 1's. The only difference is that the MS bit will have same weight as in the case of unsigned number, but with a negative sign. That means 1101 as a signed number will have the value –8 + 4 + 0 + 1 = –3.

Therefore, the largest magnitude negative number using 4 bits is 1000 = –8 + 0 + 0 + 0 = –8 = -2^3. Thus, if numbers are represented using N bits, the largest magnitude negative number would be -2^{N-1}, (this method of representing numbers is also termed as Fixed point notation).

The 8087 supports signed integers of 16, 32 and 64 bit size.

(*a*) *16 bit signed integer (word integer).* In this case the range of numbers that can be represented is from -2^{15} to $+(2^{15}-1)$. In other words the range is from –32768 to +32767. The number has to be stored in memory with the LS byte at a lower address, and the MS byte at the next higher address. Range of number representation is as shown :

sign	15 bit magnitude for positive numbers
0 = +	(in 2's comp. notation for negative numbers)
1 = –	

(*b*) *32 bit signed integer (short integer).* It should not be confused with a byte of data. In fact, 8087 does not support byte data. In this case the range of numbers that can be represented is from -2^{31} to $+(2^{31}-1)$. That means the range is approximately from $-2\times10^{+9}$ to $+2\times10^{+9}$. The number has to be stored in memory in consecutive locations with the LS byte at the lowest address. Range of number representation is as shown :

sign	31 bit magnitude for positive numbers
0 = +	(in 2's comp. notation for negative numbers)
1 = –	

(*c*) *64 bit signed integer (Long integer).* In this case the range of numbers that can be represented is from -2^{63} to $+(2^{63}-1)$. In other words the range is approximately from $-8\times10^{+18}$ to $+8\times10^{+18}$. The number has to be stored in memory in consecutive locations, with the LS byte at the lowest address. The range of number representation is as shown :

sign	63 bit magnitude for positive numbers
0 = +	(in 2's comp. notation for negative numbers)
1 = –	

11.9.2 Packed Decimal Number (BCD) Type

This format is very useful for working with decimal numbers, such as Financial Accounting programs, which is a 80 bit data type. Here the MS bit indicates the sign of the number, and next 7 bits are not having any meaning, which are ignored by the 8087.

The remaining 72 bits are used to represent 18 BCD digits, using 4 bits for each BCD digit. The number has to be stored in memory in consecutive locations, with the LS byte at the lowest address. For the convenience of the user, the LS 72 bits, which represent 18 BCD digits, always provides the magnitude of the number immaterial of the sign of the number.

Here the range of numbers that can be represented is from $-(10^{18} - 1)$ to $+(10^{18} - 1)$. In other words the range is approximately from -10^{+18} to $+10^{+18}$, the format is as shown :

sign	7 bits	72 bits for 18 packed BCD digits
0 = +	don't	
1 = −	cares	

11.9.3 The Floating Point or Real Types

Real numbers are used very frequently which will have an integer part and a fractional part, such as 23.456. But this notation is not convenient for representing very large magnitude numbers, such as − 123456789.4567. Same number can be more conveniently represented in scientific notation as $-1.2345678 \times 10^{-08}$. But this actually stands for −123456780. So there is an error of 9.4567, which forms a very small percentage error. Here, 1.2345678 is called the Mantissa, and + 08 is the Exponent to base 10.

But, the advantage of this scientific notation is that we can represent a very much larger number than the $-1.2345678 \times 10^{-08}$ using same number of digits. For example, $+9.9999999 \times 10^{+99}$ is the largest magnitude positive number using same number of digits.

So also, the way we represent real numbers in our daily life is not convenient for representing very small numbers, like + 0.00000045678912. This same number can be more conveniently represented in scientific notation as $+4.56789 \times 10^{-07}$. But this actually stands for + 0.000000456789. So there is an error of 0.00000000000012, which forms a very small percentage error.

The advantage of this scientific notation is that we can represent a very much smaller magnitude number than the $+4.56789 \times 10^{-07}$ using same number of digits. For instant, -1.00000×10^{-99} is the smallest magnitude negative number using same number of digits.

Similarly to represent very large, or very small signed numbers, a computer uses floating point notation. This may result in small errors, but the error will be negligible, if large number of digits are reserved for the fractional part. But the important thing is that we will be able to represent much larger range of numbers than is possible with fixed point notation.

A floating point number like $+1.22 \times 10^4$ can be written in several equivalent ways such as :

$$+12.4 \times 10^3$$

$$+.124 \times 10^5$$

There are several ways of representing floating point numbers in a computer also, the 8087 supports floating point numbers of 32, 64 and 80 bit size, which are discussed here. with examples :

(*a*) Short real number or 32-bit floating point number

This notation makes use of 32 bits to represent a number. It is normally called a Short Real number, or a Single precision number. Here the MS bit indicates the sign of the number and the next 8 bits provide the exponent to base 2 of the number in excess 7FH notation. In the next bit position a 1 is always assumed to be present. But in reality, the bit is not physically present. The remaining 23 bits provide the fractional part of the mantissa. To expalin this notation, let us see few examples :

sign	8 bits	23 bits for fractional part
0 = +	exponent	with implied '1.' before
1 = -	in Ex 7FH	the fractional part

Ex (*i*) : Assume that, we want to represent 23.25 in this notation. First let us we represent 23.25 in binary as 10111.01. Then we represent this as $+1.011101 \times 2^{+4}$, which is called the Normalized form of representation. In the normalized form, the mantissa will always have an integer part with value 1. Floating point notations supported by 8087 always represent a number in the normalized form. In 32 bit and 64 bit floating point notations the integer part of the mantissa, of value 1, is just implied to be present, but not explicitly indicated in the bit pattern for the number. That means the LS 23 bits are used to indicate only the fractional part of the mantissa and so will be 011 1010 0000 0000 0000 0000. The MS bit will be 0 to indicate that the number is positive and the next 8 bits provide the exponent in excess 7FH format. Therefore next 8 bits will be 4 + 7F = 83H = 1000 0011. Hence the 32 bit floating point representation for 23.25 is as below.

sign	exp. in Ex 7FH	23 bit fractional part of mantissa
0	1000 0011	011 1010 0000 0000 0000 0000

Ex (*ii*) : Now let us observe what is the value of the 32 bit floating point number 1 0111 1100 100 0000 0000 0000 0000 0000. It has its MS bit as a 1. Thus the number is negative. The next 8 bits are 0111 1100 = 7CH. Thus 7CH is the exponent in excess 7FH format. Hence, the actual exponent is 7CH – 7FH = –03. The actual mantissa is obtained by appending 1 to the LS 23 bits. Thus the actual mantissa is 1.100 0000 0000 0000 0000 0000. Therefore the value of the given 32 bit floating point number would be

$$\begin{aligned} & -1.100\ 0000\ 0000\ 0000\ 0000\ 0000 \times 2^{-03} \\ = & -1.1 \times 2^{-03} \\ = & -0.0011 \times 2^{0} \\ = & -0.0011 \\ = & -0.1875 \end{aligned}$$

Hence the given 32 bit number represents the value –0.1875.

Now let us analyse for two cases with short real number :

(*i*) Largest magnitute number possible

(*ii*) Smallest magnitude number possible.

(*i*) *Largest magnitude number possible (with Short Real notation).* The largest possible positive number in this notation will be 0 1111 1111 111 1111 1111 1111 1111 1111. It is a positive number as the MS bit is a 0. The next 8 bits correspond to FFH. This is the exponent in excess 7FH. Thus the actual exponent is FFH – 7FH = 80H. So this bit pattern stands for the number $+1.111\ 1111\ 1111\ 1111\ 1111\ 1111 \times 2^{+80H}$. This is approximately $+2^{+81H}$, or $+5.12 \times 10^{+38}$.

The largest magnitude negative number will have the MS bit as a 1, with no change in the other bits. Its value will be approximately $-5.12 \times 10^{+38}$. It is to be noted that the range for a 32 bit floating point number is very much larger than the range for a 32 bit integer.

(*ii*) *Smallest magnitude number possible (with Short Real notation).* The smallest number that can be represented indicates the possible error in the representation for a number. The smallest possible positive number in this notation is 0 0000 0000 000 0000 0000 0000 0000 0000. It is a positive number as the MS bit is a 0. The next 8 bits correspond to 00H. THis is the exponent in excess 7FH. Thus the actual exponent is 00H – 7FH = –7FH. So this bit pattern stands for the number + 1.000 0000 0000 0000 0000 0000 × 2^{-7FH}. This is + 2^{-7FH}, or approximately + 8 × 10^{-39}.

Note. For approximation, 210 range taken as 10^3, where actual value is 1.024 × 10^3. And the smallest magnitude negative number will have the MS bit as a 1, with no change in the other bits. Its value will be approximately –8 × 10^{-39}.

(*b*) Long real number or 64-bit floating point number

It is also called a **Double precision number,** which uses 64 bits to represent a number. This notation is similar to short real notation, however, it provides much larger range as well as precision.

Here in this notation, the MS bit indicates the sign of the number and the next 11 bits provide the exponent to base 2 of the number is excess 3FFH notation. In the next bit position a 1 is always assumed to be present. But in reality, the bit is not physically present. The remaining 52 bits provide the fractional part of the mantissa, as shown below :

sign	11 bits	52bits for fractional part
0 = +	exponent	with implied '1.' before
1 = –	in Ex3FFH	the fractional part

This notation is still made clear with example below :

Ex (*i*) : (Assume that it is required). To represent 23.25 in this notation, first we represent 23.25 in binary as 10111.01, and then we represent this as +1.011101 × 2^{+4}, which is called the normalized form of representation. In the normalized form, the mantissa will always have an integer part with value 1. (The floating point notations supported by 8087 always represent a number in the normalized form.) In the 32 bit and 64 bit floating point notations the integer part of the mantissa, of value 1, is just implied to be present, but not explicitly indicated in the bit pattern for the number. That means the LS 52 bits are used to indicate only the fractional part of the mantissa and so will be 0111 0100 0000 0000 0000 0000 00000 0000 0000 0000 0000 0000 0000. The MS bit will be 0 to indicate that the number is positive. The next 11 bits provide the exponent in excess 3FFH format. So the next 11 bits will be 4 + 3FF = 403H = 100 0000 0011. Thus the 64 bit floating point representation for 23.25 will be

sign	exp. in Ex 3FFH	52 bit fractional part of mantissa
0	100 0000 0011	0111 0100 00...............00

Ex (*ii*) : Now let us observe what is the value of the 64 bit floating point number 1 100 0000 0011 0100 0000 0000 0000 0000 0000 0000 0000 0000 0000 0000 0000 0000, which has its MS bit as a 1. That means the number is negative. The next 11 bits are 100 0000 0011 = 403H. Thus 403H is the exponent in excess 3FFH format. In other words, the actual exponent is 403H – 3FFH = +04. The actual mantissa is obtained by appending 1, to the LS 52 bits. Hence the actual mantissa is 1.0100 0000 0000 0000 0000 0000 0000 0000 0000 0000 0000 0000 0000. Thus the value of the given 64 bit floating point number would be

$$\begin{aligned} & -1.0100\ 0000\ \ldots\ 0000 \times 2^{+04} \\ &= -1.01 \times 2^{+04} \\ &= -10100 \times 2^{0} \\ &= -10100 \\ &= -20 \end{aligned}$$

Finally, the given 64 bit number represents the value –20, as discussed above.

Now let us analyse for two cases with Long Real Number, *i.e.,*

(*i*) Largest magnitude number possible, and

(*ii*) Smallest magnitude number possible.

(*i*) *Largest magnitude number possible (with Long Real notation).* The largest possible positive number in this notation will be 0 111 1111 1111 1111.....1111, which is a positive number as the MS bit is a 0. The next 11 bits correspond to 7FFH, which is the exponent in excess 3FFH. Therefore the actual exponent is 7FFH – 3FFH = 400H. So this bit pattern stands for the number +1.1111.....1111 × 2^{+400H}. This is approximately $+2^{+401H}$, or +3.2 × 10^{+307}.

The largest magnitude negative number will have the MS bit as a 1, with no change in the other bits. Its value will be approximately –3.2 × 10^{+307}. (It is observed that the range for a 64 bit floating point number is very much larger than the range for a 32 bit floating point number.)

(*ii*) *Smallest magnitude number possible (with Long Real notation).* The smallest possible positive number in this notation will be 0 000 0000 0000 0000......0000, which is a positive number as the MS bit is a 0. The next 11 bits correspond to 000H, which is the exponent in excess 3FFH. Thus the actual exponent is 000H – 3FFH = –3FFH. Hence this bit pattern stands for the number +1.0000.....0000 × 2^{-3FFH}. This is $+2^{-3FFH}$, or approximately +1.25 × 10^{-307}.

Note. The smallest magnitude negative number will have the MS bit as a 1, with no change in the other bits, whose value will be approximately –1.25 × 10^{-307}.

(*c*) Temporary real number or 80-bit floating point number

This notation uses 80 bits to represent a number, which is similar to long real notation. But is provides a much larger range as well as precision. This is the format to which all numbers are converted by the 8087, after reading them from memory. (Internally all computations are done using this format).

(In this notation MS bit indicates the sign of the number, and next 15 bits provide the exponent to base 2 of the number in excess 3FFFH notation. The remaining 64 bits provide the mantissa in the normalized form. (Note that there is no implied '1', before the fractional part of the mantissa.) The integer value of 1 in the mantissa is explicitly given. Only the binary point after this integer value of 1 is implied. The bits allocation is as shown below :

sign	15 bits	64 bits for mantissa with
0 = +	exponent	explicit '1.' before
1 = –	in Ex3FFFH	the fractional part

This notation will be still more clear after observing following examples.

Ex (*i*) : Imagine that, we want to represent 23.25 in this notation. First we represent 23.25 in binary as 10111.01, and then we represent this as $+1.011101 \times 2^{+4}$. This is called the Normalized form of representation. In the normalized form, the mantissa will always have an integer part with value 1. (The floating point notations supported by 8087 always represent a number in the normalized form.)

Therefore LS 64 bits are used to indicate the mantissa and so will be 1011 1010 0000 0000 0000 0000 0000 0000 0000 0000 0000 0000 0000 0000 0000 0000. The MS bit will be 0 to indicate that the number is positive. The next 15 bits provide the exponent in excess 3FFFH format. Hence the next 15 bits will be 4 + 3FFF = 4003H = 100 0000 0000 0011. Now the 80 bit floating point representation for 23.25 will be as below :

sign	exp. in Ex. 3FFFH	64 bit mantissa
0	100 0000 0000 0011	1011 1010 00............00

Ex (*ii*) : Now let us realize what is the value of the 64 bit floating point number 1 100 0000 0000 0011 1010 0000.....0000. It has its MS bit as a 1. Thus the number is negative. The next 15 bits are 100 0000 0000 0011 = 4003H. Thus 4003H is the exponent in excess 3FFFH format. In other words, the actual exponent is 4003H – 3FFFH = +04. The actual mantissa is 1.010 0000......0000, where the binary point is implied to be present after the MS bit of the mantissa. Hence the value of the given 80 bit floating point number would be

$$
\begin{aligned}
& -1.010\ 0000\ \ldots\ 0000 \times 2^{+04} \\
&= -1.01 \times 2^{+04} \\
&= -10100 \times 2^{0} \\
&= -10100 \\
&= -20
\end{aligned}
$$

Therefore the given 80 bit number represents the value-20, as explained above.

Now let us observe the notation for the cases with temporary real number which are :

(*i*) Largest magnitude number possible, and

(*ii*) Smallest magnitude number possible.

(*i*) *Largest magnitude number possible (with Temporary Real notation).* The largest possible positive number in this notation will be 0 111 1111 1111 1111 1111.....1111. It is a positive number as the MS bit is 0. The next 15 bits correspond to 7FFFH. This is the exponent in excess 3FFFH. Thus the actual exponent is 7FFFH – 3FFFH = 4000H. Hence this bit pattern stands for the number $+1.111....1111 \times 2^{+4000H}$. This is approximately $+2^{+4001H}$, or $+3.2 \times 10^{+4915}$.

Note. The largest magnitude negative number will have the MS bit as a 1, with no change in the other bits, whose value will be approximately $-3.2 \times 10^{+4915}$. It is to be noted that the range for a 80 bit floating point number is very much larger than the range for a 64 bit floating point number.

(*ii*) *Smallest magnitude number possible (with Temporary Real notation).* The smallest possible positive number in this notation will be 0 000 0000 0000 0000 1000....0000. It is a positive number as the MS bit is a 0. The next 15 bits correspond to 0000H. This is the exponent in excess 3FFFH. That means the actual exponent is 0000H – 3FFFH = – 3FFFH. So this bit pattern stands for the number + 1.000.....0000 × 2^{-3FFFH}. This is + 2^{-3FFFH}, or approximately + 1.25 × 10^{-4915}. In other words, this is the possible error in the representation of a number.

Note. The smallest magnitude negative number will have the MS bit as a 1, with no change in the other bits. Its value will be approximately –1.25 × 10^{-4915}.

11.10 8087 INTERNAL ARCHITECTURE

The overal organization of 8087 numeric data processor is as illusted in Fig. 11.5. It is fundamentally having two register *i.e.,* control register and status register, both are of 16 bits length as shown in Fig. 11.5.

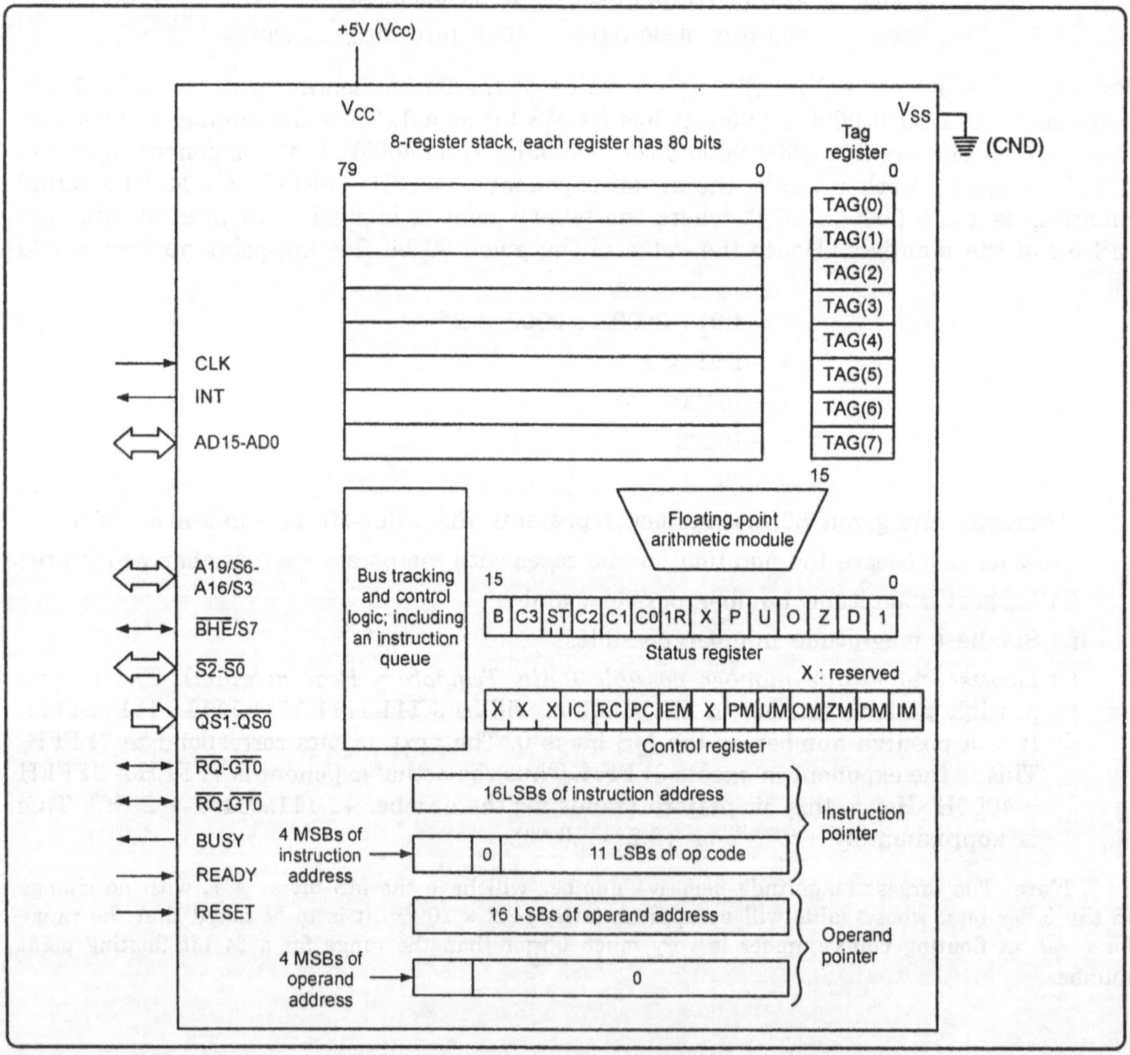

Fig. 11.5. Internal block diagram of 8087.

The important functional registers of 8087 are :

1. The Control Register,
2. The TAG Register,
3. The Status Register,
4. Instruction pointer/Queue and
5. Data Registers.

11.10.1 The Control Register

It is a 16 bit register in 8087. The 8087 provides several processing options which are selected by loading a word from memory into the control register, which gives the facility to mask each error type individually from causing an interrupt. It can be used to set precision levels, rounding type and infinity representation. The Fig. 11.6 shows bit definition of control register of 8087 as below :

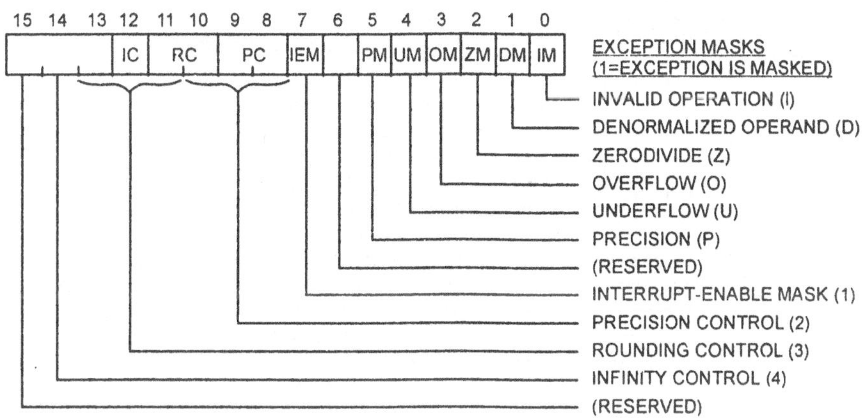

(1) Interrupt - Enable Mask
  0 = Interrupts Enable
  1 = Interrupts Disabled (Masked)

(2) Precision Control
  00 = 24 bits
  01 = (reserved)
  10 = 53 bits
  11 = 64 bits

(3) Rounding Control
  00 = Round to Nearest or Even
  01 = Round Down (toward %)
  10 = Round Up (toward %)
  11 = Chop (Truncate Toward Zero)

(4) Infinity Control
  0 = Projective
  1 = Affine

DEFAULT AFTER FINITE

Fig. 11.6. Bit pattern of Control Register in 8087.

Here the low order byte of this control register configures 8087 interrupts and exception masking. The Bits 5-0 of the control register contain individuals masks for each of the six exceptions that the 8087 recognizes and bit 7 contains a general mask bit for all 8087 interrupts.

The high order byte of the control register configures the 8087 operating mode including precision, rounding, and infinity controls.

Once reset or initialization of the 8087 is done, the condition of bits will be PC = 11, RC = 00, IC = 0, IEM = 0 and all error mask bits are 1.

11.10.2 The TAG Register

TAG register holds the status of the contents of data register, which include

0 0 – Data valid
0 1 – Zero
1 0 – A special value *i.e.,* NAN (Not a Number), infinity or denormal
1 1 – Empty

11.10.3 The Status Register

This register is also of 16 bit, which reflects the overall state of the 8087 and indicates various errors, stores condition code for certain instructions, specifies which register is top of the stack and indicates the BUSY status. The Fig. 11.7 illustrates the bit definitions of the Status Register as below :

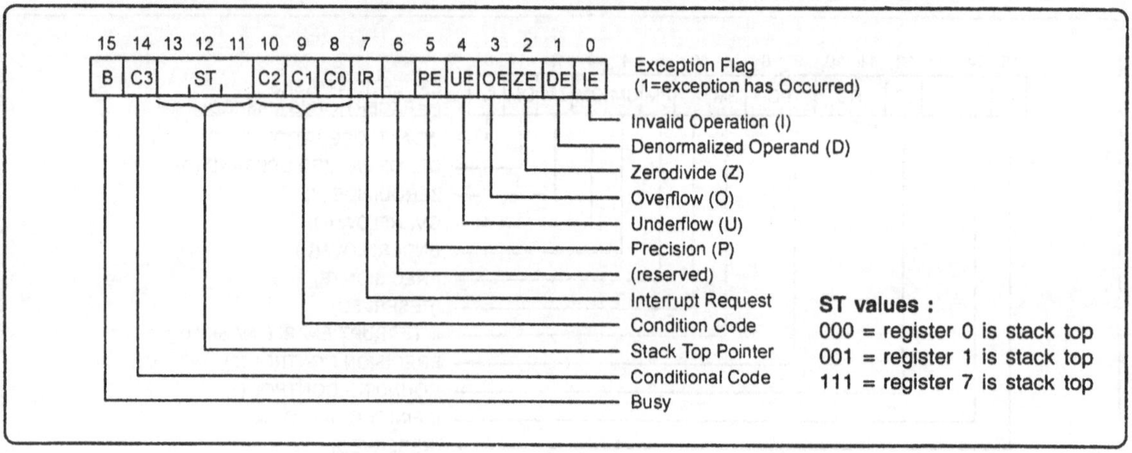

Fig. 11.7. Bit pattern of status register in 8087.

The error flags are explained in the Table 5 below :

Table 5

Error Flags	Explanation
IE	An invalid operation such as stack overflow, stack underflow, invalid operand, square root of a negative number etc.
DE	The operand is not normalized.
ZE	A divide by zero erro.
OE	An exponent overflow error, *i.e.,* the biased exponent is too small.
PE	A precision error, *i.e.,* the result cannot be represented in the designed format and hence is rounded off.

The interrupt flag IR : will indicate the existance of interrupt request.

The Condition Code : C_0 - C_3 indicate the condition code. The condition code is set by the compare and examine instructions, (which are discussed later).

The Stack Bits : ST : $\overline{S_0}$ – $\overline{S_2}$ indicate the top of stack.

The Busy Status : B : Indicates current operation is not completed.

Once 8087 is reset or initialized, all status bits except the condition code are cleared.

11.10.4 Instruction Pointer or Queue

It maintains a 6 byte instruction queue and tracks a execution sequence of the host. When the current host instruction is an ESC instruction, 8087 decodes the external opcode to perform the specified operation and captures the operand address, (other instructions are ignored by 8087).

11.10.5 Data Registers

The data register is 8 data registers. Each register is a 80-bit and it is accessed as a stack. An operand may be pushed or popped from top of stack. A 'push' operation decrements TOP of stack by 1 and loads a value into the new top register. A 'pop' operation stores the value from the current top register and then increments TOP by 1. The top stack element is pointed by ST bits, *i.e.,* bits 13, 12 and 11 of the status register.

Note. The control words are transferred to the 8087 coprocessor by writing them to a memory location using the memory mapped I/O. The 8087 reads the control word from the memory by executing an instruction. Similarly, to transfer a status word from the coprocessor to the memory, the 8087 instruction has to be executed, which writes the status to the memory and can be processed through the 8086 instruction, *i.e.,* the communication of data between the main processor and the coprocessor takes place through the main memory only.

11.11 STACK OPERATION OF 8087

8087 internally operates on all numbers in the 80-bit temporary real format, which has eight, 80-bit registers used as a Last-In-First-Out (LIFO) stack. The register stack holds the operands on which the coprocessor instruction operates. The 8087 has a 3-bit stack pointer to hold a number of registers which currently points to the top of stack known as the coprocessor stack pointer. The stack operates as a circular stack of fixed size (8 elements) on the basis of LIFO access.

Stack pointer is initialized to 000B, when the 8087 coprocessor is initialized, so that the register 0 (R0) becomes the top of the stack (TOS) as shown in Fig. 11.7 (*a*). The operation of reading data from memory and writing to the register stack is analogous to PUSH operation of the 8086 processor. It first decrements the stack pointer by one and then writes data from the memory to the stack register pointed to by the stack pointer. The data is converted to temporary real format before copying to the stack register, when required. Dring the first push (data load to stack register), the TOS has value 000B, the push operation decrements TOS by one which becomes 111B and places the data in register 111B (R7). The operation of TOS has wraparound effect, *i.e.,* if TOS is 000B, then TOS ← TOS-1 becomes 111B, if TOS is 111B, then TOS ← TOS + 1 becomes 000B. The condition of the coprocessor stack after the Push(X), Push(Y), Push(Z), Pop(Z), Pop(Y) and Pop(X) are shown in Fig. 11.8 below in detail.

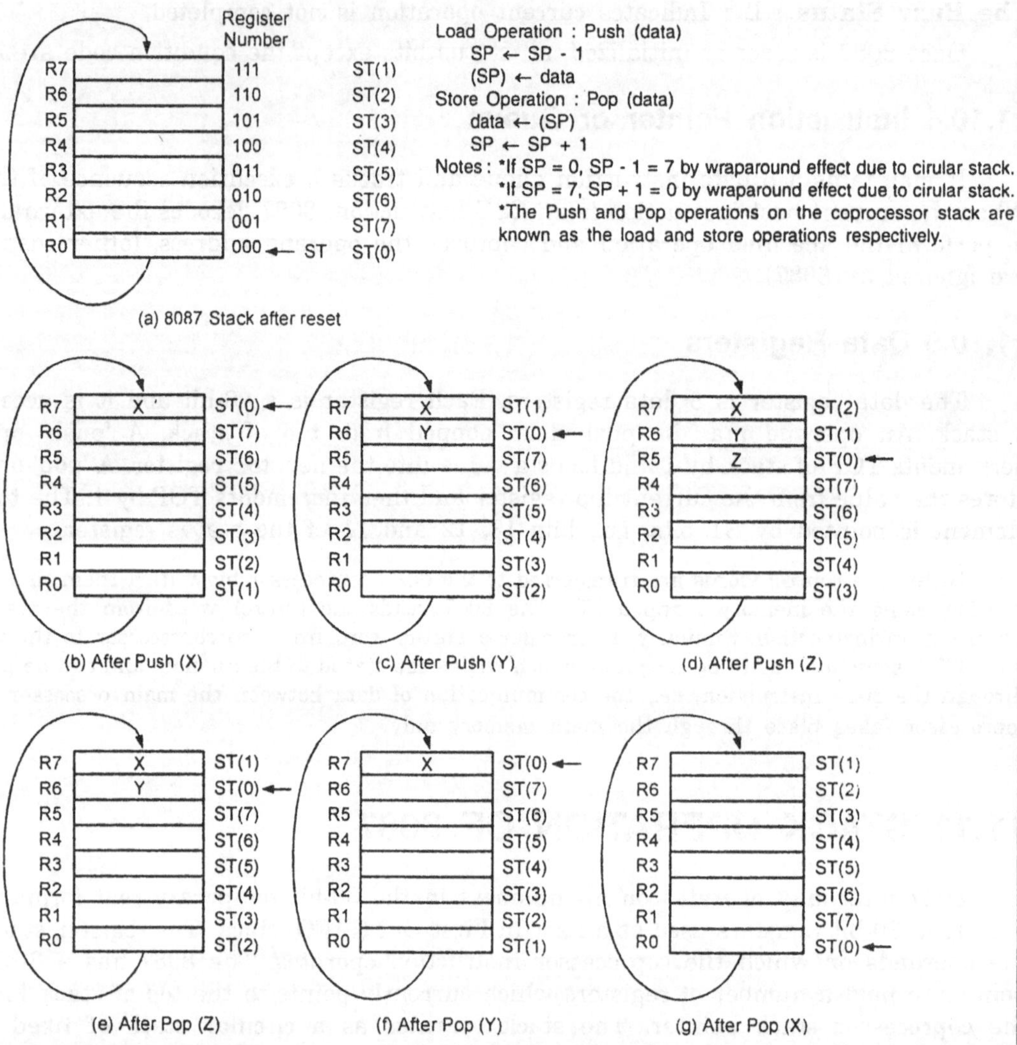

Fig. 11.8. 8087 Coprocessor Stack Operations.

The register which is currently the top of stack is referred as ST(0) in the 8087 instruction [or simply Stack Top (ST)]. The immediate register below ST is referred as ST(1) and the next to it as ST(2) and so on upto ST(7). It implies that, by popping a value from stack, the TOS will be pointing to earlier ST(1) which becomes new ST(0) or ST, because pop will decrement the stack pointer by one. The push operation increments stack pointer by one and hence the old ST becomes the new ST(1) and ST points to the register above the old ST. The register which is on the top of stack can be determined in the program by transferring the coprocessor status word to memory and examining the bits labelled ST in the status word.

11.12 8087 INSTRUCTION SET

The instruction mnemonics of 8087 starts with F, stands for floating point. The instruction of 8087 numeric data processor can be classified into following six groups :

1. Data transfer instructions
2. Arithmetic instructions
3. Compare instructions
4. Transcendental instructions
5. Load constant instructions
6. Processor control instructions

Now in each classification different instruction set can be observed, which are illustrated in the following Tables (*i.e.*, 6 to 11).

11.12.1 Data Transfer Instruction

Table 6

S. No.	Instruction	Description with example
1	FLD source	Decrements stack pointer by one and copies a real number from a stack element or memory location to the new ST. A short-real or long-real number from memory is automatically converted to temporary real format by the 8087 before it is put in ST. Exceptions : I, D. Examples : FLD ST(2) ; Copies ST(2) to ST FLD [BX] ; Number from memory pointed ; by BX copied to ST
2	FST destination	Copies ST to a specified stack position or to a specified memory location. Exceptions : I, O, U, P. Examples : FST ST(3) ; Copy ST to ST(3) FST [BX] ; Copy ST to memory pointed by [BX]
3	FSTP destination	Copies ST to a specified stack element or memory location and increments stack pointer by one to point to the next element on the stack. This is a stack POP operation.
4	FXCH destination	Exchanges contents of ST with the contents of a specified stack element. If no destination is specified, then ST(1) is used. Exception : I. Example : FXCH ST(4) ; Swap ST and ST(4)

5	FILD source	Integer load. Converts integer number from memory to temporary-real format and pushes converted number on 8087 stack. Exception : I. Example : FILD DWORD PTR [BX] ; Short integer from ; memory location pointed ; by [BX]
6	FIST destination	Integer store. Converts number from ST to integer form, and copies to memory. Exceptions : I, P Example : FIST INT_NUM ; ST to memory locations named ; INT_NUM
7	FISTP destination	Integer store and pop. Similar to FIST except that stack pointer is incremented after copy.
8	FBLD source	Packed decimal (BCD) load. Convert number from memory to temporary-real format and push on top of 8087 stack. Exception : I. Example : FBLD AMOUNT ; Ten byte BCD number from ; memory location AMOUNT to ST
9	FBSTP destination	BCD store in memory and pop 8087 stack. Pops temporary-real from stack, converts to 10-byte BCD, and stores result to memory. Exception : I. Example : FBSTP MONEY ; Contents from top of stack are ; converted to BCD, and stored ; in memory.

Note. In Table 6, *i.e.*, in data transfer instruction, serial number 1 to 4 belongs to real transfers, 5 to 7 belongs to integer transfers and 8 to 9 belongs to packed decimal transfer types.

11.12.2 Arithmetic Instruction

Table 7

S. No.	Instruction	Description with example
1	FADD destination, soure	Will add real number from specified source to real number at specified destination. Source can be stack element or memory location. Destination must be a stack element. If no source or destination is specified, then ST is added to ST(1) and the stack pointer is incremented so that the result of the addition is at ST. Exceptions : I, D, O, U, P.

		Examples : FADD ST(2), ST ; Add ST to ST(2), result in ST(2) FADD ST, ST(5) ; Add ST(5) to ST, result in ST FADD SUM ; Real number from memory + ST FADD ; ST + ST(1), pop stack-result at ST
2	FADDP destination, source	Adds ST to specified stack element and increments stack pointer by one. Exceptions : I, D, O, U, P. Example : FADDP ST(2) ; Add ST(2) to ST ; Increment stack pointer so ; ST(2) becomes ST
3	FIADD source	Adds integer from memory to ST, Stores the result in ST. Exceptions : I, D, O, P. Example : FIADD CARS_SOLD ; Integer number from memory + ST
4	FSUB destination, source	Subtracts the real number at the specified source from the real number at the specified destination and puts the result in the specified destination. Exceptions : I, D, O, U, P. Examples : FSUB ST(3), ST ; ST(3) ← ST(2) − ST FSUB DIFFERENCE; ST ← ST − real from memory FSUB ; ST ← (ST(1) − ST)
5	FSUBP destination, source	Subtracts ST from specified stack element and puts result in specified stack element. Then increments stack pointer by one. Exceptions : I, D, O, U, P. Examples : FSUBP ST(2) ; ST(2) − ST . ST(1) becomes new ST.
6	FISUB source	Subtracts integer number stored in memory from ST and stores result in ST. Exceptions : I, D, O, P. Example : FISUB DIFFERENCE ; ST ← ST − integer ; from memory
7	FSUBR destination, source	
8	FSUBRP destination, source	
9	FISUBR source	These instructions operate same as FSUB instructions discussed earlier except that these instructions subtract the contents of the specified destination from the contents of the specified source and put the difference in the specified destination. [Normal FSUB instruction subtracts source from destination.]

10	FMUL destination, source	Multiply real number from source by real number from specified destination, and put result in specified stack element. Exceptions : I, D, O, U, P. FMUL ST(2), ST ; Multiply ST(2) and ST, result ; in ST(2) FMUL ST, ST(5) ; Multiply ST(5) to ST, result in ST FMULP destination, source-Multiplies the real number from specified source by real number from specified destination, puts result in specified stack element, and increment stack pointer by one. With no specified operands FMULP multiplies ST(1) by ST and Pops stack to leave result at ST. Exceptions : I, D, O, U, P. Example : FMULP ST(2) ; Multiply ST(2) to ST. Increment stack ; pointer so STI (1) becomes ST
11	FIMUL source	Multiply integer from memory ST and put result in ST. Exceptions : I, D, O, P. Example : FIMUL DWORD PTR [BX] ; Integer number from memory ; pointed by BX × ST and result ; in ST
12	FDIV destination, source	Divides destination real by source real, stores result in destination. Exceptions : I, D, Z, O, U, P. Example : FDIV ST(2), ST ; Divides ST by ST(2) ; stores result in ST
13	FDIVP destination, source	Same as FDIV, but also increments stack pointer by one after DIV. Exception : I, D, Z, O, U, P. Example : FDIV ST(2), ST ; Divides ST by ST(2), stores ; result in ST and increments ; stack pointer
14	FIDIV source	Divides ST by integer from memory, stores result in ST. Exceptions : I, D, Z, O, U, P. Example : FIDIV PERCENTAGE ; ST ← ST/integer number
15	FDIVR destination, source	
16	FDIVP destination, source	
17	FIDIVR source	These three instructions are identical in format to the FDIV, FDIVP and FIDIV instructions above except that they divide the source operand by the destination operand and put the result in the destination.

18	FSQRT	Contents of ST are replaced with its square root. Exceptions : I, D, P. Example : FSQRT.
19	FSCALE	Scales the number in ST by adding an integer value in ST(1) to the exponent of the number in ST. Fast way of multiplying by integral powers of two. Exceptions : I, O, U.
20	FPREM	Partial remainder. The contents of ST(1) are subtracted from the contents of ST over and over again until the contents of ST are smaller than the contents of ST(1). Exceptions : I, D, U. Example : FPREM.
21	FRNDINT	Round number in ST to an integer. The round-control (RC) bits in the control word determine how the number will be rounded. Exceptions : I, P.
22	FXTRACT	Separates the exponent and the significant parts of a temporary-real number in ST. After the instruction executes, ST contains a temporary-real representation of the significant of the number and ST (1) contains a temporary-read representation of the exponent of the number. Exception : 1.
23	FABS	Replaces ST by its absolute value. Instruction simply makes sign positive. Exception : 1.
24.	FCHS	Complements the sign of the number in ST. Exception : 1.

Note. In Table 7, serial no. 1 to 3 belongs to Addition, 4 to 6 Substraction, 7 to 9 Reversed Substraction, 10 to 11 Multiplication, 12 to 14 Division, 15 to 17 Reversed Division and 18 to 24 other Arithmetic Operations.

11.12.3 Compare Instructions

These instruction compares the contents of ST with contents of specified or default source. The source may be another stack element or real number in memory. Such compare instructions set the condition code bits C3, C2 and C0 of the status words use as shown in Table (*i*), (*ii*) and (*iii*) below :

Table (*i*)

Order	C3	C0
(ST) > (SRC)	0	0
(ST) < (SRC)	0	1
(ST) = (SRC)	1	0
Not comparable	1	1

Table (*ii*)

Order	C3	C0
(ST) > 0 . 0	0	0
(ST) < 0 . 0	0	1
(ST) = 0 . 0	1	0
Not comparable	1	1

Table (*iii*)

(ST)	C3	C2	C1	C0
+ Unnormal	0	0	0	0
+ NAN	0	0	0	1
– Unnormal	0	0	1	0
– NAN	0	0	1	1
+ Normal	0	1	0	0
+ ∞	0	1	0	1
– Normal	0	1	1	0
– ∞	0	1	1	1
+ 0	1	0	0	0
Empty	1	0	0	1
– 0	1	0	1	0
Empty	1	0	1	1
+ Denormal	1	1	0	0
Empty	1	1	0	1
– Denormal	1	1	1	0
Empty	1	1	1	1

Note. It is possible to transfer status word to memory with 8087 FSTSW instruction and then we can use 8086 instructions to determine the results of comparision.

Now Table 8 shows the different compare instructions :

Table 8

S. No.	Instruction	Description with example
1	FCOM source	Compares ST with real number in another stack element or memory. Exceptions : I, D. Examples : FCOM ; Compares ST with ST(1) FCOM ST(4) ; Compares ST with ST(4) FCOM VALUE ; Compares ST with real number ; from memory
2	FCOMP source	Identical to FCOM except that the stack pointer is incremented by one after the compare operation.
3	FCOMPP	Compares ST with ST(1) and increments stack pointer by 2 after compare. Exceptions : I, D.
4	FICOM source	Compares ST to a short or long integer from memory. Exceptions : I, D.

5	FICOMP source	Identical to FICOM except stack pointer is incremented by one after compare.
6	FTST	Compares ST with zero. Exceptions : I, D.
7	FXAM	Tests ST to see if it is zero, infinity, unnormalized, or empty. Sets bits C_3, C_2, C_1 and C_0 to indicate result. Exceptions : None.

11.12.4 Transcendental Instructions (Trigonometric and Exponential Instructions)

Table 9

S. No.	Instruction	Description
1	FPTAN	Computes the values for a ratio of Y/X for an angle in ST. The angle must be expressed in radians, and the angle must be in the range of $0 < \text{angle} < \pi/4$. (FPTAN does not work correctly for angles of exactly 0 and $\pi/4$. Exceptions : I, P.)
2	FPATAN	Computes the angle whose tangent is Y/X. The X value must be in ST, and the Y value must be in ST(1). Also X and Y must satisfy the inequality $0 < Y < X < \infty$. The resulting angle expressed in radians replaces Y in the stack. After the operation the stack pointer is incremented so the result is then ST. Exceptions : U, P.
3	F2XM1	Computes the function $Y = 2^x - 1$ for an X value in ST. The result, Y, replaces X in ST . X must be in the range $0 \leq X \leq 0.5$. Exceptions : U, P.
4	FYL2X	Calculates Y times the log to the base 2 of X or Y ($\log_2 X$). X must be in the range of $0 < X < \infty$ and Y must be in the range $-\infty < Y < +\infty$. X must initially be in ST and Y must be in ST(1). The result replaces Y and then the stack is popped so that the result is then at ST. Exception : P.
5	FYL2XP1	Computes the function Y times the log to the base 2 of (X + 1) or Y ($\log_2$ (X + 1)). This instruction is almost identical to FYL2X except that it gives more accurate results when computing the log of a number very close to one.

11.12.5 Load Constant Instructions

Table 10

S. No.	Instruction	Description
1	FLDZ	– Push 0.0 onto stack
2	FLD1	– Push + 1.0 onto stack
3	FLDPI	– Push the value of π onto stack
4	FLD2T	– Push log of 10 to the base 2 onto stack ($log_2 10$)
5	FLDL2E	– Push log of e to the base 2 onto stack ($log_2 e$)
6	FLDLG2	– Push log of 2 to the base 10 onto stack ($log_{10} 2$)

Note. The load constant instruction will just push indicated constant into the stack.

11.12.6 Processor Control Instructions

Table 11

S. No.	Instruction	Description
1	FINIT/FNINT	Initializes 8087. Disables interrupt output, sets stack pointer to register 7, sets default status.
2	FDISI/FNDISI	Disables the 8087 interrupt output pin so that it cannot cause an interrupt when an exception (error) occurs.
3	FENI/FNENI	Enables 8087 interrupt output so it can cause an interrupt when an exception occurs.
4	FLDCW source	Loads a status word from a memory location into the 8087 status register. This instruction should be preceded by the FCLEX instruction to prevent a possible exception response if an exception bit in the status word is set.
5	FSTCW/FNSTCW destination	Copies the 8087 control word to a memory location. You can determine its current value with 8086 instructions.
6	FSTSW/FNSTW destination	Copies the 8087 status word to a memory location. You can check various status bits with 8086 instructions and take further action on the state of these bits.
7	FCLEX/FNCLEX	Clears all of the 8087 exception flag bits in the status register. Unasserts BUSY and INT outputs.
8	FSAVE/FNSAVE destination	Copies the 8087 control word, status word, pointers and entire register stack to 94-byte area of memory. After copying all of this the FSAVE/FNSAVE instruction initializes the 8087.
9	FRSTOR source	Copies a 94-byte area of memory into the 8087 control register, status register, pointer registers, and stack registers.
10	FSTENV/FNSTENV destination	Copies the 8087 control register, status register, tag words, and exception pointers to a series of memory locations. This instruction does not copy the 8087 register stack to memory as the FSAVE/FNSAVE instruction does.

11	FLDENV source	Loads the 8087 control register, status register, tag word and exception pointers from a named area in memory.
12	FINCSTP	Increment the 8087 stack pointer by one.
13	FDECSTP	Decrement stack pointer by one.
14	FFREE destination	Changes the tag for the specified destination register to empty.
15	FNOP	Performs no operation. Actually copies ST to ST.
16	FWAIT	This instruction is actually an 8086 instruction which makes the 8086 wait until it receives a not busy signal from the 8087 to its $\overline{\text{TEST}}$ pin.

Note. The processor control instructions actually do not perform computations but they are made used to perform tasks like initializing 8087, enabling intempty, etc.

11.13 8087 PROGRAMMING GUIDELINES

The coprocessor 8087, offers a wide variety of instructions for writing programs which operate on floating point data. It offers extended data types, registers and instructions to the microprocessor. The coprocessor register stack space allows constants and temporary results to be held in the registers during calculations, thus reducing the memory access and improving the speed as well as the availability of system bus.

Coprocessor programming is a distinct discipline, because the coprocessor works in parallel with the microprocessor. The parallel operation decreases the operating time by allowing the coprocessor to do mathematical calculations while the microprocessor continues to do other operations. The main processor and the coprocessor must be properly synchronized during the execution of program.

Instruction FWAIT is used to synchronize the processor (8086 or 80286) with the math coprocessor (8087 or 80287). It is normally used while switching from the coprocessor instructions to the processor instructions, but not used in switching from the processor instruction to the coprocessor instruction since every coprocessor instruction is prefixed with the FWAIT instruction by the assembler, as such.

While programming with 8087 coprocessor, these steps are to be followed as discussed below :

1. Coprocessor Initialization,
2. Synchronization, and
3. Reading status of coprocessor.

11.13.1 Coprocessor Initialization

The coprocessor should be initialized by using FINIT instruction in the beginning which sets the top of stack of 000B (register R0). The coprocessor control word register has to be initialized to a suitable value depending on the requirement using FLDCW instruction.

For Example :

The instruction,

```
FINIT                          ; initialize coprocessor
```

initializes the coprocessor with default settings.

Control Word Setting :

Taking a memory variable *controlword* defined as,

```
CONTROLWORD    DW    ?         ; controlword
```

1. Initialize control word to some value as required.

   For example, validate all flags,

```
MOV    CONTROLWORD  03ffh      ; set control word to check for
                               ; the validity of all flags
```

2. Load control word : It is performed by,

```
FLDCW  CONTROLWORD             ; load control word into the control register
```

The initialization and control word setting is performed by the following sequence of instruction :

```
FINIT                          ; Initialize the Numeric Coprocessor
MOV  CONTROLWORD,  03ffh       ; set control world to check for the
                               ; validity of all flags
FLDCW  CONTROLWORD             ; load control word into the control register
```

11.13.2 Synchronization

Synchronize 8087 with 8086 while switching from coprocessor (8087) instruction execution to the main processor (8086) instruction execution, which can be achieved by inserting the FWAIT instruction while switching.

As an Examples :

Consider the following instruction sequence.

```
. .
FADD     ST, NUM1              ; ST ← ST+NUM1
MOV      AX, 100               ; AX ← 100
. .
```

The above instruction sequence should be coded as,

```
. .
FADD     ST, NUM1              ; ST ← ST+NUM1
FWAIT                          ; Synchronization of 8086 and 8087
MOV      AX, 100               ; AX ← 100
```

11.13.3 Reading Status of the Coprocessor

The status of execution of the coprocessor instruction (success or failure) is determined by writing the status word from the coprocessor to the memory location and observing it with the 8086 instructions. Which must be performed after the execution of procedures or while terminating the program to make sure of correct coprocessor operation.

For instant, following sequence of codes are used to test the status of the coprocessor activity.

```
        FSTSW   STATUSWORD          ; Get status word
        FWAIT                       ; Synchronize processors
        MOV     AX, STATUSWORD
        AND     AH, 0bfh            ; Are any error hits set
        JNZ     FAIL                ; if yes, the coprocessor activity fail
SUCC :
        ; coprocessor activity successful
        . .
FAIL :
        ; coprocessor activity fails
```

Now let us observe few program examples using 8087, after analysing the programming guidelines.

Ex : Prg. 1 : To calculate the area of a circle : *i.e.*, A = πR^2 where R is radius and A area of the circle.

```
Label        Mnemonics                         Comments
        ; Program calculates the area of a circle.
        ; The radius must be stored at memory location RADIUS.
        ; The result is found in memory location AREA after the program execution.
.MODEL SMALL
.DATA
        RADIUS  DD  3 . 4
        AREA    DD  ?
.CODE
START : MOV AX, @DATA                       ; [Initialise
        MOV DS, AX                          ; data segment]
        FINIT                               ; Initialise 8087
        FLD RADIUS                          ; Loads radius (R) in ST
        FMUL ST, ST(0)                      ; Square radius (R × R)
        FLDPI                               ; π to ST
        FMUL                                ; Multiply ST = ST × ST(1) i.e. (π × R × R)
        FSTP AREA                           ; Save area
        END START
```

Ex : Prg 2 : To find the roots of quadratic equation where roots of equation $ax^2 + bx + c = 0$ are $\frac{-b \pm \sqrt{b^2 - 4ac}}{2c}$. In the procedure it must find roots R_1 and R_2 for constant stored in memory locations A, B and C and the directions .286 and .287 must identify 80286 as processor and 80287 as an coprocessor.

```
         .286                               ; tells assembler that processor is 80286
         .287                               ; tells assembler that coprocessor is 80287
.MODEL SMALL
.DATA
         TWO      DD    2.0
         FOUR     DD    4.0
         A        DD    1.0
         B        DD    – 16.0
         C        DD    + 39.0
         R1       DD    ?
         R2       DD    ?

.CODE
START:   MOV AX, @DATA                      ; [Initialise
         MOV DS, AX                         ; data segment]
         FINIT                              ; Initialise 8087
         FLD      TWO
         FMUL     A                         ; form 2a
         FLD      FOUR
         FMUL     A                         ; form 4a
         FMUL     C                         ; form 4ac
         FLD      B
         FMUL     B                         ; form b²
         FSUBR                              ; form b² – 4ac
         FSQRT                              ; form square root of b² – 4ac
         FLD      B
         FSUB     ST, ST(1)
         FDIV     ST, ST(2)
         FSTP     R1                        ; save root 1
         FLD      B
         FADD
         FDIVR
         FSTP     R2                        ; save root 2
         END      START
         END
```

Ex : Prg. 3 : Program to calculate hypotenuse of a right angle triangle, whose sides A and B are stored in memory named SIDE 1 and SIDE 2.

```
        .287                          ; tells assembler that coprocessor is 80287
        .286                          ; tells assembler that processor is 80286
.MODEL  SMALL
.DATA
        SIDE1       DD  3.0
        SIDE2       DD  4.0
        HYPOTENUSE  DD  0
        CW          DW  0             ; space for control word
        SW          DW  0             ; space for status word
.CODE
START:  MOV AX, @DATA                 ; [Initialise
        MOV DS, AX                    ; data segment]
        FINIT                         ; Initialise 8087 coprocessor
        MOV    CW,    03FFII          ; Put the CW in memory which
                                      ; masks the interrupts and
                                      ; sets round to even
        FLDCW  CW                     ; Load control world in 8087
        FLD    SIDE1                  ; Put value of SIDE1 on top of stack
        FMUL   ST, ST(0)              ; Square SIDE1
        LD     SIDE2                  ; Put value of SIDE2 on top of stack
        FMUL   ST, ST(0)              ; Square SIDE2
        FADD   ST, ST(0)              ; A² + B² result at the top of stack
        FSQRT                         ; √(A²+B²) = hypotenuse in top of stack
        FSTSW  SW                     ; Copy status word to memory
                                      ; so 8086 can access it
        MOV    AX, SW                 ; Bring status to AX to check
                                      ; for errors
        AND    AL, 0BFH               ; Mask unneeded bits
        JNZ    LAST                   ; handle error if found
        FSTP   HYPOTENUSE             ; No error, copy result from 8087 to
                                      ; memory.
LAST :  NOP
        END    START
```

Ex : Prg : 4 : Program to find resonant frequency of an LC circuit whose equation is $F = \frac{1}{2\pi\sqrt{LC}}$.

```
.MODEL SMALL
.DATA
        RESOF   DD    ?             ; resonant frequency
        L       DD    .000001       ; inductance
        C       DD    .000002       ; capacitance
        TWO           2.0           ; constant
.CODE
START:  MOV AX,@DATA                ; [Initialise
        MOV DS,AX                   ; data segment]
        FINIT                       ; Initialise 8087
        FLD     L                   ; Get L .
        FMUL    C                   ; Find LC
        FSQRT                       ; Find √LC
        FMUL    TWO                 ; Find 2√LC
        FLDPI                       ; Get π
        FMUL                        ; Get 2π√LC
        FLD1                        ; Get 1
        FDIVR                       ; Form 1/(2π√LC)
        FSTP    RESOF               ; Save frequency
        END     START
        END
```

REVIEW QUESTIONS

1. Explain the importance of numeric data processor in a microcomputer system.
2. With the help of functional pin diagram of 8087 briefly explain the functions of pins.
3. What are the features of 8087 ?
4. Explain 8086-8087 operation.
5. Draw and explain the internal block diagram of 8087.
6. What are data types in 8087 ?
7. Draw and explain 8087 coprocessor data formates.
8. Draw and explain bit pattern of control register in 8087.
9. Briefly explain the stack operations of 8087.
10. Describe with example the packed BCD data types available in 8087.
11. Write the code for the following numbers in short real notation in 8087 :

    (*a*) +13.875 (*b*) −25.0625.

12 INTRODUCTION TO MICROCONTROLLERS

12.0 OBJECTIVE

At the end of this chapter reader will be familiar with Microcontroller, its Architectural block diagram and its applications. Apart, the comparison of Microcontroller with Micro-processor will be made clear. The reader will also know the details of 8051 and 8048/8748 series microcontrollers.

12.1 INTRODUCTION

The rapidly expanding technology with area of microcontrollers have contributed to many development in computer applications, industrial instrumentation and controls. With the advancement in the semiconductor technology, the whole of CPU of a digital computer is now fabricated on a single chip using LSI and VLSI technology.

12.2 MICROPROCESSOR V/S MICROCONTROLLER

A microcontroller differs form a microprocessor in several important ways. Early name for a microcontroller was micro computer. The main difference between a microprocessor and microcontroller is the completeness of the machine each represents. In order to put a microprocssor into use the designer required memory peripheral chips and serial and parallel ports to make completely funtional computer. On the other hand a complete computer based system could be build using a single chip microcontrollers, with a minimum of external components. Fig 12.1 shows simplified block diagram of a microprocessor, which consists of an arithmetic and logic unit (ALU), general purpose registers, stack pointer (SP), program counter (PC), clock timing circuit and interrupt circuit.

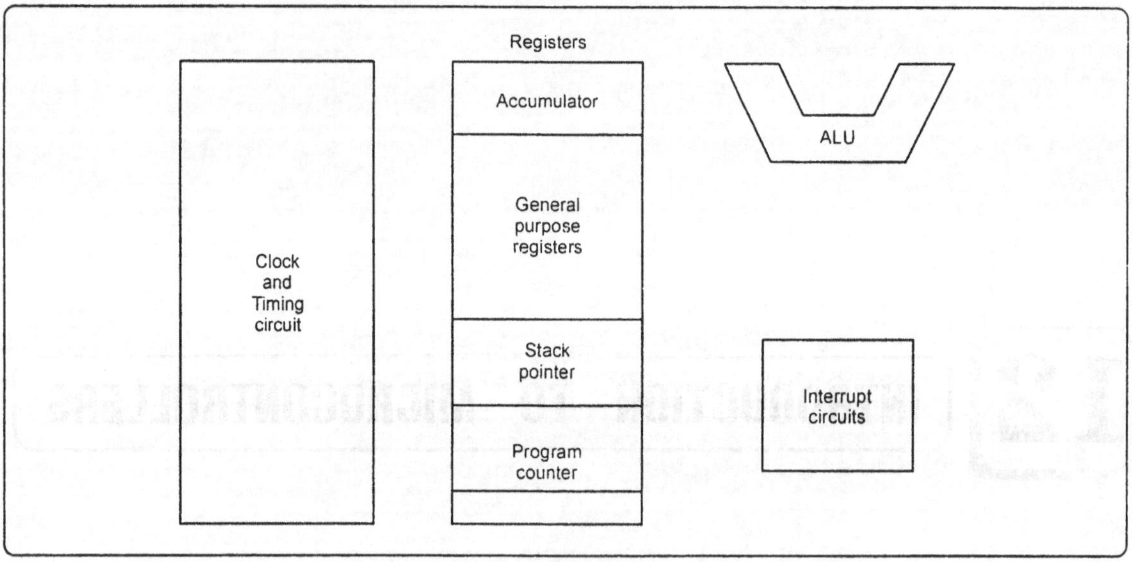

Fig. 12.1 Block diagram of a microprocessor.

To make a complete microcomputer system only microprocessor is not sufficient, but it is required to add other peripherals such as read only memory (ROM), read/write memory (RAM), decoders, drivers, number of input/output devices to make a complete microcomputer system. In addition, special purpose devices, such as interrupt controller, programmable timers, programmable I/O devices, DMA controllers may be added to improve the capacity and performance and flexibility of a microcomputer system.

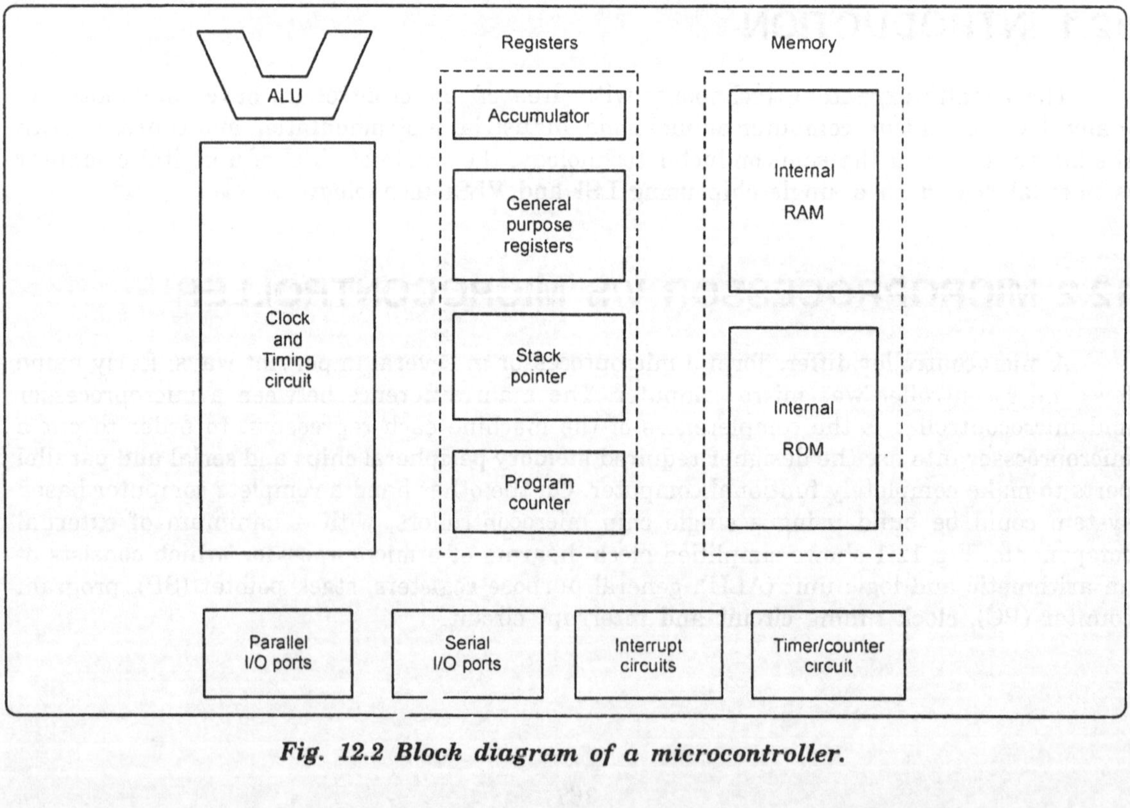

Fig. 12.2 Block diagram of a microcontroller.

The microcontroller incorporates all the features that found in microprocessor, apart it has also added features to make a complete microcomputer system on its own. The microcontroller has built-in ROM, RAM, parallel I/O, serial I/O, counters and a clock circuit. The Fig. 12.2 shows the simplified block diagram of a microcontroller.

Now the Table 1 gives few comparisions between Microprocessor and Microcontroller as below :

Table 1. Comparision Between Microprocessor and Microcontroller

Sl. No.	Microprocessor	Microcontroller
1.	Microprocessor contains ALU, general purpose register, stack pointer, program counter, clock timing circuit and interrupt circuit.	Microcontroller contains the circuitry of microprocessor and in addition it has built-in ROM, RAM, I/O devices, timers and counters.
2.	It has many instructions to move data between memory and CPU.	It has one or two instructions to move data between memory and CPU.
3.	It has one or two bit handling instructions.	It has many bit handling instructions.
4.	Acess times for memory and I/O devices are more.	Less access times for built-in memory and I/O devices.
5.	Microprocessor based system requires more hardware.	Microcontroller based system requires less hardware reducing PCB size and increasing the reliability.
6.	Microprocessor based system is more flexible in design point of view.	Less flexible in design point of view.
7.	It has single memory map for data and code.	It has separate memory map for data and code.
8.	Less number of pins are multi-functioned.	More number pins are multifunctioned.

12.3 HISTORY OF MICROCONTROLLER

Intel introduced 8048 series chip micro controller in 1978, known as MCS-48. It contains 8-bit CPU, 1/2, 4 k ROM/EPROM, 64/128, 256 byte RAM, timer/counter, parallel I/ Q ports, and 8-bit ADC.

In 1980 Intel introdueced more powerful microcontrollers series 8051. It has 8-bit CPU, 4/8 K ROM/EPROM, 128/256 byte RAM, timer/Counter, Parallel I/O ports, serial I/O ports etc., It has provision to increase its memory size by using extend EPROMs and Rams, 8031 has no on chips ROM/RPROM.

In 1983 Intel introduced 16-bit microcontrole series as MCS-96. It contains 16-bit CPU, 8K ROM, 232 byte RAM, timer/counter, parallel I/O ports, serial I/O Ports, 10-bit ADC, high speed I/O, pulse width modulated output, watch dog timer etc.

12.4 APPLICATIONS OF MICROCONTROLLER

There are many benefits by using a Microcontroller. Because all the standard features of a microcomputer embedded on a single chip with its performance equal to that of a multiple chip devices. When the device is used as a standard one controller, it needs interface only with its I/O peripherals. That means the execution speed of the processing is limited only by the speed of the chip, because there is no slowdown from transferring data between memory and CPU as in multi chip design. In microcontroller, execution will take at least 8-states even for 0 shift.

There are many applications using microcontroller, in different fields, such as :

- **HAND HELD INSTRUMENTS**
  - -small pagers
  - -Electronic planemeter
  - -Digital level meter
  - -IC Tester
- **PERIPHERAL DEVICES**
  - -keyboard controller
  - -modem
  - -printer buffer
  - -color plotter
- **STAND ALONE DEVICES**
  - -color copier
  - -typewriter
  - -cable TV terminal
  - -Fish finder
  - -Lawn sprinkler
  - -Charge card phones
- **INSTRUMENT SUB FUNCTIONS**
  - -Front panel handler for a digitizing oscilloscope
  - -Touch sensitive liquid crystal display
  - -Interface option for a microwave counter
  - -High performance multimeter
  - -Modular spectrum analyser
- **AUTOMOTIVE APPLICATIONS**
  - -Engine Control Modules
  - -Antilock braking system
  - -Dynamic ride control
  - -Electronic instrument panels
- **OTHER APPLICATIONS**
  - -CRT displays
  - -Navigation systems
  - -Transmission controllers
  - -Multiplexed wiring systems
  - -Cellular Telephones
  - -Electronic entry and security systems
  - -Electronic power storing

-Collision avoidance systems
-DC Motor control system
-PLC
-PID control system
-Automatic Belling system
-Stepper control system
-Voice synthesiser

12.5 COMMERCIAL MICROCONTROLLERS

Commercially available microcontrollers are 8-bit and 16-bit single chip microcontrollers. The 8048, 8051 service are 8-bit microcontrollers and 8096, 8097 are of 16-bit microcontrollers. These are constant improvement over one to another. However depending upon the particular requirement suitable Microcontroller can be selected. The Table 2 depicts 8-bit and 16-bit single chip Microcontrollers, which are available.

Table 2. 8-bit and 16-bit Single Chip Microcontrollers

Sl. No.	ROM/EPROM	RAM	I/O Ports	Ext. Interrupts	Timer	ADC
1	8048 1 K ROM	64	3 × 8 bit	1	1	NO
2	8035 0	64	3 × 8 bit	1	1	NO
3	8039 0	128	3 × 8 bit	1	1	NO
4	8748 1K EPROM	64	3 × 8 bit	1	1	NO
5	8049 2K EPROM	64	3 × 8 bit	1	1	NO
6	8741 1K EPROM	64	3 × 8 bit	1	1	NO
7	8022 2K ROM	64	3 × 8 bit	1	1	NO
8	8744 4K EPROM	64	3 × 8 bit	5	2	NO
9	8751 4K EPROM	128	4 × 8 bit	5	2	NO
10	8096 0	252	5 × 8 bit	8	4	NO
11	8097 0	252	5 × 8 bit	8	4	YES

12.6 ARCHITECTURE OF INTEL 8051 MICROCONTROLLER

Intel 8051 is a powerful microcontroller series having 8-bit CPU, 4/8k ROM/EPROM, 128/256 byte RAM timer/counter, parallel I/O ports, serial I/O ports etc. Its memory size can be increased if needed. Its special feature are :

- clock frequency 12 MHZ
- Programs memory security
- 128 * 8 RAM
- 32-I/O lines (4 numbers 8-bit port)
- 2-numbers of 16-bit timer/counter,
- Programmable Full-Duplex Serial Channel,
- 128K Accesible External Memor,

- Boolean Processor
- 218- user Bit-addressable location

Intel 8751 H is pin compatable EPROM version of 8051 AH microcontroller. It is available in hermetically sealed, ceranmic 40-pin Dual in line package (DIP). The Fig 12.3 depicts the architectural block diagram of Intel-8051 microcontroller.

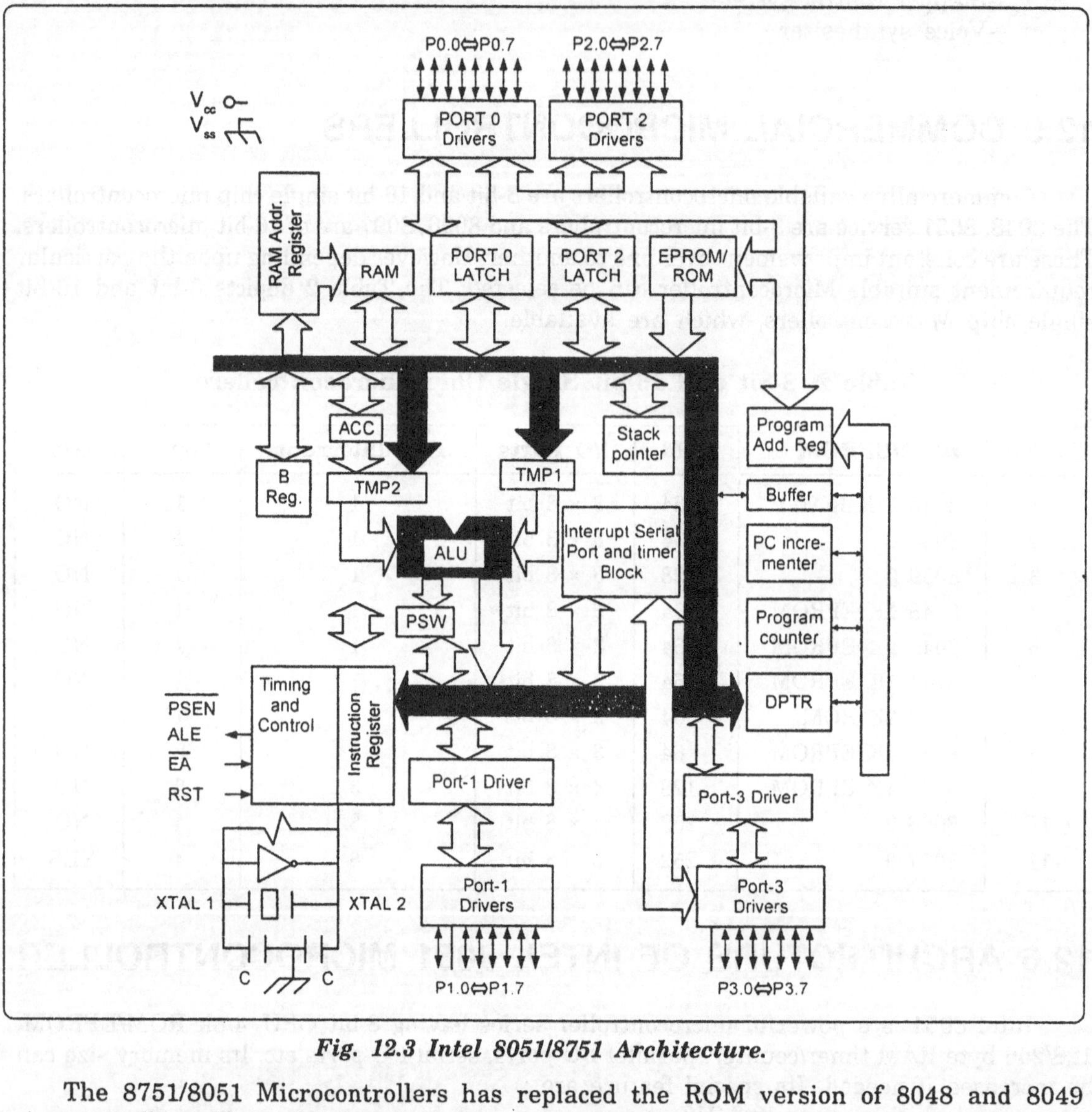

Fig. 12.3 Intel 8051/8751 Architecture.

The 8751/8051 Microcontrollers has replaced the ROM version of 8048 and 8049 microcontrollers. These chips operate from a single +5 volt power supply, however, 8748/ 8751 requires additional 25 volts supply for EPROM programming.

12.6.1 Pin-out of 8051 Microcontroller

The Fig. 12.4 depicts the pin-out of 8051 microcontroller, in which V_{ss} is circuit ground potential and V_{cc} is supply voltage during programming, verification and nomal operation.

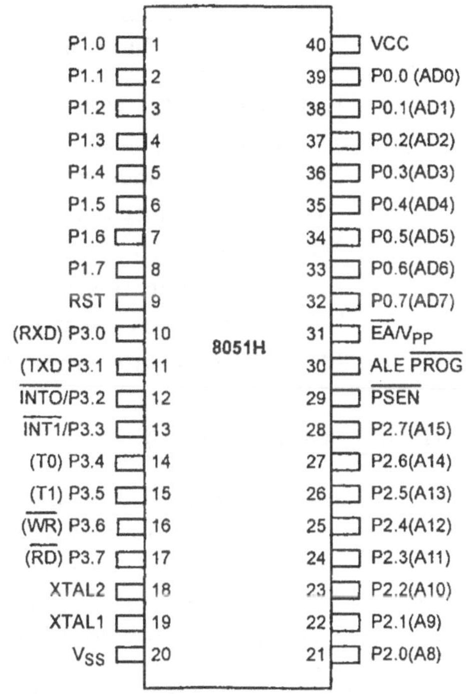

Fig. 12.4. Pin-out of 8051/8751 microcontroller.

12.6.2 I/O Ports of 8051

The Input/Output Ports of 8051 microcontroller has different identifications, which is explained as :

- **Port O :** It is an 8-bit open drain bidirectional I/O port. It is also the multiplexed low order address and data bus during acesses to external memory. Port-O can sink 8 LS TTL inputs.
- **Port 1 :** This is an 8-bit I/O port with internal pull-ups. Port-1 can sink/source 4 LS TTL inputs.
- **Port 2 :** It is also 8-bit I/O port with internal pullups. It emits the higher order address byte during accesses to external memory. Port-2 can sink/source 4 LS TTL inputs.
- **Port 3 :** It is again an 8-bit bidirectional port with internal pullups. Port-3 can sink/ source four LS TTL inputs. It also serves the functions of various special features listed in Table 3.

Table 3

Port Pin	Function
P3.0	RXD (Serial input port)
P3.1	TXD (Serial output port)
P3.2	INTO (External interupt)
P3.3	INT1 (Internal interupt)
P3.4	TO (Timer/Counter 0 external input)
P3.5	T1 (Timer/Counter 1 external input)
P3.6	WR (External data memory write strobe)
P3.7	RD (External data memory read strche)

12.6.3 Internal RAM and Registers of 8051

The register of internal RAM have 128 bytes, and there are sixteen special function registers. Next R0 to R7 are eight registers at 00 to 07 locations in RAM.

These registers are used for temporary data storage, movement, increment and decrement and compare operations with byte opcode. Four Sixty register banks could be selected, which occupy internal RAM locations, 00-07, 08-0F, 10-17, 18-1F. The selection is done by a single instruction. R0 and R1 can be used for indirect addressing. The Fig 12.5 shows 4 register bank arrangement, with lower 128-bytes of internal RAM.

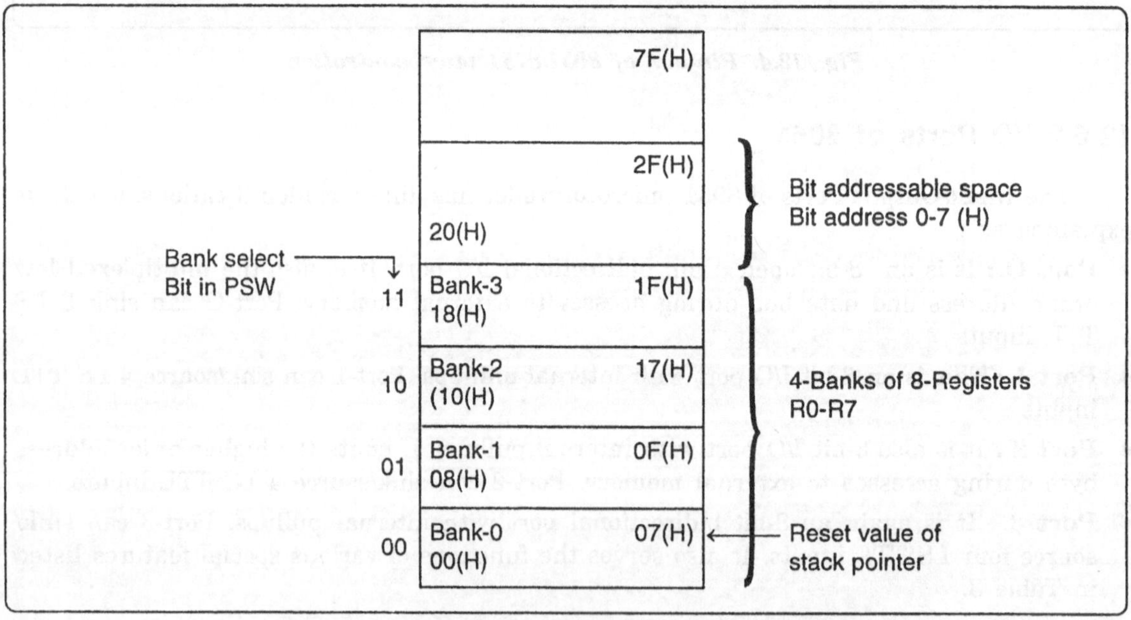

Fig. 12.5 The lower 128-bytes of Internal RAM.

12.6.4 8051-INTERRUPTS

Interrupts are used whenever it is necessary for a program to respond as quickly as possible to external events. Such an event might be the arrival of a measurment value that is to be processed as fast as possible. Interrupts from various sources can occur at any time during the program execution, which makes their programming and debugging complex.

The interrupt capabilities of the 8051 family are briefly discussed here. External interrupts are applied via IC input pins INTO/(pin 12) and INT1 (pin 13). The user can select the interrupt mode whether transition sensitive or level sensitive. Further interrupt sources are Timer Flag0 and Timer Flag1 (TFO and TF1), Receiver Interrupt (RI and Transmitter Interrupt (TI). When an interrupt occurs, the processor executes a kind of LCALL instruction, which results in a jump to address associated with the interrupt routine. Interrupt sources and associated addresses are listed below in Table 4.

Table 4

Interrupt	Interrupt address
IEO	0003H
TFO	000BH
TE1	0013H
TF1	001BH
RI + TI	0023H

12.7 INTEL 8748 MICROCONTROLLER

Intel 8048 family consists of 8048, 8748, 8049, 8749, 8035 and 8039. The 8035 and 8039 chips has no on-chip memory, however 8035 has execution speed same that of 8048. The Intel 8748 and 8749 are the two versions of chip microcontroller having 1K and 2K bytes

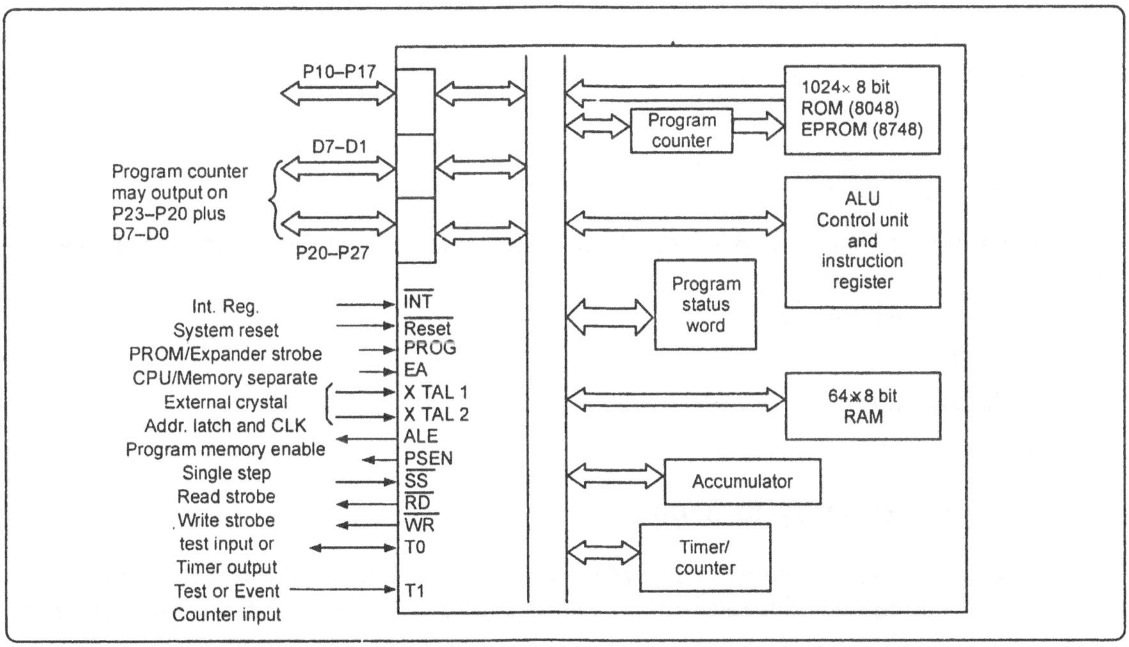

Fig. 12.6 Functional Blcok diagram of Intel 8748.

of EPROM respectively, which operate from a single +5 volts power supply. The Fig 12.6 depicts the functional block diagram of 8748 microcontroller.

Now few details of 8748 chip which are in Fig 12.6 are briefly discussed as below :

- Intel 8748 has 1024 bytes of Erasable Programmable Read Only Memory (EPROM), 12-bit program counter, 8-bit Interval Timer/Even counter, two single level interrupts, 27-I/O lines and over 96-instruction. About 80% of instructions are of single byte type.
- The arithmetic logic unit (ALU), the control unit and instruction register are all inaccessible to the users. The constants of program counter/timer, accumulator, and program status word (PSW) can be outputed on the Bus Port and few other pins of Port-2 (*i.e.,* on P20-P23)
- The Bus Port of single chip Microcontroller is a true bidirectional I/O Port with I/O strobes. The outputs can be statically latched whereas inputs are not latched. Thus the external logic must hold input data stable at bus port till the data has been read, (All 8-ports of the Bus Port must be assigned to either input or output).
- The I/O port-1 and 2 are secondary I/O ports with characteristics different from bus port. A 8-bit parallel data output at I/O port-1 or 2 by pulling individual pins from high to low.

12.7.1 8748 Processor's Architecture

Intel 8748 Processor's architecture may be divided into nine sections as :

1. Arithmetic Section
2. Program Memory
3. Data Memory
4. I/O Ports
5. Program Counter and Stac.
6. Program Status Word
7. Timer/Counter
8. Reset
9. Programming/Verification.

The arithmetic section of 8748 has basic data manipulation functions with following four blocks :

1. Instruction Decoder
2. ALU
3. Accumulator
4. Carry flag.

12.7.2 Program Memory in 8748

The resident program memory in 8748 consists of 1K byte in the form of EPROM 8-bit of wide, which can be addressed by program counter address of 1KB resident program memory is from 000(H) to 3FF(H) in Hex (or 0000(D) to 1023(D) in Decimal). Apart 3K additional momory can be interfaced as it has 12-bit program counter. The address of external memory will be 400(H) to FFF(H). The address of 2K words or less is done in the conventional manner. The address beyond 2KB can be reached by executing a program memory bank switch instruction (SEL MB0 SEL MB1) followed by a branch (JMP or CALL). (The bank switch feature extends the range of branch instructions beyond their normal 2K range and at the same time prevents the user from inadvertently crossing the 2K boundary). There are three important program Memory locations, which are :

(*a*) 000 (H) The reset line of the processor causing the first instruction to be fetched from this locations.

(*b*) 003 (H) Activates the Interrupt input line of the processor and causes a jump to the interrupt subroutine.

(*c*) 007 (H) A timer/counter interrupt resulting from timer counter overflow causes a jump to subroutine.

Note. The first instruction to be executed after reset is stored in location 000(H). The external vector interrupt subroutine or a jump instruction to a subroutine is stored in location 003(H), and the first instruction of a timer/counter service routine is stored in location 007(H), (Instruction such as MOVP and MOVP3 allows easy access to data look-up tables).

12.7.3 Data Memory of 8748

Intel 8748 has resident 64 × 8 bits data memory, whose locations are indirectly addressable through two RAM Pointer Registers which resides at address 0 and 1 of register array (R0 and R1 (RO" and R1"). There are two set of register banks, each of 8-bytes, known as Resigter Bank-0 and Register Bank-1. Location-0 to 7 is known as register Bank (RBO). Which are directly addressable by several instructions. Fig. 12.7 depicts the data memory map of 8748 microcontroller.

One more set of 8-working registers (R0-R7') is form Register Bank-1 (RB1). The RAM loctions 24-31 are known as bank-1 working register. The RAM has 16 reserved locations from 08 to 23 for 8-stack level which is used during subroutine jump, since there are only 8-stack levels, maximum subroutine nesting is fixed to eight. (If the nesting is less than 8, then remaining RAM locations can be used for general purpose). These locations are addressed through R0 or R1 of RBO or through R0' or R1' of RB1, as the case may be.

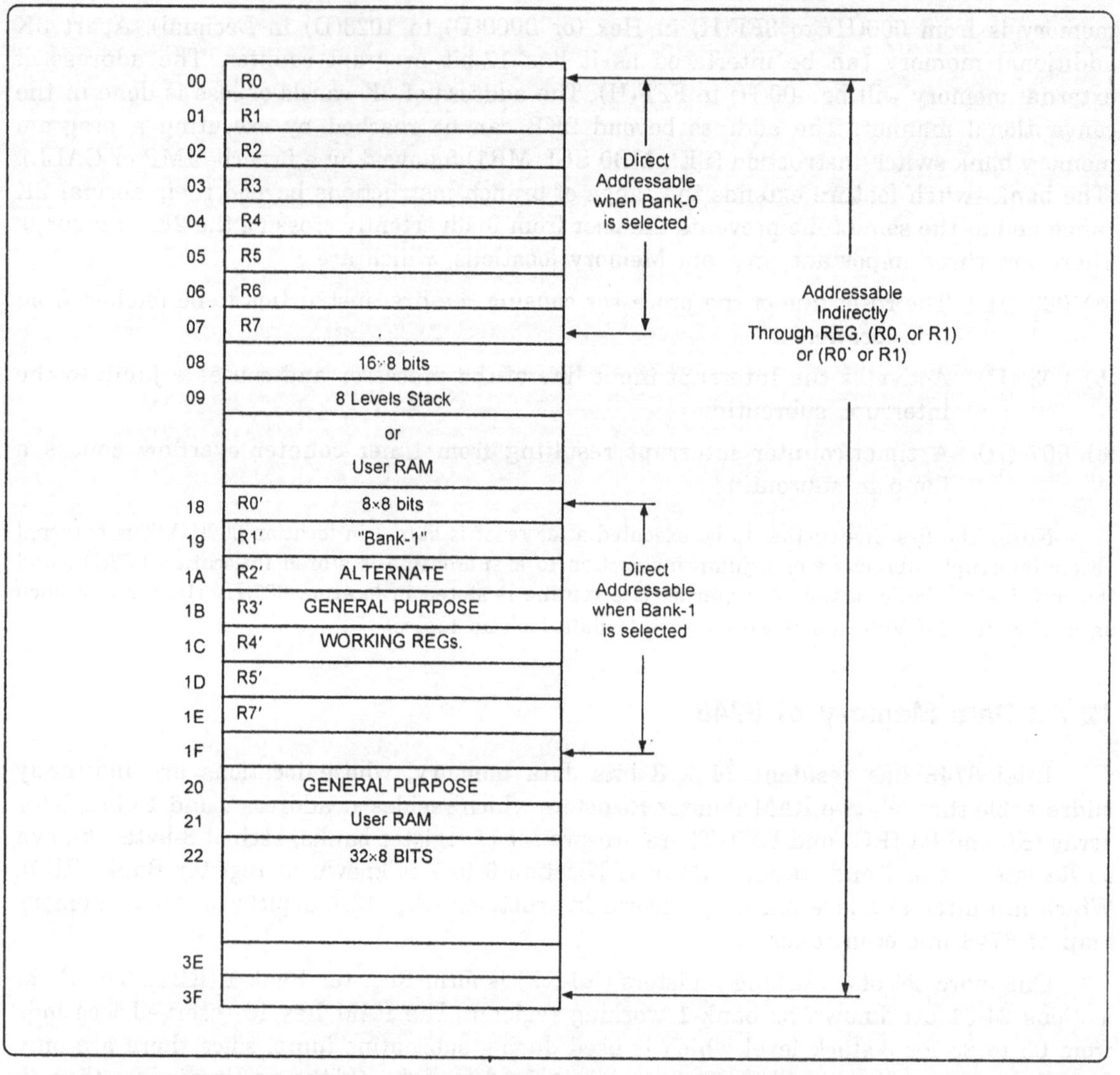

Fig. 12.7 Scratch Pad Memory of 8748 chip.

12.7.4 Input/Output lines of 8748

Intel 8748 has 27-I/O lines which can be used for input or output functions. These lines are grouped as three ports of eight-lines each of them serve as either inputs, outputs or bidirectonal ports and 3-test inputs, which can alter program sequences when tested by conditional jump instruction. The Table 5 shows the condition of three I/O ports along with other three input lines.

Table 5 Condition of Three I/O Ports

Pin Names	Function	Remarks
P10-P17	Port-1 Input and/or Output	Individual bit can be input or output
P20-P27	Port-2 Input and/or Output	
D0-D7	Bus Port Input Output Bidirectional	All 8-bits be input or output
T0	Test input, optional clock output and program/verification mode select	
T1 $\overline{INT}$	Test input, optional clock output Input Interrupts Request	

A brief explation of the three ports can be observed, *i.e.*, Port-1 and Port-2 and Bus Port :

- **Port-1 and Port-2 :**

  Each of these ports are 8-bits wide and have identical characteristics and the data written to these ports is statically latched and remains unchanged until new data is written. As input ports, these lines are non latching *i.e.* inputs may be present until read by an input instruction. The line of port-1 and 2 are called quasi-bidirectional, because of a special output circuit structure which allow each line to serve as an input and outputs or both even though the outputs are statically latched.

- **Bus Port :**

  This is also an 8-bit port, which is a true Bidirectional port with associated input and output strobes. If the Bidirectional nature is not needed it can serve as either statically latched output port or non latching input port, (input and Output lines on this port cannot be mixed).

- **Test and INT inputs :**

  There are three pins which serve as inputs and are testable with the conditional jump instructions, *i.e.,* TO, T1 and $\overline{INT}$. Without the necessity to load an input port into the accumulator, these pins allow input to cause program branches.

12.7.5 Pin out of 8748 microcontroller

Intel 8748 microcontroller is pakaged in 40-pin-Dual-In-Line Package (DIP). The Fig 12.8 (*a*) and Fig 12.8 (*b*) depicts the pinout of 8748 microcontroller and its logic symbol respectively.

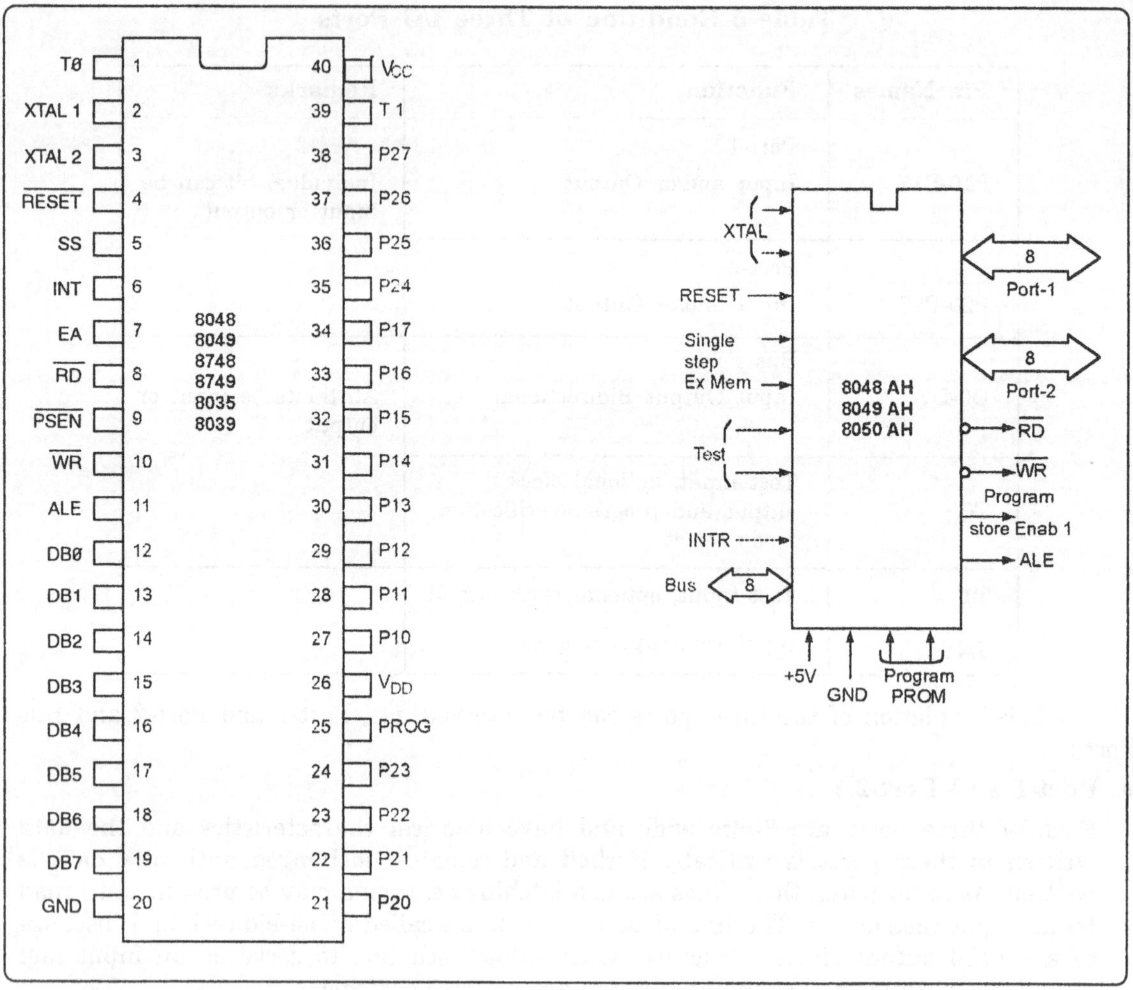

Fig. 12.8 (a) Pinout of 8048/8748 microconroller. ***Fig. 12.8 (b) Logic Symbol of 8048/8748.***

The Table 6 provides the summary of the functions of each pins of 8048/8748 microcontroller shown in Fig 12.8 (*a*)

Table 6. Pin Details of 8048/8748 Microcontroller

Pin Number	Designation	Function
20	V_{SS}	Circuit GND Potential
26	V_{DD}	Programming Power Supply; 21 V during program for the 8748H; + 5 V during operation for both ROM and EPROM.
40	V_{CC}	Main power supply + 5 V during operation.
25	PROG	Program Pulse ; + 18 V input pin during 8748H programing.

27-34	P10-P17 (PORT-1)	8 bit quasi bidirectional port
21-24 35-38	P20-P27 (PORT-2)	8 bit quasi bidirectional port
12-19	D0-D7	True bidirectional port which can be written or read synchronously using the RD, WR strobes. The port can also be statically latched.
1	T0	Input pin testable using the conditional transfer instruction, JTO and JNTO.
39	T1	Input pin testable using the JTI and JNTI instruction.
6	$\overline{INT}$	Interrupt input. Initiates and Interrupts if interrupts is enabled. Interrupts is disabled after a reset.
8	$\overline{RD}$	Output strobe activated during a BUS read. It can be used to enable data onto the bus from an external device.
4	$\overline{RESET}$	Input which is used to initialize in the processor. Also used during programming as verification.
10	$\overline{WR}$	Output store during a BUS write.
11	ALE	Address latch enable. This signal occurs once during each cycle and is useful as a clock output.
9	$\overline{PSEN}$	Programm store Enable. This output occurs only during a fetch to external program memory.
5	$\overline{SS}$	Single step input can be used in conjunction with ALE to "single step" the processor through each instruction.
7	EA	External access input which forces all program memory fetched to external (reference) memory.
2	XTAL 1	One side of crystal input for internal oscillator.
3	XTAL 2	Other side of crystal/ext. source input.

13 TYPES OF MICROCONTROLLERS

13.0 OBJECTIVE

After completion of this chapter, the reader will be familiar with

- Embedded Microcontroller
- External Memory Microcontroller
- Princeton Architecture
- Harvard Architecture
- CISC Vs RISC Architecture
- Microcontroller Memory types

13.1 INTRODUCTION

Similar to microprocessors microcontrollers also have their family, depending on the type of microcontrollers, the supporting chips also differ to develop a particular system using microcontrollers. Now let us discuss about following types of microcontollers so as to be familiar to select a suitable device to meet system requirements :

1. Embedded Microcontrollers
2 External Memory Microcontrollers

13.1.1 Embedded Microcontrollers

The Embedded Microcontroller is refferred when a complete hardware required to run a particular application, provided on a microcontroller chip. The Fig 13.1 depicts a typical microcontroller system with Embedded Microcontroller.

The embedded microcontrollers can communicate with external devices with its digital Input/Output pins. The Fig 13.1 shows brief features of embedded microcontroller.

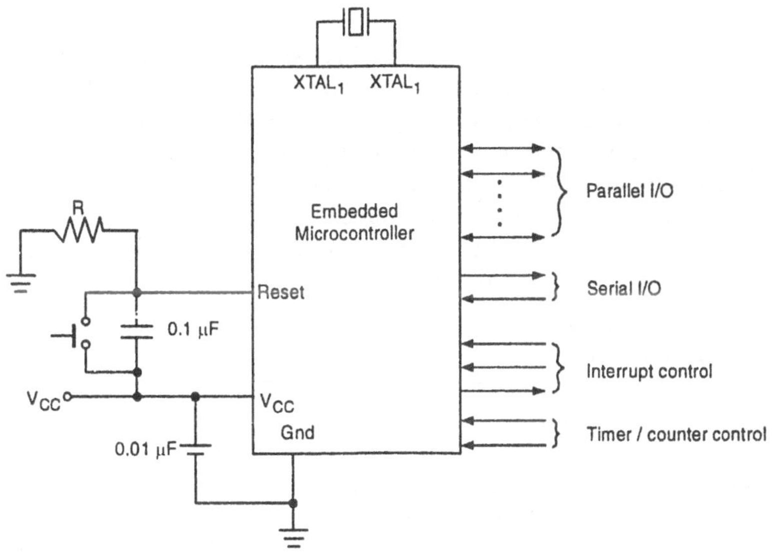

Fig. 13.1 Microcontroller system with embedded Microcontroller.

13.1.2 External Memory Microcontrollers

Normally in large systems, buit-in program memory and data memory will not be sufficient. Hence to overcome such problem some microcontrollers allow the connection of external memory. For the connection of external memory some parallel port pins are used as address and data lines. Thus connecting external memory to the microcontroller reduces its parallel input/output capabilities. Now the Fig 13.2 shows the microcontroller with external memory connections.

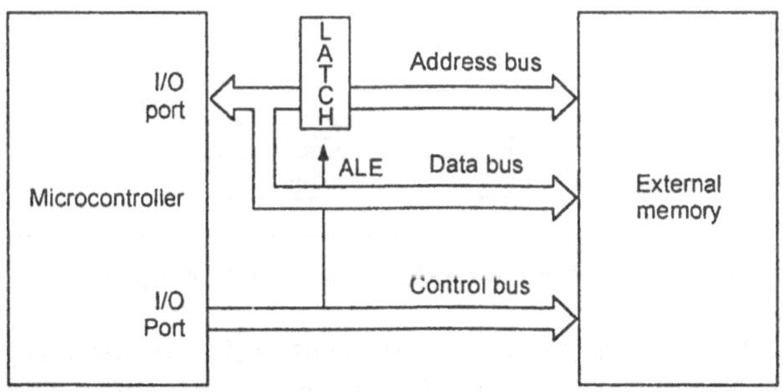

Fig. 13.2 External memory arrangement to microcontroller.

As observed in Fig 13.2, address and data lines are multiplexed and separated by external latch and ALE signal from the Microcontroller, as and when required. The 8051 Microcontroller can work very effectively as an Embeded device or with External Memory also.

13.2 PROCESSOR ARCHITECTURES

Under the direction of Eckert and Mauchly at Moora School of Engineering first electronic computer ENIAC *i.e.,* Electronic Numerical Integrator And Computer, was developed and constructed. At that time it was required more that 1800 Vacuum tubes and about 1500 relays, whose primary function was to compute ballistic trajectories. The ability of ENIAC was to perform about 5000 additions/subtraction per sec. Later in the development process, Harvard and Princeton Universities of US brought up a Computer Architecture to which can be used in computing tables of Naval Artillary Shell distance to vary elevations and enviroment conditions. Now let us observe few details of :

1. Princeton Architecture
2. Harvard Architecture.

13.2.1 Princeton Architecture

Main feature of the computer architecture developed by Princeton university was that the computer had common memory for storing the control program as well as variables and other data structures. The chief scientist for this project was Von Neumann. As a name of developer this architecture is best known as Von Neumann architecture. The Fig 13.3 depicts the block diagram of Princeton architecture.

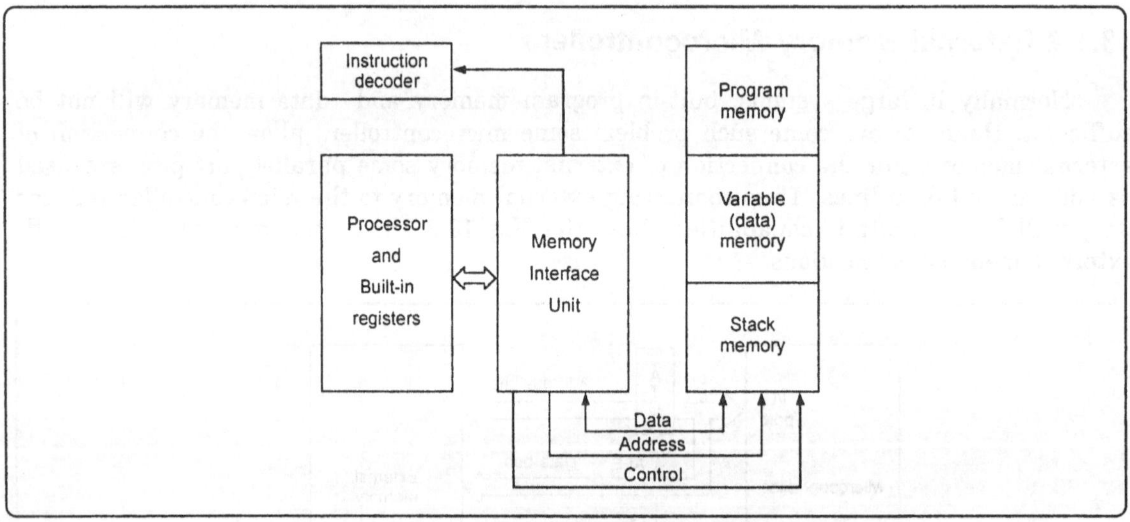

Fig. 13.3 Princeton Architecture Block diagram.

As observed in Fig. 13.3. It consists of processor with built-in registers, instruction decoder, memory and memory interface unit. The memory interface unit is responsible for arbitrating access to the memory space between reading the code, reading and writing the data with the processor and its internal registers. This architecture will not support to fetch the instruction code and data simultaneously. But, it is possible to fetch next instruction when convert instruction is executing, without requiring memory access. Such feature is known as per-fetching.

13.2.2 Harvard Architecture

When compaired to Princeton Architecture, Harvard's architecture suggested to have separate memory banks for program storage (code memory), the processor stack, and variable RAM (data Memory), as shown in Fig. 13.4, below :

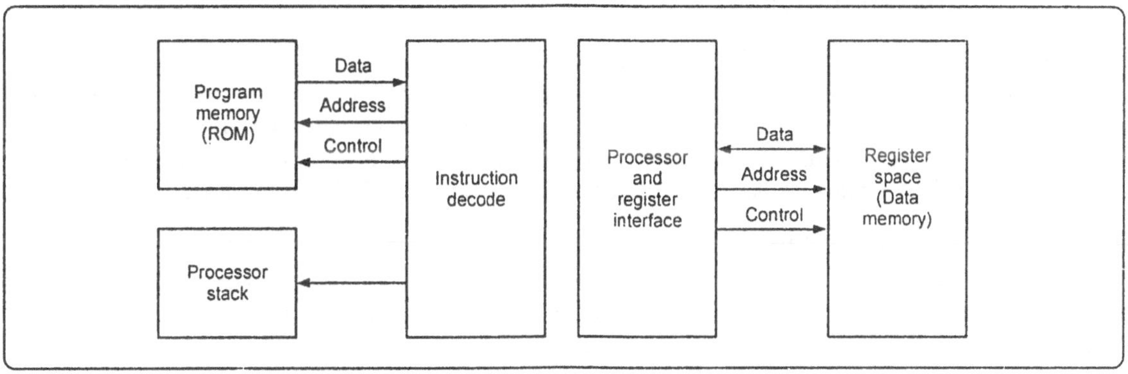

Fig. 13.4 Harvard Architecture Block diagram.

As observed in Fig. 13.4, Harvard Architecture has an advantage of executing instructions in fewer instructions cycles than Princeton architecture, which is because a much greater amount of instruction parallelism can be achieved in the Harvard architecture due to separate memory banks.

Apart from the advantage of parallelism in Harvard architecture, Princeton architecture was preferred by scientist because of the unreliability of technology of that time. They found one memory architecture more suitable of computers. One memory architecture allows greater flexibility in developing software, primarily in the area of real-time operating systems because program counter and stack contents are accessible to the programmer along with the data.

13.3 CISC V/S RISC ARCHITECTURES

RISC machines *i.e.*, Reduced Instruction Set Computer are the newly developed architectures, which is intended as a constrast with CISC machine, *i.e.*, Complex Instruction Set Computers. The examples of CISC machines are, IBM 360, DEC-VAX, Intel 80386 and Motorola 68030. Certain points are followed while designing RISC architecture which are

(*a*) Limited and simple instruction set.

(*b*) Large number of general purpose registers, or the use of compiler technology to optimize register usage.

(*c*) Emphasis on optimizing instruction pipeline.

13.3.1 Characteristics of RISC

The important characteristics of RISC, which are to be discussed are :

(*a*) One Instruction per cycle,

(*b*) Hardwired Instructions,

(*c*) Register to Register Operations,

(*d*) Reduced Number of Instruction,

(*e*) Simple Addressing Modes,

(*f*) Simple Instruction Format, and

(*g*) Instruction Pipelining.

(*a*) One Instruction per cycle

There is an one instruction per machine cycle, in RISC processor, where a machine cycle is defined to be the time taken to fetch two operands from register to perform an ALU operation and stores result in a register. With this, RISC machine instructions are not complicated and can execute as fast as microinstruction on CISC machines.

(*b*) Hardwired Instruction

With a simple, one cycle instruction, there is no need for microinstructions by which machine instructions can be hardwired. These instructions are executed faster than the instructions implemented with microinstructions, since it is not necessary to access a microprogram control memory during instruction execution.

(*c*) Register to Register Operations

This is the most important characteristics of RISC processor, whose architecture encourages the optimization of register use, so that frequently accessed operands remain in high-speed storage. To implement register to register operations, RISC processors provide multiple sets of registers. These register sets are organized into overlapped windows and act as a small, fast buffer for holding a subset of all variables that are most likely to be used. The Fig. 13.5 depicts overlapping register windows.

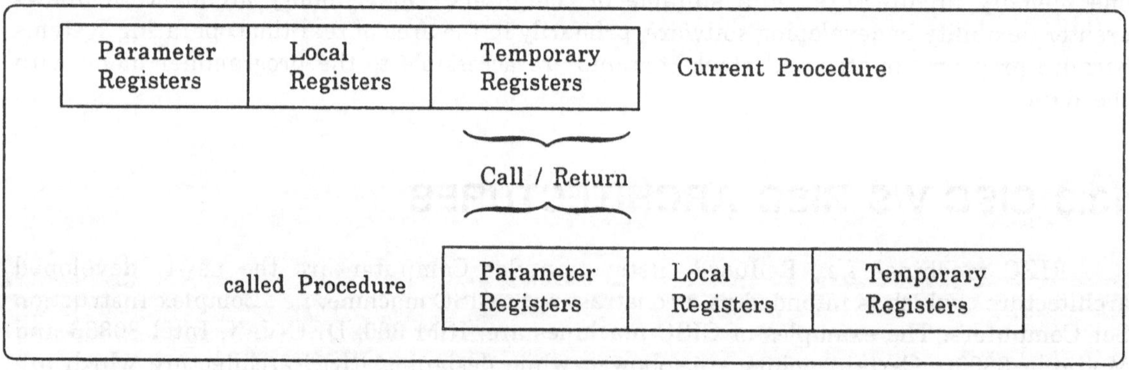

Fig. 13.5 Overlapping Register Window.

As observed in Fig. 13.5, the window is divided into three fixed-size areas. Parameter registers hold parameters passed down from the procedure that called the current procedure and results to be passed back up, local registers are used for local variables, as assigned by the compiler. Temporary registers are used to exchange parameters and results with the procedure called by current procedure. The temporary registers of current procedure are physically same as parameter registers of the called procedure, such overlap permits parameters to be passed without actual movement of data.

(d) Reduced Number of Instruction

The RISC processor can provide limited number of instruction, which will make the design of control unit simpler.

(e) Simple Addressing Modes

The RISC processor uses simple addressing modes, where almost all instructions use simple register addressing. Several other addressing modes are also used, such as displacement and PC-relative. It does not support complexed addressing modes. They can be synthesized in software from the simple ones. Such architecture of RISC processor simplified the instruction set and the control unit.

(f) Simple Instruction Format

The RISC processor makes use of simple instruction formats with fixed instruction length. The instruction length is aligned on word boundaries. Field locations, especially opcodes are fixed. The benefits of RISC architecture are :

(i) With fixed fields, opcode decoding and register operands addressing can occur simultaneously.

(ii) It simplifies the design of control unit.

(iii) Instruction fetching is optimized since word-length units are fetched.

(g) Instruction Pipelining

The concept of pipelining where in the CPU contains several independent units that work in parallel. One of them fetches the instructions, and other ones decode and execute them. At any instant, several instructions are in various stages of processing. All RISC processor will provide this pipelining feature.

13.3.2 Comparisons between RISC and CISC

The RISC and CISC architectury have some different features, the same is listed in the Table 1.

Table 1. RISC Vs CISC

S. No.	RISC	CISC
1.	Simple instructions taking one cycle.	Complex instructions taking multiple cycles.
2.	Very few instructions refer memory.	Most of instructions may refer memory.
3.	Instructions are executed by hardware.	Instructions are executed by microprogram.
4.	Fixed format instructions.	Variable format instructions.
5.	Few instructions only.	Many instructions.
6.	Few addressing mode, and most instructions have register to register addressing mode.	Many addressing modes.
7.	Complexed addressing modes are synthesized in software.	Supports complex addressing modes.

8.	Multiple register sets.	Single register set.
9.	Highly pipelined.	Not pipelined (or less pipelined).
10.	Complexity is in the compiler.	Complexity is in the microprogram.
11.	Conditional jump can be based on a bit anywhere in memory.	Conditional jump is usually based on status register bit.

Intel 8051 chip combines the features of RISC and CISC processors, whose architecture encourage the register to register operation by providing large register set, (as such, it also has the feature of CISC processor of having large instruction set).

13.4 MICROCONTROLLER MEMORY TYPES

To execute program in a microcontroller, programmer has to write program from scratch; he has to write a code to input variables and to display the result along with the actual program unlike in PC. No built-in subroutines are available like operating system calls (DOS calls) in PC systems.

However similar to microprocessors, microcontrollers require memory to store program and the variable data. The memory used to store program is known as **program memory** or **control storage** and the memory used to store variable data is known as **variable memory** or **data memory.**

13.4.1 Control Storage

The microcontrollers uses non-volatile memory, which does not lose its contents when power is taken away from the chip. Follow non-volatile memory can be discussed :

(*a*) ROM
(*b*) PROM
(*c*) EPROM
(*d*) EEPROM
(*e*) Flash Memory.

(*a*) ROM (Read Only Memory)

The Fig. 13.6 shows a very simple four byte diode ROM. Which consists of only diodes and a decoder. As observed in the Fig. 13.6, address lines A_0 and A_1 are decoded by 2 : 4 decoder and used to select one of the four rows. As decoder output is active low, it places a logic 0 on the selected row. Each output data line goes to logic 0 if a diode connects the output data column to the selected row. Data is available on the output data lines only when output enable $(\overline{OE})$ signal is low.

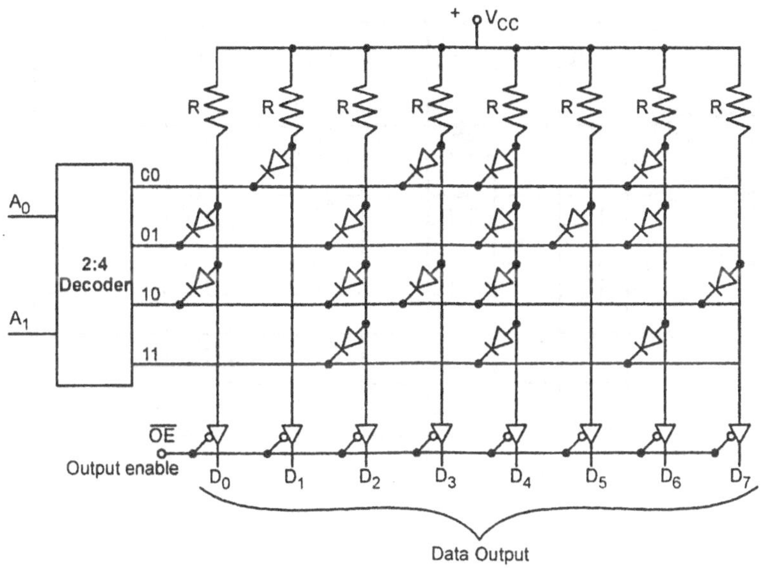

Fig. 13.6. Four byte diode ROM.

Now the Table 2 provides the contents of ROM at four locations.

Table 2. ROM contents

Address in binary	Binary Data								Data in Hex
	D_0	D_1	D_2	D_3	D_4	D_5	D_6	D_7	
00	1	0	1	0	0	1	0	1	A5
00	0	1	0	1	0	0	0	1	51
10	0	1	0	0	0	1	1	0	46
11	1	1	0	1	0	1	0	1	D5

Normally, in these days ROM's uses MOS technology instead of diode. The Fig. 13.7 depicts a four nibble ROM using MOS transistors.

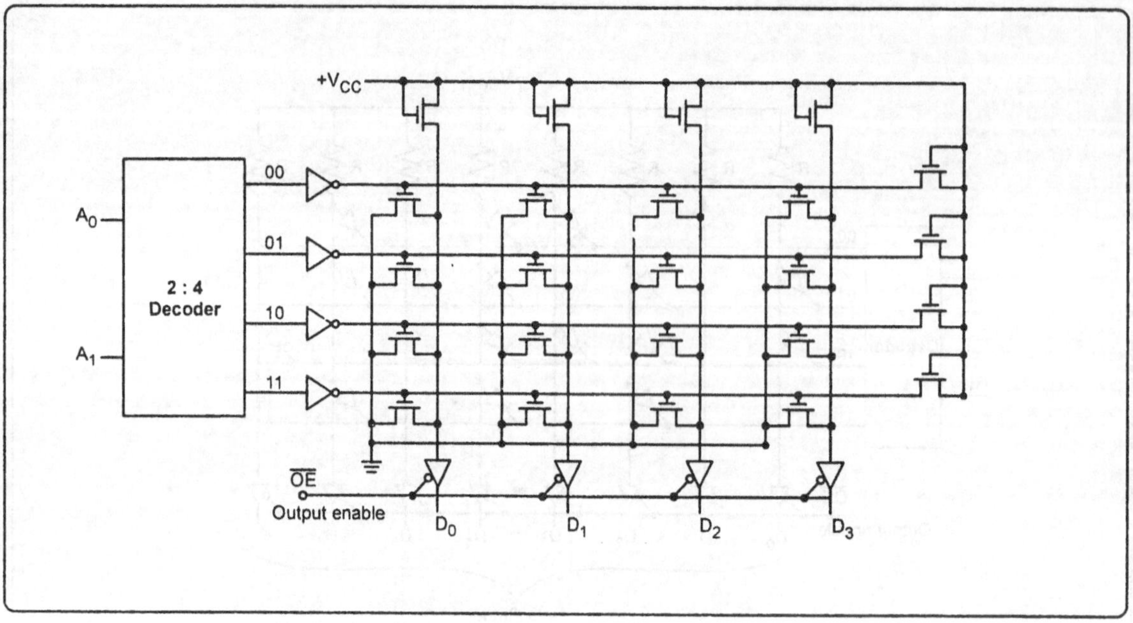

Fig. 13.7 Four half-byte ROM.

As observed in Fig. 13.7 the diodes and pull up resistors are replaced by MOS transistors. The address on the address lines (A_0 and A_1) is decoded by 2 : 4 decoder. Decoder selects one of the four rows making it logic 0. The inverter connected at the output of decoder inverts the state of selected row (*i.e.,* logic 1). Therefore, each output data line goes to logic 0 if a gate of MOS transistor is connected to row select lines. When gate of the MOS transistor is connected there selected row, MOS transistor is turned on, which pulls the corresponding column data line to logic 0.

Normally in integrated circuits, a thin metallized layer connects the gates of some transistors to the row select lines. The gate connections of MOS transistors depend on the data to be stored in the ROM, hence, according to the user truth table, manufacturer can deposit thin layer of metal to connect gates of the transistors. Once the pattern/ mask is decided, it is possible to make thousands of such ROMs, called **Mask-programmed ROMs.**

(*b*) PROM (Programmable Read Only Memory)

The Fig 13.8, depicts a four byte PROM, where diode is used in every bit position, hence initially all output will be Os, with each diode having fusible link in series with it.

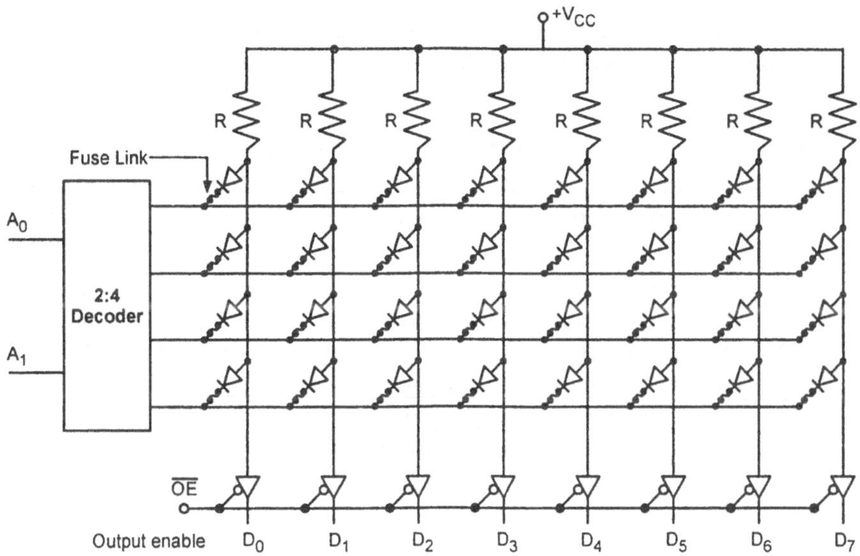

Fig. 13.8 Four byte PROM.

Here by addressing bit and applying proper current pulse at the corresponding output, we can blow out the fuse, storing logic 1 at that bit position, (fuse uses material like nichrome and polycrystalline). For blowing the fuse it is necessary to pass around 20 to 50 mA of current for a period of 5 to 20 μs. The blowing of fuses according to the truth table is called **programming of ROM.** The user can program PROMs with special PROM programmer. The PROM programmer selectively burns the fuses according to the bit pattern to be stored, (process is also known as **burning of PROM**). The PROMs are one time programmable, once programmed, the information stored is permanent.

(c) EPROM (Erasable Programmable Read Only Memory)

The erasable programmable ROMs use MOS circuitry, which can store 1s and 0s as a packet of charge in a buried layer of the IC chip. EPROMs can be programmed by the user with a special EPROM programmer. The important point is that we can erase the stored data in the EPROMs by exposing the chip to ultraviolet light through its quartz window for 15 to 20 minutes, as depicted in Fig. 13.9.

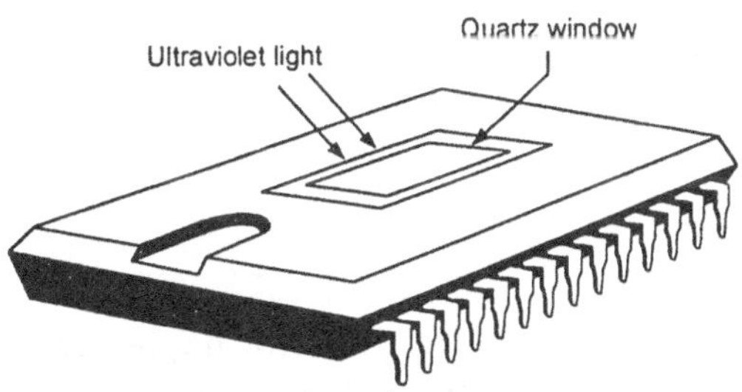

Fig. 13.9 EPROM chip.

In EPROM's it is not possible to erase selective information, when erased the entire information is lost, however the chip can be reprogrammed. This memory is ideally suitable for product development, experimental projects and college laboratories, since this chip can be fused many times.

(*d*) EEPROM (Electrically Erasable Read Only Memory)

The EEPROM's also make use of MOS circuitry similar to that of EPROM, and the data stored as charge or no charge on an insulated layer, (or an insulated floating) gate in the device.

Such insulating layer is made very thin (< 200 Å), hence, a voltage as low as 20 to 25 V can be used to move charges across the thin barrier in either direction for programming or erasing. EEPROM allows selective erasing at the register level rather than erasing all the information since the information can be changed by using electrical signals. The EEPROM memory also has a special chip erase mode by which entire chip can be erased in 10 ms, this time is quite small as compared to time required to erase EPROM, and it can be erased and reprogrammed with device right in the circuit. However, EEPROMs are most expensive and the least dense than ROMs.

(*e*) Flash Memory

The flash memory is a type of constantly-powered nonvolatile memory that can be erased and reprogrammed in units of memory called blocks. It is a variation of electrically erasable programmable read-only memory (EEPROM) which, unlike flash memory, is erased and rewritten at the byte level, which is slower than flash memory updating. The flash memory can be written in block (rather than byte) sizes, making it to update easy.

Flash memory gets its name as the microchip is organized so that a section of memory cells are erased in a single action or "flash". It uses bussed circuit for erasing the cells. The erasure is caused by Fowler-Nordheim tunneling in which electrons pierce through a thin dielectric material to remove an electronic charge from a floating gate associated with each memory cell. Intel offers a form of flash memory that holds two bits (rather than one) in each memory cell, thus doubling the capacity of memory without a corresponding increase in price. The flash memory has other features such as :

(*i*) High speed
(*ii*) Operating voltage low
(*iii*) Good durability.

13.4.2 The Variable Area

The microcontrollers makes use of four types of internal variable data storages :

- bits,
- registers,
- variable RAM, and
- program counter stack.

All variable storage in microcontroller 8051 are implemented as static random-access memory (SRAM), which will retain the current contents as long as power is applied to it. When power is taken off SRAM lose the contents, therefore SRAMs are referred to as volatile memory.

Now the Fig. 13.10, depicts simplified schematic of MOS static RAM cell.

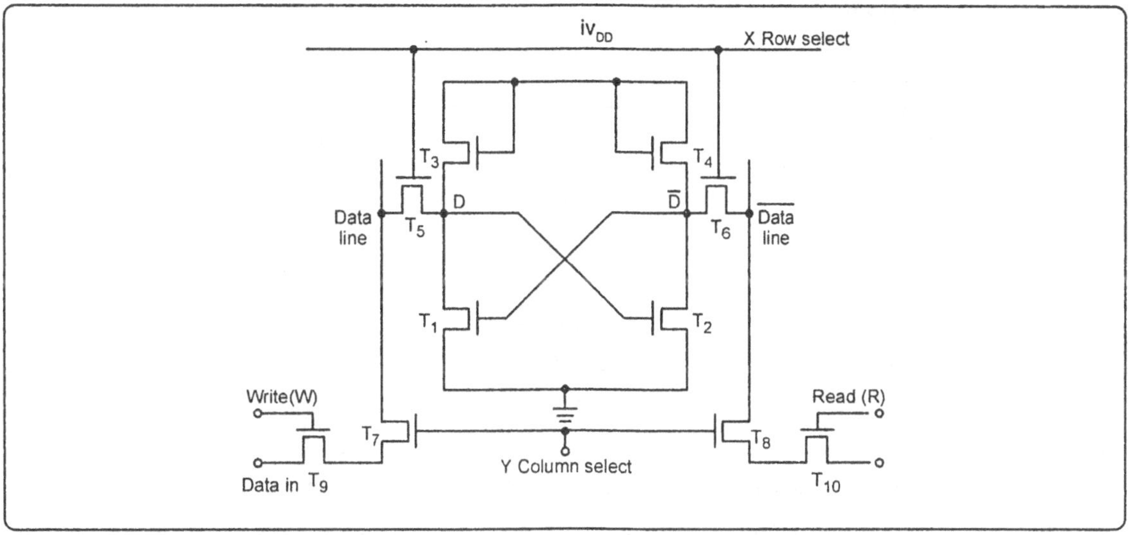

Fig. 13.10 Schematic of MOS static RAM cell.

In Fig. 13.10, the enhancement mode MOSFET transistors are used to make such RAM cell, where T_1 and T_2 form the basic cross coupled inverters and T_3 and T_4 act as load resistors for T_1 and T_2. X and Y lines are used for addressing the cell. When X and Y both are high cell is selected. For X = 1, T_5 and T_6 get ON and the cell is connected to the data and $\overline{\text{Data}}$ line, and for Y = 1, T_7 and T_8 are made ON. Due to this, either read or write operation is possible.

- **Write Operation.** Write operation may be enabled by making W signal high. With write operation enabled, if data-in signal is logic 1, node D is also at logic 1. This turns ON T_2 and T_1 is cutoff. If new data on data-in pin is logic 0, T_2 will be cutoff and T_1 will be turned ON.
- **Read Operation.** Read operation may be enabled by making R signal high. With read operation enabled, T_{10} becomes ON. This connects the data output $\left(\overline{\text{Data}}\right)$ line to the data out and thus the complement of the bit stored in the cell is available at the output.
- **The Program Counter Stack.** The processor stacks are a simple and fast way of saving data during program execution. Stack in microcontroller operates in same as in microprocessor. It operates as last in first out (LIFO) fashion.

13.4.3 Hardware Interface Registers (I/O Space)

There are different ways to implement the hardware registers or I/O space in the microcontrollers according the architecture used. According to Von Neumann architecture the hardware register space can either be a part of memory space or in a separately addressed I/O space, as dipicted in Fig. 13.11 below :

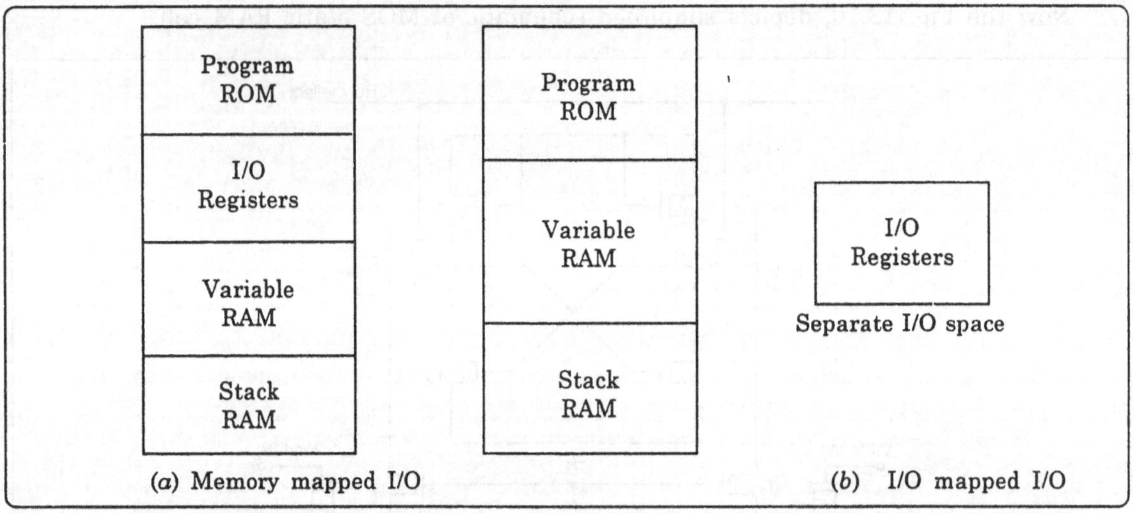

Fig. 13.11 I/O registers as per Princeton Architecture.

The I/O architecture is known as memory mapped I/O, when hardware registers are the part of memory, while the I/O architecture is known as I/O mapped I/O when hardware registers are separately addressed I/O spare as observed in Fig. 13.11 (*a*) and Fig. 13.11 (*b*).

Now as per Harvard architecture I/O registers can be in program ROM in register space or in separate space as depicted in Fig. 13.12

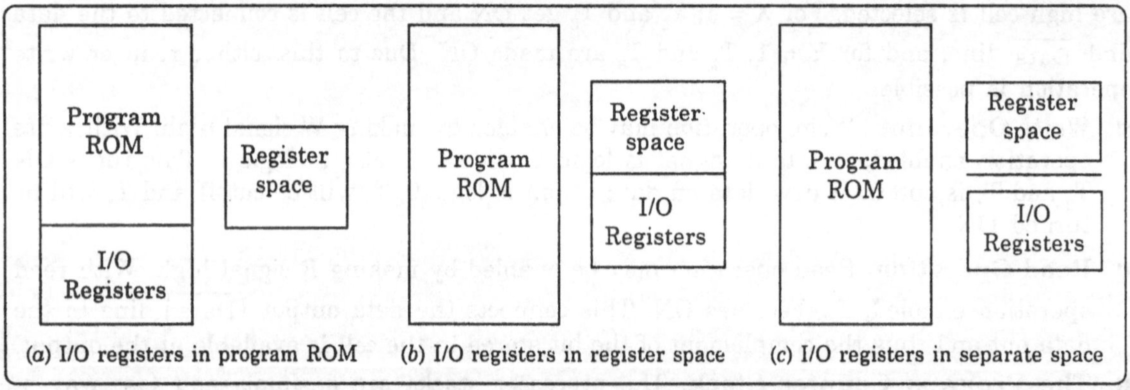

Fig. 13.12 I/O registers as per Harvard Architecture.

In most of the Harvard Architecture microcontrollers, including 8051, the I/O registers are located in the variable memory space.

13.5 MICROCONTROLLER FEATURES

Before using any Microprocessor or Microcontroller chips it is suitable to analyze their features. Now few important features of Microcontrollers to be discussed are :

1. Clocking,
2. I/O Pins,

3. Interupts,
4. Timers, and
5. Peripherals.

13.5.1 Clocking

Almost all microcontrollers provide built-in clock circuitry to allow a simple connection of a crystal or other hardware such as RC and LC circuits to generate a clock signal. But few microcontrollers have an internal ring oscillator which allows it to generate clock signal without any external component other than power.

Similar to microprocessor, in microcontroller each instruction cycle is made up of number of clock cycles. In 8051, each instruction takes 12 clock cycles to execute. Requiring 12 cycles per instruction. At 2 MHz clock, 8051 executes around 150,000 instructions per second. The instuction cycle time varies for high speed microcontrollers. The high speed microcontrollers require 4 clock cycles per instruction cycle. For high speed microcontrollers power requirements are more and they also more likely to be suffered from electrical noise, (hence if speed is not the prime factor then it suggested to operate microcontroller at its slowest possible clock speed).

13.5.2 I/O Pins of Microcontroller

The microcontroller supports built-in I/O ports, where each port pin has a output driver and it is controlled by a single bit in the control register. Many microcontroller I/O pins are dedicated to specialized functions that further complicates them, along with input/ output pins. Such specialized functions include serial ports, analog I/O etc. However in few applications it is required to drive single output using multiple outputs, which can be satisfied by open collector outputs. Now Fig 13.13 depicts the bus which is pulled up with a number of MOS transistors, controlled by control signals, (any of which can pull the line low).

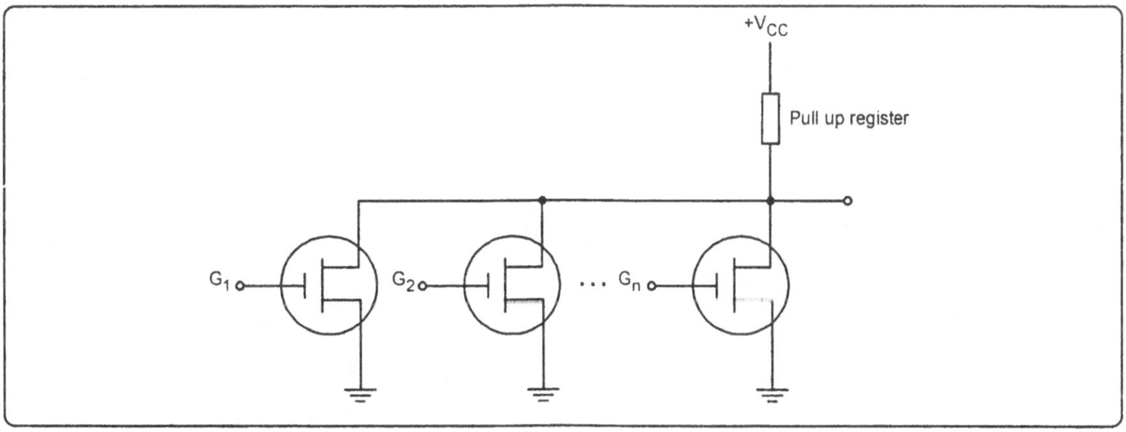

Fig. 13.13 Open collector bus arrangement.

Now while any of the MOS transistors in this circuit are turned on, the bus pin will go low,but when all the MOS transistors are turned off the bus will go high. The bus will only be high when all the transistors are off, or their functions are high, therefore this is known as a wired AND or dotted AND.

MCS-51 family I/O port structure is extremely versatile, which has 32 I/O pins configured as four eight unparallel ports (P_0, P_1, P_2 and P_3). All four ports are bidirectional, *i.e.* each pin will be configured as input or output (or both) under software control, (each port consists of a latch, an output driver, and an input buffer).

The input bufferes of Port 0, and output drives of Port 0 and Port 2 are used in access to external memory. As discussed earlier the Port 0 outputs the low order byte of the external memory address, time multiplexed with the data being written or read, and Port 2 outputs the high order byte of the external memory address when the address is 16 bits wide. Otherwise Port 2 gives the contents of special function register P_2.

All the port-pins of Port 3 are multifunctional, having special functions including two external interrupt, two counter inputs, two special data lines and two timing control strobes. Now the Fig 13.14 depicts the functional diagram of a typical latch and I/O buffers, in each of the four ports.

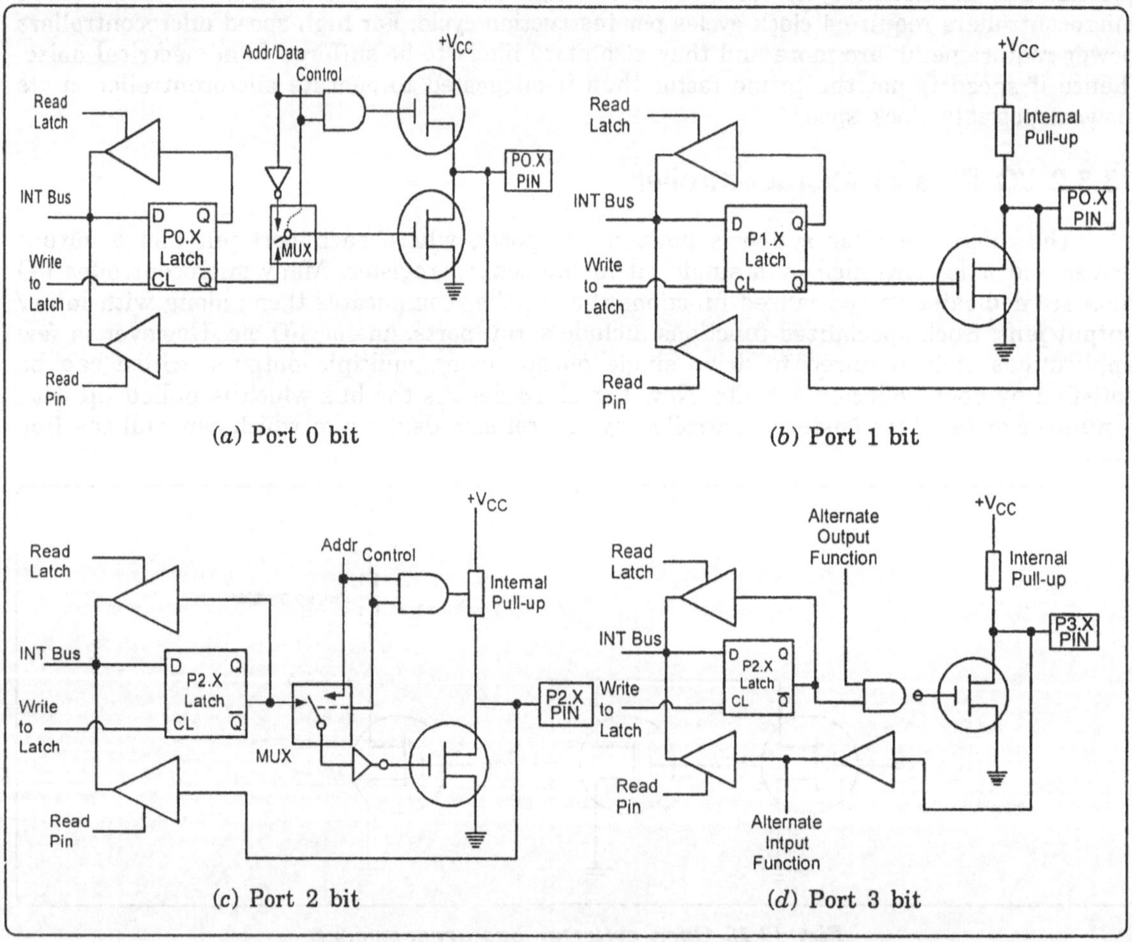

Fig. 13.14 Functional diagram of bit latch and I/O buffers for 8051.

In Fig 13.14, the bit latch is nothing but the one bit in port's SFR, represented as D-FF. The signal 'Write to latch' acts as clock input for D-FF. The data from the internal bus is clock-in,in response to a 'Write-to-latch' signal from the CPU.

The ($\bar{Q}$/Q) Q output after inversion from D flip-flop is connected at the gate input of the drive FET. The ON and OFF state of the drive FET due to the data available at the output of latch decides the status of the output pin. It is possible to read Q output of latch by activating "read latch" signal from the CPU, (the actual port status can be read by activating "read pin" signal, as shown in Fig 13.14)

As observed in Fig 13.14 for Port 0 and Port 2 drivers are switchable to internal ADDR/ DATA and ADDR bus, respectively, by internal CONTROL signal. The switching is required to access external memory. During external memory access, the P2 SFR remains unchanged, but P0 SFR gate 1s written to it. As mentioned earlier, Port 3 has multifunction pins, hence each pin of Port 3 can be programmed to use as I/O or as one of the alternate function. This is achieved by the another control input, "Alternate output function", as shown in Fig 13.14. When latch bit of Port 3 contains 1, the output level is controlled by control input, "alternate output function."

The port pin can be configured as an input by writing 1 in the latch bit of the corresponding pin. It turns OFF the output driver FET. For, Ports 1, 2 and 3, the pin is pulled high by the internal pull-up, but can be pulled low by an external source. There is no internal pull-up for port 0, hence its output pin floats when 1 is written in the latch bit, and pin can be used as a high impedance input. The port 0 is said to be **"true bidirectional"**, because when configured as an input it floats.

Note. The output of Ports 1, 2 and 3 are pulled high with pull-up registers, when configured as an input. Thus, they are sometimes called **"quasi bidirectional"** ports.

13.5.3 Interrupts

About interrupts and its conceps were discussed in Chapter 8 for 8086 microprocessor and in Chapter 12 for 8051 microcontroller. So now we can say that there are three different ways that processor, can handle the interrupt. The first method of handling the interrupt is to ignore it, the second method interrupts are kept pending by setting certain flags and processed them when processor completes its current task. 8051 microcontoller is ideal for implementing such a type of interrupt. The third method of handling an interrupt is to process it as it comes in.

For an interrupt request acknowledged by any processor, the processor will do the following task to serve the interrupt by it.

- It saves the context register information.
- It resets the hardware requesting the interrupt.
- It resets the interrupt controller.
- It processes the interrupt.
- It restores the context information.
- It returns to the previously executing code.

The context registers are the registers used by the interrupt service routine. For instant, if interrupt occurs when processor was about to run the instruction : **subb A, @R1** then the context, registers that would need to be protected by the interrupt service routine before it changed those registers would be as mentioned below :

- The accumulator
- The status register (which saves the carry/borrow flag)
- The contents of "R1", which is pointing to the data to be subtracted from the accumulator.

Now if any of these three registers are changed without being saved and restored, then the result of the instruction will be wrong. The context information is normally saved on the stack as a series of push instructions. Normally, all registers in a processor are saved ; however, in an 8051, the number of registers are large so only those registers that are modified within the interrupt service routine are saved.

Once the context registers are saved, in 8051, interrupt requesting hardware and the interrupt controller are often reset by just the execution of the RETI (return from interrupt) instruction.

Note. Interrupts can occur at different priority levels, in 8051 and most other processors, different interrupt sources have different priorities. A higher priority interrupt will be services before a lower priority one.

13.5.4 Timers

The timers are required by microcontrollers also similar to microprocessors so as to implement real-time systems. Along with providing real-time information the timers also provide a number of other tasks helpful in the operation of the application, the Fig 13.15 depicts the typicl time circuit.

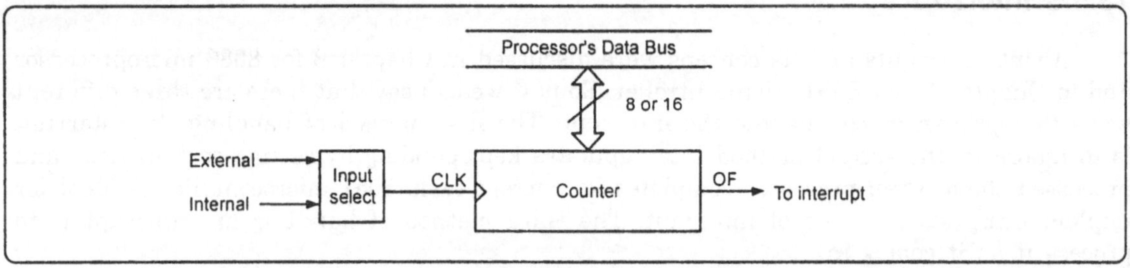

Fig. 13.15 Timer Circuit.

As observed in Fig 13.15 it consists of a counter that can be read from or written to by the processor and is driven by some constant frequency source, (a frequency source can be internal (microcontroller's clock) or an external source). In many cases a counter overflow signal is used as an interrupt request. Normally counter is 8 or 16 bits wide and the value in the counter can be read from or written to during the processor's operation. When the counter overflow occurs, an interrupt request is made of the processor.

Modification of basic timer circuit is easy to generate repeating interrupt at a specific interval, for instant :

(*a*) Pulse width measurement
(*b*) Generation of repeating interrupts
(*c*) Generation PWM
(*d*) Timer/counters in 8051.

(a) Pulse Width M easurement

The timer circuit to measure pulse width is as shown in Fig 13.16, where an external control or gate input is used to mask the clock source except when the pulse is active.

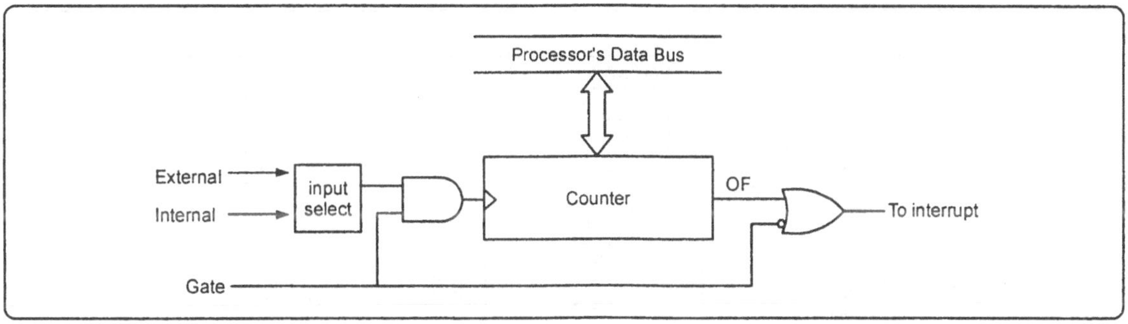

Fig. 13.16 Timer circuit for Pulse meagurement.

Now here when the gate becomes high (the pulse is active), the conter's clock source is allowed to increment the counter. When the pulse becomes inactive (low), the counter's clock source is masked and an interrupt is generated. The generated interrupt allows the application program to read the counter value that is proportional to the width of the pulse. The counter is reset when the pulse becomes active, so the value is correct for each pulse, (the interrupt is also generated to indicate count overflow during the active pulse).

(b) Generation of Repeating Interrupts

To generate the repeating interrupts after a specific time interval, the contents of the counter are reloaded with a set value, where the counter overflow indicates the end of interval. After counter overflow, the contents of the counter are again reloaded with a set value and process is repeated, the arrangements is as shown in Fig 13.17.

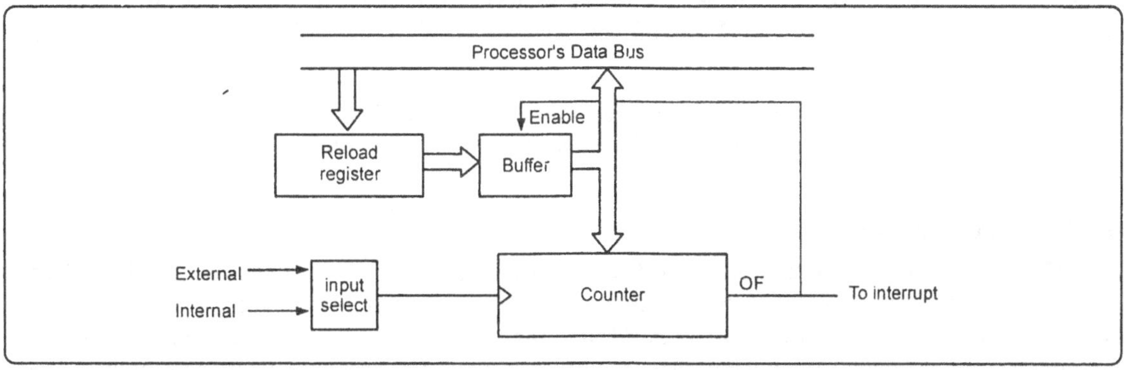

Fig. 13.17 Timer arrangement for Generation of Repeating Interrupts.

The interval between two overflows is given by :

$$\text{Interval} = \frac{[(\text{Max\_Count}) - (\text{Reload\_Count})]}{\text{Frequency of clock source}}$$

Where Max-Count is the maximum count of the counter (*i.e.* overflow value of the counter)

Its value for 8-bit counter will be 255 and for 16-bit counter it will be 65, 535.

(c) Generation of PWM

The circuit for generation of PWM or variable duty cycle signal is as shown in Fig 13.18, where the current value of the counter is compared with the duty cycle value.

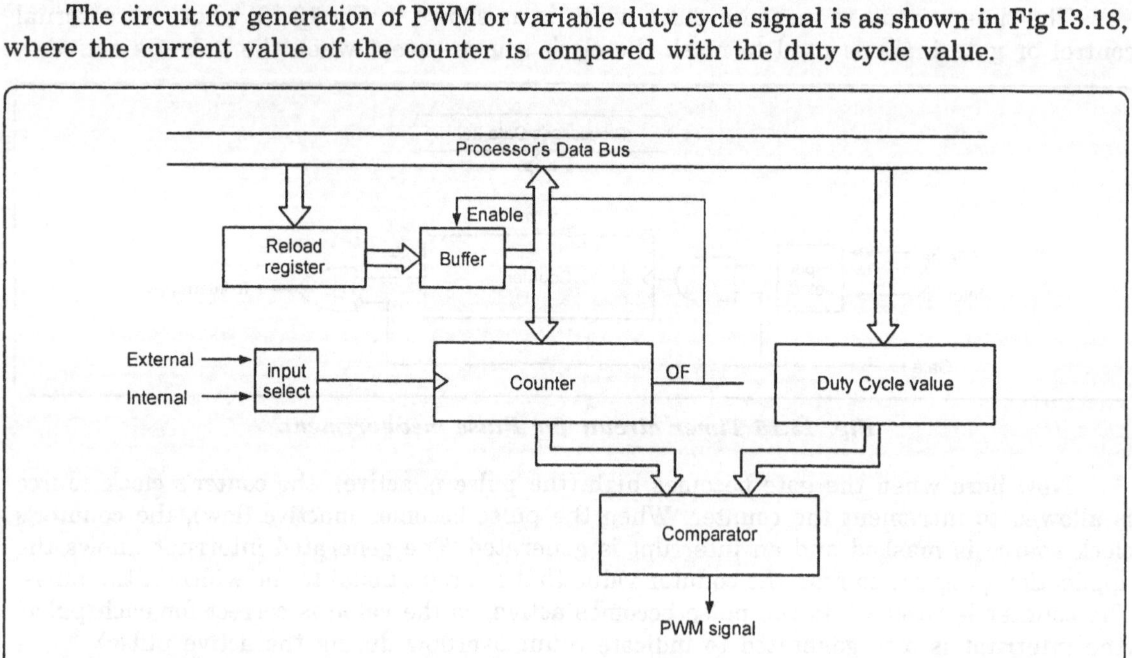

Fig. 13.18 Timer arrangement with PWM signal output.

If the counter value is less than the set value (the duty-cycle value), the comparator will output a high value. When the counter is equal to or greater than this set value, the comparator will output a low value. During successive counter overflows, count to be reloaded is incremented from 0 to set value and decremented from set value to 0 value to get the variable duty cycle, *i.e.*, the PWM waveform, (the PWM signal is often used for controlling the speed of a DC motor).

Note. Timers of 8051 do not support the pulse input measuring or PWM outputs, but does support a number of operating modes that allows us to implement these features in the applications very easily.

(d) Timers/Counters in 8051

8051 has two 16 bit Timers/Counters registers : Timer 0 and Timer 1 where both registers can be independently configured to operate as timers or event counters. When used as a Timer, the register is incremented after each machine cycle, because a machine cycle consists of 12 oscillator periods, the count rate is 1/12 of the oscillator frequency. When used as a Counter, the register in incremented in response to a high to low transition of the corresponding external input pin, T_0 and T_1.

Apart from the "Timer" or "Counter" selection, Timer 0 and Timer 1 have four operating modes. The Timer or Counter operation, and the mode of operation can be configured using TMOD register in the special function registers, as :

- **MODE 0 :** In this mode, the Timer register is configured as a 13-bit register (TL-5 bits and TH-8 bits). The 5-bits in the TL sets the divide by 32 prescale to the TH as an 8-bit counter.

- **MODE 1 :** Mode 1 is the same as the Mode 0, except that the Timer register is being run with all 16-bits.
- **MODE 2 :** In this mode, timer register is configured as an 8-bit counter (TL) with autoload facility. The overflow from TL not only sets timer interrupt flag (TF), but also reloads TL with the contents of TH. The reloading operation does not affect the contents of TH.
- **MODE 3 :** In mode 3, timer 1 simply hold its count, where as Timer 0 registers T_{L0} and T_{H0} are used as two separate 8-bit counters.

13.5.5 Peripherals

The microcontroller consists of a processor, program memory, RAM, I/O pins, interrupts, and timers. Along with these features, many microcontroller manufactures add additional peripherals to create devices suited for particular applications.

Such additional peripherals provide interfaces to microcontroller network devices to simplify wiring applications and not depend on using an external memory interface. These interfaces allow a microcontroller to be connected to digital devices-such as a EEPROM memory, thermometers, analog-to-digital converters (ADCs), etc.–and other microcontrollers. These peripherals are built directly onto the chip to avoid new buses with the microcontroller.

Note. The Microcontrollers with additional peripherals may appear to be more costly than basic microcontrollers and external hardware peripherals, but there are advantages in using an integrated device in terms of costs (the PCB used for the application will be smaller and assemble costs will be less), wiring complexity, power requirements, quality, reliability, performance etc.

REVIEW QUESTIONS

1. Briefly explain the embedded Microcontroller system.
2. Explain the differences between Princeton and Harvard Architecture.
3. List out the characteristics of RISC.
4. List out the comparisons between RISC and CISC.
5. Write short notes on :
   (*a*) EPROM
   (*b*) EEPROM
   (*c*) Flash Memory
6. Explain the following :
   (*a*) Hardware interface Registers
   (*b*) Program Counter stack
   (*c*) Peripherals.

14 8051 ARCHITECTURE

14.0 OBJECTIVE

At the end of this chapter the reader will be able to know the features of 8051, its architecture, memory organisation, interrupts and 8051's instruction execution.

14.1 INTRODUCTION

The microcontrollers are well suited for specific applications, unlike microprocessor. The microcontroller was called microcomputer earlier. Such microcontroller is making great impact on every activity of the mankind and is playing and expected to play very important role in the daily functioning of the developed and developing societies. A single chip microcontroller contains all essential elements of a microcomputer on a single chip. The introduction to microcontroller family is explained in chapter 12, however now here let us deal with 8051/MCS51 microcontroller's architecture, and its features.

14.2 FEATURES OF 8051

The major features of 8051 microcontroller or MCS51 are as listed below :

- 8-bit CPU.
- Single +5 volts operating supply.
- On-chip oscillator and clock circuitry.
- 128 bytes data memory on-chip.
- Four register bank.
- 32 bidirectional I/O lines.
- 64 K address space for external data memory.
- 64 K address space for external program memory.
- Two 16-bit timer/counters (three on 8032/8052)
- A five-source interrupt structure (six sources on 8032/8052) with two priority levels.
- Full duplex serial port.
- Boolean processor.

All three microcontroller devices 8031, 8051 and 8751 are available in a standard 40-pin Dual In-Line Package with the same pin-out, the same timing and the same electrical characteristics. The primary difference between the three is the on-chip program memory-different types are offered to satisfy different user requirements. The Fig 14.1 depicts the pin diagram of 8051.

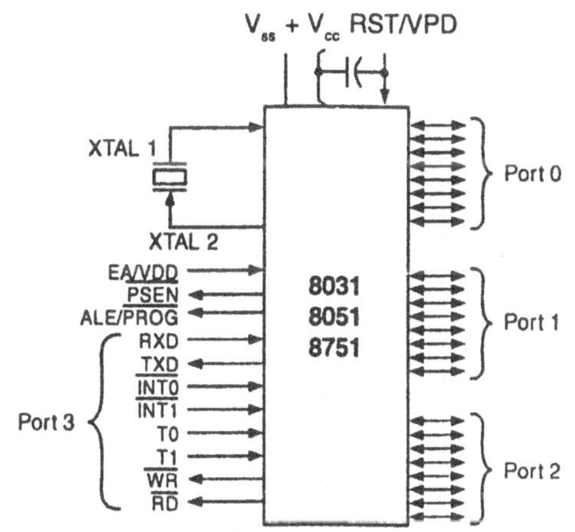

Fig. 14.1 Pin diagram of 8051.

The pin diagram shown in Fig 14.1 for all the three microcontrollers is same, however few difference among 8751, 8051 and 8031 are :

- 8751 provides 4 Kbytes of Ultraviolet-Erasable Programmable Read Only Memory (EPROM) for program development and prototyping. As we know, in development and prototyping stages due to change in design specification or software bugs, it is necessary to change or erase the software. With EPROM it is a matter of few minutes, by exposure to unlraviolet light and reprogrammed with the modified code.
- 8051 has 4 Kbytes of ROM which is programmed by manufacturer when the chip is built. This part is considerably less expensive, but cannot be erased or altered after fabrication, hence it is used to store final version of the program.
- 8031 does not have any program memory on chip, but may be used with upto 64 Kbytes of external standard or multiplexed ROMs PROMs or EPROMs. The 8031 fit well in applications requiring significantly larger or smaller amounts of memory that the 4 Kbytes provided by 8051 and 8751.

Apart 8051 and 8751 can automatically access external program memory for all addresses greater than 4096 bytes on chip.

14.3 ARCHITECTURE OF 8051

The Fig 14.2 depicts the architectural block diagram of 8051/MCS-51 Chip.

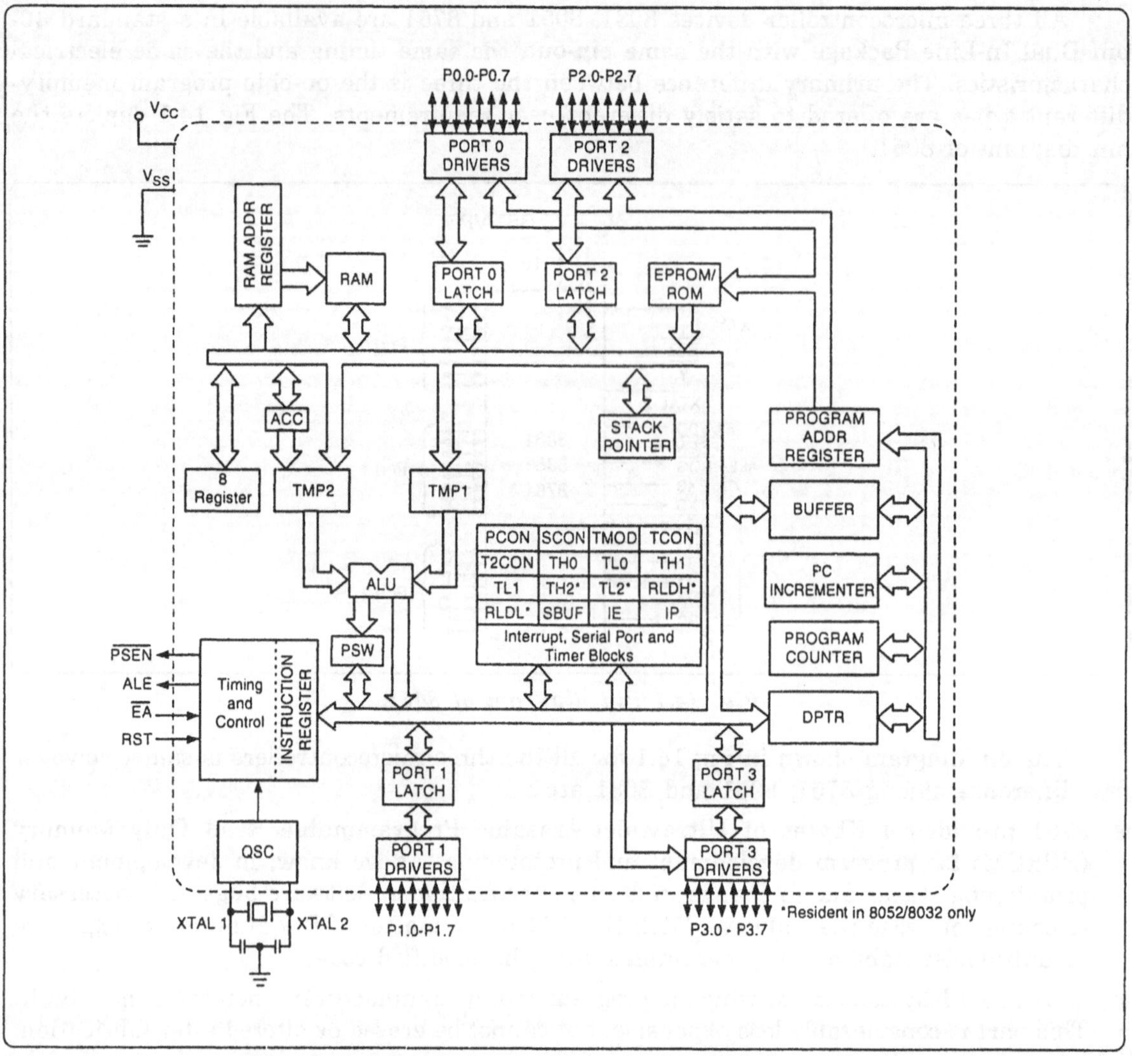

Fig. 14.2 Architectural Block Diagram of MCS-51/8051.

The architectural block diargam of Fig 14.2 consists of a CPU (Central Processing Unit) two kinds of memories (data RAM plus program ROM/EPROM) Input/Output ports, the mode status, and data registers and random logic needed for a variety of peripheral functions. These elements communicate through an eight bit data bus which runs throughout the chip referred to as internal data bus, which is buffered to the outside world through an I/O port when memory or I/O expansion is desired. The Table 1 lists few commercial grade members of 8051.

Table 1. 8051 Family Members

PART	Technology	ON-Chip Program Memory	ON-Chip Data Memory
8051 AH	HMOS II	4K-ROM	128
8031 AH	HMOS II	None	128
8751 H	HMOS I	4K-EPROM	128

14.3.1 CPU

MCS-51/8051 consists of eight bit Arithmetic and logic unit with associated registers like A, B, PSW, SP, the sixteen bit program counter and DPTR register, (Data pointer).

CPU Timing

A machine cycle consists of 6 states (12 oscillator periods). Each state is divided into a Phase 1 half, during which the Phase 1 clock is active, and a Phase 2 half, during which the Phase 2 clock is active. Thus, a machine cycle consists of 12 oscillator periods, numbered S1P1 (State1, Phase 1), through S6P2 (State 6, Phase 2). Each phase lasts for one oscillator period. Each state lasts for two oscillator periods. Typically, arithmetic and logical operations take place during Phase 1 and internal register-to-register transfers take place during Phase 2. The timing diagram shown in Fig 14.3 shows the fetch/execute timing referred to the internal states and phases.

However since these internal clock signals are not user accessible, the XTAL2 oscillator signal and the ALE (Address Latch Enable) signal are shown for external reference. ALE is normally activated twice during each machine cycle : once during S1P2 and S2P1, and again during S4P2 and S5P1.

Execution of a one-cycle instruction begins at S1P2 when the opcode is latched into the Instruction Register. If it is a two-byte instruction, the second byte is read during S4 of the same machine cycle. If it is a one-byte instruction, there is still a fetch at S4, but the byte read (which would be the next opcode), is ignored, and the Program Counter is not incremented. In any case, execution is complete at the end of S6P2. Figures 14.3A and 14.3B show the timing for a 1-byte, 1-cycle instruction and for a 2-byte, 1-cycle instruction.

Most 8051 instructions execute in one cycle. MUL (multiply) and DIV (divide) are the only instruction that take more than two cycles to complete, they take four cycles.

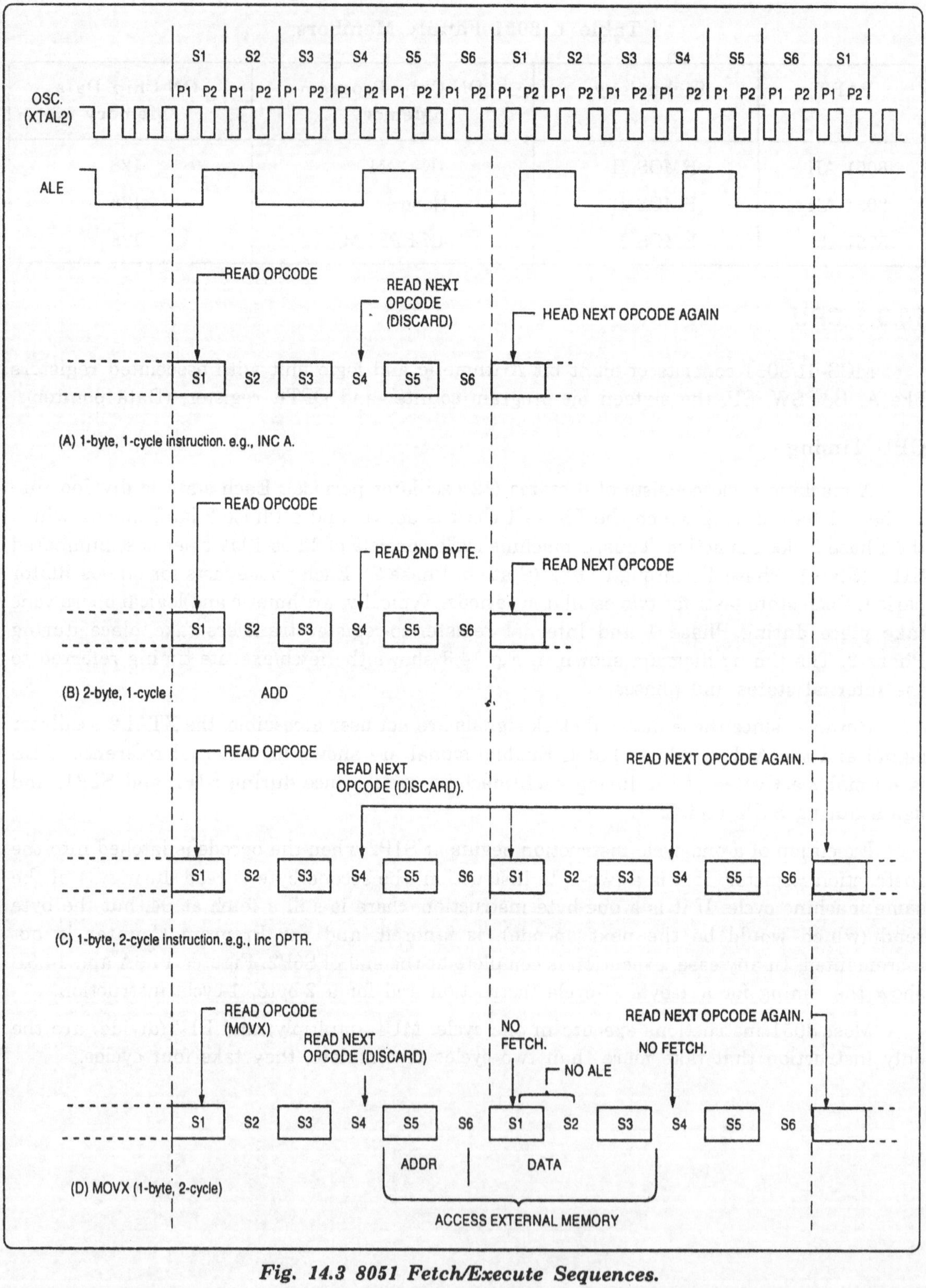

Fig. 14.3 8051 Fetch/Execute Sequences.

14.3.2 ALU

The ALU in 8051 performs arithmetic and logic functions on 8-bit variables. The arithmetic unit can perform addition, subtraction, multiplication and division and the logic unit can perform logical operations such as AND, OR and Exclusive-OR, as well as rotate, clear and complement, apart ALU looks after the branching decisions. An important and unique feature of the 8051 architecture is that the ALU can also manipulate one bit as well as eight-bit data types. Individual bits may be set, cleared, complemented, moved, tested, and used in logic computation.

14.3.3. Accumulator

The accumulator is also written as ACC or simply A is an 8-bit register. ACC holds a source operand and receives the result of the arithmetic instructions (addition, subtraction, multiplication, and division). The accumulator can be the source or destination for logical operations and a number of special data movement instructions, including look-up tables and external RAM expansion. Several functions apply exclusively to the accumulator : rotate, parity computation, testing for zero, and so on. Apart from the accumulator, B register is available, which is used during multiply and divide operations. For other instructions it can be treated as another scratch pad register.

14.3.4 Program Status Word (PSW)

The PSW register contains program status information, which is of 8-bit word. The Fig 14.4 illustrates the detailed PSW.

(MSB)							(LSB)
CY	AC	F0.	RS1	RS0	OV	—	P

Symbol	Position	Name and Significance
CY	PSW.7	Carry flag. Set/cleared by hardware or software during certain arithmetic and logical Instructions.
AC	PSW.6	Auxillary Carry flag. Set/cleared by hardware during addition or subtraction instructions to indicate carry or borrow out of bit 3.
F0	PSW.5	Flag 0 Set/cleared/tested by software as a user-define, status flag.
RS1	PSW.4	Register bank Select control bits 1 and 0. Set/cleared by software to determine working register bank (see Note)
RS0	PSW.3	
OV	PSW.2	Overflow flag. Set/cleared by hardware during arithmetic instructions to indicate overflow conditions.
—	PSW.1	(reserved)
P	PSW.0	Parity flag. Set/cleared by hardware each Instruction cycle to Indicate an odd/even number of "one" bits in the accumulator, *i.e.*, even parity.

Note. the contents of (RS1, RS0) enable the working register banks as follows:

(0.0)-Bank0	(00H-07H)
(0.1)-Bank1	(08H-0FH)
(1.0)-Bank2	(10H-17H)
(1.1)-Bank3	(18H-1FH)

Fig. 14.4 Program Status Word Register.

14.3.5 Other CPU Registers of MCS-51

The MCS-51 has two more CPU Registers (*a*) Stack Pointer and (*b*) Data Pointer (DPTR).

(*a*) **STACK POINTER**

The Stack Pointer register is 8 bits wide. It is incremented before data is stored during PUSH and CALL executions. While the stack may reside anywhere in onchip RAM, the Stack Pointer is initialized to 07H after reset. This causes the stack to begin at location 08H.

(*b*) **DATA POINTER**

The Data pointer (DPTR) consists of a high byte (DPH) and a low byte (DPL). Its function is to hold a 16-bit address. It may be manipulated as a 16-bit register or as two independent 8-bit registers.

RAM Byte	(MSB)							(LSB)	
7FH									127
2FH	7F	7E	7D	7C	7B	7A	79	78	47
2EH	77	76	75	74	73	72	71	70	46
2DH	6F	6E	6D	6C	6B	6A	69	68	45
2CH	67	66	65	64	63	62	61	60	44
2BH	5F	5E	5D	5C	5B	5A	59	58	43
2AH	57	56	55	54	53	52	51	50	42
29H	4F	4E	4D	4C	4B	4A	49	48	41
28H	47	46	45	44	43	42	41	40	40
27H	3F	3E	3D	3C	3B	3A	39	38	39
26H	37	36	35	34	33	32	31	30	38
25H	2F	2E	2D	2C	2B	2A	29	28	37
24H	27	26	25	24	23	22	21	20	36
23H	1F	1E	1D	1C	1B	1A	19	18	35
22H	17	16	15	14	13	12	11	10	34
21H	0F	0E	0D	0C	0B	0A	09	08	33
20H	07	06	05	04	03	02	01	00	32
1FH	Bank 3								31
16H									24
17H	Bank 2								23
10H									16
0FH	Bank 1								15
08H									8
07H	Bank 0								7
00H									0

Fig 14.5 (a) Special Function Bit Addressable Locations.

Direct Byte Address	(MSB)	Bit Addresses						(LSB)	Hardware Register Symbol
F0 240	F7	F6	F5	F4	F3	F2	F1	F0	8
E0 224	E7	E6	E5	E4	E3	E1	E1	E0	ACC
	CY	AC	F0	R81	R80	OV		P	
DC 204	D7	D6	D5	D4	D3	D2	D1	D0	P&W
	TF2	EXF2	RCLK	TCLK	EXEN2	TR2	C/$\overline{T2}$	CP/$\overline{RL2}$	
C3 200	CF	CE	CD	CC	CB	CA	C9	C8	T2CON
			PT2	P8	PT1	PX1	PT0	PX0	
B8 164	—	—	BD	BC	BB	BA	B9	B8	IP
B0 176	B7	B6	B5	B4	B3	B2	B1	B0	P3
	EA		ET2	ES	ET1	EX1	ET0	EX0	
A8 168	AF	—	AD	AC	AB	AA	A9	A8	IE
A0 160	A7	A6	A5	A4	A3	A2	A1	A0	P2
	SM0	SM1	SM2	REN	TS4	R04	TI	R1	
93 162	OF	OE	OD	OC	OB	OA	O9	O8	SCON
90 144	07	96	95	94	93	92	91	90	P1
	TF1	TR1	TF0	TRO	IE1	IT1	IE0	IT0	
88 138	8F	8E	8D	8C	8B	8A	89	88	TCON
80 128	87	86	85	84	83	82	81	80	P0

Fig 14.5 (b) Special Function Register Bit Address.

14.3.6 Special Function Register (SFR)

8051 uses memory mapped I/O through a set of special function registers that are implemented in the address space immediately above the 128 bytes of RAM. Fig 14.5 illustrates special function bit address and special function Register bit addresses.

In Fig 14.5 all access to the four I/O ports, the CPU registers, interrupt-control registers, the timer/counter, UART, and power control are performed through registers between 80H and FFH.

14.3.7 RAM

The RAM consists of 4 registers banks and 144 general purpose bits forming Boolean Processor RAM. The Fig 14.6 depicts RAM bit address.

RAM Byte	(MSB)							(LSB)
7FH								
2FH	7F	7E	7D	7C	7B	7A	79	78
2EH	77	76	75	74	73	72	71	70
2DH	6F	6E	6D	6C	6B	6A	69	68
2CH	67	66	65	64	63	62	61	60
2BH	5F	5E	5D	5C	5B	5A	59	58
2AH	57	56	55	54	53	52	51	50
29H	4F	4E	4D	4C	4B	4A	49	48
28H	47	46	45	44	43	42	41	40
27H	3F	3E	3D	3C	3B	3A	39	38
26H	37	36	35	34	33	32	31	30
25H	2F	2E	2D	2C	2B	2A	29	28
24H	27	26	25	24	23	22	21	20
23H	1F	1E	1D	1C	1B	1A	19	18
22H	17	16	15	14	13	12	11	10
21H	0F	0E	0D	0C	0B	0A	09	08
20H	07	06	05	04	03	02	01	00
1FH–18H	Bank 3							
17H–10H	Bank 2							
0FH–08H	Bank 1							
07H–00	Bank 0							

Fig. 14.6 RAM Bit Address.

In Fig 14.6 the register bank-select bits refer to four 8 byte regions at the bottom of memory. Certain instructions can access these registers in RAM directly with 1 byte instructions, since 3 bits of the instruction specify the desired register. The RS_0 and RS_1 bits in PSW must be set to the appropriate bank before execution of the instruction. This method of addressing saves one byte per instruction and the 144 general purpose bits can be used as software flags or program variables.

14.3.8 I/O Ports

MCS-51 family I/O port structure is extremely versatile, it has 32 I/O pins configured as four eight-bit parallel ports (P_0,P_1,P_2 and P_3). All four ports are bidirectional, *i.e.*, each pin will be configured as input or output (or both) under software control. Each port consists of a latch, an output driver, and an input buffer.

The output drives of Ports 0 and 2 and the input buffers of Port 0, are used to access external memory. As mentioned earlier, Port 0 outputs the low order byte to the external memory address, time multiplexed with the data being written or read, and Port 2 outputs the high order byte of the external memory address when the address is 16 bits wide, else Port 2 gives the contents of special function register P_2. Now the Fig 14.7 depicts the alternate special functions of Port 3.

(MSB) (LSB)

$\overline{RD}$	$\overline{WR}$	T_1	T_0	$\overline{INT1}$	$\overline{INT0}$	TXD	RXD

Symbol	Position	Name and Significance
$\overline{RD}$	P3.7	Read data control output. Active low pulse, generated by hardware when external data memory is read.
$\overline{WR}$	P3.6	Write data control output. Active low pulse generated by hardware when external data memory is written.
T_1	P3.5	Timer/counter 1 external input or test pin.
T_0	P3.4	Timer/counter 0 external input or test pin.
$\overline{INT1}$	P3.3	Interrupt 1 input pin. Low-level or falling-edge triggered.
$\overline{INT0}$	P3.2	Interrupt 0 input pin. Low-level or falling-edge triggered.
TXD	P3.1	Transmit Data pin for serial port in UART mode. Clock output in shift register mode.
RXD	P3.0	Receive Data pin for serial port in UART mode. Data I/O pin in shift register mode.

Fig. 14.7 Alternate special functions of port 3.

In Fig 14.7 all port pins of Port 3 are multifuntional and have special functions including two external interrupts, two counter inputs, two special data lines and two timing control strobes.

14.3.9 Special Peripheral Functions

A microcontroller or microcomputer system has to fulfil certain common needs so as to make the system more versatile, such as.

(*a*) Keeping track of elasped real-time.

(*b*) Maintaining a count of signal transitions.

(*c*) Measuring the precise width of input pulses.

(*d*) Communicating with ohter systems or people.

(*e*) Closely monitoring asynchronous external events.

The microprocessor systems require peripheral chips such as timer, counters, USARTs etc so as to meet, above said needs. However MCS-51 family integrates such capabilities on-chip like :

(*a*) Timer/Counter.

(*b*) Serial Interface.

(*a*) Timer/Counter

MCS-51/8051 has two 16 bit Timer/Counter registers-Timer 0 and Timer 1, which can be independently configured to operate as timer or counters.

- When used as a Timer, the register is incremented after every machine cycle. Since a machine cycle consists of 12 oscillator periods, the count rate is 1/12 of the oscillator frequency.
- When used as a Counter, the register in incremented in response to a high to low transition of the corresponding external input pin T_0 and T_1.

Apart from the "Timer" or "Counter" selection, Timer 0 and Timer 1 have four operating modes. The Timer of Counter operation, and the mode of operation can be configured using TMOD register in the special function registers. The bit pattern for Timer/Counter Mode Control Register (TMOD) is as illustrated in Fig 14.8.

(MSB) (LSB)

GATE	C/T	M1	M0	GATE	C/T	M1	M0
Timer 0				Timer 1			

GATE Gating control when set. Timer/Counter "X" is enabled only while "$\overline{INTx}$" pin is high and "TRx" control bit is set. When cleared Timer "x" is enabled whenever "TRx" control bit is set.

$C/\overline{T}$ Timer or Counter selector cleared for Timer operation (input from internal system clock). Set for Counter operation (input from "Tx" input pin).

M1	M0	Operating Mode
0	0	8-bit Timer/Counter "THx" with "TLx"s 5-bit prescaler.
0	1	16-bit Timer/Counter "THx" with "TLx" are cascaded ; there is no prescaler.
1	0	8-bit auto-reload Timer/Counter "THx" holds a value which is to be reloaded into "TLx" each time it overflows.
1	1	(Timer 0) TL0 is an 8-bit Timer/Counter controlled by the standard Timer 0 control bits. TH0 is an 8-bit timer only controlled by Timer 1 control bits.
	1	1(Timer 1) Timer/Counter 1 stopped.

Fig. 14.8 Timer/Counter mode control register (TMOD).

Here the control bit $C/\overline{T}$ acts as a Timer or Counter selector. When this bit is 0 register works as a timer ; otherwise it will operate as a counter. Bits M_0 and M_1 decide the mode of operation. Amongst the four operating modes, first three modes, Mode 0, 1 and 2 are same for both Timers/Counters, where Mode 3 is different.

■ Mode 0

In this mode, the Timer register is configured as a 13-bit register (TL-5 bits and TH-8 bits), the 5-bits in the TL sets the divide by 32 prescale to the TH as an 8-bit counter as observed in Fig 14.9.

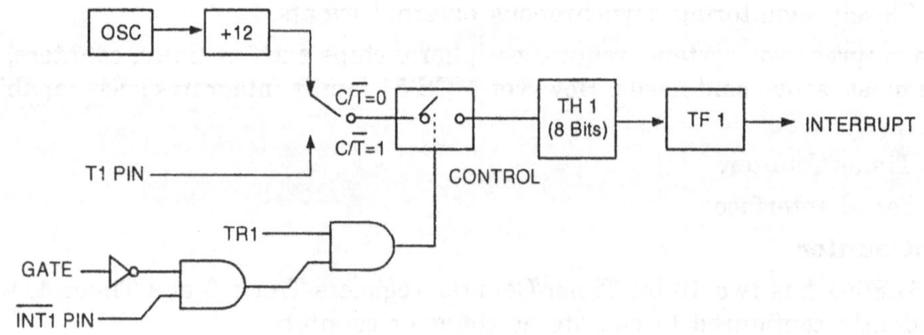

Fig. 14.9 Timer/Counter 1-mode 0 : 13-bit Counter.

The Timer is enabled when TR = 1 and Gate = 0 or $\overline{INT}$ = 1. Setting GATE = 1 allows the Timer to be controlled by external input $\overline{INT}$, to facilitate pulse width measurements. TR is control bit in the special function Register TCON, as observed in Fig 14.10.

(MSB) (LSB)

TF1	TR1	TF0	TR0	IE1	IT1	IE0	IT0

Symbol	Position	Name and Significance
TF1	TCON.7	Timer 1 overflow Flag. Set by hardware on timer/counter overflow. Cleared when interrupt processed.
TR1	TCON.6	Timer 1 Run control bit. Set/cleared by software to turn timer/counter on/off.
TF0	TCON.5	Timer 0 Overflow Flag. Set by hardware on timer/counter overflow. Cleared when interrupt processed.
TR0	TCON.4	Timer 0 Run control bit. Set/cleared by software to turn timer/counter on/off.
IE1	TCON.3	Interrupt 1 Edge Flag. Set by hardware when external interrupt edge detected. Cleared when interrupt processed.
IT1	TCON.2	Interrupt 1 Type control bit. Set/cleared by software to specify falling edge/low level triggered external interrupts.
IE0	TCON.1	Interrupt 0 Edge Flag. Set by hardware when external interrupt edge detected. Cleared when interrupt processed.
IT0	TCON.0	Interrupt 0 Type control bit. Set/cleared by software to specify falling edge/low level triggered external interrupts.

Fig. 14.10 TCON-timer/counter control/status register.

As observed in Fig 14.10 the timer interrupt flag TF is set as soon as count rolls over from all 1s to all 0s.

- **Mode 1**

  The Mode 1 is just simillar to that of Mode 0, except that the Timer register is being run with all 16-bits, here.

- **Mode 2**

  In this mode, timer register is configured as an 8-bit counter (TL) with autoload facility, as shown in Fig 14.11, below.

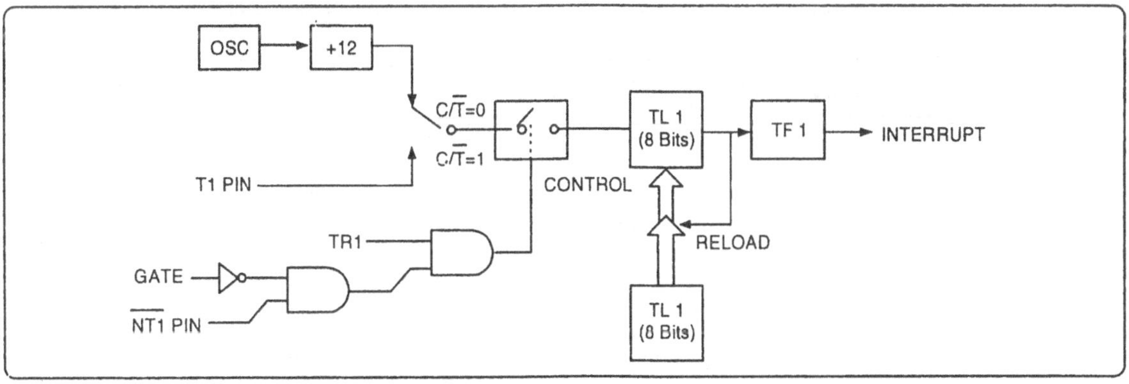

Fig. 14.11 Timer/Counter 1 in Mode 2 : 8-bit auto reload.

Here the overflow from TL not only sets timer interrupt flag (TF), but also reloads TL with the contents of TH. The reloading operation does not affect the contents of TH.

- **Mode 3**

  In this mode, Timer 1 simply hold its count, whereas timer 0 registers T_{L0} and T_{H0} are used as Two separate 8-bit counters, as shown in Fig 14.12 below :

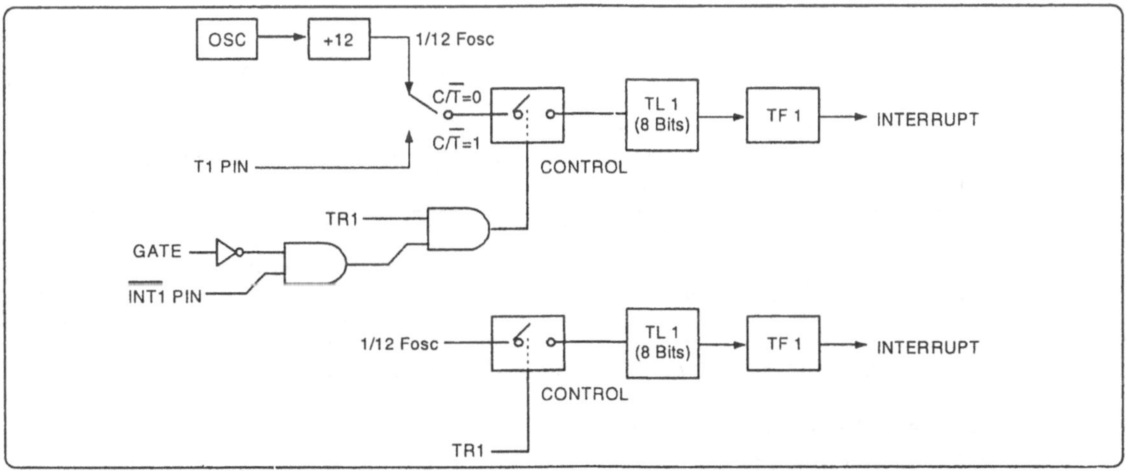

Fig. 14.12 Timer/Counter 0 in Mode 3 : two 8-bit counters.

As observed in Fig 14.12 the T_{L0} uses the Timer 0 control bits : $C/\overline{T}$ **Gate,** TR0, $\overline{INT0}$, and TF0. TH0 counts machine cycles and takes over the use of TR1 and TF1 from Timer1.

(b) Serial Interface

Full-duplex asynchronous serial interface is possible to implement on chip, using serial interface control circuitry. The special feature of this circuit is a double buffered receiver, due to which it can receive a second character while the first one is being held in an intermediate register until the software fetches it. This double buffering reduces the time complexity of servicing a serial interface and reduces the chances of an overrun error. The serial interface can be operated in four different modes, which can be configured using serial port control/status register. The Fig. 14.13 below depicts the Bit pattern for serial port Control/status register.

(MSB)	1	0	1				(LSB)
SM0	SM1	SM2	REN	TB8	RB8	TI	RI

Symbol	Position	Name and Significance
SM0	SCON.7	Serial port Mode control bit 0. Set/cleared by software (see note).
SM1	SCON.6	Serial port Mode control bit 1. Set/cleared by software (see note).
SM2	SCON.5	Serial port Mode control bit 2. Set by software to disable reception of frames for which bit 8 is zero.
REN	SCON.4	Receiver Enable control bit. Set/Cleared by software to enable/disable serial data reception.
TB8	SCON.3	Transmit Bit 8. Set/Cleared hardware to determine state of ninth data bit transmitted in 9-bit UART mode.
RB8	SCON.2	Receive Bit 8. Set/cleared by hardware to indicate state of ninth data bit received.
TI	SCON.1	Transmit Interrupt flag. Set by hardware when byte transmitted. Cleared by software after servicing.
RI	SCON.0	Receive Interrupt flag. Set by hardware when byte received. Cleared by software after servicing.

Fig. 14.13 Serial port control/status register (SCON).

Note. The state of (SM0, SM1) selects :

(0,0)-Shift register I/O expansion.

(0,1)-8-bit UART, variable data rate.

(1,0)-9-bit UART, fixed data rate.

(1,1)-9-bit UART, variable data rate.

Now the four Modes are briefly discussed below :

- **Mode 0** : It is half duplex mode. The serial data is received and transmitted through the RXD line. Shift clock is output through the TXD line.The baud rate is fixed. It is one twelfth of the oscillator frequency. The start and stop bits are not required, since only eight bits are transferred at a time.
- **Mode 1** : It is full duplex mode. Here 1 is transmitted when there is change in data, otherwise zero is transmitted. This is referred as non return to zero (NRZ) operation. In this mode 10 bits (1-start bit, 8-data bits, LSB first, and 1-stop bit) are transmitted at a time. The baud rate is variable.
- **Mode 2** : It is full duplex with 11 bits (1-start bit, 8 data bits, a programmable 9th bit and 1-stop bit) transmission in NRZ mode. The baud rate is programmable which is either 1/32 or 1/64 of the microcontroller clock frequency.

Note. On transmit, the 9th data bit (TB8 in SCON) can be assigned the value of 0 or 1 as a parity bit. On receive, the 9th data bit goes into RB8 in special function register (SCON).

- **Mode 3** : It is same as Mode 2 with a variable baud rate, (it is determined by timer1 overflow rate.)

14.4 MEMORY ORGANIZATION

The MCS-51 architecture provides on-chip memory as well as off-chip memory expansion capabilities. Several addressing mechanisms are incorporated to allow for an optimal instruction set.

The 8051 has three basic memory address spaces :

- 64K-byte Program Memory ;
- 64K-byte External Data Program Memory ; and
- 256-byte (384 bytes for the 8032/8052) Internal Data Memory.

Now Fig 14.14 illustrates MCS-51 memory maps.

14.4.1 Program Memory Address Space

The 64K-byte Program Memory space consists of an internal and an external memory portion. If the $\overline{EA}$ pin is held high, the 8051 executes out of internal program memory unless the address exceeds 0FFFH (1FFFH for the 8052). Locations 1000H through 0FFFFH (2000H through 0FFFFH for the 8052), are then letched from external Program memory. If the $\overline{EA}$ pin is held low, the 8051 fetches all instructions from external Program Memory. In either case, the 16-bit Program Counter is the addressing mechanism.

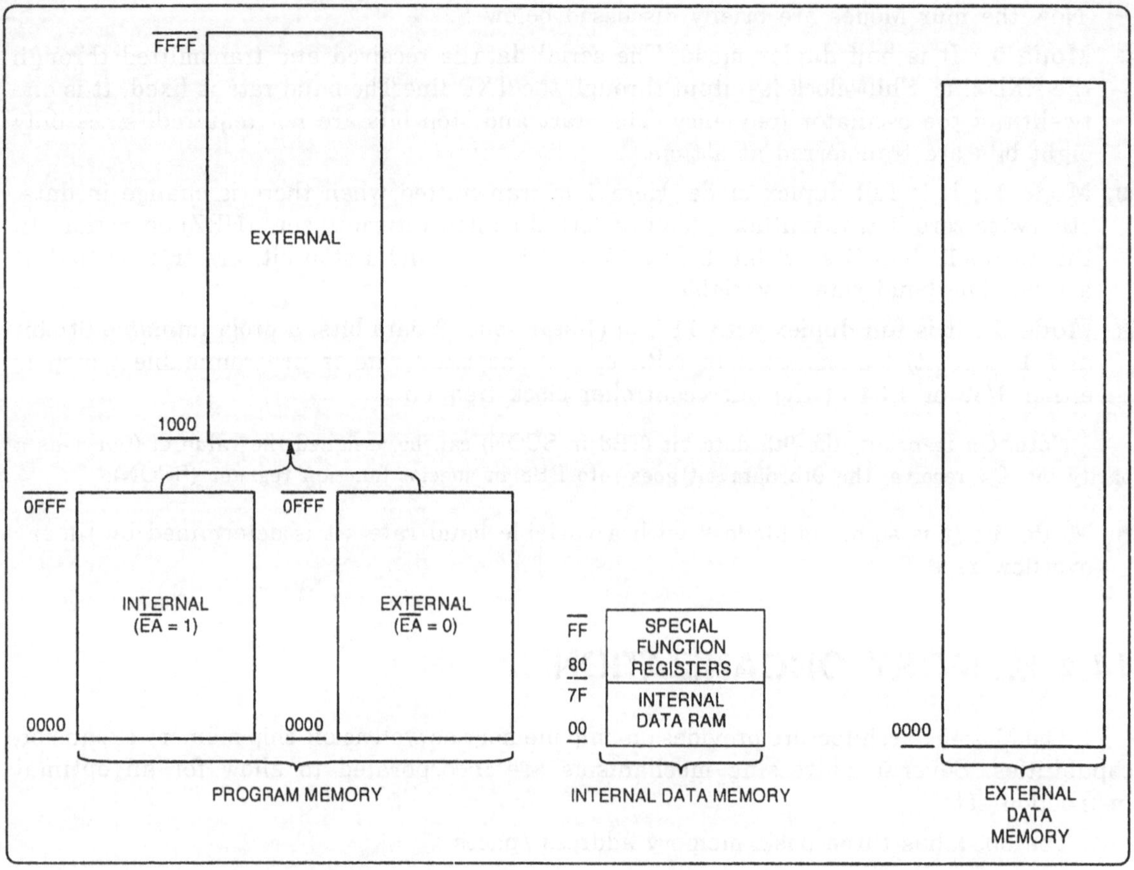

Fig. 14.14 MCS-51 MEMORY MAP

The locations 00 through 42H (00 through 2BH for the 8032/8052) in Program Memory are reserved for interrupt service routines as indicated in Table 2, below.

Table 2. Addressing Method and Associated Memory Spaces

Register Addressing

— R0-R7

— ACC, B, CY(bit), DPTR

Direct Addressing

— Lower 128 bytes of internal RAM

— Special Function Registers

Register Indirect Addressing

— Internal RAM (@R1, @R0, SP)

— External Data Memory (@R1, R0, @DPTR)

Immediate Addressing

— Program Memory

Base-Register plus Index-Register Indirect Addressing

— Program Memory (@DPTR +A, @PC +A)

14.4.2 Data Memory Address Space

The Data Memory address space consists of an internal and an external memory space. External Data Memory accessed when a MOVX instruction is executed.

Internal Data Memory is divided into three physically separate and distinct blocks : the lower 128 bytes of RAM ; the upper 128 bytes of RAM (accessable in the 8032/8052 only); and the 128-byte Special Function Register (SFR) area. While the upper RAM area and the SFR area share the same address locations, (they are accessed through different addressing modes). Now the Fig 14.15 illustrates mapping of Internal Data Memory.

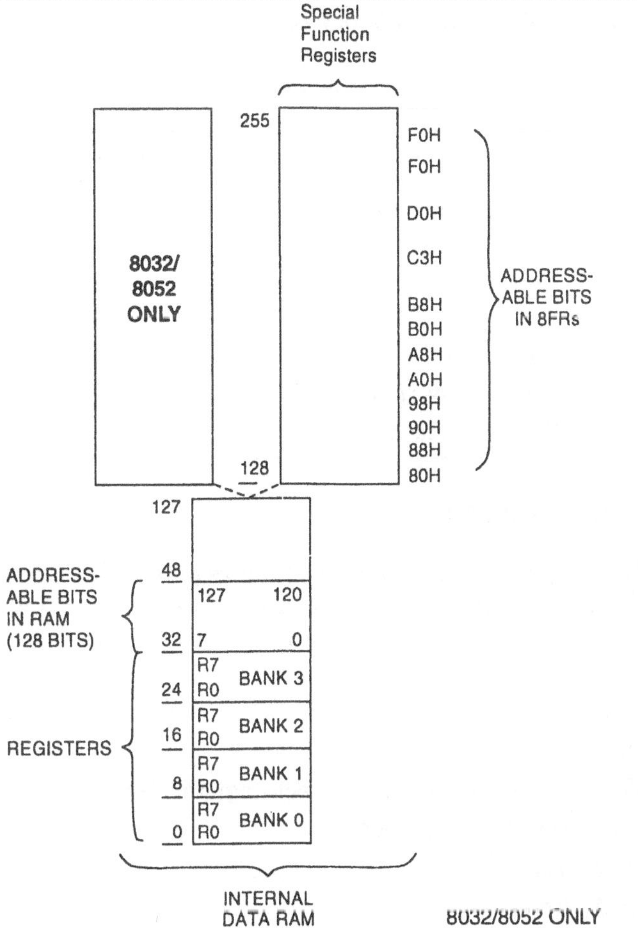

Fig. 14.15 Internal Data Memory Address Space.

As observed in Fig 14.15, the four 8-Register Banks occupy locations 0 through 31 in the lower RAM area. Only one of these banks may be enabled at a time (through a two-bit field in the PSW). The next sixteen bytes, locations 32 through 47 contain 128 bit addressable locations. The RAM bit addresses and SFR bit address are already shown in Fig 14.5 (*a*) and 14.5 (*b*) respectively.

Note. Reading from unused locations in Internal Data Memory will yield random data.

14.5 ADDRESSING MODES

The 8051 uses five addressing modes : viz ;

- Register ;
- Direct ;
- Register Indirect ;
- Immediate ; and
- Base-Register plus Index-Register Indirect.

The Table 3 summarizes which memory spaces may be accessed by each of the addressing modes.

Table 3

Source	Address
External Interrupt 0	0003H
Timer 0 Overflow	000BH
External Interrupt 1	0013H
Timer 1 Overflow	001BH
Serial Port	0023H
Timer 2 Overflow/TZEX	002BH
Negative Transition	

14.5.1 Register Addressing

Register Addressing access the eight working registers (R0-R7) of the selected Register Bank. The least significant three bits of the instruction op code indicate which register is to be used. ACC, B, DPTR and CY, the Boolean Processor accumulator, can also be addressed as registers.

Ex : Add the contents of register R3 and R4 from bank 2

Step 1 Select register bank,

```
MOV PSW, # 00001000B ; select register Bank 2
```

Step 2 Add the contents of R3 and R4

```
MOV A, R3
ADD A, R4
```

14.5.2 Direct Addressing

Direct Addressing is the only method of accessing the Special Function Registers. The lower 128 bytes of Internal RAM are also directly addressable.

Ex : Add the contents of locations 50H and 51H

```
MOV A, 50H      ; load byte from address 50H into A
ADD A, 51 H     ; Add the contents of A and the contents ; at memory location 51H.
```

14.5.3 Register - Indirect Addressing

Register-Indirect Addressing uses the contents of either R0 or R1 (in the selected Register Bank) as a pointer to locations in a 256-byte block : the lower 128 bytes of internal RAM ; the upper 128 bytes of internal RAM (8032/8052 only); or the lower 256 bytes of external Data memory.

It may be remembered that the Special Function Registers are not accessable by this method. Access to the full 64K external Data Memory address space is accomplished by using the 16-bit Data Pointer.

Execution of PUSH and POP instructions also use Register-Indirect addressing. The Stack Pointer may reside anywhere in Internal RAM.

Ex : Add the contents of memory location addressed by register 1 to the contents of RAM location pointed by register 0.

```
MOV A, @ R0      ; load the contents pointed by R0 in A
ADD A, @ R1      ; Add the contents of A and the contents pointed by R1
```

14.5.4 Immediate Addressing

Immediate Addressing allows constants to be part of the op code instruction in Program Memory. In this addressing mode source operand is a constant rather than a variable. So the constant can be incorporated in the instruction. Sing "#" indicates it is a immediate addressing mode.

Ex : Add the constant 52 Decimal in Accumulator

```
MOV A, # 52
```

14.5.5 Base-Register Plus Index-Register Indirect Addressing

Base-Register plus Index Register-Indirect Addressing allows a byte to be accessed from Program Memory via an indirect move from the location whose address is the sum of a base register (DPTR or PC) and index register, ACC. (This mode facilitates look-up-table accesses).

Ex : Read data from the program memory (Index).

```
MOVC A, @ A + DPTR      ; This instruction adds the unsigned 8-bit
                        ; and accumulator contents into sixteen bit Data.
                        ; pointer, and uses the sum as an address from which
                        ; byte to be moved into accumulator.
```

14.6 BOOLEAN PROCESSOR

The Boolean Processor is an integrated bit processor within the 8051. It has its own instruction set, accumulator (the carry flag), and bit addressable RAM and I/O.

The bit-manipulation instructions allow a bit to be set, cleared, complimented, jump-if-set, jump-if-not-set, jump-if-set-then-cleared and moved to/from the carry. Addressable bits, or their compliments, may be logically ANDed or ORed with the contents of the carry flag. The result is returned to the carry register.

14.7 MEMORY ORGANIZATION AND EXTERNAL ADDRESSING

As mentioned in article 14.4, 8051 can access upto 64K program memory and 64K Data memory, the Fig 14.16 depicts the Basic Memory structure of MCS-51.

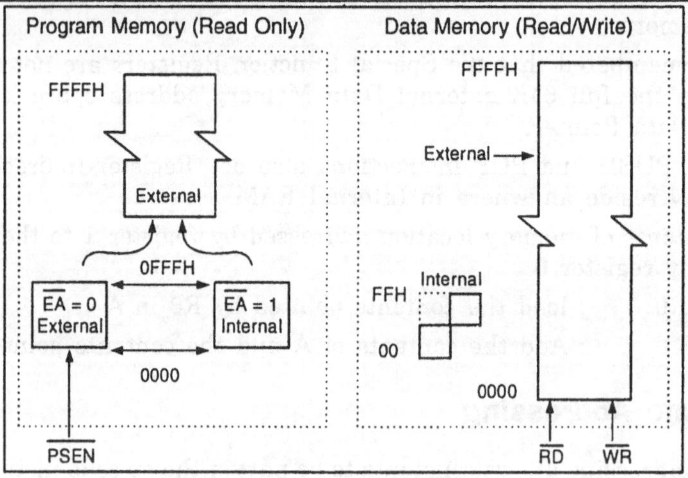

Fig. 14.16 Memory structure of MCS-51.

MCS-51 has 4K bytes of internal program memory and 256 bytes of internal data memory, which is also mentioned earlier. Each memory type has different addressing mechanism, different control signals, and a different function. Each may be added independently, and each uses the same address and data buses, but with different control signals. Now let us discuss briefly about :

(1) Program Memory, and

(2) Data Memory.

Both the memories can be observed for accessing external memory in their individual cases.

14.7.1 Program Memory

The program memory map for MCS-51 or 8051 is as shown in Fig 14.17.

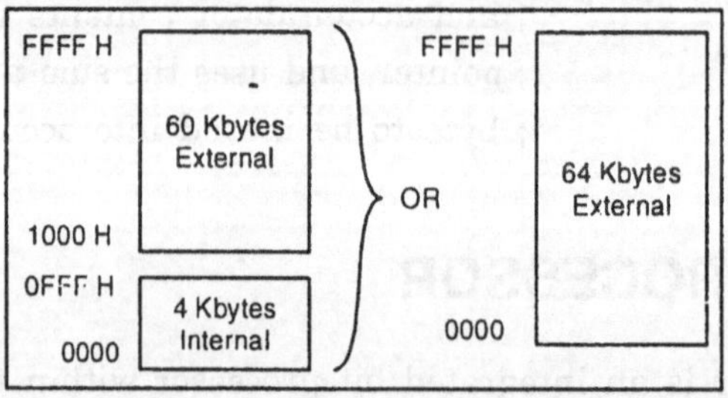

Fig. 14.17 Program Memory for MCS-51.

In MCS-51/8051 when the $\overline{EA}$ pin is connected to V_{CC}, program fetches to addresses 0000H through 0FFFH are directed to the internal ROM and program fetches to addresses 1000H through FFFFH are directed to external ROM/EPROM. On the other hand when

$\overline{EA}$ pin is grounded, all addresses (0000H to FFFFH) fetched by program are directed to the external ROM/EPROM. The $\overline{PSEN}$ signal is used to activate output enable signal of the external ROM/EPROM, which can be observed in Fig 14.18 below.

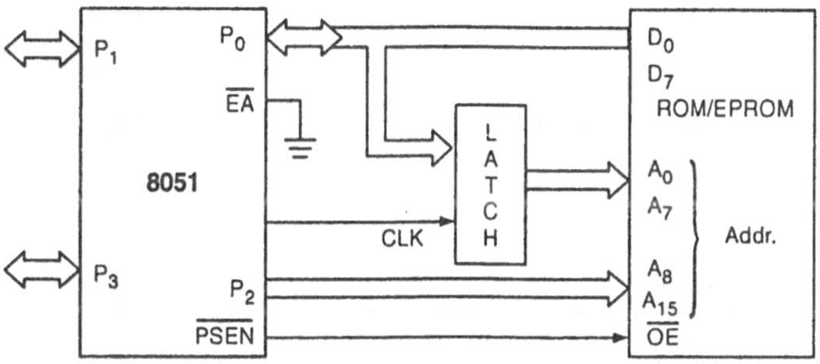

Fig. 14.18 External Program Memory Accessing.

As, observed in Fig 14.18 the port 0 is used as a multiplexed address/bus. It gives lower order 8-bit address in the initial T-cycle and later it is used as a data bus. The 8-bit address is latched using external latch and ALE signal generated by 8051. The port 2 provides the higher order 8-bit address. The timing waveforms for external program memory read cycle, is as shown in Fig 14.19.

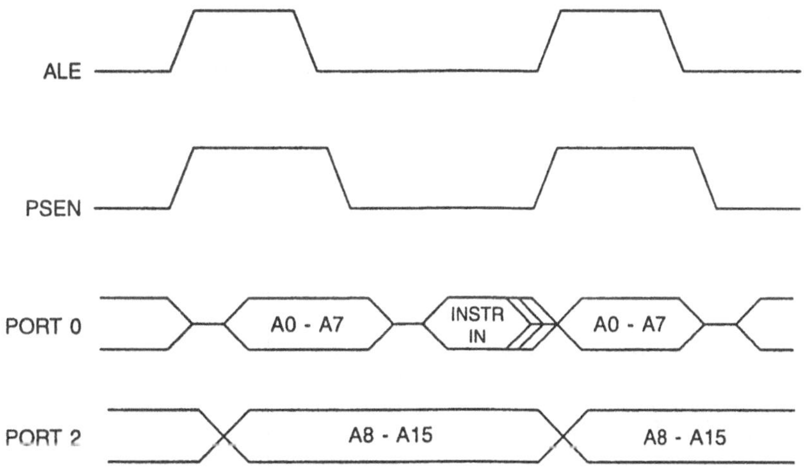

Fig. 14.19 Timing waveforms for External program memory read cycle.

Now the Fig 14.20 shows vector address map, where the lower part of program memory stores the vector address for various interrupt service routines.

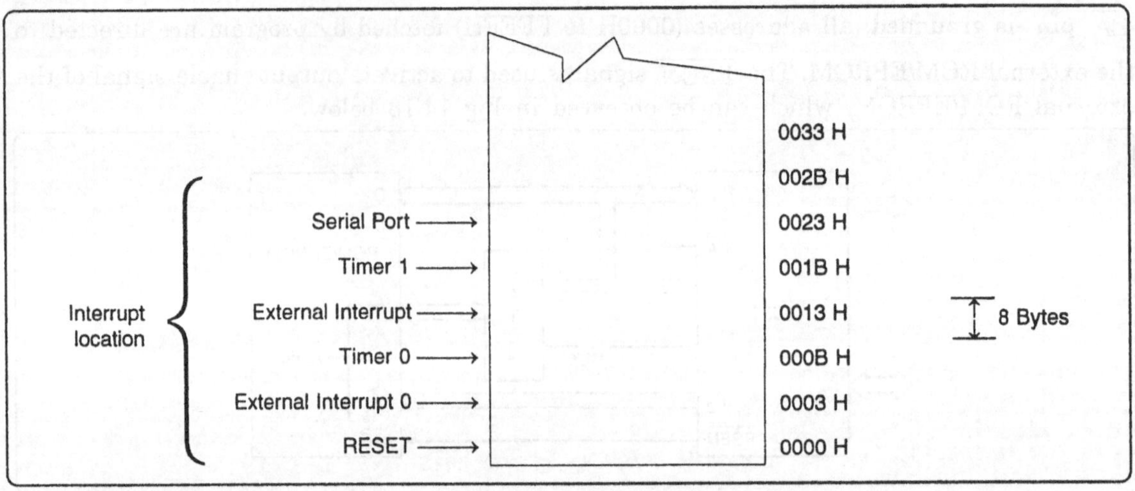

Fig. 14.20 Interrupt/Vector locations (in the lower part of program memory).

Here each interrupt is assigned with a fixed location in program memory. For example, external interrupt 0 is assigned to location 0003H. The interrupt service locations are spaced at 8-byte intervals such as 0003H for External Interrupt 0,000BH for Timer 0, 0013H for External Interrupt 1, 001BH for Timer1, etc. When interrupt is to be used, its service routine must begin at corresponding location. If the interrupt is not to be used, its service location is available as general purpose program memory.

14.7.2 Data Memory

The Data memory map for 8051 microcontroller is as shown in Fig 14.21.

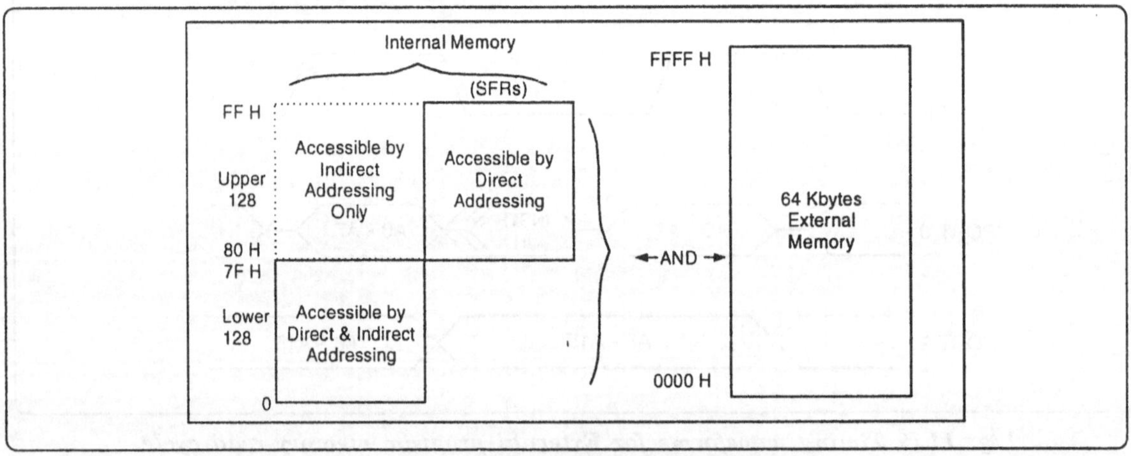

Fig. 14.21 Data Memory map for 8051.

8051 can address upto 64 Kbytes of external data memory. The "MOVX" instruction is used to access the external data memory. The internal data memory space for 8051 is divided into three blocks : Lower 128 bytes, Upper 128 bytes and SFRs. The upper addresses and SFRs occupy the same block of address space, 80H through FFH, although they are physically separate entities.

As observed in Fig 14.21 the upper address sace is accessible by indirect addressing only and SFRs are accessible by direct addressing only. On the other hand, lower address space can be accessed either by direct addressing or by indirect addressing. The circuit arrangement for connecting external data memory is as shown in Fig. 14.22.

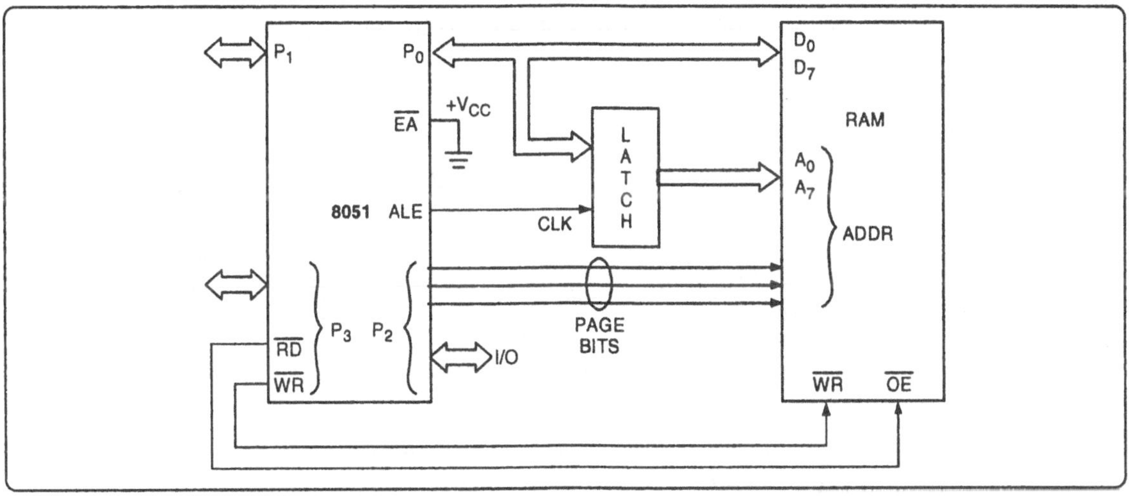

Fig. 14.22 External Data Memory Accessing.

In Fig 14.22 the multiplexed address/data bus provided by port 0 is demultiplexed by external latch and ALE signal. Port 2 gives the higher order address bus. The $\overline{RD}$ and $\overline{WR}$ signals from 8051 selects the memory read and memory write operation, respectively.

Now the timing waveforms of external Data Memory for read and write cycles are as shown in Fig 14.23 (*a*) and Fig 14.23 (*b*) respectively.

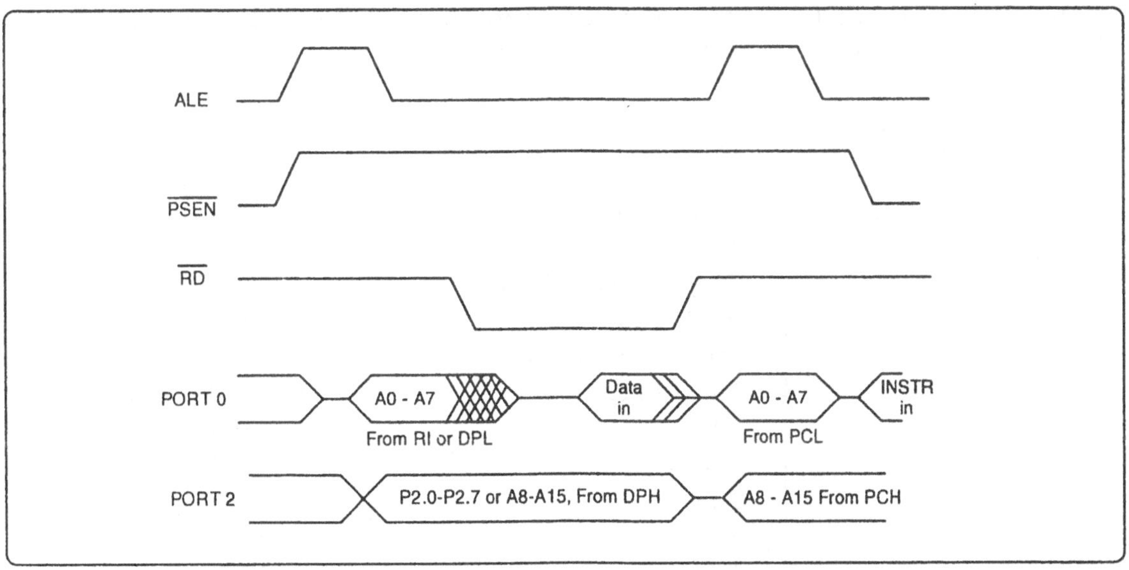

(a) Timing waveforms for external data memory read cycle

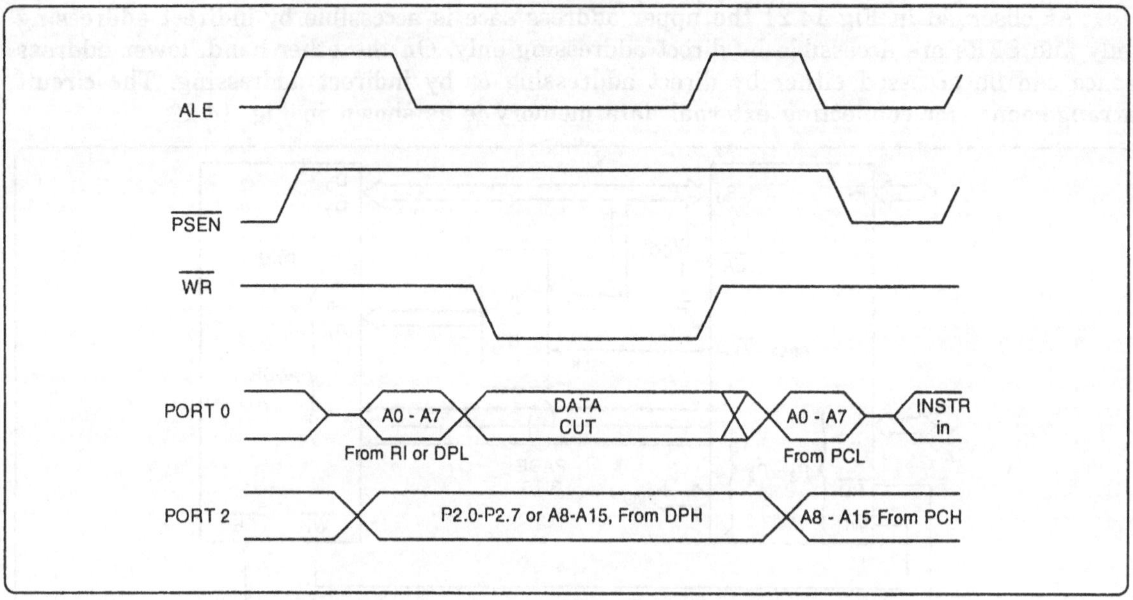

(b) Timing waveform for external data memory write cycle

Fig. 14.23 Timing waveforms of External Data Memory.

14.7.3 Locations of SFRs in Memory Organization

As mentioned earlier, 8051 has separate address spaces for Program memory and Data memory. The Program memory can be upto 64K bytes long, the lower 4K of which may reside on-chip. The Data memory has 128-byte of on-chip RAM, a 128-byte Special Function Register (SFR) area and 64 K bytes of external data memory. The SFR area contains the following registers in the indicated locations, in Table 4.

Table 4

Symbol	Name	Location
* ACC	Accumulator	0E0H
* B	B Register	0F0H
* PSW	Program Status Word	0D0H
SP	Stack Pointer	81H
DPTR	Data Pointer (consist-ing of DPH)	83H and
	DPL	82H
* P0	Port 0	80H
* P1	Port 1	90H
* P2	Port 2	0A0H
* P3	Port 3	0B0H
* IP	Interrupt Priority Contol	0B8H
* IE	Iterrupt Enable Control	0A8H

TMOD	Timer/Counter Mode Control	89H
+* T2CON	Timer/Counter Control	88H
TCON	Timer/Counter 2 Control	0C8H
TH0	Timer/Counter 0 (high byte)	8CH
TL0	Timer/Counter 0 (low byte)	8AH
TH1	Timer/Counter 1 (high byte)	8DH
TL1	Timer/Counter 1 (low byte)	8BH
+ TH2	Timer/Counter 2 (high byte)	0CDH
+ TL2	Timer/Counter 2 (low byte)	0CCH
+ RLDH	Timer/Counter 2 Auto-Reload (high byte)	0CBH
+ RLDL	Timer/Counter 2 Auto-Reload (low byte)	0CAH
* SCON	Serial Control	98H
SBUF	Serial Data Buff	99H
PCON	Power Control	87H

Theastericked (*) registers are both byte and bit addressable. The Timer/Counter 2 related registers, those marked with (+), are present in the 8032/8052 only.

14.8 INTERRUPTS

8051 has five hardware interrupt sources, one from the serial port when a transmission or reception is completed, two from the timers when overflow occurs, and two from the input pins INT0 and INT1. Each interrupt has its own vector location which is depicted in Table 5 below.

Table 5

Interrupt Source	Vector Address
External 0	0003H
Timer/counter 0	000BH
External 1	0013H
Timer/counter 1	001BH
Serial port	0023H

Apart each of the five sources can be individually enabled or disabled by setting or resetting bits in the interrupt enable register. One of the two priorities can be assigned to each source with the help of interrupt priority control register.

During requests of the same priority level are received simultaneously, an internal polling sequence determines which request is serviced. Thus within each priority level there is a second priority structure determined by polling sequence, which is depicted in Table 6 below.

Table 6

No.	Source	Priority level
1	IE 0	Highest
2	TF 0	
3	IE 1	
4	TF 1	
5	RI + TI	Lowest

Now Fig 14.24 gives details of IE, *i.e.* Interrupt enable register, where in the symbol position of IE and its control word is also shown.

(MSB) (LSB)

EA	—	—	ES	ET1	EX1	ET0	EX0

Symbol	Position	Name and Significance
EA	IE.7	Enable All control bit. Cleared by Software to disable all interrupts, independent of the state of IE.4-IE.0.
—	IE.6	(Reserved)
—	IE.5	(Reserved)
ES	IE.4	Enable Serial port control bit. Set/cleared by software to enable/disable interrupts from TI or RI flags.
ET1	IE.3	Enable Timer 1 control bit. Set/cleared by software to enable/disable interrupts from timer/counter 1
EX1	IE.2	Enable External interrupt 1 control bit. Set/cleared by software to enable/disable interrupts from INTI.
ET0	IE.1	Enable Timer 0 control bit. Set/cleared by software to enable/disable interrupts from timer/counter 0.
EX0	IE.0	Enable external interrupt 0 control bit. Set/cleared by software to enable/disable interrupts from INT0.

Fig. 14.24 Interrupt enable register.

The Interrupt Priority Control register (IP), is as shown in Fig 14.25. Here also the symbol, its position and their significance with name are shown.

(MSB)							(LSB)
—	—	—	PS	PT1	PX1	PT0	PX0

Symbol	Position	Name and Significance
—	IP.7	(Reserved)
—	IP.6	(Reserved)
—	IP.5	(Reserved)
PS	IP.4	Serial port Priority control bit. Set/cleared by software to specify high/low priority interrupts for Serial port.
PT1	IP.3	Timer 1 Priority control bit Set/cleared by software to specify high/low priority interrupts for timer/counter 1.
PX1	IP.2	External interrupt 1 Priority control bit. Set/cleared by software to specify high/low priority interrupts for INT1.
PT0	IP.1	Timer 0 Priority control bit. Set/cleared by software to specify high/low priority interrupts for timer/counter 0.
PX0	IP.0	External interrupt 0 Priority control bit. Set/cleared by software to specify high/low priority interrupts for INT0.

Fig. 14.25 Interrupt priority control register.

During an event when enabled for interrupt occurs, the CPU automatically executes an internal subroutine call to the corresponding address. A user subroutine starting at this location then performs the instruction to service that particular source. After completing the interrupt service routine, execution returns to the background program as usual.

14.9 8051 INSTRUCTION EXECUTION

Control unit in the microprocessor/microcontroller is responsible for generating control signal required for execution of particular instruction. The two different ways to design control unit, are :

(*a*) Hardwired control

(*b*) Microprogrammed control

In the hardwired control, the control units use fixed logic circuits to interpret instructions and generate contol signals from them, a typical hardwired control unit is as shown in Fig 14.26.

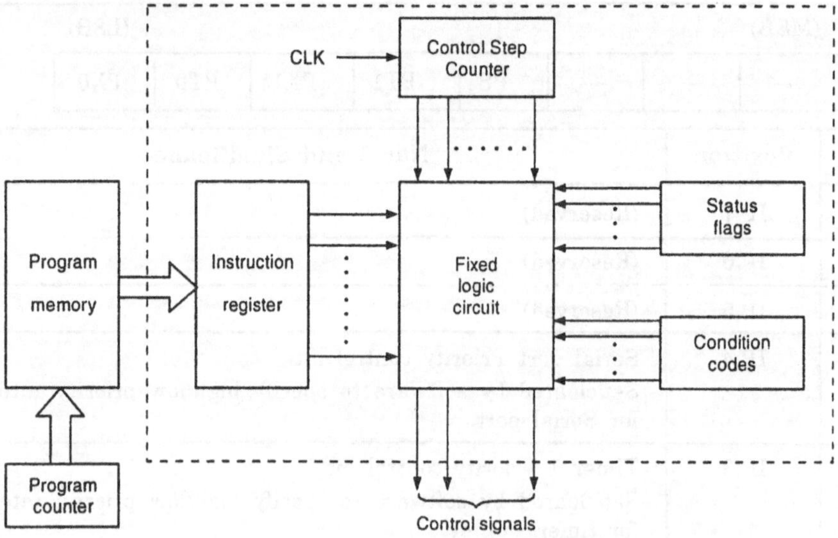

Fig. 14.26 Typical hardwired control unit.

8051 has microprogrammed control unit, unlike other microcontrollers. In this 'microprogrammed control', every instruction in a CPU is implemented by a sequence of one or more sets of concurrent microoperations. Each microoperation is associated with a specific set of control lines which, when activated, causes that microoperation to take place. Since the number of instructions and control lines is often in the hundreds, the complexity of hardwired control unit is very high, (*i.e.* it is costly and difficult to design). Further, the hardwired control unit is relatively inflexible because it is difficult to change the design, if one wishes to correct design error or modify the instruction set.

Microprogramming

It is a method of control unit design in which the control signal selection and sequencing information is stored in a ROM or RAM called a **control memory** CM. The control signals to be activated at any time are specified by a **microinstruction,** which is fetched from CM in much similar way an instruction is fetched from main memory. Each microinstruction also explicitly or implicitly specifies the next microinstruction to be used, thereby providing the necessary information for microoperation sequencing. A sequence of one or more microoperations designed to control specific operation, such as addition, multiplication is called a **microprogram**. The microprograms for all instructions are stored in the control memory. The address where these microinstructions are stored in CM is generated by microprogram sequencer/microprogram controller.

Note : The microprogram sequencer generates the address for microinstruction according to the instruction stored in the IR.

Now Fig 14.27 depicts the microprogrammed control unit. It consists of control memory, control address register, micro instruction register and microprogram sequencer.

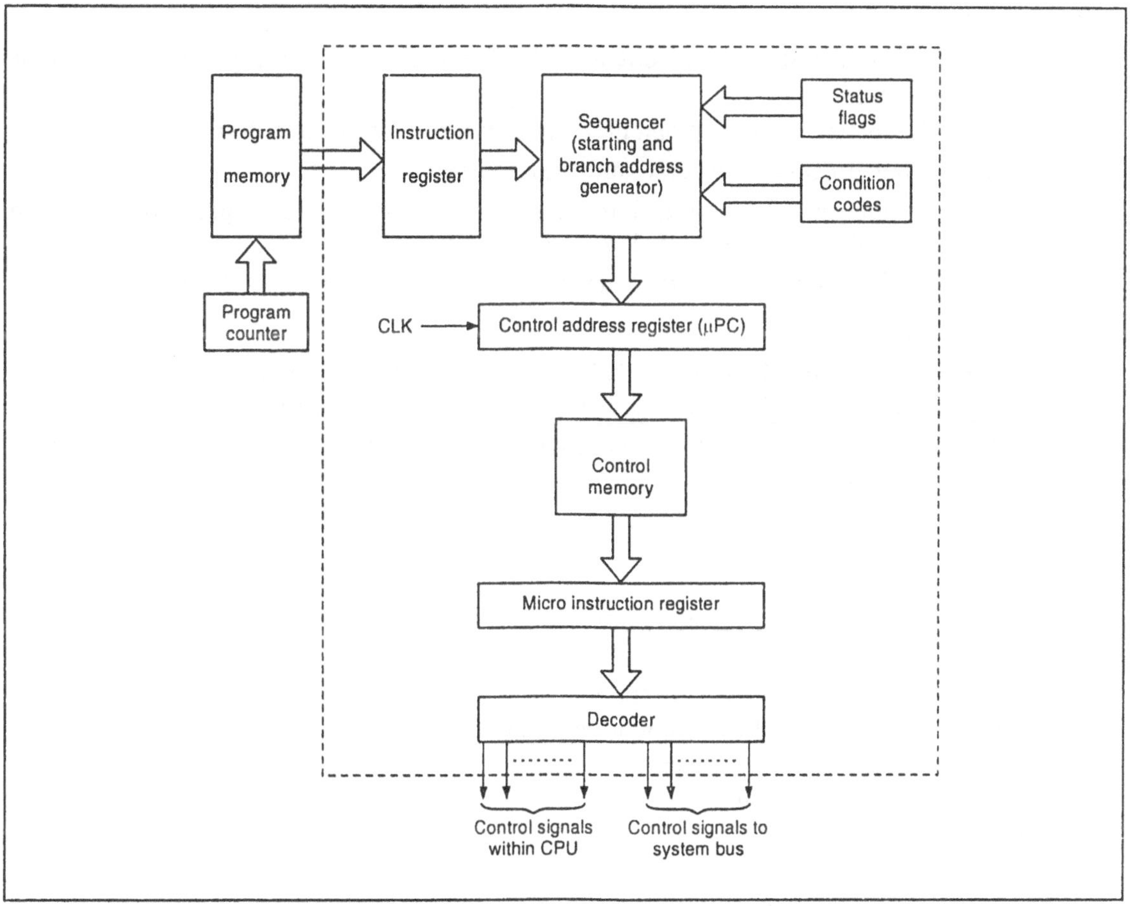

Fig. 14.27 8051 microprogrammed control unit.

The sequence by which the components of control unit works are :

(*a*) The control address register holds the address of the next microinstruction to be read.

(*b*) When address is available in control address register, the sequencer issues READ command to the control memory.

(*c*) After issue of READ command, the word from the addressed location is read into the micro instruction register.

(*d*) Now the content of the micro instruction register generates control signals and next address information for the sequencer.

(*e*) The sequencer loads a new address into the control address register based on the next address information.

14.9.1 Advantages and Disadvantages of Microprogrammed Control

There are few advantages as well as few disadvantages of Microprogrammed Control, which are briefly listed below :

■ Advantages :

(*i*) It simplifies the design of control unit. Thus it is both cheaper and less error prone to implement.

(*ii*) Control functions are implemented in software rather than hardware.

(*iii*) The design process is orderly and systematic.

(*iv*) More flexible, can be changed to accommodate new system specifications or to correct the design errors quickly and cheaply.

(*v*) Complex function such as floating point arithmetic can be realised efficiently.

■ Disadvantages :

(*i*) A microprogrammed control unit is somewhat shower than the hardwired control unit, because time is required to access the microinstructions from CM.

(*ii*) The flexibility is achieved at some extra hardware cost due to the control memory and its access circuitry.

REVIEW QUESTIONS

1. List out the main features of 8051 microcontroller.
2. Draw the architecture of 8051 and listout the main blocks.
3. Briefly explain Fetch/Execute sequences in 8051.
4. Explain Program Status Word.
5. Briefly explain the following
   (*a*) I/O Ports
   (*b*) Special peripheral functions.
6. Draw the Memory map of MCS-51 and explain.
7. Write short notes on following Addressing Modes
   (*a*) Register Addressing
   (*b*) Direct Addressing.

15 8051 INSTRUCTION SET

15.0 OBJECTIVE

At the end of this chapter the reader will be familliar with the different instruction set of 8051. Different functional groups of 8051 instructions will be made clear. Finaly reader can develop few ALP in 8051 microcontroller.

15.1 INTRODUCTION

8051 instruction set includes 111 instructions, 49 of which are single-byte, 45 two-byte and 17 three byte. The instruction op code format consists of a function mnemonic followed by a "destination, source" operand field. This field specifies the data type and addressing method(s) to be used.

By Intel convention, multiple byte address and data operands are stored least significant byte in the high-order address and most significant byte in the low-order address.

15.2 FUNCTIONAL OVERVIEW

The 8051 instruction set is divided into four functional groups, as listed below

1. Data transfer/Data movement instructions
2. Arithmetic Instructions
3. Logic
4. Control Transfer.

15.2.1 Data Transfer Instructions

The Data transfer operations are again divided into three classes :

(*a*) General Purpose

(*b*) Accumulator-specific

(*c*) Address-object

None of these operations affect the PSW flag settings except a POP or MOV directly to PSW.

(a) GENERAL PURPOSE TRANSFER

MOV performs a bit or a byte transfer from the source operand to the destination operand.

The data pointer can be loaded with a 16-bit value using the instruction MOV DPTR. #data 16. The data used is stored in the second and third instruction bytes, high-order byte first.

The MOVC (Move Constant) instructions (MOVC A.@ A+DPTR and MOVC A.@A+PC) read into the accumulator bytes of data from the program memory logical address space. Both use a form of indexed addressing : the former adds the unsigned eight-bit accumulator contents with the sixteen-bit data pointer register, and uses the resulting sum as the address from which the byte is fetched. A sixteen-bit addition is performed : a carry-out from the low-order eight bits may propagate through higher-order bits. but the contents of the DPTR are not altered (The latter form uses the incremented program counter as the base value instead of the DPTR again, neither version affects the flags).

- PUSH increments the SP register and then transfers a byte from the source operand to the stack location currently addressed by SP.
- POP transfer a byte operand from the stack location addressed by SP to the destination operand and then decrements SP.

The PUSH instruction increments the stack pointer by one, then transfers the contents of the single byte variable indicated (direct addressing only) into the internal RAM location addressed by the stack pointer. Conversely, POP copies the contents of the internal RAM location addressed by the stack pointer to the byte variable indicated, then decrements the stack pointer by one. Interrupt service routines must not change any variable or hardware registers modified by the main program, or else the program may not resume correctly. The Fig 15.1, shows the contents of stack during interrupt.

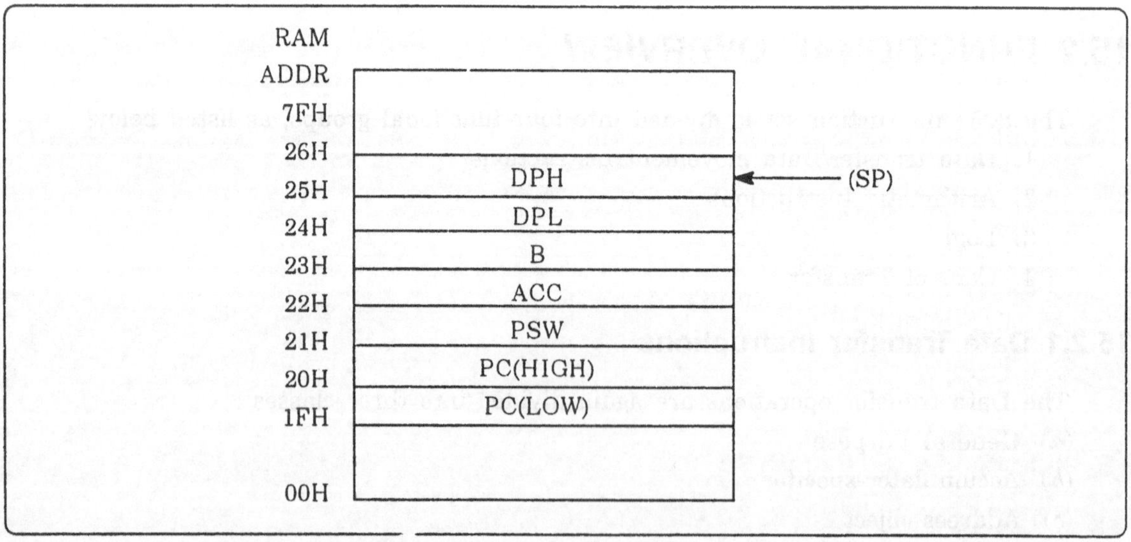

Fig. 15.1 Contents of Stack during interrupt.

The PUSH and POP operations provide an efficient and convenient way to save register states on the stack. The following example shows how to use stack for states saving on Interrupts :

Label	Opcode		Comments
	LOC_TMP	EQU	REMEMBER LOCATION COUNTER
	ORG	0003H	STARTING ADDRESS FOR INTERRUPT ROUTINE
	LJMP	SERVER	JUMP TO ACTUAL SERVICE ROUTINE LOCATED ELSEWHERE
	ORG	LOC_TMP	RESTORE LOCATION COUNTER
SERVER	PUSH	PSW	
	PUSH	ACC	SAVE ACCUMULATOR (NOTE DIRECT ADDRESSING NOTATION)
	PUSH B		SAVE B REGISTER
	PUSH	DPL	SAVE DATA POINTER
	PUSH	DPH	
	MOV	PSW,#00001000B	SELECT REGISTER BANK 1
	POP	DPH	RESTORE REGISTERS IN REVERSE ORDER
	POP	DPL	
	POP	B	
	POP	ACC	
	POP	PSW	RESTORE PSW AND RE-SELECT ORIGINAL REGISTER BANK
	RETI		RETURN TO MAIN PROGRAM AND RESTORE INTERRUPT LOGIC

(*b*) ACCUMULATOR SPECIFIC TRANSFERS

- XCH exchanges the byte source operand with register A (accumulator).
- XCHD exchanges the low-order nibble of the byte source operand with the low-order nibble of A.
- MOVX performs a byte move between the External Data Memory and the accumulator. The external address can be specified by the DPTR register (16-bit) or the R1 or R0 register (8-bit).
- MOVC moves a byte from Program memory to the accumulator. The operand in A is used as an index into a 256-byte table pointed to by the base register (DPTR or PC). The byte operand accessed is transferred to the accumulator.

(*c*) ADDRESS-OBJECT TRANSFER

- MOV DPTR, #data loads 16-bits of immediate data into a pair of destination registers, DPH and DPL.

  It is already mentioned in General purpose Transfer.

15.2.2 Arithmetic Instructions

The 8051 has four basic mathematical operations. Only 8-bit operations using unsigned arithmetic are supported directly. The overflow flag, however, permits the addition and subtraction operation to serve for both unsigned and signed binary integers. Arithmetic can also be performed directly on packed decimal (BCD) representations. The four mathematical operations are :

(*a*) Addition

(*b*) Subtraction

(*c*) Multiplication

(*d*) Division

(*a*) ADDITION

- INC (increment) adds one to the source operand and puts the result in the operand.
- ADD adds A to the source operand and returns the result to A.
- ADDC (add with Carry) adds A and the source operand, then adds one (1) if CY is set, and puts the result in A.
- DA (decimal-add-adjust for BCD addition) corrects the sum which results from the binary addition of two two-digit decimal operands. The packed decimal sum formed by DA is returned to A. CY is set if the BCD result is greater than 99; otherwise, it is cleared.

ADD instruction adds a byte variable with the accumulator, leaving the result in the accumulator. The carry flag is set if there is an overflow from bit 7 and cleared otherwise. The AC flag is set to the carry-out from bit 3 for use by the DA instruction described later. ADDC adds the previous contents of the carry flag with the two byte variables, but otherwise is the same as ADD.

Decimal addition is possible by using DA instruction in conjunction with ADD and / or ADDC. The eight-bit binary value in the accumulator resulting from an earlier addition of two variables (each a packed BCD digit-pair) is adjusted to form two BCD digits of four bits each. If the contents of accumulator bits 3-0 are greater than nine or if the AC flag had been set, six is added to accumulator producing the proper BCD digit in the low-order nibble. (This addition might itself set—but would not clear—the carry flag.) If the carry flag is set, or if the four high-order bits now exceed nine these bits are incremented by six.

Note. The carry flag is left set if originally set or if either addition of six produces a carry out of the highest-order bit, indicating the sum of the original two BCD variables is greater than or equal to decimal 100.

Ex : Program to add two byte decimal with registers and constants.

BCDADD : ADD THE CONSTANT 1,234 (DECIMAL) TO
THE CONTENTS OF REGISTER PAIR <R3><R2>

Label	Opcode	
BCDADD	MOV	A,R2
	ADD	A,#34
	DA	A
	MOV	R2, A
	MOV	A, R3
	ADDC	A, #12H
	DA	A
	MOV	R3, A
	RET	

(*b*) SUBTRACTION

- SUBB (subtract with borrow) subtracts the second source operand from the first operand (the accumulator), subtracts one (1) if CY is set and returns the result to A.
- DEC (decrement) subtracts one (1) from the source operand and returns the result to the operand.

SUBB (subtract with borrow) instruction subtracts the byte variable indicated and the contents of the carry flag together from the accumulator, and puts the result back in the accumulator. The carry flag serves as a "Borrow Required" flag during subtraction operations when a greater value is subtracted from a lesser value (as in subtracting 5 from 1) requiring a borrow into the highest order bit, the carry flag is set ; otherwise it is cleared.

(*c*) MULTIPLICATION

- MUL performs an unsigned multiplication of the A register by the B register, returning a double-byte result. A receives the low-order byte, B receives the high-order byte. OV is cleared if the top half of the result is zero and is set if it is non-zero. CY is cleared. AC is unaffected.

Instruction "MUL AB" multiplies the unsigned eight-bit integer values held in the accumulator and B-register. The low-order byte of the sixteen-bit product is left in the accumulator, the higher-order byte in B. If the high-order eight-bits of the product are all zero, the overflow flag is cleared ; otherwise it is set. The programmer can check OV to determine when the B register is non-zero and must be processed.

(*d*) DIVISION

- DIV performs an unsigned division of the A register by the B register and returns the integer quotient to A and returns the fractional remainder to the B register. Division by zero leaves indeterminate data in registers A and B and sets OV; otherwise OV is cleared. CY is cleared, AC is unaffected.

  Unless otherwise stated in the above descriptions, the flags of PSW are affected as follows :

- CY is set if the operation causes a carry to or from the resulting high-order bit. Otherwise CY is cleared.

- AC is set if the operation results in a carry from the low-order four bits of the result (during addition), or a borrow from the high-order bits to the low-order bits (during subtraction); otherwise AC is cleared.
- OV is set if the operation results in a carry to the high-order bit of the result but not a carry from the high-order bit, or vice versa ; otherwise OV is cleared. OV is used in two's-complement arithmetic, because it is set when the signed result cannot be represented in 8 bits.
- P is set if the modulo 2 sum of the eight bits in the accumulator is 1 (odd parity); otherwise P is cleared (even parity). When a value is written to the PSW register, the P bit remains unchanged, as it always reflects the parity of A.

Divide instructions is also useful for purposes such as radix conversion or separating bit fields of the accumulator. A short subroutine can convert an eight-bit unsigned binary integer in the accumulator (between 0 and 255) to a three-digit (two byte) BCD representation. The hundred's digit is returned in one register (HUND) and the ten's and one's digits returned as packed BCD in another (TENONE).

Ex : Program using DIV instruction for Radix Conversion.

BINBCD : CONVERT B-BIT BINARY VARIABLE IN ACC
TO 3-DIGIT PACKED BCD FORMAT
HUNDREDS PLACE LEFT IN VARIABLE 'HUND'
TENS AND ONES PLACES IN 'TENONE'

Label	Opcode		Comments
	HUND	EQU 21H	
	TENONE	EQU 22H	
BINBCD	MOV	B,#100	; DIVIDE BY 100 TO
	DIV	AB	; DETERMINE NUMBER OF HUNDREDS
	MOV	HUND,A	
	MOV	A,#10	; LOAD DIVIDER
	XCH	A,B	; COUNT
	DIV	AB	; DIVIDE
			; REMAINDER BY 10
	SWAP	A	
	ADD	A,B	; PACK BCD DIGITS IN ACC
	MOV	TENONE,A	
	RET		

15.2.3 Logic Instructions

The 8051 performs basic logic operations on both bit and byte operands. The two operations logic instruction can perform are :

(*a*) Single-operand operations.

(*b*) Two-operand operations.

(a) SINGLE-OPERAND OPERATIONS

- CLR sets A or any directly addressable bit to zero(0).
- SET B sets any directly addressable bit to one (1).
- CPL is used to compliment the contents of the A register without affecting any flags, or any directly addressable bit location.
- RL, RLC, RR, RRC, SWAP are the five rotate operations that can be performed on A. RL, rotate left, RR, rotate right, RLC, rotate left through C, RRC, rotate right through C, and SWAP, rotate left four. For RLC and RRC the CY flag becomes equal to the last bit rotated out. SWAP rotates A left four places to exchange bits 3 through 0 with bits 7 through 4.

(b) TWO-OPERAND OPERATIONS

- ANL performs bitwise logical and of with two source operands (for both bit and byte operands) and returns the result to the location of the first operand.
- ORL performs bitwise logical or of two source operands (for both bit and byte operands) and returns the result of the location of the first operand.
- XRL performs bitwise logical or of two source operands (byte operands) and returns the result to the location of the first operand.

15.2.4 Control Transfer/Program Control

There are three classes of control transfer operations : unconditional calls, returns and jumps ; conditional jumps ; and interrupts. All control transfer operations cause, some upon a specific condition, the program execution to continue at a non-sequential location in program memory. The three classes of control transfer operations are :

(a) Unconditional Calls, Returns and Jumps.

(b) Conditional Jumps

(c) Interrupt Returns.

(a) UNCONDITIONAL CALLS, RETURNS AND JUMPS

Unconditional calls, returns and jumps transfer control from the current value of the Program Counter to the target address. Both direct and indirect transfers are supported.

- ACALL and LCALL push the address of the next instruction onto the stack and then transfer control to the target address. ACALL is a 2-byte instruction used when the target address is in the current 2K page. LCALL is a 3-byte instruction that addresses the full 64K program space. In ACALL, immediate data (*i.e.*, an 11 bit address field) is concatenated to the five most significant bits of the PC (which is pointing to the next instruction). If ACALL is in the last 2 bytes of a 2K page then the call will be made to the next page since the PC will have been incremented to the next instruction prior to execution.
- RET transfers control to the return address saved on the stack by a previous call operation and decrements the SP register by two (2) to adjust the SP for the popped address.
- AJMP, LJMP and SJMP transfer control to the target operand. The operation of AJMP and LJMP are analogous to ACALL and LCALL. The SJMP (short jump) instruction

provides for transfers within a 256 byte range centered about the starting address of the next instruction (–128 to +127).

- JMP @A+DPTR performs a jump relative to the DPTR register. The operand in A is used as the offset (0-255) to the address in the DPTR register. Thus the effective destination for a jump can be anywhere in the Program Memory space.

(b) CONDITIONAL JUMPS

Conditional jumps perform a jump contingent upon a specific condition. The destination will be within a 256-byte range centered about the starting address of the next instruction (–128 to +127)

- JZ performs a jump if the accumulator is zero.
- JNZ performs a jump if the accumulator is not zero.
- JC performs a jump if the carry flag is set.
- JNC performs a jump if the carry flag is not set.
- JB performs a jump if the Direct Addressed bit is set.
- JNB performs a jump if the Direct Addressed bit is not set.
- JBC performs a jump if the Direct Addressed bit is set and then clears the Direct Addressed bit.
- CJNE compares the first operand to the second operand and performs a jump if they are not equal. CY is set if the first operand is less than the second operand ; otherwise it is cleared. Comparisons can be made between A directly addressable bytes in Internal Data Memory or between an immediate value and either A, a register in the selected Register Bank, or a Register-Indirect addressed byte of Internal RAM.
- DJNZ decrements the source operand and returns the result to the operand. A jump is performed if the result is not zero. The source operand of the DJNZ instruction may be any byte in the Internal Data Memory. Either Direct or Register Addressing may be used to address the source operand.

(c) INTERRUPT RETURNS

- RET1 transfers control as does RET, but additionally enables interrupts of the current priority level.

15.3 8051 INSTRUCTION SET SUMMARY

The Interrupt Response time is the time to finish execution of current instruction, respond to the interrupt request push the PC and to vector to the first instruction of the interrupt service program requires 38 to 81 oscillator periods, (3 to 7 μsec@12 MHz).

The following Table 1 provides the Instructions that affect flag settings.

Table 1. Instruction that Affect Flag Settings

INSTRUCTION	FLAG		
	C	OV	AC
ADD	X	X	X
ADDC	X	X	X
SUBB	X	X	X
MUL	O	X	
DIV	O	X	
DA	X		
RRC	X		
RLC	X		
SETBC	1		
	C	OV	AC
CLR C	O		
CPL C	X		
ANL C,bit	X		
ANL C,/bit	X		
ORL C,bit	X		
ORL C,bit	X		
MOV C,bit	X		
CJNE	X		

Note. The operations on SFR byte address 208 or bit addresses 209-215 (*i.e.*, the PSW or bits in the PSW) will also affect flag settings.

Now the Table 2 provides the notes on Instruction set and addressing modes :

Table 2. Notes on instruction set and addressing modes

Instruction Set	Addressing Modes
Rn	— Register R7-R0 of the currently selected Register Bank.
diroct	— 8-bit internal data location's address. This could be an Internal Data RAM location (0-127) or a SFR [*i.e.*, I/O port, control register, status register, etc. (128-255)].
@Ri	— 8-bit internal data RAM location (0-255) addressed indirectly through register R1 to R0.
#data	— 8-bit constant included in instruction.
#data16	— 16-bit constant included in instruction.
addr 16	— 16-bit destination address. Used by LCALL and LJMP. A branch can be anywhere within the 64K-byte Program Memory address space.

addr 11	— 11-bit destination address. Used by ACALL and AJMP. The branch will be within the same 2K-byte page of program memory as the first byte of the following instruction.
rel	— Signed (two's complement) 8-bit offset byte. Used by SJMP and all conditional jumps. Range is –128 to +127 bytes relative to first byte of the following instruction.
bit	— Direct Addressed bit in Internal Data RAM or Special Function Register.
	— New operation not provided by 8048AH/8049AH.

15.4 INSTRUCTION DICTIONARY OF 8051

The instruction set of 8051 can be divided into five sub sets as mentioned below :

(1) Arithmetic Operations.

(2) Logical Operations.

(3) Data Transfer

(4) Boolean variable Manipulation

(5) Program Branching

15.4.1 Arithmetic Operation

Mnemonic		Description	Byte	Oscillator Period
ADD	A.Rn	Add register to Accumulator	1	12
ADD	A.direct	Add direct byte to Accumulator	2	12
ADD	A.@Ri	Add indirect RAM to Accumulator	1	12
ADD	A.#data	Add immediate data to Accumulator	2	12
ADDC	A.Rn	Add register to Accumulator with carry	1	12
ADDC	A.direct	Add direct byte to Accumulator with carry	2	12
ADDC	A.@Ri	Add indirect RAM to Accumulator with carry	1	12
ADDC	A.#data	Add immediate data to Acc with carry	2	12
SUBB	A.Rn	Subtract register from Acc with borrow	1	12
SUBB	A.direct	Subtract direct byte from Acc with borrow	2	12
SUBB	A.@Ri	Subtract indirect RAM from Acc with borrow	1	12
SUBB	A.#data	Subtract immediate data from Acc with borrow	2	12
INC	A	Increment Accumulator	1	12
INC	Rn	Increment register	1	12
INC	direct	Increment direct byte	2	12
INC	@Ri	Increment indirect RAM	1	12

DEC	A	Decrement Accumulator	1	12
DEC	Rn	Decrement Register	1	12
DEC	direct	Decrement direct byte	2	12
DEC	@Ri	Decrement Indirect RAM	1	12
INC	DPTR	Increment Data Pointer	1	24
MUL	AB	Multiply A and B	1	48
DIV	AB	Divide A by B	1	48
DA	A	Decimal Adjust Accumulator	1	12

15.4.2 Logical Operations

Mnemonic		Description	Byte	Oscillator Period
ANL	A.Rn	AND register Accumulator	1	12
ANL	A.direct	AND direct byte to Accumulator	2	12
ANL	A.@Ri	AND indirect RAM to Accumulator	1	12
ANL	A.#data	AND immediate data to Accumulator	2	12
ANL	direct.A	AND accumulator to direct byte	2	12
ANL	direct.#data	AND immediate data to direct byte	3	24
ORL	A.Rn	OR register to accumulator	1	12
ORL	A.direct	OR direct byte to accmulator	2	12
ORL	A.@Ri	OR indirect RAM to Accumulator	1	12
ORL	A.#data	OR immediate data to Accumulator	2	12
ORL	direct.A	OR accumulator to direct byte	2	12
ORL	direct.#data	OR immediate data to direct byte	3	24
XRL	A.Rn	Exclusive-OR register to accumulator	1	12
XRL	A.direct	Exclusive-OR direct byte to accumulator	2	12
XRL	A.@Ri	Exclusive-OR indirect RAM to Accumulator	1	12
XRL	A.#data	Exclusive-OR immediate data to Accumulator	2	12
XRL	direct A	Exclusive-OR Accumulator to direct byte	2	12
XRL	direct.#data	Exclusive-OR immediate data to direct byte	3	24
CLR	A	Clear Accumulator	1	12
CPL	A	Complement Accumulator	1	12
RL	A	Rotate Accumulator Left	1	12
RCL	A	Rotate Accumulator Left through the carry	1	12

RR	A	Rotate Accumulator Right	1	12
RRC	A	Rotate Accumulator Right through the carry	1	12
SWAP	A	Swap nibbles within the Accumulator	1	12

15.4.3 Data Transfer

Mnemonic		Description	Byte	Oscillator Period
MOV	A.Rn	Move register to Accumulator	1	12
MOV	A.direct	Move direct byte to Accumulator	2	12
MOV	A.@Ri	Move indirect RAM to Accumulator	1	12
MOV	A.#data	Move immediate data to Accumulator	2	12
MOV	Rn.A	Move accumulator to register	1	12
MOV	Rn.direct	Move direct byte to register	2	24
MOV	Rn.#data	Move immediate data to register	2	12
MOV	direct.A	Move accumulator to direct byte	2	12
MOV	direct.Rn	Move register to direct byte	2	24
MOV	direct.direct	Move direct byte to direct	3	24
MOV	direct.@Ri	Move indirect RAM to direct byte	2	24
MOV	direct.#data	Move immediate data to direct byte	3	24
MOV	@Ri.A	Move accumulator to indirect RAM	1	12
MOV	@Ri.direct	Move direct byte to indirect RAM	2	24
MOV	@Ri.#data	Move immediate data to indirect RAM	2	12
MOV	DPTR,#data16	Load Data Pointer with a 16-bit Constant	3	24
MOVC	A,@A+DPTR	Move Code byte relative to DPTR to Acc	1	24
MOVC	A.@A+PC	Move Code byte relative to PC and Acc	1	24
MOVX	A.@Ri	Move external RAM (8-bit addr) to Acc	1	24
MOVX	A.@DPTR	Move External RAM (16-bit addr) to Acc	1	24
MOVX	@Ri.A	Move Acc to external RAM (8-bit addr)	1	24
MOVX	@DPTR.A	Move Acc to external RAM (16-bit addr)	1	24
PUSH	direct	Push direct byte onto stack.	2	24
POP	direct	Pop direct byte from stack.	2	24
XCH	A.Rn	Exchange register with Accumulator	1	12

XCH	A.direct	Exchange direct byte with accumulator	2	12
XCH	A.@Ri	Exchange indirect RAM with accumulator	1	12
XCHD	A.@Ri	Exchange low order Digit indirect RAM with Acc	1	12

15.4.4 Boolean Variable Manipulation

Mnemonic		Description	Byte	Oscillator Period
CLR	C	Clear Carry	1	12
CLR	bit	Clear direct bit	2	12
SETB	C	Set Carry	1	12
SETB	bit	Set direct bit	2	12
CPL	C	Complement carry	1	12
CPL	bit	Complement direct bit	2	12
ANL	C.bit	AND direct bit to carry	2	24
ANL	C./bit	AND complement of direct bit to carry	2	24
ORL	C.bit	OR direct bit to carry	2	24
ORL	C./bit	OR complement of direct bit to carry	2	24
MOV	C.bit	Move direct bit to carry	2	12
MOV	bit.C	Move Carry to direct bit	2	24
IC	rel	Jump if Carry is set	2	24
JNC	rel	Jump if Carry not set	2	24
JB	bit.rel	Jump if direct Bit is set	3	24
JNB	bit.rel	Jum if direct Bit is Not set	3	24
JBC	bit.rel	Jump if direct Bit is set and clear bit	3	24

15.4.5 Program Branching

Mnemonic		Description	Byte	Oscillator Period
ACALL	addr11	Absolute Subroutine Call	2	24
LCALL	addr16	Long Subroutine Call	3	24
RET		Return from Subroutine	1	24
RET1		Return from interrupt	1	24
AJMP	addr11	Absolute Jump	2	24
LJMP	addr16	Long Jump	3	24
SJMP	rel	Short Jump (relative addr)	2	24
JMP	@A+DPTR	Jump indirect relative to the DPTR	1	24
JZ	rel	Jump if Accumulator is Zero	2	24

JNZ	rel	Jump if Accumulator is Not Zero	2	24
CJNE	A.direct.rel	Compare direct byte to Acc and Jump if Not Equal	3	24
CJNE	A.#data.rel	Compare immediate to Acc and Jump if Not Equal	3	24
CJNE	Rn.#data.rel	Compare immediate to register and Jump if Not Equal	3	24
CJNE	@Ri.#data.rel	Compare immediate to indirect and Jump if Not Equal	3	24
DJNZ	Rn.rel	Decrement register and Jump if Not Zero	3	24
DJNZ	direct.rel	Decrement direct byte and Jump if Not Zero	3	24
NOP		No Operation	1	12

15.5 SAMPLE PROGRAMS OF 8051 MICROCONTROLLER

Once the microcontroller is discussed for its instruction set and its features, now few assembly level sample programs are added to make it still more clear.

15.5.1 Program 1

Write an ALP in 8051 to add two 8 bit numbers.

Mnemonics		**Comments**
MOV A, #30H	;	(A)←30
ADD A, #50H	;	(A)←(A) + 50H

15.5.2 Program 2

Write an ALP in 8051 to substract two 8-bit numbers and exchange digits.

Mnemonics		**Comments**
MOV A, #9F	;	Get First number in A
MOV R0, #40	;	Get the second number in R0
SUBB A, R0	;	A←[A-(R0)]
SWAP A	;	Exchange digits.

15.5.3 Program 3

Write ALP in 8051 to bad accumulator A, DPH and DPL with 40H.

Mnemonics		**Comments**
MOV A, #40H	;	Loads 40H in Accumulator
MOV DPH, A	;	(DPH)←(A)
MOV PPL,A	;	(DPL)←(A)

15.5.4 Program 4

Write ALP in 8051 to multiply two 8-bit numbers.

Mnemonics		Comments
MOV A, #8F	;	Get the first number in A
MOV B, #77	;	Get the second number in B
MUL B	;	(A×B), Higher by of result
	;	in B and lower byte of
	;	result in A.

15.5.5 Program 5

Write ALP in 8051 for division of two 8-bit number.

Mnemonics		Comments
MOV A, #80	;	Get the first number in A
MOV B, #40	;	Get the second number in B
DIV AB	;	(A÷B), Remainder in B
	;	and quotient in A.

15.5.6 Program 6

Write ALP in 8051 to find 2's complement of a number in R0.

Mnemonics		Comments
MOV A, R0	;	(A)←(R0)
CPL A	;	1's Complement A
ADD A, #1	;	Add 1 to A to get 2's complement

15.5.7 Program 7

Write ALP in 8051 to subtract the contents of R1 of Bank 0 from the contents of R0 of Bank 2.

Mnemonics		Comments
MOV PSW, #10	;	Select the Bank 2.
MOV A, R0	;	(A)←(R0) from Bank 2.
MOV PSW, #00	;	Select Bank 0
SUBB A, R	;	[A←A—(R1)] from Bank 0

15.5.8 Program 8

Write ALP in 8051 to add the contents of 9000 and 9001 addresses and store the result in address 9002.

```
; Sample program to add the contents of
; 9000 and 9001, Write the result at 9002.
```

```
Address  Code     Mnemonics           Comments
8000              ORG 8000H         ; Starting from address 8000.
8000     90 9000  MOV DPTR, #9000H  ; load data pointer value
8003     E0       MOVX A, @DPTR     ;
8004     FSF0     MOV B,A           ; Store the first value in B.
8006     A3       INC DPTR          ; take next value
8007     E0       MOVXA, @DPTR      ;
8008     A3       INC DPTR          ; now DPTR points to result field.
8009     25F0     ADD A,B           ;
800B     D4       DA A              ; decimal adjust for addition.
800C     F0       MOVX @DPTR,A      ; Store result
800D     020000   JMP MONITOR       ; Here it is monitor.
0000              END.
```

15.5.9 Program 9

Write an ALP in 8051 microcontroller, to add two 16 bit numbers.

```
Mnemonics             Comments
MOV DPTR, #2040H    ; (DPTR)←2040H (16-bit number)
MOV A, #2BH         ; (A)←2BH (lower byte of second 16-bit number)
MOV B, #20H         ; (B)←20H (Higher byte of second 16-bit numer)
ADD A, DPL          ; Add lower bytes
MOV DPL,A           ; Save result of lower byte addition
MOV A, B            ; Get higher byte of second number in A
ADDC A, DPH         ; Add higher bytes with any carry
                    ; from lower byte addition.
MOV DPH, A          ; Save result of higher byte addition
```

15.5.10 Program 10

Write ALP in 8051 to generate BCD up Counter and send each count to port A.

```
Label    Mnemonics           Comments
         MOV A, #00      ;   Initialize counter
BACK:    MOV P1, A       ;   Send count to port A
         ADD A, #01      ;   Increment counter
         DA A            ;   Decimal adjust the counter
         AJMP BACK       ;   Jump BACK
```

15.5.11 Program 11

Write ALP in 8051 to arrange the given ten 8-bit numbers in ascending order.

Label	Mnemonics		Comments
	MOV R0, #09	;	Initialize counter1
AGAIN:	MOV DPTR,#2000H	;	Initialize memory pointer.
	MOV R1,#09	;	Initialize counter2
BACK:	MOV R2, DPL	;	Save lower byte of memory address
	MOVX A,@DPTR	;	Get the number
	MOV B,A	;	Increment memory pointer
	MOVX A,@DPTR	;	Get the next number
	CJNE A,B,NE		
	AJMP SKIP		
NE:	JC SKIP		
	MOV DPL,R2	;	[Exchange
	MOV @DPTR,A	;	the contents
	INC DPTR	;	of two
	MOV A,B	;	memory
	MOV @DPTR,A	;	locations]
SKIP:	DJNZ R1,BACK	;	If R1 not equal to 0 go to BACK
	DJNZ R0, AGAIN	;	If R0 not equal to 0 go to AGAIN.

15.5.12 Program 12

Write ALP in 8051 to find the maximum number from a given 8-bit ten numbers.

Label	Mnemonics		Comments
	MOV DPTR, #2000	;	Initialize pointer to memory where
		;	numbers are stored
	MOV R0, #0A	;	Initialize counter
	MOV 30, #00	;	Maximum = 0
AGAIN:	MOVX A, @DPTR	;	Get the number from memory
	CJNE A, 30, NE	;	Compare number with maximum number
	AJMP SKIP	;	If equal go to SKIP
NE:	JC SKIP	;	If not equal check for carry,
		;	If carry go to SKIP
	MOV 30,A	;	Otherwise maximum = number
SKIP:	INC DPTR	;	Increment memory pointer
	DJNZ R0,AGAIN	;	Decrement count, if count = 0 stop
		;	otherwise go to AGAIN

15.5.13 Program 13

Write ALP to find minimum value of data in memory block 9000 to 90FF and store the result in 9100H.

```
Label       Mnemonics                Comments
            MOV  DPTR, #9000H  ;     load data pointer value
            MOV  R0, #00H      ;     number of locations
            MOV  B, #0FFH      ;     to start with B will have
                               ;     maximum value
LOOP:
            MOVX A, @DPTR
            CJNE  A, B NEXT    ;     this instruction is for Comparision
NEXT:
            JNC PROCEED        ;     if A is equal or bigger than B
                               ;     no swapping
            MOV B,A            ;     if A is smaller, transfer its value
                               ;     to B
PROCEED:
            INC   DPTR
            DJNZ  R0,LOOP      ;     be in loop till 100 location are completed.
            MOV   A, B         ;
            MOVX @DPTR,A       ;     store minimum value
            END
```

15.5.14 Program 14

Write a program to convert 8-bit binary number to its equivalent BCD.

```
Mnemonics          Comments
MOV A, #76H    ;   Load the binary number in A
MOV B, #100    ;   Load B with 100 decimal
DIV AB         ;   Divide number with 100
MOV R0, A      ;   Save the hundreds of the number
               ;   (Quotient of the previous division)
MOV A,B        ;   Get the remainder
MOV B, #10     ;   Load B with 10 decimal
DIV AB         ;   Divide number with 10
MOV R1,A       ;   Save the tens of the number
MOV R3,B       ;   Save the ones of the number
```

APPENDIX - I

DOS Function Calls

In order to use DOS function calls, always place the function number into register AH and load all other pertinent information into registers as described in the table as entry data. Once this is accomplished, follow with an INT 21H to execute the DOS function.

Ex. To display ASCII on CRT at the current cursor position with DOS instruction call.

```
0000  B4  06     MOV AH, 6        ;load function 06H
0002  B2  41     MOV DL, 'A'      ;select letter 'A'
0004  CD  21     INT 21H          ;call DOS function
```

List of DOS function calls

00H	TERMINATE A PROGRAM
Entry	AH = 00H
	CS = program segment prefix address
Exit	DOS is entered.
01H	READ THE KEYBOARD
Entry	AH = 01H
Exit	AL = ASCII character
Notes	If Al = 00H the function call must be invoked again to read an extended ASCII character. This function call automatically echoes whatever is typed to the video screen.
02H	WRITE TO STANDARD OUTPUT DEVICE
Entry	AH = 02H
	DL = ASCII character to be displayed
Notes	This function call normally displays data on the video display
03H	READ CHARACTER FROM COM1
Entry	AH = 03H
Exit	AL = ASCII character read from the communications port
Notes	This function call reads data from the serial communications port
04H	WRITE TO COM1
Entry	AH = 04H
	DL = character to be sent out of COM1
Notes	This function transmits data through the serial communications port. The COM port assignment can be changed to use other COM ports with functions 03H and 04H by using the DOS MODE command to reassign COM1 to another COM port.

05H	WRITE TO LPT1
Entry	AH = 05H
	DL = ASCII character to be printed
Notes	Prints DL on the line printer attached to LPT1. Note that the line printer port can be changed with the DOS MODE command.
06H	DIRECT CONSOLE READ/WRITE
Entry	AH = 06H
	DL = 0FFH or DL = ASCII character
Exit	AL = ASCII character
Notes	IF DL = 0FFH on entry, then this function reads the console. If DL = ASCII character, then this function displays the ASCII character on the console (CON) video screen. If a character is read from the console keyboard, the zero flag (ZF) indicates whether a character was typed. A zero condition indicates no key was typed, and a not-zero condition indicates that AL contains the ASCII code of the key or a 00H. If AL = 00H, the function must again be invoked to read an extended ASCII character from the keyboard. Note that the key does not echo to the video screen.
07H	DIRECT CONSOLE INPUT WITHOUT ECHO
Entry	AH = 07H
Exit	AL = ASCII character
Notes	This functions exactly as function number 06H with DL = 0FFH, but it will not return from the function until the key is typed.
08H	READ STANDARD INPUT WITHOUT ECHO
Entry	AH = 08H
Exit	AL = ASCII character
Notes	Performs as function 07H, except that it reads the standard input device. The standard input device can be assigned as either the keyboard or the COM port. This function also responds to a control break, where function 06H and 07H do not. A control-break causes INT 23H to execute. By default, this functions as does function 07H.
09H	DISPLAY A CHARACTER STRING
Entry	AH = 09H
	DS:DX = address of the character string
Notes	The character string must end with an ASCII (24H). The character string can be of any length and may contains control characters such as carriage return (0DH) and line feed (0AH)
0AH	BUFFERED KEYBOARD INPUT
Entry	AH = 0AH
	DS:DX = address of keyboard input buffer
Notes	The first byte of the buffer contains the size of the buffer (up to 255) The second byte is filled with the number of characters typed upon return. The third byte through the end of the buffer contains the character string typed followed by a carriage return (0DH). This function continues to read the keyboard (displaying data as typed) until either the specified number of characters are typed or until the enter key is typed.

0BH	TEST STATUS OF THE STANDARD INPUT DEVICE
Entry	AH = 0BH
Exit	AL = status of the input device
Notes	This function tests the standard input device to determine if data are available. If AL = 00, no data are available. If AL = 0FFH then data are available that must be input using function number 08H.
0CH	CLEAR KEYBOARD BUFFER AND INVOKE KEYBOARD FUNCTION
Entry	AH = 0CH
	AL = 01H, 06H, 07H, or 0AH
Exit	See exit for functions 01H, 06H, 07H or 0AH
Notes	The keyboard buffer holds keystrokes while programs execute other tasks. This function empties or clears the buffer and then invokes the keyboard function located in register AL
0DH	FLUSH DISK BUFFERS
Entry	AH = 0DH
Notes	Erases all file names stored in disk buffers. This function does not close the files specified by the disk buffers, so care must be exercised in its usage.
0EH	SELECT DEFAULT DISK DRIVE
Entry	AH = 0EH
	DL = desired default disk drive number
Exit	AL = the total number of drives present in the system
Notes	Drive A = 00H, drive B = 01H, drive C = 02H, and so forth
0FH	@ OPEN FILE WITH FCB
Entry	AH = 0FH
	DS:DX = address of the unopened file control block (FCB)
Exit	AL = 00H if file found
	AL = 0FFH if file not found
Notes	The file control block (FCB) is only used with early DOS software and should never be used with new programs. File control blocks do not allow path names as do the newer file access function codes presented later. To open a file, the file must either be present on the disk or be created with function call 16H.
10H	@ CLOSE FILE WITH FCB
Entry	AH = 10H
	DS:DX = address of the opened file control block (FCB)
Exit	AL = 00H if the closed
	AL = 0FFH if error found
Notes	Errors that occurs usually indicate either that the disk is full or the media is bad

11H	@SEARCH FOR FIRST MATCH (FCB)
Entry	AH = 11H
	DS:DX = address of the file control block to be searched
Exit	AL = 00H if file found
	AL = 0FFH if file is not found
Notes	Wild card character (? or *) may be used to search for a file name. The? wild card character matches any character and the *matches any name or extension.
12H	@ SEARCH FOR NEXT MATCH (FCB)
Entry	AH = 12H
	DS:DX = address of the file control block to be searched
Exit	AL = 00H if file found
	AL = 0FFH if file not found
Notes	This function is used after function 11H finds the first matching file name.
13H	@DELET FILE USING FCB
Entry	AH = 13H
	DS:DX = address of the file control block to be deleted.
Exit	AL = 00H if file deleted
	AL = 0FFH if error occurred
Notes	Errors that most often occur are defective media errors
14H	@SEQUENTIAL READ (FCB)
Entry	AH = 14H
	DS:DX = address of the file control block to be read
Exit	AL = 00H if read successful
	AL = 01H if end of file reached
	AL = 02H if DTA had a segment wrap
	AL = 03H if less than 128 bytes were read
15H	@SEQUENTIAL WRITE (FCB)
Entry	AH = 15H
	DS:DX = address of the file control block to be written
Exit	AL = 00H if write successful
	AL = 01H if disk is full
	AL = 02H if DTA had a segment wrap
16H	@CREATE A FILE (FCB)
Entry	AH = 16H
	DS:DX = address of an unopened file control block
Exit	AL = 00H if file created
	AL = 01H if disk is full

17H	@RENAME A FILE (FCB)
Entry	AH = 17H
	DS:DX = address of a modified file control block
Exit	AL = 00H if file renamed
	AL = 01H if error occurred
19H	RETURN CURRENT DRIVE
Entry	AH = 19H
Exit	AL = current drive
Notes	AL = 00H for drive A, 01H for drive B, and so forth
1AH	SET DISK TRANSFER AREA
Entry	AH = 1AH
	DS:DX = address of new DTA
Notes	The disk trasfer area is normally located within the program segment prefix at offset address 80H. The DTA is used by DOS for all disk data transfers using file control blocks.
1BH	GET DEFAULT DRIVE FILE ALLOCATION TABLE (FAT)
Entry	AH = 1BH
Exit	AL = number of sectors per cluster
	DS:BX = address of the media-descriptor
	CS = size of a sector in bytes
	CX = number of clusters on drive
Notes	The DS register is chagned by this function, so make sure to save it before using this function.
1CH	GET ANY DRIVE FILE ALLOCATION TABLE (FAT)
Entry	AH = 1CH
	DL = disk drive number
Exit	AL = number of sectors per cluster
	DS:BX = address of the media-descriptor
	CX = size of a sector in bytes
	DX = number of clusters on drive
21H	@RANDOM READ USING FCB
Entry	AH = 21H
	DS:DX = address of opened FCB
Exit	AL = 00H if read successful
	AL = 01H if end of file reached
	AL = 02H if the segment wrapped
	AL = 03H if less than 128 bytes read

22H	@RANDOM WRITE USING FCB
Entry	AH = 22H
	DS:DX = address of opened FCB
Exit	AL = 00H if write successful
	AL = 01H if disk full
	AL = 02H if the segment wrapped
23H	@RETURN NUMBER OF RECORDS (FCB)
Entry	AH = 23H
	DS:DX = address of FCB
Exit	AL = 00H number of records
	AL = 0FFH if file not found
24H	@SET RELATIVE RECORD SIZE (FCB)
Entry	AH = 24H
	DS:DX = address of FCB
Notes	Sets the record field to the value contained in the FCB
25H	SET INTERRUPT VECTOR
Entry	AH = 25H
	AL = interrupt vector number
	DS:DX = address of new interrupt procedure
Notes	Before changing the interrupt vector, it is suggested that the current interrupt vector first be saved using DOS function 35H. This allows a back-link so the original vector can later be restored.
26H	CREATE NEW PROGRAM SEGMENT PREFIX
Entry	AH = 26H
	DX = segment address of new PSP
27H	@RANDOM FILE BLOCK READ (FCB)
Entry	AH = 27H
	CX = the number of records
	DS:DX = address of opened FCB
Exit	AL = 00H if read successful
	AL = 01H if end of file reached
	AL = 02H if the segment wrapped
	AL = 03H if less than 128 bytes read
	CX = the number of records read
28H	@RANDOM FILE BLOCK WRITE (FCB)
Entry	AH = 28H
	CX = the number of records
	DS:DX = address of opened FCB
Exit	AL = 00H if write successful
	AL = 01H if disk full
	AL = 02H if the segment wrapped
	CX = the number of records written

29H	@PARSE COMMAND LINE (FCB)
Entry	AH = 29H
	AL = parse mask
	DS:SI = address of FCB
	DS:DI = address of command line
Exit	AL = 00H if no file name characters found
	AL = 01H if file name characters found
	AL = 0FFH if drive specifier incorrect
	DS:SI = address of character after name
	DS:DI = address first byte of FCB
2AH	READ SYSTEM DATE
Entry	AH = 2AH
Exit	AL = day of the week
	CX = the year (1980-2099)
	DH = the month
	DL = day of the month
Notes	The day of the week is encoded as Sunday = 00H through Saturday = 06H. The year is a binary number equal to 1980 through 2099.
2BH	SET SYSTEM DATE
Entry	AH = 2BH
	CX = the year (1980-2099)
	DH = the month
	DL = day of the month
2CH	READ SYSTEM TIME
Entry	AH = 2CH
Exit	CH = hours (0-23)
	CL = minutes
	DH = seconds
	DL = hundredths of seconds
Notes	All times are returned in binary form, and hundredths of second may not be available.
2DH	SET SYSTEM TIME
Entry	AH = 2DH
	CH = hours
	CL = minutes
	DH = seconds
	DL = hundredths of seconds
2EH	DISK VERIFY WRITE
Entry	AH = 2EH
	AL = 00H to disable verify on write
	AL = 01H to enable verify on write
Notes	By default, disk verify is disabled

2FH	READ DISK TRANSFER AREA ADDRESS
Entry	AH = 2FH
Exit	ES:BX = contains DTA address
30H	READ DOS VERSION NUMBER
Entry	AH = 30H
Exit	AH = fractional version number
	AL = whole number version number
Notes	For example, DOS version number 3.2 is returned as a 3 in AL and a 14H in AH.
31H	TERMINATE AND STAY RESIDENT (TSR)
Entry	AH = 31H
	AL =.the DOS return code
	DX = number of paragraphs to reserve for program
Notes	A paragraph is 16 bytes and the DOS return code is read at the batch file level with ERRORCODE.
33H	TEST CONTROL-BREAK
Entry	AH = 33H
	AL = 00H to request current control-break
	AL = 01H to change control-break
	DL = 00H to disable control-break
	DL = 01H to enable control-break
Exit	DL = current control-break state
34H	GET ADDRESS OF InDOS FLAG
Entry	AH = 34H
Exit	ES : BX = address of InDOS flag
Notes	The InDOS flag is available in DOS versions 3.2 or newer and indicates DOS activity. If InDOS = 00H, DOS if inactive or 0FFH if DOS is active and pursuing another operation.
35H	READ INTERRUPT VECTOR
Entry	AH = 35H
	AL = interrupt vector number
Exit	ES : BX = address stored at vector
Notes	This DOS function is used with function 25H to install/remove interrupt handlers.
36H	DETERMINE FREE DISK SPACE
Entry	AH = 36H
	DL = drive number
Exit	AX = FFFFH if drive invalid
	AX = number of sectors per cluster
	BX = number of free clusters
	CX = bytes per sector
	DX = number of clusters on drive
Notes	The default disk drive is DL = 00H, drive A = 01H, drive B = 02H, and so forth

38H	RETURN COUNTRY CODE
Entry	AH = 38H
	AL = 00H for current country code
	BX = 16-bit country code
	DS:DX = data buffer address
Exit	AX = error code if carry set
	BX = counter code
	DS : DX = data buffer address
39H	CREATE SUB-DIRECTORY
Entry	AH = 39H
	DS : DX = address of ASCII-Z string subdirectory name
Exit	AX = error code if carry set
Notes	The ASCII-Z string is the name of the cubdirectory in ASCII code ended with a 00H instead of a carriage return/line feed.
3AH	ERASE SUB-DIRECTORY
Entry	AH = 3AH
	DS : DX = address of ASCII-Z string sub-directory name
Exit	AX = error code if carry set
3BH	CHANGE SUB-DIRECTORY
Entry	AH = 3BH
	DS : DX = address of new ASCII-Z string sub-directory name
Exit	AX = error code if carry set
3CH	CREATE A NEW FILE
Entry	AH = 3CH
	CX = attribute word
	DS : DX = address of ASCII-Z string file name
Exit	AX = error code if carry set
	AX = file handle if carry cleared
Notes	The attribute word can contain any of the following *(added together):* 01H read-only access, 02H = hidden file or directory, 04H = system file, 08H = volume label, 10H - subdirectory, and 20H = archive bit. In most cases, a file is created with 0000H.
3DH	OPEN A FILE
Entry	AH= 3DH
	AL = access code
	DS : DX = address of ASCII-Z string file name
Exit	AX = errror code if carry set
	AX = file handle if carry cleared
Notes	The access code in AL = 00H for a read-only access, AL = 01H for a write-only access, and AL = 02H for a read/write access.
	For shared file in a network environment, bit 4 of AL = 1 will deny read/write access, bit 5 of AL = 1 will deny a write access, bits 4 and 5 of AL = 1 will deny read access, bit 6 of AL = 1 denies none, bit 7 of AL = 0 causes the file to be inherited by child, if bit 7 of AL = 1, file is restricted to current process.

3EH	CLOSE A FILE
Entry	AH = 3EH
	BX = file handle
Exit	AX = error code if carry set
3FH	READ A FILE
Entry	AH = 3FH
	BX = file handle
	CX = number of bytes to be read
	DS:DX = address of file buffer to hold data read
Exit	AX = error code if carry set
	AX = number of bytes read if carry cleared
40H	WRITE A FILE
Entry	AH = 40H
	BX = file handle
	CX = number of bytes to write
	DS:DX = address of file buffer that holds write data
Exit	AX = error code if carry set
	AX = number of bytes written if carry cleared
41H	DELETE A FILE
Entry	AH = 41H
	DS:DX = address of ASCII-Z string file name
Exit	AX = error code if carry set
42H	MOVE FILE POINTER
Entry	AH = 42H
	AL = move technique
	BX = file handle
	CX:DX = number of bytes pointer moved
Exit	AX = error code if carry set
	AX:DX = bytes pointer moved
Notes	The move technique causes the pointer to move from the start of the file if AL = 00H, from the current location if AL = 01H, and from the end of the file if AL = 02H. The count is stored so DX contains the least-significant 16-bits and either CX or AX contains the most significant 16-bits.
43H	READ/WRITE FILE ATTRIBUTES
Entry	AH = 43H
	AL = 00H to read attributes
	AL = 01H to write attributes
	CX = attribute word (see function 3CH)
	DS:DX = address of ASCII-Z file name
Exit	AX = error code if carry set
	CX = attribute word of carry cleared

44H	I/O DEVICE CONTROL (IOTCL)
Entry	AH = 44H
	AL = sub-function code (see notes)
Exit	AX = error code (see function 59H) if carry set
Notes	The sub-function codes found in AL are as follows :
	00H = read device status
	Entry : BX = handle
	Exit : DX = status
	01H = write device status
	Entry : BX = file handle, DH = 0, DL = device information
	Exit : AX = error code if carry set
	02H = read control data from character device
	Entry : BX = file handle, CX = number of bytes, DS:DX = I/O buffer address
	Exit : AX = number of bytes read
	03H = write control data to character device
	Entry : BX = file handle, CX = number of bytes, DS:DX = I/O buffer address
	Exit : AX = number of bytes written
	04H = read control data from block device
	Entry : BL = drive number (0 = default, 1 = A, 2 = B etc.) CX = number of bytes, DS:DX = I/O buffer address
	Exit : AX = number of bytes read
	05H = write control data to block device
	Entry : BL = drive number, CX = number of bytes, DS:DX = I/O buffer address
	Exit : AX = number of bytes written
	06H = check input status
	Entry : BX = file handle
	Exit : AL = 00H ready or FFH not ready
	07H = check output status
	Entry : BX = file handle
	Exit : AL = 00H ready or FFH not ready
	08H = removable media ?
	Entry : BL = drive number
	Exit : AL = 00H removable, 01H fixed
	09H = netword block device?
	Entry : BL = drive number
	Exit : bit 12 of DX set for network block device
	0AH = local or network character device ?
	Entry : BX = file handle
	Exit : bit of 15 of DX set for network character device

	0BH = change entry count (must have SHARE.EXE loaded) Entry : CX = delay loop count, DX = retry count Exit : AX = error code if carry set 0CH = generic I/O control for character devices Entry : BX = file handle, CH = category, CL = function Categories : 00H = unknown, 01H = COM port, 02H = CON, 05H = LPT ports Function : CL = 45H; set iteration count CL = 4AH; select code page CL = 4CH; start code page preparation CL = 4DH; end code page preparation CL = 5FH; set display information CL = 65H; get iteration count CL = 6AH; query selected code page CL = 6BH; query preparation list CL = 7FH; get display information 0DH = generic I/O control for block devices Entry : BL = drive number, CH = category, CL = function, DS:DX = address of parameter block Category : 08H = disk drive Function : CL = 40H; set device parameters CL = 41H; write track CL = 42H; format and verify track CL = 46H; set media ID code CL = 47H; set access flag CL = 60H; get device parameters CL = 61H; read track CL = 62H; verify track CL = 66H; get media ID code CL = 67H; get access code 0EH = return logical device map Entry : BL = drive number Exit : AL = number of last device 0FH = change logical device map Entry : BL = drive number Exit : AL = numbr of last device
45H	DUPLICATE FILE HANDLE
Entry	AH = 45H BX = current file handle
Exit	AX = error code if carry set AX = duplicate file handle

46H	FORCE DUPLICATE FILE HANDLE
Entry	AH = 46H
	BX = current file handle
	CX = new file handle
Exit	AX = error code if carry set
Notes	This function works like function 45H except that function 45H allows DOS to select the new handle, while this function allows the user to select the new handle.
47H	READ CURRENT DIRECTORY
Entry	AH = 47H
	DL = drive number
	DS:SI = address of 64 byte buffer for directory name
Exit	DS:SI address current directory name if carry cleared
Notes	Drive A = 00, drive B = 01, and so forth
48H	ALLOCATE MEMORY BLOCK
Entry	AH = 48H
	BX = number of paragraphs to allocate
	CX = new file handle
Exit	BX = largest block available if carry cleared
49H	RELEASE ALLOCATED MEMORY BLOCK
Entry	AH = 49H
	ES = segment address of block to be released
	CX = new file handle
Exit	Carry indicates an error if set
4AH	MODIFY ALLOCATED MEMORY BLOCK
Entry	AH = 4AH
	BX = new block size in paragraphs
	ES = segment address of block to be modified
Exit	BX = largest block available if carry cleared
4BH	LOAD OR EXECUTE A PROGRAM
Entry	AH = 4BH
	AL = function code
	ES:BX = address of parameter block
	DS:DX = address ASCII-Z string command
Exit	Carry indicates an error if set
Notes	The function codes are AL = 00H to load and execute a program, AL = 01H to load a program but not execute it, AL = 03H to load a program overlay, and AL = 05H to enter the EXEC state.
4CH	TERMINATE A PROCESS
Entry	AH = 4CH
	AL = error code

Exit	Returns control to DOS
Notes	This function returns control to DOS with the error code saved so it can be obtained using DOS ERROR LEVEL batch processing system. We normally use this function with an error code of 00H to return to DOS.
4DH	READ RETURN CODE
Entry	AH = 4DH
Exit	AX = return error code
Notes	This function is used to obtain the return status code created by executing a program with DOS function 4BH. The return codes are AX = 0000H for a normal-no error-termination, AX = 0001H for a control-break termination, AX = 0002H for a critical device error, and AX = 0003H for a termination by an INT 31H.
4EH	FIND FIRST MATCHING FILE
Entry	AH = 4EH
	CX = file attributes
	DS:DX = address ASCII-Z string file name
Exit	Carry is set for file not found
Notes	This function searches the current or named directory for the first matching file. Upon exit, the DTA contains the file information.
4FH	FIND NEXT MATCHING FILE
Entry	AH = 4FH
Exit	carry is set for file not found
Notes	This function is used after the first file is found with function 4EH
50H	SET PROGRAM SEGMENT PREFIX (PSP) ADDRESS
Entry	AH = 50H
	BX = offset address of the new PSP
Notes	Extreme care must be used with this function because no error recovery is possible.
51H	GET PSP ADDRESS
Entry	AH = 51H
Exit	BX = current PSP segment address
54H	READ DISK VERIFY STATUS
Entry	AH = 54H
Exit	AL = 00H if verify off
	AL = 01H if verify on
56H	RENAME FILE
Entry	AH = 56H
	ES:DI = address of ASCII-Z string containing new file name
	DS:DX = address of ASCII-Z string containing file to be renamed.
Exit	Carry is set for error condition

57H	READ FILE'S DATE AND TIME STAMP
Entry	AH = 57H
	AL = function code
	BX = file handle
	CX = new time
	DX = new date
Exit	Carry is set for error condition
	CX = time if carry cleared
	DX = date if carry cleared
Notes	AL = 00H to read date and time or 01H to write date and time
59H	GET EXTENDED ERROR INFORMATION
Entry	AH = 59H
	BX = 0000H for DOS version 3.X
Exit	AX = extended error code
	BH = error class
	BL = recommended action
	CH = locus
Notes	Following are the extended error codes found in AX :
	0001H = invalid function number
	0002H = file not found
	0003H = path not found
	0004H = no file handles available
	0005H = access denied
	0006H = file handle invalid
	0007H = memory control block failure
	0008H = insufficiet memory
	0009H = memory block address invalid
	000AH = environment failure
	000BH = format invalid
	000CH = access code invalid
	000DH = data invalid
	000EH = unknown unit
	000FH = disk drive invalid
	0010H = attempted to remove current directory
	0011H = not same device
	0012H = no more files
	0013H = disk write-protected
	0014H = unknown unit
	0015H = drive not ready
	0016H = unknown command
	0017H = data error (CRC check error)
	0018H = bad request structure length

0019H = seek error
001AH = unknown media type
001BH = sector not found
001CH = printer out of paper
001DH = write fault
001EH = read fault
001FH = general failure
0020H = sharing violation
0021H = lock violation
0022H = disk change invalid
0023H = FCB unavailable
0024H = sharing buffer exceeded
0025H = code page mismatch
0026H = handle end of file operation not completed
0027H = disk full
0028H-0031H reserved
0032H = unsupported network request
0033H = remote machine not listed
0034H = duplicate name on network
0035H = network name on network
0036H = network busy
0037H = device no longer exists on network
0038H = netBIOS command limit exceeded
0039H = error in network adapter hardware
003AH = incorrect response from network
003BH = unexpected network error
003CH = remote adapter is incompatible.
003DH = print queue is full
003EH = not enough room for print file
003FH = print file was deleted
0040H = network name deleted
0041H = network access denied
0042H = incorrect network device type
0043H = network name not found
0044H = network name exceeded limit
0045H = netBIOS session limit exceeded
0046H = temporary pause
0047H = network request not accepted
0048H = print or disk redirection pause
0049H-004Fh reserved
0050H = file already exists
0051H = duplicate FCB
0052H = cannot make directory
0053H = failure in INT 24H (critical error)

	0054H = too many re-directions 0055H = duplicate redirection 0056H = invalid password 0057H = invalid parameter 0058H = network write failure 0059H = function not supported by network 005AH = required system component not installed 0065H = device not selected Following are the error class codes as found in BH : 01H = no resources available 02H = temporary error 03H = authorization error 04H = internal software error 05H = hardware error 06H = system failure 07H = application software error 08H = item not found 09H = invalid format 0AH = item blocked 0BH = media error 0CH = item already exists 0DH = unknown error Following in the recommended action as found in BL : 01H = retry operation 02H = delay and retry operation 03H = user retry 04H = abort processing 05H = immediate exit 06H = ignore error 07H = retry with user intervention Following is a list of locus in CH : 01H = unknown source 02H = block device error 03H = network area 04H = serial device error 05H = memory error
5AH	CREATE UNIQUE FILE NAME
Entry	AH = 5AH CX = attribute code DS:DX = address of the appended directory name
Exit	Carry is set for error condition AX = file handle if carry cleared DS:DX = address of the appended directory name

Notes	The ASCII-Z file directory path must end with a backslash (\). On exit, the directory name is appended with a unique file name.
5BH	CREATE A DOS FILE
Entry	AH = 5BH
	CX = attribute code
	DS:DX = address of the ASCII-Z string contain the file name
Exit	Carry is set for error condition
	AX = file handle if carry cleared
Notes	The function only works in DOS version 3.X or higher. It is almost identical to function 3CH, except that function 3CH erases the file; if it already exists, while function 5BH reports that the file exists without erasing it.
5CH	LOCK/UNLOCK FILE CONTENTS
Entry	AH = 5CH
	BX = file handle
	CX:DX = offset address of locked/unlocked area
	SI:DI = number of bytes to lock or unlock beginning at offset
Exit	Carry is set for error condition
5DH	SET EXTENDED ERROR INFORMATION
Entry	AH = 5DH
	AL = 0AH
	DS:DX = address of the extended error data structure
Notes	This function is used by DOS version 3.1 or higher to store extended error information.
5EH	NETWORK/PRINTER
Entry	AH = 5EH
	AL = 00H (get network name)
	DS:DX = address of the ASCII-Z string containing network name
Exit	Carry is set for error condition
	CL = netBIOS number if carry cleared
Entry	AH = 5EH
	AL = 02H (define network printer)
	BX = redirection list
	CX = length of setup string
	DS:DX = address of printer setup buffer
Exit	Carry is set for error condition
Entry	AH = 5EH
	AL = 03H (read network printer setup string)
	BX = redirection list
	DS:DX = address of printer setup buffer
Exit	Carry is set for error condition
	CX = length of setup string if carry cleared
	ES:DI = address of printer setup buffer

62H	GET PSP ADDRESS
Entry	AH = 62H
Exit	BX = segment address of the current program
Notes	The function only works in DOS version 3.0 or higher.
65H	GET EXTENDED COUNTRY INFORMATION
Entry	AH = 65H AL = function code ES:DI = address of buffer to receive information
Exit	Carry is set for error condition CX = length of country information
Notes	The function only works in DOS version 3.3 or higher
66H	GET/SET CODE PAGE
Entry	AH = 65H AL = function code BX = code page number
Exit	Carry is set for error condition BX = active code page number DX = default code page number
Notes	A function code in AL of 01H gets the code page number, and a code of 02H sets the code page number
67H	SET HANDLE COUNT
Entry	AH = 67H BX = number of handles desired
Exit	Carry is set for error condition
Notes	This function is available for DOS version 3.3 or higher.
68H	COMMIT FILE
Entry	AH = 68H BX = handle number
Exit	Carry is set for error condition; otherwise, the date and time stamp is written to directory
Notes	This function is available for DOS version 3.3 or higher
6CH	EXTENDED OPEN FILE
Entry	AH = 6CH AL = 00H BX = open mode CX = attributes DX = open flag DS,SI = address of ASCII-Z string file name
Exit	AX = error code if carry is set AX = handle if carry is cleared CX = 0001H file existed and was opened CX = 0002H file did not exist and was created
Notes	This function is available for DOS version 4.0 or higher.

APPENDIX - II

BIOS Function Calls

Apart from DOS function call INT 21H, some other BIOS function calls are useful in controlling the I/O the computer. Unlike INT 21H, which exists in the DOS program, the BIOS function calls are found stored in the system and video BIOS ROMs. These BIOS ROM functions directly control the I/O devices with or without DOS loaded into a system.

INT 10H

The INT 10H BIOS interrupt is often called the video services interrupt because it directly controls the video display in a system. The INT 10H instruction uses register. AH to select the video service provided by this interrupt. (*The video BIOS ROMS is located on the video board and varies from one video card to another).*

Table I : Video display modes

Mode	Type	Columns	Rows	Resolution	Standard	Colours
00H	Text	40	25	320 × 200	CGA	2
00H	Text	40	25	320 × 250	EGA	2
00H	Text	40	25	360 × 400	VGA	2
01H	Text	40	25	320 × 200	CGA	16
01H	Text	40	25	320 × 350	EGA	16
01H	Text	40	25	360 × 640	VGA	16
02H	Text	80	25	640 × 200	CGA	2
02H	Text	80	25	640 × 350	EGA	2
02H	Text	80	25	720 × 400	VGA	2
03H	Text	80	25	640 × 200	CGA	16
03H	Text	80	25	640 × 350	EGA	16
03H	Text	80	25	720 × 400	VGA	16
04H	Graphics	80	25	320 × 200	CGA	4
05H	Graphics	80	25	320 × 350	CGA	2
06H	Graphics	80	25	640 × 200	CGA	2
07H	Text	80	25	720 × 350	EGA	4
07H	Text	80	25	720 × 400	VGA	4
ODH	Graphics	80	25	320 × 200	CGA	16
OEH	Graphics	80	25	640 × 200	CGA	16
OFH	Graphics	80	25	640 × 350	EGA	4
10H	Graphics	80	25	640 × 350	EGA	16
11H	Graphics	80	30	640 × 480	VGA	2
12H	Graphics	80	30	640 × 480	VGA	16
13H	Graphics	40	25	320 × 200	VGA	256

The VGA can use any mode listed, while the other displays are more restrictive in use. Additional higher resolution modes are explained later in this section.

Ex : Short sequence of instruction to place video display into mode 03H operation.

(This mode is available on CGA, EGA and VGA displays. This mode allows the display to draw test data with 16 colours at various resolutions dependent upon the display adapter)

```
0000  B4  00      MOV AH, 0      ;select mode
0002  B0  03      MOV AL, 3      ;mode is 03H
0004  CD  10      INT 10H
```

Cursor Control and Other Standard Features

Appendix II illustrates the function codes (*placed in AH*) used to control the cursor on the video display. *(These cursor control functions will work on any video display from the CGA display to the latest super VGA display. It also lists the functions used to display data and change to a different character set).*

Video BIOS (INT 10H) functions

00H	SELECT VIDEO MODE
Entry	AH = 00H
	AL = mode number
Exit	Most changed and screen cleared

01H	SELECT CURSOR TYPE
Entry	AH = 02H
	CH = starting line number
	CL = ending line number
Exit	Cursor size changed
02H	SELECT CURSOR POSITION
Entry	AH = 02H
	BH = page number (usually 0)
	DH = row number (beginning with 0)
	DL = column number (beginning with 0)
Exit	Changes cursor to new position
03H	READ CURSOR POSITION
Entry	AH = 03H
	BH = page number
Exit	CH = starting line (cursor size)
	CL = ending line (cursor size)
	DH = current row
	DL = current column
04H	READ LIGHT PEN
Entry	AH = 04H (not supported in VGA)

Exit	AH = 0, light pen triggered BX = pixel column CX = pixel row DH = character row DL = character column
05H	SELECT DISPLAY PAGE
Entry	AH = 05H AL = page number
Exit	Page number selected, Following are the valid page numbers: Mode 0 and 1 support pages 0-7 Mode 2, and 3 support pages 0-7 Mode 4,5 and 6 support page 0 Mode 7 and D support pages 0-7 Mode E supports pages 0-3 Mode F and 10 support pages 0-1 Mode 11, 12 and 13 support page 0
Notes	Most modern displays use page 0 for most operations.

06H	SCROLL PAGE UP
Entry	AH = 06H AL = number of lines to scroll (0 clears window) BH = character attribute for new lines CH = top row of scroll window CL = left column of scroll window DH = bottom row of scroll window DL = right column of scroll window
Exit	Scrolls window from the bottom toward the top of the screen. Blank lines fill the bottom using the character attribute in BH
07H	SCROLL PAGE DOWN
Entry	AH = 07H AL = number of lines to scroll (0 clears window) BH = character attribute for new lines CH = top row of scroll window CL = left column of scroll window DH = bottom row of scroll window DL = right column of scroll window
Exit	Scrolls window from the top toward the bottom of the screen. Blank lines fill from the top using the character attribute in BH.

08H	READ ATTRIBUTE/CHARACTER AT CURRENT CURSOR POSITION
Entry	AH = 08H
	BH = page number
Exit	AL = ASCII character code
	AH = character attribute
Notes	This function does not advance the cursor
09H	WRITE ATTRIBUTE/CHARACTER AT CURRENT CURSOR POSITION
Entry	AH = 09H
	AL = ASCII character code
	BH = page number
	BL = character attribute
	CX = number of characters to write
Notes	This function does not advance the cursor

0AH	WRITE CHARACTER AT CURRENT CURSOR POSITION
Entry	AH = 0AH
	AL = ASCII character code
	BH = page number
	CX = number of characters to write
Notes	This function does not advance the cursor
0FH	READ VIDEO MODE
Entry	AH = 0FH
Exit	AL = current video mode
	AH = number of character columns
	BH = page number
10H	SET VGA PALETTE REGISTER
Entry	AH = 10H
	AL = 10H
	BX = colour number (0-255)
	CH = green (0-63)
	CL = blue (0-63)
	DH = red (0-63)
Exit	Palette register colour is changed Note : The first 16 colors (0-15) are used in the 16-color, VGA text mode and other modes
10H	READ VGA PALETTE REGISTER
Entry	AH = 10H
	AL = 15H
	BX = color number (0-255)
Exit	CH = green
	CL = blue
	DH = red

11H	GET ROM CHARACTER SET
Entry	AH = 11H
	AL = 30H
	BH = 2 = ROM 8 × 14 character set
	BH = 3 = ROM 8 × 8 character set
	BH = 4 = ROM 8 × 8 extended character set
	BH = 5 = ROM 9 × 14 character set
	BH = 6 = ROM 8 × 16 character set
	BH = 7 = ROM 9 × 16 character set
Exit	CX = bytes per character
	DL = rows per character
	ES:BP = address of character set

APPENDIX - III

8086 Instruction set summary

Mnemonic	Description	Clock cycles	Number of bytes
AAA	ASCII adjust for addition	4	1
AAD	ASCII adjust for division	60	2
AAM	ASCII adjust for multiplication	83	2
AS	ASCII adjust for subtraction	4	1
ADC	Add with carry		
	Register to register	3	2
	Memory to register	9 + EA	2-4
	Register to memory	16 + EA	2-4
	Immediate to register	4	3-4
	Immediate to memory	17 + EA	3-6
	Immediate to accumulator	4	2-3
ADD	Addition		
	Register to register	3	2
	Memory to register	9 + EA	2-4
	Register to memory	16 + EA	2-4
	Immediate to register	4	3-4
	Immediate to memory	17 + EA	3-6
	Immediate to accumulator	4	2-3
AND	Logical AND		
	Register to register	3	2
	Memory to register	9 + EA	2-4
	Register to memory	16 + EA	2-4
	Immediate to register	4	3-4
	Immediate to memory	17 + EA	3-6
	Immediate to accumulator	4	2-3
CALL	Call a procedure		
	Intrasegment direct	19	3
	Intrasegment indirect through register	19	2
	Intrasegment indirect through memory	21 + EA	2-4
	Intersegment direct	28	5
	Intersegment indirect	37 + EA	2-4
CBW	Convect. byte to word	2	1
CLC	Clear carry flag	2	1

Mnemonic	Description	Clock cycles	Number of bytes
CLD	Clear direction flag	2	1
CLI	Clear interrupt flag	2	1
CMC	Complement carry flag	2	1
CMP	Compare		
	Register to register	3	2
	Memory to register	9 + EA	2-4
	Register to memory	9 + EA	2-4
	Immediate to register	4	3-4
	Immediate to memory	17 + EA	3-6
	Immediate to accumulator	4	2-3
CMPS/ CMPSB/ CMPSW	Compare string/ Compare byte string/ Compare word string		1
	Not repeated	22	
	Repeated	9 + 22/rep	
CWD	Convert word to double word	5	1
DAA	Decimal adjust for addition	4	1
DAS	Decimal adjust for subtraction	4	1
DEC	Decrement by 1		
	16-bit register	2	1
	8-bit register	3	2
	Memory	15 + EA	2-4
DIV	Unsigned division		
	8-bit register	80-90	2
	16-bit register	144-162	2
	8-bit memory	(86-96) + EA	2-4
	16-bit memory	(150-168) + EA	2-4
ESC	Escape		
	Register	2	2
	Memory	8 + EA	2-4
HLT	Halt	2	1
IDIV	Integar division		
	8-bit register	101-112	2
	16-bit register	165-184	2
	8-bit memory	(107-118) + EA	2-4
	16-bit memory	(171-190) + EA	2-4
IMUL	Integer multiplication		
	8-bit register	80-98	2
	16-bit register	128-154	2
	8-bit memory	(86-104) + EA	2-4
	16-bit memory	(134-160) + EA	2-4

Mnemonic	Description	Clock cycles	Number of bytes
IN	Input from I/O port		
	Fixed port	10	2
	Variable port	8	1
INC	Increment by 1		
	16-bit register	2	1
	8-bit register	3	2
	Memory	15 + EA	2-4
INT	Interrupt		
	Type-3	52	1
	Type-3	51	2
INTO	Interrupt if overflow	1	
	Interrupt is taken	53	
	Interrupt is not taken	4	
IRET	Return from interrupt	24	1
JA/	Jump if above/	16/4	2
JNBE	Jump if not below or equal		
JAE/	Jump if above or equal/	16/4	2
JNB/	Jump if not below/		
JNC/	Jump if not carry/		
JB/	Jump if below/	16/4	2
JNAE/	Jump if not above or equal/		
JC	Jump if carry		
JBE/	Jump if below or equal/	16/4	2
JNA	Jump if not above		
JCXZ	Jump if CX is zero	18/6	2
JE/	Jump if equal/	16/4	2
JZ	Jump if zero		
JG/	Jump if greater/	16/4	2
JNLE	Jump if not less or equal		
JGE/	Jump if greater or equal/	16/4	2
JNL	Jump if not less		
JL/	Jump if less/	16/4	2
JNGE	Jump if not greater or equal		
JLE/	Jump if less or equal/	16/4	2
JNG	Jump if not greater		
JMP	Jump		
	Intrasegment direct short	15	2
	Intrasegment direct	15	3
	Intersegment direct	15	5
	Intrasegment indirect through memory	18 + EA	2-4

Mnemonic	Description	Clock cycles	Number of bytes
	Intrasegment indirect through register	11	2
	Intersegment indirect	24 + EA	2-4
JNE/	Jump if not equal/	16/4	2
JNZ	Jump if not zero		
JNO	Jump if not overflow	16/4	2
JNP/	Jump if not parity/	16/4	2
JPO	Jump if parity odd		
JNS	Jump if not sign	16/4	2
JO	Jump if overflow	16/4	2
JP/	Jump if parity/	16/4	2
JPE	Jump if parity even		
JS	Jump if sign	16/4	2
LAHF	Load AH from flags	4	1
LDS	Load pointer using DS	16 + EA	2-4
LEA	Load effective address	2 + EA	2-4
LES	Load pointer using ES	16 + EA	2-4
LOCK	Lock bus	2	1
LODS/	Load string/	1	
LODSB	Load byte string/		
LODSW	Load word string		
	Not repeated	12	
	Repeated	9 + 13/rep	
LOOP	Loop	17/5	2
LOOPE/	Loop if equal/	18/6	2
LOOPZ	Loop if zero		
LOOPNE/	Loop if not equal	19/5	2
LOOPNZ	Loop if not zero		
MOV	Move		
	Accumulator to memory	10	3
	Memory to accumulator	10	3
	Register to register	2	2
	Memory to register	8 + EA	2-4
	Register to memory	9 + EA	2-4
	Immediate to register	4	2-3
	Immediate to memory	10 + EA	3-6
	Register to SS, DS, or ES	2	2
	Memory to SS, DS or ES	8 + EA	2-4
	Segment register to register	2	2
	Segment register to memory	9 + EA	1

Mnemonic	Description	Clock cycles	Number of bytes
MOVS/	Move string/		1
MOVSB/	Move byte string/		
MOVSW	Move word string		
	Not repeated	18	
	Repeated	9 + 17/rep	
MUL	Unsigned multiplication		
	8-bit register	70-77	2
	16-bit register	118-133	2
	8-bit memory	(76-83) + EA	2-4
	16-bit memory	(124-139) + EA	2-4
NEG	Negate		
	Register	3	2
	Memory	16 + EA	2-4
NOP	No operation	3	1
NOT	Logical NOT		
	Register	3	2
	Memory	16 + EA	2-4
OR	Logical OR		
	Register to register	3	2
	Memory to register	9 + EA	2-4
	Register to memory	16 + EA	2-4
	Immediate to accumulator	4	2-3
	Immediate to register	4	3-4
	Immediate to memory	17 + EA	3-6
OUT	Output to I/O port		
	Fixed port	10	2
	Variable port	8	1
POP	Pop word off stack		
	Register	8	1
	Segment register SS, DS, or ES	8	1
	Memory	17 + EA	2-4
POPF	Pop flags off stack	8	1
PUSH	Push word onto stack		
	Register	11	1
	Segment register	10	1
	Memory	16 + EA	2-4
PUSHF	Push flags onto stack	10	1
RCL	Rotate left through carry		
	Register with single-shift	2	2
	Register with variable-shift	8 + 4/bit	2

Mnemonic	Description	Clock cycles	Number of bytes
	Memory with single-shift	15 + EA	2-4
	Memory with variable-shift	20 + EA + 4/bit	2-4
RCR	Rotate right through carry		
	Register with single-shift	2	2
	Register with variable-shift	8 + 4/bit	2
	Memory with single-shift	15 + EA	2-4
	Memory with variable-shift	20 + EA + 4/bit	2-4
REP	Repeat string operation	2	1
REPE/	Repeat operation while equal/	2	1
REPZ	Repeat operation while zero		
REPNE/	Repeat operation while not equa/	2	1
REPNZ	Repeat operation while not zero		
RET	Return from procedure		
	Intrasegment	8	1
	Intrasegment with constant	12	3
	Intersegment	18	1
	Intersegment with constant	17	3
ROL	Rotate last		
	Register with single-shift	2	2
	Register with variable-shift	8 + 4/bit	2
	Memory with single-shift	15 + EA	2-4
	Memory with variable-shift	20 + EA + 4/bit	2-4
ROR	Rotate right		
	Register with single-shift	2	2
	Register with variable-shift	8 + 4/bit	2
	Memory with single-shift	15 + EA	2-4
	Memory with variable-shift	20 + EA + 4/bit	2-4
SAHF	Store AH into flags	4	1
SAL/	Shift arithmetic left/		
SHL	Shift logical left		
	Register with single-shift	2	2
	Register with variable-shift	8 + 4/bit	2
	Memory with single-shift	15 + EA	2-4
	Memory with variable-shift	20 + EA + 4/bit	2-4
SAR	Shift arithmetic right		
	Register with single-shift	2	2
	Register with variable-shift	8 + 4/bit	2

Mnemonic	Description	Clock cycles	Number of bytes
	Memory with single-shift	15 + EA	2-4
	Memory with variable-shift	20 + EA + 4/bit	2-4
SBB	Subtract with borrow		
	Register from register	3	2
	Memory from register	9 + EA	2-4
	Register from memory	16 + EA	2-4
	Immediate from accumulator	4	2-3
	Immediate from register	4	3-4
	Immediate from memory	17 + EA	3-6
SCAS/	Scan string/		1
SCASB/	Scan byte string/		
SCASW	Scan word string		
	Not repeated	15	
	Repeated	9 + 15/rep	
SHR	Shift logical right		
	Register with single-shift	2	2
	Register with variable-shift	8 + 4/bit	2
	Memory with single-shift	15 + EA	2-4
	Memory with variable-shift	20 + EA + 4/bit	2-4
STC	Set carry flag	2	1
STD	Set direction flag	2	1
STI	Set interrupt flag	2	1
STOS/	Store string/		1
STOSB/	Store word string		
	Not repeated	11	
	Repeated	9 + 10/rep	
SUB	Subtraction		
	Register from register	3	2
	Memory from register	9 + EA	2-4
	Register from memory	16 + EA	2-4
	Immediate from accumulator	4	2-3
	Immediate from register	4	3-4
	Immediate from memory	17 + EA	3-6
TEST	Test		
	Register with register	3	2
	Memory with register	9 + EA	2-4
	Immediate with accumulator	4	2-3
	Immediate with register	5	3-4
	Immediate with memory	11 + EA	3-6

Mnemonic	Description	Clock cycles	Number of bytes
WAIT	Wait while $\overline{TEST}$ pin not asserted	3 + 5n	1
XCHG	Exchange		
	Register with accumulator	3	1
	Register with memory	17 + EA	2-4
	Register with register	4	2
XLAT/ XLATB	Translate	11	1
XOR	Logical exclusive OR		
	Register with register	3	2
	Memory with register	9 + EA	2-4
	Register with memory	16 + EA	2-4
	Immediate with accumulator	4	2-3
	Immediate with register	4	3-4
	Immediate with memory	17 + EA	3-6

APPENDIX - IV

8086 Instruction formates

Mnemonic and Description	Instruction Code

I DATA TRANSFER

MOV = MOVE :

7 6 5 4 3 2 1 0 7 6 5 4 3 2 1 0 7 6 5 4 3 2 1 0 7 6 5 4 3 2 1 0

1 0 0 0 1 0 dw	mod reg/m

Register/Memory to/from Register

Immediate to Register/Memory

1 1 0 0 0 1 1 w	mod 0 0 0 r / m	data	data if w = 1

Immediate to Register

1 0 1 1 w reg	data	data if w = 1

Memory to Accumulator

1 0 1 0 0 0 0 w	addr—low	addr—high

Accumulator to Memory

1 0 1 0 0 0 1 w	addr—low	addr—high

Register/Memory to Segment Register

1 0 0 0 1 1 1 0	mod 0 reg r/m

Segment Register to Register/Memory

1 0 0 0 1 1 0 0	mod 0 reg r/m

PUSH = Push

Register/Memory

1 1 1 1 1 1 1 1	mod reg r/m

Register

0 1 0 1 0 reg

Segment Register

0 0 0 reg 1 1 1

POP = Pop :

Register/Memory

1 0 0 0 1 1 1 1	mod 0 0 0 r/m

Register

0 1 0 1 1 reg

Segment Register

0 0 0 reg 1 1 1

XCHG = Exchange :

Register/Memory with Register	1 0 0 0 0 1 1 w	mod reg r/m
Register with Accumulator	1 0 0 1 0 reg	

IN = Input from :

Fixed Port	1 1 1 0 0 1 0 w	port
Variable Port	1 1 1 0 1 1 0 w	

OUT = Output to

Fixed Port	1 1 1 0 0 1 1 w	port
Variable Port	1 1 1 0 1 1 1 w	
XLAT = Translate Byte to AL	1 1 1 0 1 1 1 w	
LEA = Load EA to Register	1 0 0 0 1 1 0 1	mod reg r/m
LDS = Load Pointer to DS	1 1 0 0 0 1 0 1	mod reg r/m
LES = Load Pointer to ES	1 1 0 0 1 1 0 0	mod reg r/m
LAHF = Load AH with Flags	1 0 0 1 1 1 1 1	
SAHF = Store AH into Flags	1 0 0 1 1 1 1 0	

Mnemonic and Description	**Instruction Code**
PUSHF = Push Flags	1 0 0 1 1 1 0 0
POPF = Pop flags	1 0 0 1 1 1 0 1

II ARITHMETIC

ADD = Add

7 6 5 4 3 2 1 0 7 6 5 4 3 2 1 0 7 6 5 4 3 2 1 0 7 6 5 4 3 2 1 0

0 0 0 0 0 0 dw	mod reg/m

Reg/Memory with Register to Either Immediate to Register/Memory

1 0 0 0 0 0 sw	mod 0 0 0 r/m	data	data if s:w = 01

Immediate to Accumulator

0 0 0 0 0 1 0 w	data	data if w = 1

ADC = Add with Carry :

Reg/Memory with Register to Either

0 0 0 1 0 0 dw	mod reg r/m

Immediate to Register/Memory

1 0 0 0 0 0 sw	mod 0 0 1 r/m	data	data if s:w = 01

Immediate to Accumulator

0 0 0 1 0 1 0 w	data	data if w = 1

INC = Increment :

Register Memory

1 1 1 1 1 1 w	mod 0 0 0 r/m

Register

0 1 0 0 0 reg

AAA = ASCII Adjust for Add

0 0 1 1 0 1 1 1

BAA = Decimal Adjust for ADD

0 0 1 0 0 1 1 1

SUB = Subtract :

Reg/Memory and Register to Either

0 0 1 0 1 0 dw	mod reg r/m

Immediate from Register/Memory

1 0 0 0 0 0 sw	mod 1 0 1 r/m	data	data if s:w = 01

Immediate from Accumulator

0 0 1 0 1 1 0 w	data	data if w = 1

SSB = Subtract with Borrow

Reb/Memory and Register to Either

0 0 0 1 1 0 dw	mod reg r/m

Immediate from Register/Memory

1 0 0 0 0 0 sw	mod 0 1 1 r/m	data	data if s:w = 01

Immediate from Accumulator

0 0 0 1 1 1 w	data	data if w = 1

DEC = Decrement :

Register/Memory

1 1 1 1 1 1 1 w	mod 0 0 1 r/m

Register

0 1 0 0 1 reg

NEG = Change sign

1 1 1 0 1 1 dw	mod 0 1 1 r/m

CMP = Compare :

Register/Memory and Register	1 1 1 1 1 1 1 w	mod 0 0 1 r/m	

Immediate to Register/Memory

1 0 0 0 0 0 sw	mod 1 1 1 r/m	data	data if s:w = 01

Immediate to Accumulator	0 0 1 1 1 1 0 w	data	data if w = 1
AAS = ASCII Adjust for Subtract	0 0 1 1 1 1 1 1		
DAS = Decimal Adjust for Subtract	0 0 1 0 1 1 1 1		
MUL = Multiply (Unsigned)	1 1 1 1 0 1 1 w	mod 1 0 0 r/m	
IMUL = Integer Multiply (Signed)	1 1 1 1 0 1 1 w	mod 1 0 1 r/m	
AAM = ASCII Adjust for Multiply	1 1 0 1 0 1 0 0	0 0 0 0 1 0 1 0	
DIV = Divide (Unsigned)	1 1 1 1 0 1 1 w	mod 1 1 0 r/m	
IDIV = Integer Divide (Signed)	1 1 1 1 0 1 1 w	mod 1 1 1 r/m	
AAD = ASCII Adjust for Divide	1 1 0 1 0 1 0 1	0 0 0 0 1 0 1 0	
CBW = Convert Byte to Word	1 0 0 1 1 0 0 0		
CWD = Convert Word to Double Word	1 0 0 1 1 0 0 1		

Mnemonic and Description **Instruction Code**

III LOGIC

	7 6 5 4 3 2 1 0	7 6 5 4 3 2 1 0	7 6 5 4 3 2 1 0	7 6 5 4 3 2 1 0
NDT = Invert	1 1 1 1 0 1 1 w	mod 0 1 0 r/m		
SRLSAL = Shift Logical/Arithmetic left	1 1 0 1 0 0 vw	mod 1 0 0 r/m		
SHR = Shift Logical Right	1 1 0 1 0 0 vw	mod 1 0 1 r/m		

SAR = Shift Arithmetic Right | 1 1 0 1 0 0 vw | mod 1 1 1 r/m |

ROL = Rotate Left | 1 1 0 1 0 0 vw | mod 0 0 0 r/m |

ROR = Rotate Right | 1 1 0 1 0 0 vw | mod 0 0 1 r/m |

RCL = Rotate Through Carry Flag Left | 1 1 0 1 0 0 vw | mod 0 1 0 r/m |

RCR = Rotate Through Carry Right | 1 1 0 1 0 0 vw | mod 0 1 1 r/m |

AND = And :

Reg/Memory with Register to Either | 0 0 0 0 0 0 w | mod reg r/m |

Immediate to Register/Memory

| 1 0 0 0 0 0 0 w | mod 1 0 0 r/m | data | data if w = 1 |

Immediate to Accumulator

| 0 0 1 0 0 1 0 w | data | data if w = 1 |

TEST = And Function to FLAGs, No Result :

Register/Memory and Register | 1 0 0 0 0 1 0 w | mod reg r/m |

Immediate Data and Register/Memory

| 1 0 0 0 0 0 0 w | mod 0 0 1 r/m | data | data if w = 1 |

Immediate Data and Accumulator | 1 0 1 0 1 0 0 w | data | data if w = 1 |

OR = Or :

Reg/Memory with Register to Either | 0 0 0 0 1 0 dw | mod reg r/m |

Immediate to Register/Memory

| 1 0 0 0 0 0 0 w | mod 0 0 1 r/m | data | data if w = 1 |

Immediate to Accumulator | 0 0 0 0 1 1 0 w | data | data if w = 1 |

XOR = Exchange or :

Reg/Memory with Register to Either | 0 0 1 1 0 0 dw | mod reg r/m |

Immediate to Resigter/Memory

1 0 0 0 0 0 0 w	mod 1 1 0 r/m	data	data if w = 1

Immediate to Accumulator

0 0 1 1 0 1 0 w	data	data if w = 1

IV STRING MANIPULATION

REF = Repeat	1 1 1 1 0 0 1 z
MOVS = Move Byte/Word	1 0 1 0 0 1 0 w
CMPS = Compare Byte/Word	1 0 1 0 0 1 1 w
SCAS = Scan Byte/Word	1 0 1 0 1 1 1 w
LODS = Load Byte/Word to AL/AX	1 0 1 0 1 1 0 w
STOS = Stor Byte/Wd from AL/A	1 0 1 0 1 0 1 w

V CONTROL TRANSFER

CALL = Call :

Direct within Segment	1 1 1 0 1 0 0 0	disp-low	disp-high
Indirect with Segment	1 1 1 1 1 1 1 1	mod 0 1 0 r/m	
Direct Intersegment	1 0 0 1 1 0 1 0	offset-low	offset-high
		seg-low	seg-high
Indirect Intersegment	1 1 1 1 1 1 1 1	mod 0 1 1 r/m	

Mnemonic and Description **Instruction Code**

JMP = Uncondition Jump

7 6 5 4 3 2 1 0 7 6 5 4 3 2 1 0 7 6 5 4 3 2 1 0

Direct within Segment	1 1 1 0 1 0 0 1	disp-low	disp-high
Direct within Segment-Short	1 1 1 0 1 0 1 1	disp	
Indirect within Segment	1 1 1 1 1 1 1 1	mod 1 0 0 r/m	

Instruction			
Direct Intersegment	1 1 1 0 1 0 0 1	offset-low	offset-high
		seg-low	seg-high
Indirect Intersegment	1 1 1 0 1 0 1 0	mod 1 0 1 r/m	

RET = Return from CALL :

Instruction			
Within Segment	1 1 0 0 0 0 1 1		
Within Seg Adding Immed to SP	1 1 0 0 0 0 1 0	data-low	data-high
Intersegment	1 1 0 0 1 0 1 1		
Intersegment Adding Immediate to SP	1 1 0 0 1 0 1 0	data-low	data-high

Instruction		
JE/JZ = Jump on Equal/Zero	0 1 1 1 0 1 0 0	disp
JL/JNGE = Jump on Less/Not Greater or Equal	0 1 1 1 1 1 0 0	disp
JLE/JNG = Jump on Less or Equal/ Not Greater	0 1 1 1 1 1 1 0	disp
JB/JNAE = Jump on Blow/Not Above or Equal	0 1 1 1 0 0 1 0	disp
JBE/JNA = Jump on Below or Equal/ Not Above	0 1 1 1 0 1 1 0	disp
JP/JPE = Jump on Parity/Parity Even	0 1 1 1 1 0 1 0	disp
JO = Jump on Overflow	0 1 1 1 0 0 0 0	disp
JS = Jump of Sign	0 1 1 1 1 0 0 0	disp
JNE/JNZ = Jump on Not Equal/Not Zero	0 1 1 1 0 1 0 1	disp
JNL/JGE = Jump on Not Less/Greater or Equal	0 1 1 1 1 1 0 1	disp

Mnemonic and Description	Instruction Code	
JNLE/JG = Jump on Not Less or Equal/Greater	0 1 1 1 1 1 1 1	disp
JNB/JAE = Jump on Not Below/Above or Equal	0 1 1 1 0 0 1 1	disp
JNBE/JA = Jump on Not Below or Equal/Above	0 1 1 1 0 1 1 1	disp
JNP/JPO = Jump on Not Par/Par Odd	0 1 1 1 1 0 1 1	disp
JNO = Jump on Not Overflow	0 1 1 1 0 0 0 1	disp
JNS = Jump on Not Sign	0 1 1 1 1 0 0 1	disp
LOOP = Loop CX Times	1 1 1 0 0 0 1 0	disp
LOOPZ/LOOPE = Loop While Zero/Equal	1 1 1 0 0 0 0 1	disp
LOOPNZ/LOOPNE = Loop While Not Zero/Equal	1 1 1 0 0 0 0 0	disp
JNXZ = Jump on CX Zero	1 1 1 0 0 0 1 1	disp

INT = Interrupt

Mnemonic and Description	Instruction Code	
Type Specified	1 1 0 0 1 1 0 1	type
Type 3	1 1 0 0 1 1 0 0	
INTo = Interrupt on Overflow	1 1 0 0 1 1 1 0	
IRET = Interrupt Return	1 1 0 0 1 1 1 1	

Mnemonic and Description	Instruction Code
VI PROCESSOR CONTROL	7 6 5 4 3 2 1 0 7 6 5 4 3 2 1 0
CLC = Clear Carry	1 1 1 1 1 0 0 0
CMC = Complement Carry	1 1 1 1 0 1 0 1
STC = Set Carry	1 1 1 1 1 0 0 1
CLD = Clear Direction	1 1 1 1 1 1 0 0

STD = Set Direction	1 1 1 1 1 1 0 1
CLI = Clear Interrupt	1 1 1 1 1 0 1 0
STI = Set Interrupt	1 1 1 1 1 0 1 1
HLT = Halt	1 1 1 1 0 1 0 0
WAIT = Wait	1 0 0 1 1 0 1 1
ESC = Escape (to External Device)	1 1 0 1 1 x x x | mod x x x r/m
LOCK = Bus Lock Prefix	1 1 1 1 0 0 0 0

APPENDIX - V

Microcontroller Instruction Set

Microcontroller instruction set is very powerful for a processor of its size. It is straight forward and very efficient in its use of program memory. All instructions are either one or two bytes in length and over 80% are only one byte long. The instruction set has been summarized, alphabetical listing including the informations :

- ☐ Mnemonic
- ☐ Hex Code
- ☐ Symbolic description
- ☐ Bytes and Cycle

For (*a*) Accumulator,
(*b*) Input/Output,
(*c*) Registers,
(*d*) Branch,
(*e*) Subroutines,
(*f*) Flags,
(*g*) Data Moves,
(*h*) Time Counters,
(*i*) Control.

(*a*) ACCUMULATOR

Mnemonic	Hex code	Description	Bytes	Cycle
ADD A, R0	68	Add Register R0 to A	1	1
ADD A, R1	69	Add Register R1 to A	1	1
ADD A, R2	6A	Add Register R2 to A	1	1
ADD A, R3	6B	Add Register R3 to A	1	1
ADD A, R4	6C	Add Register R4 to A	1	1
ADD A, R5	6D	Add Register R5 to A	1	1
ADD A, R6	6E	Add Register R6 to A	1	1
ADD A, R7	6F	Add Register R7 to A	1	1
ADD A, @R0	60	Add data memory pointed by R0 to A	1	1
ADD A, @R1	61	Add data memory pointed by R1 to A	1	1
ADD A, #data	03	Add immediate to A	2	2
ADDC A, R0	78	Add Register R0 with carry	1	1
ADDC A, R1	79	Add Register R1 with carry	1	1
ADDC A, R2	7A	Add Register R2 with carry	1	1
ADDC A, R3	7B	Add Register R3 with carry	1	1
ADDC A, R4	7C	Add Register R4 with carry	1	1

Mnemonic	Hex code	Description	Bytes	Cycle
ADDC A, R5	7D	Add Register R5 with carry	1	1
ADDC A, R6	7E	Add Register R6 with carry	1	1
ADDC A, R7	7F	Add Register R7 with carry	1	1
ADDC A, @R0	70	Add data memory pointed by R0 with carry	1	1
ADDC A, @R1	71	Add data memory pointed by R1 with carry	1	1
ANL A, R0	58	AND Register R0 to A	1	1
ANL A, R1	59	AND Register R1 to A	1	1
ANL A, R2	5A	AND Register R2 to A	1	1
ANL A, R3	5B	AND Register R3 to A	1	1
ANL A, R4	5C	AND Register R4 to A	1	1
ANL A, R5	5D	AND Register R5 to A	1	1
ANL A, R6	5E	AND Register R6 to A	1	1
ANL A, R7	5F	AND Register R7 to A	1	1
ANL A, @R0	50	Add data memory pointed by R0 to A	1	1
ANL A, @R1	51	Add data memory pointed by R1 to A	1	1
ANL A, #data	53	AND Immediate to A	2	2
ORL A, R0	48	OR Register R0 to A	1	1
ORL A, R1	49	OR Register R1 to A	1	1
ORL A, R2	4A	OR Register R2 to A	1	1
ORL A, R3	4B	OR Register R3 to A	1	1
ORL A, R4	4C	OR Register R4 to A	1	1
ORL A, R5	4D	OR Register R5 to A	1	1
ORL A, R6	4E	OR Register R6 to A	1	1
ORL A, R7	4F	OR Register R7 to A	1	1
ORL A, @R0	40	OR data memory pointed by R0 to A	1	1
ORL A, @R1	41	OR data memory pointed by R1 to A	1	1
ORL A, #data	43	OR Immediate to A	2	2
XRL A, R0	D8	Ex-OR Register R0 to A	1	1
XRL A, R1	D9	Ex-OR Register R1 to A	1	1
XRL A, R2	DA	Ex-OR Register R2 to A	1	1
XRL A, R3	DB	Ex-OR Register R3 to A	1	1
XRL A, R4	DC	Ex-OR Register R4 to A	1	1
XRL A, R5	DD	Ex-OR Register R5 to A	1	1
XRL A, R6	DE	Ex-OR Register R6 to A	1	1
XRL A, R7	DF	Ex-OR Register R7 to A	1	1
XRL A, @R0	D0	Ex-OR data memory pointed by R0 to A	1	1
XRL A, @R1	D1	Ex-OR data memory pointed by R1 to A	1	1
XRL A, #data	D3	Ex-OR Immediate to A	2	2
INC A	17	Increment A	1	1
DEC A	07	Decrement A	1	1

Mnemonic	Hex code	Description	Bytes	Cycle
CLR A	27	Clear A	1	1
CPL A	37	Complement A	1	1
DAA	57	Decimal adjust	1	1
SWAP A	47	Swap nibbles of A	1	1
RL A	E7	Rotate A left	1	1
RLC A	F7	Rotate A left through carry	1	1
RR A	77	Rotate A left through carry	1	1
RRC A	67	Rotate A right through carry	1	1

(*b*) INPUT/OUTPUT

Mnemonic	Hex code	Description	Bytes	Cycle
IN A, P1	09	Input Port-1 to A	1	2
IN A, P2	0A	Input Port-2 A	1	2
OUTL P1, A	39	Output A to port-1	1	2
OUTL P2, A	3A	Output A to port-2	1	2
ANL P1, #data	99	AND immediate to port-1	2	2
ANL P2, #data	9A	AND immediate to port-1	2	2
ORL P1, #data	89	OR immediate to port-1	2	2
ORL P2, #data	8A	OR immediate to port-2	2	2
INS A, BUS	08	Input Bus port to A	1	2
OUTL BUS, A	02	Output A to Bus port	1	2
ANL BUS, #data	98	AND immediate to Bus port	2	2
ORL BUS, #data	88	OR immediate to Bus	2	2
Operation with Expander Port :				
MOVD A, P4	0C	Input Expander port-4 to A	1	2
MOVD A, P5	0D	Input Expander port-5 to A	1	2
MOVD A, P6	0E	Input Expander port-6 to A	1	2
MOVD A, P7	0F	Input Expander port-7 to A	1	2
MOVD P4, A	3C	Output A to expander port-4	1	2
MOVD P5, A	3D	Output A to expander port-5	1	2
MOVD P6, A	3E	Output A to expander port-6	1	2
MOVD P7, A	3F	Output A to expander port-7	1	2
ANLD P4, A	9C	AND A to expander port-4	1	2
ANLD P5, A	9D	AND A to expander port-5	1	2
ANLD P6, A	9E	AND A to expander port-6	1	2
ANLD P7, A	9F	AND A to expander port-7	1	2
ORLD P4, A	8C	OR A to expander port-4	1	2
ORLD P5, A	8D	OR A to expander port-5	1	2
ORLD P6, A	8E	OR A to expander port-6	1	2

(c) REGISTERS

Mnemonic	Hex code	Description	Bytes	Cycle
INC R0	18	Increment Register R0	1	1
INC R1	19	Increment Register R1	1	1
INC R2	1A	Increment Register R2	1	1
INC R3	1B	Increment Register R3	1	1
INC R4	1C	Increment Register R4	1	1
INC R5	1D	Increment Register R5	1	1
INC R6	1E	Increment Register R6	1	1
INC R7	1F	Increment Register R7	1	1
INC @R0	10	Increment data memory pointed by R0	1	1
INC @R1	11	Increment data memory pointed by R1	1	1
DEC R0	C8	Decrement Register R0	1	1
DEC R1	C9	Decrement Register R1	1	1
DEC R2	CA	Decrement Register R2	1	1
DEC R3	CB	Decrement Register R3	1	1
DEC R4	CC	Decrement Register R4	1	1
DEC R5	CD	Decrement Register R5	1	1
DEC R6	CE	Decrement Register R6	1	1
DEC R7	CF	Decrement Register R7	1	1

(d) BRANCH

Mnemonic	Hex code	Description	Bytes	Cycle
JMP Page-0	04	Jump Unconditional to page-0	2	2
JMP Page-1	24	Jump Unconditional to page-1	2	2
JMP Page-2	44	Jump Unconditional to page-2	2	2
JMP Page-3	64	Jump Unconditional to page-3	2	2
JMP Page-4	84	Jump Unconditional to page-4	2	2
JMP Page-5	A4	Jump Unconditional to page-5	2	2
JMP Page-6	C4	Jump Unconditional to page-6	2	2
JMP Page-7	E4	Jump Unconditional to page-7	2	2
JMP @A	B3	Jump Indirect within page	1	1
Conditional jump if RN Zero :				
DJNZ R0, Addr	E8	Decrement Reg. R0 and jump to addr if not zero	2	2
DJNZ R1, Addr	E9	Decrement Reg. R1 and jump to addr if not zero	2	2
DJNZ R2, Addr	EA	Decrement Reg. R2 and jump to addr if not zero	2	2
DJNZ R3, Addr	EB	Decrement Reg. R3 and jump to addr if not zero	2	2

Mnemonic	Hex code	Description	Bytes	Cycle
DJNZ R4, Addr	EC	Decrement Reg. R4 and jump to addr if not zero	2	2
DJNZ R5, Addr	ED	Decrement Reg. R5 and jump to addr if not zero	2	2
DJNZ R6, Addr	EE	Decrement Reg. R6 and jump to addr if not zero	2	2
DJNZ R7, Addr	EF	Decrement Reg. R7 and jump to addr if not zero	2	2
Conditional Jump on Flags :				
JC Addr	F6	Jump to addr. if Carry=1	2	2
JNC Addr	E6	Jump to addr. if Carry=0	2	2
JZ Addr	C6	Jump to addr. if A zero	2	2
JNZ Addr	96	Jump to addr. if A not zero	2	2
JTO Addr	26	Jump to addr. if T0=1	2	2
JNTO Addr	36	Jump to addr. if T0=0	2	2
JT1 Addr	56	Jump to addr. if T1=1	2	2
JNT1 Addr	46	Jump to addr. if T1=0	2	2
JF0 Addr	B6	Jump to addr. if F0=1	2	2
JF1 Addr	76	Jump to addr. if F1=1	2	2
JTF Addr	16	Jump to addr. if timer flag=1	2	2
JNI Addr	86	Jump to addr. if INT=0	2	2
JB0 Addr	12	Jump to addr. if Accumulator Bit-0=1	2	2
JB1 Addr	32	Jump to addr. if Accumulator Bit-1=1	2	2
JB2 Addr	52	Jump to addr. if Accumulator Bit-2=1	2	2
JB3 Addr	72	Jump to addr. if Accumulator Bit-3=1	2	2
JB4 Addr	92	Jump to addr. if Accumulator Bit-4=1	2	2
JB5 Addr	B2	Jump to addr. if Accumulator Bit-5=1	2	2
JB6 Addr	D2	Jump to addr. if Accumulator Bit-6=1	2	2
JB7 Addr	F2	Jump to addr. if Accumulator Bit-7=1	2	2

(e) SUBROUTINES

Mnemonic	Hex code	Description	Bytes	Cycle
CALL Page-0	14	Subroutine Call to page-0	2	2
CALL Page-1	34	Subroutine Call to page-1	2	2
CALL Page-2	54	Subroutine Call to page-2	2	2
CALL Page-3	74	Subroutine Call to page-3	2	2
CALL Page-4	94	Subroutine Call to page-4	2	2
CALL Page-5	B4	Subroutine Call to page-5	2	2

Mnemonic	Hex code	Description	Bytes	Cycle
CALL Page-6	D4	Subroutine Call to page-6	2	2
CALL Page-7	F4	Subroutine Call to page-7	2	2
RET	83	Return without PSW restore	1	2
RETR	93	Returned with PSW restore	1	2

(*f*) FLAGS

Mnemonic	Hex code	Description	Bytes	Cycle
CLR C	97	Clear Carry	1	1
CPL C	A7	Complement Carry	1	1
CLR F0	85	Clear Flag 0	1	1
CPL F0	95	Complement Flag 0	1	1
CLR F1	A5	Clear Flag 1	1	1
CPL F1	B5	Complement Flag 1	1	1

(*g*) DATA MOVES

Mnemonic	Hex code	Description	Bytes	Cycle
MOV A,R0	F8	Move Register R0 content to A	1	1
MOV A,R1	F9	Move Register R1 content to A	1	1
MOV A,R2	FA	Move Register R2 content to A	1	1
MOV A,R3	FB	Move Register R3 content to A	1	1
MOV A,R4	FC	Move Register R4 content to A	1	1
MOV A,R5	FD	Move Register R5 content to A	1	1
MOV A,R6	FE	Move Register R6 content to A	1	1
MOV A,R7	FF	Move Register R7 content to A	1	1
MOV A,@R0	F0	Move data from memory pointed by R0 to A	1	1
MOV A,@R1	F1	Move data from memory pointed by R1 to A	1	1
MOV A,#data	23	Move Immediate data to A	2	2
MOV R0,A	A8	Move A to Register R0	1	1
MOV R1,A	A9	Move A to Register R1	1	1
MOV R2,A	AA	Move A to Register R2	1	1
MOV R3,A	AB	Move A to Register R3	1	1
MOV R4,A	AC	Move A to Register R4	1	1
MOV R5,A	AD	Move A to Register R5	1	1
MOV R6,A	AE	Move A to Register R6	1	1
MOV R7,A	AF	Move A to Register R7	1	1

Mnemonic	Hex code	Description	Bytes	Cycle
MOV @R0,A	A0	Move A to data memory pointed by R0	1	1
MOV @R1,A	A1	Move A to data memory pointed by R1	1	1
MOV R0,#data	B8	Move immediate data to Reg. R0	2	2
MOV R1,#data	B9	Move immediate data to Reg. R1	2	2
MOV R2,#data	BA	Move immediate data to Reg. R2	2	2
MOV R3,#data	BB	Move immediate data to Reg. R3	2	2
MOV R4,#data	BC	Move immediate data to Reg. R4	2	2
MOV R5,#data	BD	Move immediate data to Reg. R5	2	2
MOV R6,#data	BE	Move immediate data to Reg. R6	2	2
MOV R7,#data	BF	Move immediate data to Reg. R7	2	2
MOV @R0#DATA	B0	Move Immediate data to data memory pointed by Register R0	2	2
MOV @R1#DATA	B1	Move Immediate data to data memory pointed by Registet R1	2	2
MOV A,PSWD	C7	Move PSW to A	1	1
MOV PSW,A	D7	Move A to PSW	1	1
XCH A,R0	28	Exchange A and Register R0	1	1
XCH A,R1	29	Exchange A and Register R1	1	1
XCH A,R2	2A	Exchange A and Register R2	1	1
XCH A,R3	2B	Exchange A and Register R3	1	1
XCH A,R4	2C	Exchange A and Register R4	1	1
XCH A,R5	2D	Exchange A and Register R5	1	1
XCH A,R6	2E	Exchange A and Register R6	1	1
XCH A,R7	2F	Exchange A and Register R7	1	1
XCH A,@R0	20	Exchange A and memory contents pointed by R0	1	1
XCH A,@R1	21	Exchange A and memory contents pointed by R1	1	1
XCHD A,@R0	30	Exchange A and data memory Nibble pointed by R0	1	1
XCHD A,@R1	31	Exchange A and data memory Nibble pointed by R1	1	1
MOVX A,@R0	80	Move external data memory to A pointed by R0	1	1
MOVX A,@R1	81	Move external data memory to A pointed by R1	1	2
MOVX @R0,A	90	Move A to external data memory pointed by R0	1	2
MOVX @R1,A	91	Move A to external data memory pointed by R1	1	1
MOVX A,@A	A3	Move current page data pointed by A to A	1	1
MOVP3 A,@A	B3	Move page-3 data pointed by A to A	1	2

(*h*) TIMER COUNTERS

Mnemonic	Hex code	Description	Bytes	Cycle
Mnemonic	**Hex code**	**Description**	**Bytes**	**Cycle**
MOV A,T	42	Move Timer/counter to A	1	1
MOV T, A	62	Move A to Timercounter	1	1
STRT T	55	Start Timer	1	1
STRT CNT	45	Start Event Counter	1	1
STOP TCNT	65	Stop Timer/Event Coutner	1	1
EN TCNTI	25	Enable Timer/Counter Interrupt	1	1
DIS TCNTI	35	Disable Timer/Counter Interrupt	1	1

(*i*) CONTROL

Mnemonic	Hex code	Description	Bytes	Cycle
EN I	05	Enable External Interrupt	1	1
DIS I	15	Disable External Interrupt	1	1
SEL RB0	C5	Select Register Bank—0	1	1
SEL RB1	D5	Select Register Bank—1	1	1
SEL MB0	E5	Select Memory Bank—0	1	1
SEL MB1	F5	Select Memory Bank—1	1	1
ENT0 CLK	75	Enable Clock Output ON T0	1	1
NOP	00	No operation	1	1

#=> Immediate Date Prefix

@=-> Indirect Address Prefix

(×)=> contents of ×

((×)) contents of location addressed by ×

APPENDIX - VI

MCS-51 OPCODES

OPCODE	MNEMONIC	OPERAND	OPCODE	MNEMONIC	OPERAND	OPCODE	MNEMONIC	OPERAND	SFR	ADD
00	NOP		2A	ADD	A,R2	54	ANL	A,#n	A	E0
01	AJMP	sadd	2B	ADD	A,R3	55	ANL	A,add	B	F0
02	LJMP	ladd	2C	ADD	A,R4	56	ANL	A,@R0	PSW	D0
03	RR	A	2D	ADD	A,R5	57	ANL	A,@R1	DPH	83
04	INC	A	2E	ADD	A,R6	58	ANL	A,R0	DPL	82
05	INC	add	2F	ADD	A,R7	59	ANL	A,R1		
06	INC	@R0	30	JNB	b,radd	5A	ANL	A,R2		
07	INC	@R1	31	ACALL	sadd	5B	ANL	A,R3		
08	INC	R0	32	RETI		5C	ANL	A,R4		
09	INC	R1	33	RLC	A	5D	ANL	A,R5		
0A	INC	R2	34	ADDC	A,#n	5E	ANL	A,R6		
0B	INC	R3	35	ADDC	A,add	5F	ANL	A,R7		
0C	INC	R4	36	ADDC	A,@R0	60	JZ	radd		
0D	INC	R5	37	ADDC	A,@R1	61	AJMP	sadd		
0E	INC	R6	38	ADDC	A,R0	62	XRL	add,A		
0F	INC	R7	39	ADDC	A,R1	63	XRL	add,#n		
10	JBC	b,radd	3A	ADDC	A,R2	64	XRL	A,#n		
11	ACALL	sadd	3B	ADDC	A,R3	65	XRL	A,add		
12	LCALL	LADD	3C	ADDC	A,R4	66	XRL	A,@R0		
13	RRC	A	3D	ADDC	A,R5	67	XRL	A,@R1		
14	DEC	A	3E	ADDC	A,R6	68	XRL	A,R0		
15	DEC	ADD	3F	ADDC	A,R7	69	XRL	A,R1		
16	DEC	@R0	40	JC	b,radd	6A	XRL	A,R2		
17	DEC	@R1	41	AJMP	sadd	6B	XRL	A,R3		
18	DEC	R0	42	ORL	add,A	6C	XRL	A,R4		
19	DEC	R1	43	ORL	add,#n	6D	XRL	A,R5		
1A	DEC	R2	44	ORL	A,#n	6E	XRL	A,R6		
1B	DEC	R3	45	ORL	A,add	6F	XRL	A,R7		
1C	DEC	R4	46	ORL	A,@R0	70	JNZ	radd		
1D	DEC	R5	47	ORL	A,@R1	71	ACALL	sadd		
1E	DEC	R6	48	ORL	A,R0	72	ORL	C,b		
1F	DEC	R7	49	ORL	A,R1	73	JP	@A+DPTR		
20	JB	b,radd	4A	ORL	A,R2	74	MOV	A,#n		
21	AJMP	sadd	4B	ORL	A,R3	75	MOV	add,#n		
22	RET		4C	ORL	A,R4	76	MOV	@R0,#n		
23	RL	A	4D	ORL	A,R5	77	MOV	@R1,#n		
24	ADD	A,#n	4E	ORL	A,R6	78	MOV	R0,@n		
25	ADD	A,add	4F	ORL	A,R7	79	MOV	R1,@n		
26	ADD	A,@R0	50	JNC	radd	7A	MOV	R2,@n		
27	ADD	A,@R1	51	ACALL	sadd	7B	MOV	R3,@n		
28	ADD	A,R0	52	ANL	add,A	7C	MOV	R4,@n		
29	ADD	A,R1	53	ANL	add,#n	7D	MOV	R5,@n		

OPCODE	MNEMONIC	OPERAND	OPCODE	MNEMONIC	OPERAND	OPCODE	MNEMONIC	OPERAND	SFR	ADD
7E	MOV	R6,@n	AA	MOV	R2,add	D6	XCHD	A,@R0		
7F	MOV	R7,@n	AB	MOV	R3,add	D7	XCHD	A,@R1		
80	SJMP	radd	AC	MOV	R4,add	D8	DJNZ	R0,radd		
81	AJMP	sadd	AD	MOV	R5,add	D9	DJNZ	R1,radd		
82	ANL	C,b	AE	MOV	R6,add	DA	DJNZ	R2,radd		
83	MOVC	A,@A+PC	AF	MOV	R7,add	DB	DJNZ	R3,radd		
84	DIV	AB	B0	ANL	C,b	DC	DJNZ	R4,radd		
85	MOV	add,add	B1	ACALL	sadd	DD	DJNZ	R5,radd		
86	MOV	add,@R0	B2	CPL	b	DE	DJNZ	R6,radd		
87	MOV	add,@R1	B3	CPL	C	DF	DJNZ	R7,radd		
88	MOV	add,R0	B4	CJNE	A,#n,radd	E0	MOVX	A,@DPTR		
89	MOV	add,R1	B5	CJNE	A,add,radd	E1	AJMP	sadd		
8A	MOV	add,R2	B6	CJNE	@R0,#n,radd	E2	MOVX	A,@R0		
8B	MOV	add,R3	B7	CJNE	@R1,#n,radd	E3	MOVX	A,@R1		
8C	MOV	add,R4	B8	CJNE	R0,#n,radd	E4	CLR	A		
8D	MOV	add,R5	B9	CJNE	R1,#n,radd	E5	MOV	A,add		
8E	MOV	add,R6	BA	CJNE	R2,#n,radd	E6	MOV	A,@R0		
8F	MOV	add,R7	BB	CJNE	R3,#n,radd	E7	MOV	A,@R1		
90	MOV	DPTR,#nn	BC	CJNE	R4,#n,radd	E8	MOV	A,R0		
91	ACALL	sadd	BD	CJNE	R5,#n,radd	E9	MOV	A,R1		
92	MOV	b,C	BE	CJNE	R6,#n,radd	EA	MOV	A,R2		
93	MOVC	A,@A+DPTR	BF	CJNE	R7,#n,radd	EB	MOV	A,R3		
94	SUBB	A,#N	C0	PUSH	add	EC	MOV	A,R4		
95	SUBB	A,add	C1	AJMP	sadd	ED	MOV	A,R5		
96	SUBB	A,@R0	C2	CLR	b	EE	MOV	A,R6		
97	SUBB	A,@R1	C3	CLR	C	EF	MOV	A,R7		
98	SUBB	A,R0	C4	SWAP	A	F0	MOVX	@DPTR,A		
99	SUBB	A,R1	C5	XCH	A,add	F1	ACALL	sadd		
9A	SUBB	A,R2	C6	XCH	A,@R0	F2	MOVX	@R0,A		
9B	SUBB	A,R3	C7	XCH	A,@R1	F3	MOVX	@R1,A		
9C	SUBB	A,R4	C8	XCH	A,R0	F4	CPL	A		
9D	SUBB	A,R5	C9	XCH	A,R1	F5	MOV	add,A		
9E	SUBB	A,R6	CA	XCH	A,R2	F6	MOV	@R0,A		
9F	SUBB	A,R7	CB	XCH	A,R3	F7	MOV	@R1,A		
A0	ORL	C,b	CC	XCH	A,R4	F8	MOV	R0,A		
A1	AJMP	sadd	CD	XCH	A,R5	F9	MOV	R1,A		
A2	MOV	C,b	CE	XCH	A,R6	FA	MOV	R2,A		
A3	INC	DPTR	CF	XCH	A,R7	FB	MOV	R3,A		
A4	MUL	AB	D0	POP	add	FC	MOV	R4,A		
A5	unused		D1	ACALL	sadd	FD	MOV	R5,A		
A6	MOV	@R0,add	D2	SETB	b	FE	MOV	R6,A		
A7	MOV	@R1,add	D3	SETB	C	FF	MOV	R7,A		
A8	MOV	R0,add	D4	DA	A					
A9	MOV	R1,add	D5	DJNZ	add,radd					

❑ ❑ ❑

Index

A

B

C

D

E

F

G

H

I

J

L

M

N

O

P

R

S

T

V

W

X